Recent Advances in Polyphenol Research

Recent Advances in Polyphenol Research

A series for researchers and graduate students whose work is related to plant phenolics and polyphenols, as well as for individuals representing governments and industries with interest in this field. Each volume in this biennial series focuses on several important research topics in plant phenols and polyphenols, including chemistry, biosynthesis, metabolic engineering, ecology, physiology, food, nutrition, and health.

Volume 8 Editors:
Juha-Pekka Salminen (University of Turku, Finland), Kristiina Wähälä (University of Helsinki, Finland), Victor de Freitas (University of Porto, Portugal), and Stéphane Quideau (University of Bordeaux, France)

Series Editor-in-Chief:
Stéphane Quideau (University of Bordeaux, France)

Recent Advances in Polyphenol Research

Volume 8

Edited by

Juha-Pekka Salminen
Professor, Natural Compound Chemistry
Department of Chemistry
University of Turku, Finland

Kristiina Wähälä
Professor, Organic Chemistry
Faculty of Medicine and Faculty of Science
University of Helsinki, Finland

Victor de Freitas
Professor, Food Chemistry
Chemistry and Biochemistry Department, Faculty of Sciences
University of Porto, Portugal

Stéphane Quideau
Professor, Organic and Bioorganic Chemistry
Institut des Sciences Moléculaires, CNRS-UMR 5255
University of Bordeaux, Talence, France
& Institut Universitaire de France, Paris, France

WILEY Blackwell

This edition first published 2023

Registered Offices
John Wiley & Sons, Inc., 111 River Street, Hoboken, NJ 07030, USA
John Wiley & Sons Ltd, The Atrium, Southern Gate, Chichester, West Sussex, PO19 8SQ, UK

For details of our global editorial offices, customer services, and more information about Wiley products visit us at www.wiley.com.

Wiley also publishes its books in a variety of electronic formats and by print-on-demand. Some content that appears in standard print versions of this book may not be available in other formats.

Library of Congress Cataloging-in-Publication Data applied for

ISBN: 9781119844761
ISSN: 2474-7696

Cover Design: Wiley
Cover Images: The River Aura and the Turku Cathedral are the landmarks of the city of Turku, the venue for the XXX International Conference on Polyphenols, hosted by the Natural Chemistry Research Group at the University of Turku, Finland – The Onagraceae plant family produces the largest ellagitannins found to date in the plant kingdom: the Natural Chemistry Research Group found the undecameric ellagitannin first time in *Oenothera biennis* – The Island of Ruissalo in Turku harbors the most famous oak forest in Finland. Together with other deciduous tree species oaks produce a great color show with their autumn coloration. The two chromatograms on the top of the oak forest show how the group-specific LC-MS methodology developed by the Natural Chemistry Research group can visualize the chromatographic fingerprints of both terminal and extension units of procyanidin and prodelphinidin units present in plant proanthocyanidins.

All rights of the photos belong to Vesa Aaltonen (first photo), Marc Johnson and Kari Loikas (second photo), and Juha-Pekka Salminen (third photo). In addition, the cover includes the official logo of the ICP2020TURKU, designed by Juha Harju.

Set in 9.5/12.5pt STIXTwoText by Straive, Pondicherry, India

Dedications

In memoriam

Prof. Takashi Yoshida (born in 1939, deceased on 1 May 2021) was a Professor of Medicinal Plants Chemistry at Okayama University, Japan, from 1993 to 2005. I worked with him as a research associate in the same laboratory. He also worked at the College of Pharmacy, Matsuyama University, Japan, as a Professor of Pharmacognosy from 2006 to 2012 after his departure from Okayama University in 2005. His research interests included the isolation and structural determination of ellagitannins and related polyphenols, as well as terpenoids in medicinal plants and foods in Japan, China, South-East Asian, and South American countries and the examination of their physiological activities. His scientific activity is documented in 261 papers on the isolation and structure elucidation of tannins and terpenoids in many medicinal plants and foods, as well as their diverse pharmacological properties, such as anti-cancer, anti-*Helicobacter pylori*, antileishmanial, and anti-methicillin-resistant *Staphylococcus aureus* activities. He received the Tannin Award at the Phytochemical Society of North America Annual Meeting in 2008 (Philadelphia, USA) and the Groupe Polyphenol Medal at the 8th Tannin Conference in 2014 (Nagoya, Japan) for his achievements in the field of polyphenolic natural products. He worked with many wonderful collaborators worldwide, including pharmacologists, biochemists, and microbiologists. He was respected and loved around the world for his caring friendship, and I have the deepest respect for him as a researcher and educator.

Professor Hideyuki Ito
Okayama Prefectural University, Japan

Prof. Hidetoshi Yamada (born in Ehime, Japan, on 4 August 1962, deceased on 23 November 2019) studied chemistry at Osaka City University, obtaining a bachelor's degree in 1985 and a master's degree in 1987. He was appointed Assistant Professor at Tokushima Bunri University, working with the late Professor Mugio Nishizawa. He received his PhD degree for the synthesis and structural revision of osladin, a sweet saponin. In 1997, he moved to Kwansei Gakuin University as Associate Professor and was promoted Full Professor in 2004. The key motif of his research was *curiosity*, centering particular attention on the conformations of carbohydrates, which were his familiar molecules from his early career. Interest in oligosaccharides led him to develop several viable glycosylation protocols, including thermal glycosylation, which allowed him to achieve the synthesis of L-rhamnose-based cyclodextrin. He and Mugio-sensei named the molecule "cycloawaodorin," a facetious trivial name associated with Tokushima locality. Through these endeavors, he had a keen interest in the conformationally flipped glucosyl donors. An ingenious design of flipped donor by bridging allowed the β-selective glycosylation without aid of the neighboring-group participation and further led to the preparation of the smallest cyclodextrin. Motivated by the presence of inverted glucose motifs, he stumbled into the synthesis of ellagitannin-class polyphenols, featuring D-glucose as a core extensively esterified by digalloyl groups. Facing such a tremendous diversity, he was as ever positive to make a slogan, "Let us synthesize these natural products all." To tackle this formidable challenge, he focused on two key motifs embedded within these complex structures: (1) hexahydroxydiphenoyl (HHDP) diesters based on two gallic acids directly C—C linked together, and (2) C—O linked digallates for which two gallic acids are linked through an oxygen atom. For the former HHDP motif, he exploited intramolecular oxidative biaryl coupling between gallates appended on a glucose scaffold, where $CuCl_2$ and *n*-$BuNH_2$ were identified as effective reagents. For the latter *O*-linked digallate motif, *ortho*-quinone mono-acetal was used as a platform to undergo an *oxa*-Michael addition/elimination sequence. His chemical synthesis study opened a door to flexible, comprehensive access to the ellagitannin molecules as pure entities.

Prof. Toshiyuki Kan (born in Hokkaido, Japan, on 15 February 1964, deceased on 24 July 2021) studied chemistry at Hokkaido University, where he obtained his bachelor's degree (1986) and master's degree (1988). He received his PhD degree in 1993 under the guidance of Prof. Haruhisa Shirahama, working on the total synthesis of grayanotoxin III, a poisonous diterpene with a highly complex structure. After working as a researcher at Suntory Institute for Bioorganic Research with Prof. Yasufumi Ohfune (1993–1996), he was appointed Assistant Professor at University of Tokyo with Prof. Tohru Fukuyama in 1996 and was promoted as Associate Professor in 2003. In 2004, he moved to the University of Shizuoka as Full Professor. His research focus was complex natural product synthesis, as represented by grayanotoxin III (diterpenoid, skeletal complexity) and ecteinascidin 743 (alkaloid, skeletal/functional complexity). In addressing the synthesis of complex natural products, he was continuously exposed to care for the management of multiple functionalities, and he became able to come up with survival tactics or even strategies to achieve the total synthesis. Among others is the design of a new protecting group, "nosyl (Ns)," which served as bases for his alkaloid syntheses, and now one of the standard choices for protecting amino groups. Upon moving to his last destination, Shizuoka, one of the major tea and orange localities in Japan, he decided to study the synthesis and chemical biology of tea- and citrus-derived polyphenols, including chafurosides A and B (black tea), epigallocatechin gallate (EGCg), and nobiletin. Here the nosyl group was proven to work for protecting phenols, endorsing a theanine synthesis via otherwise difficult biomimetic oxidative dimerization. He designed molecular probes for catechins and nobiletin, fluorescein, and PET probes, serving for in vivo imaging and antibody generation as well. He further addressed the syntheses of sesamin and sesaminol by an organocatalytic process and biomimetic construction of the furofuran skeleton and of hybrid-type polyphenols, hedyotol A, princepin, and sophoraflavanone H by exploiting C—H insertion reactions. His study highlighted and gained insights into the traditional products indigenous to Shizuoka through chemical synthesis and chemical biology. He delivered a plenary lecture on "Total Synthesis of Hybrid Type Polyphenols" at the XXX International Conference on Polyphenols in Turku, Finland, in July 2021, which turned out to be his last lecture. The initial seven slides were a tribute to his friend, Hidetoshi Yamada. In many people's memory, they both would be remembered by their energy, courage, spirit in research, determination and leadership in society, and friendship and warm-hearted attitude toward everyone. They were intimate friends, sharing genuine scientific interests in complex natural product synthesis. Here they are (Hidetoshi on the left and Toshiyuki on the right) with their relaxed smiles like naughty kids, enjoying fine food and drinks!?! 合掌

Keisuke Suzuki
Professor Emeritus, Tokyo Institute of Technology, Japan

Contents

Contributors

Sara M. Abdou
Department of Agricultural Sciences, Viikki Plant Science Centre, University of Helsinki, Helsinki, Finland

Sharmin Ahamed
Department of Agricultural Sciences, Viikki Plant Science Centre, University of Helsinki, Helsinki, Finland

Nick W. Albert
Plant & Food Research, Palmerston North, New Zealand

Emilia Alfaro-Viquez
Reed Research Group, Department of Animal and Dairy Sciences, University of Wisconsin-Madison, Madison, WI, USA

Daisuke Ando
Great Lakes Bioenergy Research Center, The Wisconsin Energy Institute, University of Wisconsin, Madison, WI, USA

Current address: Institute of Wood Technology, Akita Prefectural University, Noshiro, Japan

Tomohiro Asakawa
Department of Fisheries-Food Science, Tokai University, Shizuoka, Japan

Enrique Báez-García
Institute of Food Science, Technology and Nutrition, Spanish Research Council (ICTAN-CSIC), Madrid, Spain

Tecnológico Nacional de Mexico/Instituto Tecnológico de Tepic, Tepic, Nayarit, Mexico

Hany Bashandy
Department of Agricultural Sciences, Viikki Plant Science Centre, University of Helsinki, Helsinki, Finland

Department of Genetics, Cairo University, Giza, Egypt

Andrew Birmingham
Complete Phytochemical Solutions LLC, Cambridge, WI, USA

Rafael Carecho
NOVA Medical School, Faculdade de Ciências Médicas, Universidade NOVA de Lisboa, Lisboa, Portugal

Diogo Carregosa
NOVA Medical School, Faculdade de Ciências Médicas, Universidade NOVA de Lisboa, Lisboa, Portugal

Frank Caruso
Department of Chemical Engineering, The University of Melbourne, Parkville, VIC, Australia

Mingjie Chen
Great Lakes Bioenergy Research Center, The Wisconsin Energy Institute, University of Wisconsin, Madison, WI, USA
Current address: Institute of Microbiology, Guangdong Academy of Sciences, Guangzhou, China

José C. del Río
Instituto de Recursos Naturales y Agrobiología de Sevilla (IRNAS), CSIC, Seville, Spain

Pieter C. Dorrestein
Collaborative Mass Spectrometry Innovation Center,
Skaggs School of Pharmacy and Pharmaceutical Sciences,
University of California, San Diego, La Jolla, CA, USA

Cláudia Nunes dos Santos
NOVA Medical School, Faculdade de Ciências Médicas, Universidade NOVA de Lisboa, Lisboa, Portugal

Madeleine Ernst
Section for Clinical Mass Spectrometry, Department of Congenital Disorders, Danish Center for Neonatal Screening, Statens Serum Institut, Copenhagen, Denmark

Daniel Esquivel-Alvarado
Reed Research Group, Department of Animal and Dairy Sciences, University of Wisconsin-Madison, Madison, WI, USA

Alexis Eugene
Great Lakes Bioenergy Research Center, The Wisconsin Energy Institute, University of Wisconsin, Madison, WI, USA

Makoto Inai
School of Pharmaceutical Sciences, University of Shizuoka, Shizuoka, Japan

Yi Ju
Department of Chemical Engineering, The University of Melbourne, Parkville, VIC, Australia

Toshiyuki Kan
School of Pharmaceutical Sciences, University of Shizuoka, Shizuoka, Japan

Kyo Bin Kang
College of Pharmacy, Sookmyung Women's University, Seoul, Korea

Iris F. Kappers
Laboratory of Plant Physiology,
Plant Sciences Group,
Wageningen University and Research
Wageningen, the Netherlands

Steven D. Karlen
Great Lakes Bioenergy Research Center, The Wisconsin Energy Institute, University of Wisconsin, Madison, WI, USA

Hoon Kim
Great Lakes Bioenergy Research Center, The Wisconsin Energy Institute, University of Wisconsin, Madison, WI, USA

Abigail Kramschuster
Complete Phytochemical Solutions LLC, Cambridge, WI, USA

Christian G. Krueger
Reed Research Group, Department of Animal and Dairy Sciences, University of Wisconsin-Madison, Madison, WI, USA

Complete Phytochemical Solutions LLC, Cambridge, WI, USA

Wu Lan
Great Lakes Bioenergy Research Center, The Wisconsin Energy Institute, University of Wisconsin, Madison, WI, USA

Current address: State Key Laboratory of Pulp and Paper Engineering, South China University of Technology, Guangzhou, China

Leta L. Landucci
Great Lakes Bioenergy Research Center, The Wisconsin Energy Institute, University of Wisconsin, Madison, WI, USA

Jie Li
Department of Biochemistry and Metabolism, John Innes Centre, Norwich Research Park, Colney, Norwich, UK

Yanding Li
Great Lakes Bioenergy Research Center, The Wisconsin Energy Institute, University of Wisconsin, Madison, WI, USA

Current address: BeiGene, Zhongguancun Life Science Park, Changping District, Beijing, China

Sarah Liu
Great Lakes Bioenergy Research Center, The Wisconsin Energy Institute, University of Wisconsin, Madison, WI, USA

Fachuang Lu
Great Lakes Bioenergy Research Center, The Wisconsin Energy Institute, University of Wisconsin, Madison, WI, USA

Gerardo G. Mackenzie
Department of Nutrition, University of California-Davis, Davis, CA, USA

Daniela Marques
NOVA Medical School, Faculdade de Ciências Médicas, Universidade NOVA de Lisboa, Lisboa, Portugal

Cathie Martin
Department of Biochemistry and Metabolism, John Innes Centre, Norwich Research Park, Colney, Norwich, UK

Roosa Matomäki
Department of Agricultural Sciences, Viikki Plant Science Centre, University of Helsinki, Helsinki, Finland

Saku Mattila
Department of Agricultural Sciences, Viikki Plant Science Centre, University of Helsinki, Helsinki, Finland

Fulvio Mattivi
Department of Cellular, Computational and Integrative Biology (CIBIO), University of Trento, Trento, Italy

Department of Food Quality and Nutrition, Fondazione Edmund Mach, Research and Innovation Centre, San Michele all'Adige, Italy

Marnix H. Medema
Bioinformatics Group, Plant Sciences Group, Wageningen University and Research, Wageningen, the Netherlands

Lorenzo Mollo
Department of Agricultural Sciences, Viikki Plant Science Centre, University of Helsinki, Helsinki, Finland

Department of Environmental and Life Sciences, Laboratory of Plant and Algae Physiology, Università politecnica delle Marche, Ancona, Italy

Nuoendagula
Great Lakes Bioenergy Research Center, The Wisconsin Energy Institute, University of Wisconsin, Madison, WI, USA

Patricia I. Oteiza
Department of Nutrition, University of California-Davis, Davis, CA, USA

Daniel Papenberg
Bioinformatics Group, Plant Sciences Group, Wageningen University and Research, Wageningen, the Netherlands

Laboratory of Plant Physiology, Plant Sciences Group, Wageningen University and Research, Wageningen, the Netherlands

Jara Pérez-Jiménez
Institute of Food Science, Technology and Nutrition, Spanish Research Council (ICTAN-CSIC), Madrid, Spain

John Ralph
Great Lakes Bioenergy Research Center, The Wisconsin Energy Institute, University of Wisconsin, Madison, WI, USA

Department of Biochemistry, University of Wisconsin, Madison, WI, USA

Sally A. Ralph
The US Forest Products Laboratory, One Gifford Pinchot Drive, Madison, WI, USA

Jess D. Reed
Reed Research Group, Department of Animal and Dairy Sciences, University of Wisconsin-Madison, Madison, WI, USA

Complete Phytochemical Solutions LLC, Cambridge, WI, USA

Jorge Rencoret
Instituto de Recursos Naturales y Agrobiología de Sevilla (IRNAS), CSIC, Seville, Spain

J.J. Richardson
Department of Chemical Engineering, The University of Melbourne, Parkville, VIC, Australia

Simon Rogers
School of Computing Science, University of Glasgow, Glasgow, UK

Sonia G. Sáyago-Ayerdi
Tecnológico Nacional de Mexico/ Instituto Tecnológico de Tepic, Tepic, Nayarit, Mexico

Canan Sener
Great Lakes Bioenergy Research Center, The Wisconsin Energy Institute, University of Wisconsin, Madison, WI, USA

Rebecca A. Smith
Great Lakes Bioenergy Research Center, The Wisconsin Energy Institute, University of Wisconsin, Madison, WI, USA

Steve Spoljaric
Department of Chemical Engineering, The University of Melbourne, Parkville, VIC, Australia

Takashi Tanaka
Graduate School of Biomedical Sciences, Nagasaki University, Nagasaki, Japan

Teemu H. Teeri
Department of Agricultural Sciences, Viikki Plant Science Centre, University of Helsinki, Helsinki, Finland

Vitaliy I. Timokhin
Great Lakes Bioenergy Research Center, The Wisconsin Energy Institute, University of Wisconsin, Madison, WI, USA

Maria M. Ulaszewska
Department of Food Quality and Nutrition, Fondazione Edmund Mach, Research and Innovation Centre, San Michele all'Adige, Italy

PROMEFA Facility, San Raffaele Scientific Institute, Center for Omics Sciences, Milan, Italy

Justin J.J. van der Hooft
Bioinformatics Group, Plant Sciences Group, Wageningen University and Research, Wageningen, the Netherlands

Koichi Yoshioka
Great Lakes Bioenergy Research Center, The Wisconsin Energy Institute, University of Wisconsin, Madison, WI, USA

Wei Zhu
Department of Nutrition, University of California-Davis, Davis, CA, USA

Lingping Zhu
Department of Agricultural Sciences, Viikki Plant Science Centre, University of Helsinki, Helsinki, Finland

Preface

Every 2 years, Groupe Polyphénols (GP) hosts the International Conference on Polyphenols (ICP). The anniversary XXX ICP was planned to be held in 2020 in Turku, Finland. Unfortunately, the COVID pandemic forced GP to postpone the conference by one year, although everything was already set and organized for the conference in 2020. After one year of pandemic, GP decided that the ICP2020TURKU should not be postponed further and it was successfully organized as a fully virtual conference from 13 to 15 July, 2021. This was the first ever virtual ICP hosted by GP since its foundation in 1972. Groupe Polyphénols is the world's premier society of scientists in the fields of polyphenol chemistry, synthesis, bioactivity, nutrition, industrial applications, and ecology. Luckily, the great success of this virtual ICP encouraged GP to plan for new types of scientific activities for its members also in between the biannual ICPs. Since the ICP2020TURKU, GP has already organized the first Webinar in Polyphenols Research that will gather polyphenol scientists virtually three to four times a year to attend presentations of both established and young scientists. This approach thus also continues one of the main aims of the ICP2020TURKU by giving good opportunities to young scientists to present their recent research findings on polyphenols.

The city of Turku is a city full of history. It was the first capital of Finland, before Helsinki, and it had the first Finnish-speaking university in Finland. The main organizers of the ICP2020TURKU, the Natural Chemistry Research Group, had planned to organize the ICP at the main campus of the University of Turku, close to the River Aura and the Turku Cathedral (see the front cover). Other history-oriented activities such as the gala dinner in the medieval Turku Castle were also planned and booked. However, now it remains for all the ICP participants to visit Turku on a later notice, once the COVID pandemic allows.

The XXX ICP was attended by 250 registrants from 36 countries, with 105 invited and contributed presentations. This great number of presentations was achieved by parallel sessions that maximized the opportunities given to young scientists to present their work. This eighth edition of Recent Advances in Polyphenol Research has 12 chapters that represent the work of the invited speakers at the XXX ICP and reflect the depth of science in this important field of natural product chemistry. The conference included 19 sessions on structure, reactivity, and synthesis; bioactivity and bioavailability; metabolomics, targeted analysis, and big data; quality control and standardization; biogenesis and functions in plants and ecosystems; and biomaterials and applied sciences.

We owe a special thanks to Tina Ahonen from the Aboa Congress and Event Services for her professional and excellent help in the organization of the conference. The great execution of the virtual ICP would not have been possible without the help of professionals of RajuLive Ltd. that made sure that all tiny details of the virtual conference, including the presentation recordings were of prime quality. The members and students of the Natural Chemistry Research Groups deserve our sincere thanks for their huge efforts in making the intermission activities in the form of entertaining videos that were also uploaded on the social media and for helping with the practical organization on site. Finally, we thank all the participants, who took active part in the conference sessions, and initiated a lot of scientific discussion in the conference chat, both after and between the presentations. You created a warm atmosphere for the conference and made it a really enjoyable event and a great learning experience during these otherwise difficult COVID times. We think that the ICP2020TURKU will always be remembered as a special conference, but luckily only for very good and positive reasons.

Juha-Pekka Salminen
Kristiina Wähälä
Victor de Freitas
Stéphane Quideau

Acknowledgments

The editors wish to thank all members of the "Groupe Polyphénols" Board Committee (2018–2021) for their guidance and assistance throughout this project.

Dr. Denis Barron
Dr. Luc Bidel
Dr. Catherine Chèze
Dr. Peter Constabel
Prof. Olivier Dangles
Dr. Kevin Davies
Prof. M Teresa Escribano Bailon
Prof. Victor de Freitas
Prof. Kazuhiko Fukushima
Dr. David Gang
Dr. Sylvain Guyot
Prof. Ann E. Hagerman
Dr. Irene Mueller-Harvey
Prof. Stéphane Quideau
Prof. Jess Reed
Dr. Erika Salas
Prof. Juha-Pekka Salminen
Prof. Kristiina Wähälä

1

Lignins and Lignification

New Developments and Emerging Concepts

John Ralph[1,2], Hoon Kim[1], Fachuang Lu[1], Rebecca A. Smith[1], Steven D. Karlen[1], Nuoendagula[1], Koichi Yoshioka[1], Alexis Eugene[1], Sarah Liu[1], Canan Sener[1], Daisuke Ando[1,], Mingjie Chen[1,**], Yanding Li[1,***], Leta L. Landucci[1], Sally A. Ralph[3], Vitaliy I. Timokhin[1], Wu Lan[1,****], Jorge Rencoret[4], and José C. del Río[4]*

[1] *Great Lakes Bioenergy Research Center, The Wisconsin Energy Institute, University of Wisconsin, Madison, WI, USA*
[2] *Department of Biochemistry, University of Wisconsin, Madison, WI, USA*
[3] *The US Forest Products Laboratory, One Gifford Pinchot Drive, Madison, WI, USA*
[4] *Instituto de Recursos Naturales y Agrobiología de Sevilla (IRNAS), CSIC, Seville, Spain*

[*] *Current address: Institute of Wood Technology, Akita Prefectural University, Noshiro, Japan*
[**] *Current address: Institute of Microbiology, Guangdong Academy of Sciences, Guangzhou, China*
[***] *Current address: BeiGene, Zhongguancun Life Science Park, Changping District, Beijing, China*
[****] *Current address: State Key Laboratory of Pulp and Paper Engineering, South China University of Technology, Guangzhou, China*

1.1 Introduction

In the previous volume in this series, Volume 7, Chapter 7 highlighted recent discoveries relating to the interactions between monolignol pathways with flavonoid and stilbenoid pathways producing monomers for lignification (del Río et al. 2021). The notion that lignification, the process of polymerization from monomers to the lignin polymer, may tolerate or even favor the use of monomers beyond the canonical monolignols (*p*-coumaryl alcohol, coniferyl alcohol, and sinapyl alcohol) is becoming mainstream even if there might not be universal agreement about what exactly constitutes lignification. Advances continue as ever more is revealed about the way cell wall polymers in various tissues in "natural plants" are derived and how the perturbations of genes, now in various interacting pathways, can affect lignification and the consequent composition and structure of the polymer. Although "structure" has little meaning for a polymer lacking defined repeating units of any length and possessing an overwhelming stereochemical complexity (Ralph et al. 2008), this newly revealed complexity to the composition and structure of lignin polymers, and the blurring of the definition of lignification, may seem alarming. It is worth emphasizing, however, that the process of lignification itself is a delightfully simple

Recent Advances in Polyphenol Research, Volume 8, First Edition. Edited by Juha-Pekka Salminen, Kristiina Wähälä, Victor de Freitas, and Stéphane Quideau.

one involving a single, purely chemical mechanism and, accordingly, lignification is a much simpler process than the ones involved in the formation of hemicellulosic polysaccharides, for example.

Findings expanding the definition of lignin continue, even as the ramifications of the COVID-19 pandemic have impeded research progress in general since 2020. To document this rapidly advancing field, we cover some new findings and a few of the emerging notions on lignification. So much is happening in this field that we cannot comprehensively cover even merely the work from our own labs; this chapter is best regarded as an update in which we concentrate on some of the research on a common theme that interests us. We also cover areas that often must be jettisoned from research papers, weaving in concepts that are in principle well-known but occasionally need to be reemphasized because they have particular importance and/or may be distinctive to lignification and unfamiliar to researchers new to the field. Along with the evaluation and contemplation of "new" pathways and mechanisms, we also include minor subsections on the value and use of lignin models to understand reaction pathways, and the continued importance of developing diagnostic analytics to provide unambiguous new insight.

The chapter has been laid out in seven sections but keeping the ideas discretely under those headings has not been fully realized. Just as an example, we decided that observations on the use of monolignol conjugates in lignification needed their own section, yet much of the material could easily fit under sections preceding it. Similarly, model, synthetic, and analytical work often accompanies any discovery, but we have chosen to split one aspect out to provide some recognition for such crucial research components. Finally, there are some concepts that we wanted to convey here that simply do not fit well under the chosen headings. We trust that this will nevertheless be a readable and useful contribution despite these limitations.

1.2 The Monolignol Pathway and Interacting Pathways – New Lignins

The monolignol biosynthetic pathway produces the three canonical monolignols for lignification, *p*-coumaryl alcohol, coniferyl alcohol, and sinapyl alcohol, differing in their degrees of methoxylation *ortho* to the phenol (Freudenberg and Neish 1968; Sarkanen and Ludwig 1971). Some of the enzymes are quite specific, whereas others are more general in nature. As summarized in Figure 1.1, and as has been reviewed (Boerjan et al. 2003; Ralph et al. 2004b; Shi et al. 2010; Weng and Chapple 2010), the major flux through the pathway was simplified from the full metabolic grid originally considered (Dixon et al. 2001; Higuchi 2006; Matsui et al. 2000) as favored substrates and pathways through the grid have been elucidated (Humphreys et al. 1999a, 1999b; Li et al. 2000; Osakabe et al. 1999), and new steps and their enzymes continue to be discovered, as reviewed (Bonawitz and Chapple 2010; Mottiar et al. 2016; Ralph et al. 2019; Vanholme et al. 2010a, 2019a). Figure 1.1 attempts to capture the modern notion of broader lignification, integrating in phenolic components beyond the strict monolignol pathway. Perturbing the various genes along the pathway from phenylalanine to the monolignols, or even from shikimic acid and further back, not only provides a rich source of insight into the pathway processes but is also capable of producing some striking lignins. With reference to the line in Michael Chrichton's original Jurassic Park book (Chrichton 1990) and in the movie that "Life will find a way," plants do not simply give up and die because they find themselves unable to synthesize a monolignol. Although the very notion was once considered heresy, plants can survive, even from "instantaneous" perturbations from which they do not have the luxury of evolving, by producing a functional polymer from other available phenolic components. This is most readily evident in plants utilizing "products of truncated monolignol biosynthesis," a term we might have first introduced in print in 2003 (Boerjan et al. 2003), to produce the lignin polymer, sometimes quite successfully, but it is also clear that nature itself has, over time, explored options well beyond just utilizing pathway intermediates and still has plenty of surprises for us. The following recent examples are illustrative but are neither unique nor exhaustive.

1.2.1 Truncated Monolignol Biosynthesis

We will update just two examples here, those of plants deficient in the last enzyme of the pathway, cinnamyl alcohol dehydrogenase (CAD), and in one of the two primary *O*-methyltransferases (OMT, i.e. CCoAOMT or COMT), Figures 1.1–1.3. The result in both cases is product monomers of incomplete monolignol biosynthesis, hydroxycinnamaldehydes in the former case, and the catecholic monomers caffeyl alcohol or 5-hydroxyconiferyl alcohol in the latter. Both produce lignin polymers that function satisfactorily in the plant and, at least in the case of the catechols, are used as significant or even sole components to fabricate natural lignins in specialized tissues such as seedcoats.

1.2.1.1 CAD Deficiency

Lignins have long been known to contain low levels of hydroxycinnamaldehyde (and hydroxybenzaldehyde) units. The characteristic lignin stain, the Wiesner or phloroglucinol stain, is somewhat specific for hydroxycinnamaldehyde endgroups (Adler et al. 1948; Pomar et al. 2002); it has always been ironic that this common stain owes its utility to the incorporation of low levels of components that are often not even acknowledged as being involved in lignification. Cinnamaldehyde endgroups might be produced in the lignin in one of three ways. First, they could result from the oxidation of cinnamyl alcohol endgroups in lignin, themselves resulting from initial dimerization reactions of monolignols, particularly coniferyl alcohol. The conundrum here is that it is quite difficult to oxidize etherified hydroxycinnamyl alcohols, the decades that the lignin might remain in a tree notwithstanding. A second is for coniferaldehyde or sinapaldehyde to be produced by oxidizing the monolignols in the lignifying region of the cell wall by the action of H_2O_2, for example – such oxidation with H_2O_2 can be demonstrated, and "always" accompanies synthetic lignin preparations in which monolignols are mixed with peroxidase and H_2O_2 (Kim et al. 2003; Zhao et al. 2013). These hydroxycinnamaldehyde monomers can then

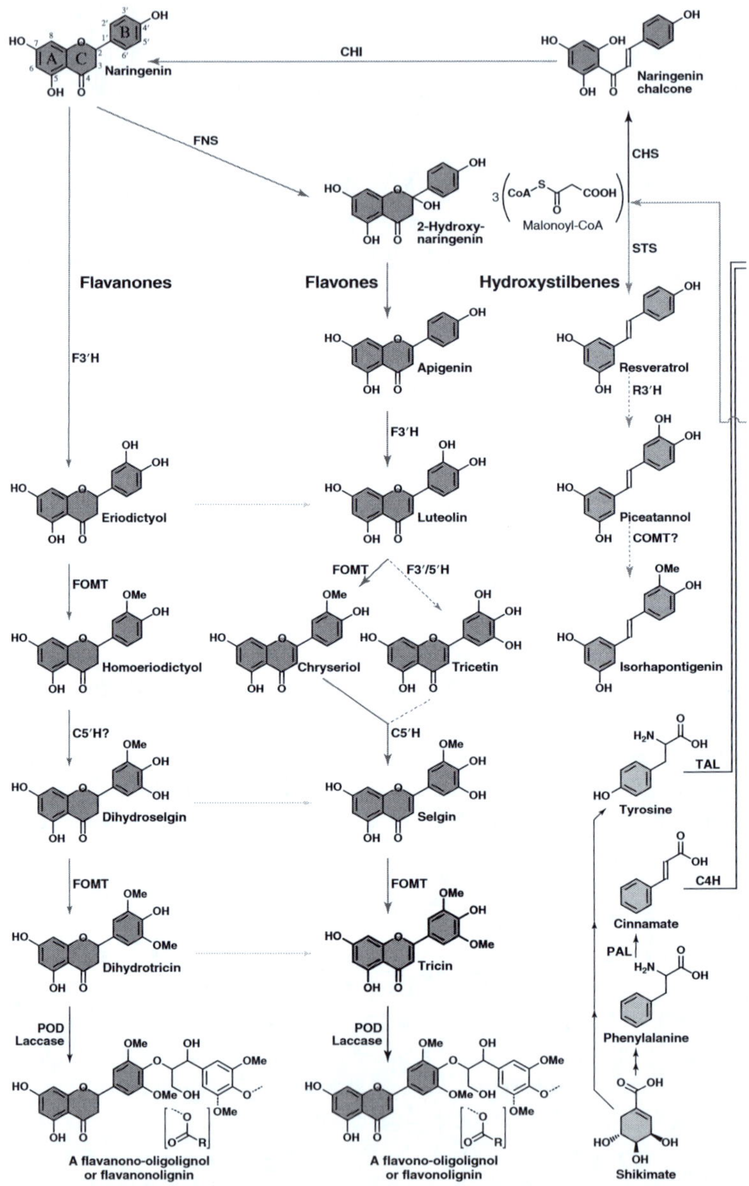

Figure 1.1 Biosynthetic pathways feeding into lignification. The primary biosynthetic pathway leading to the monolignols and the two related hydroxycinnamyl alcohols, bolded and highlighted in yellow (in the colored version of this figure), is to the right and is shown as a metabolic grid as in the old days (Dixon et al. 2001; Higuchi 2006; Matsui et al. 2000); unfortunately, since modifications to the pathway flux have been discovered, it is not possible to order the intermediates in a linear fashion vertically – conversions jump rows in a way that is simply not pleasing (and, were this a circuit-board, would represent bad design but seems unavoidable here). Where possible, the primary pathways are shown with black arrows, and the major pathways are slightly bolder, whereas minor pathways, or those that might not be completely demonstrated are in a light gray; dashed arrows represent pathways (and genes/enzymes) that are not yet known. Note that we do not try to show the pathway from *p*-hydroxybenzoate, via its CoA thioester

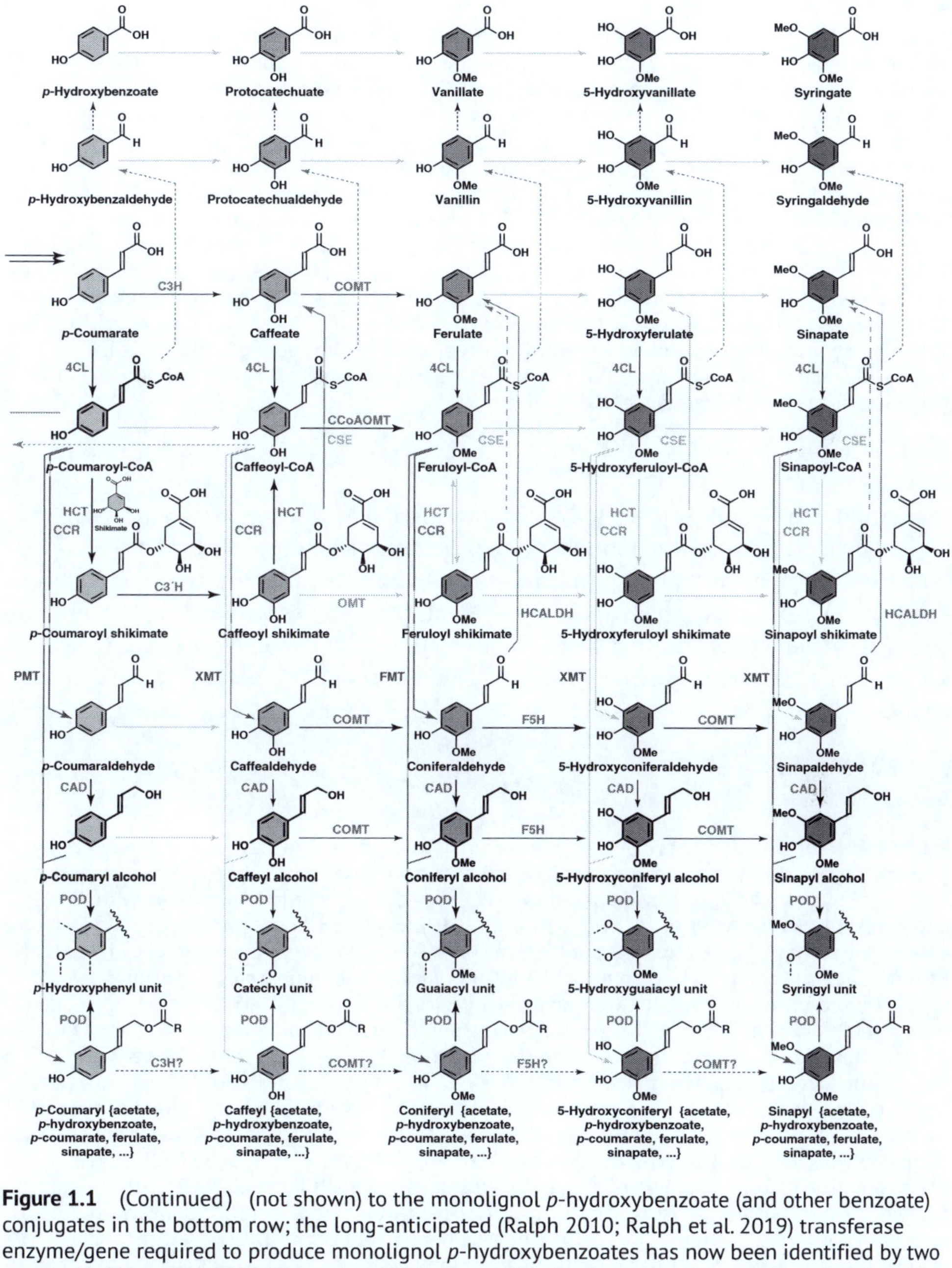

Figure 1.1 (Continued) (not shown) to the monolignol *p*-hydroxybenzoate (and other benzoate) conjugates in the bottom row; the long-anticipated (Ralph 2010; Ralph et al. 2019) transferase enzyme/gene required to produce monolignol *p*-hydroxybenzoates has now been identified by two groups (de Vries et al. 2022; Zhao et al. 2021). For clarity, the five types of aromatic nuclei, i.e. with the aromatic ring substitution at the *p*-hydroxyphenyl, catechyl, guaiacyl, 5-hydroxyguaiacyl, and syringyl level, in their columns are color-coded in common. To the left are the pathways to the hydroxystilbenes and flavonoids that may also be monomers used in lignification. The gray-ringed structures to the left of center at the bottom are the primary inputs into these pathways with the aromatics deriving from shikimic acid, via the amino-acids phenylalanine and tyrosine (Yoo et al. 2013). A major branchpoint for all three pathways is *p*-coumaroyl-CoA, also bolded, and highlighted in pink. Enzyme abbreviations: Monolignol pathway: PAL, phenylalanine

Figure 1.1 (Continued) ammonia-lyase; TAL, tyrosine ammonia-lyase (in monocots); C4H, cinnamate 4-hydroxylase; C3H, *p*-coumarate (or *p*-coumaroyl-CoA) 3-hydroxylase; C3′H, *p*-coumaroyl shikimate 3′-hydroxylase; CCoAOMT, caffeoyl-CoA *O*-methyltransferase; 4CL, 4-coumarate: CoA ligase; CCR, cinnamoyl-CoA reductase; HCT, hydroxycinnamoyl-CoA: quinate/shikimate hydroxycinnamoyl transferase; CSE, caffeoyl shikimate esterase; F5H, ferulate 5-hydroxylase; COMT, caffeic acid *O*-methyltransferase; CAD, cinnamyl alcohol dehydrogenase; HCALDH, hydroxycinnamaldehyde dehydrogenase; PMT, *p*-coumaroyl-CoA: monolignol transferase; FMT, feruloyl-CoA: monolignol transferase; XMT, a general hydroxycinnamoyl-CoA: monolignol transferase; POD, peroxidase (or laccase, not shown except in flavonoid pathway); Hydroxystilbene pathway: STS, stilbene synthase; R3′H, resveratrol 3′-hydroxylase; COMT?, the monolignol pathway COMT or another OMT; Flavonoid pathway: CHS, chalcone synthase; CHI, chalcone isomerase; FNS, flavone synthase; F3′H, flavonoid 3′-hydroxylase; F5′H, flavonoid 5′-hydroxylase or a dual-function 3′/5′-hydroxylase; FOMT, flavonoid-*O*-methyltransferase; C5′H, chrysoeriol 5′-hydroxylase. We heartily thank Ruben Vanholme (VIB) for helping clarify the nuances of the monolignol pathway. (*See insert for color representation of the figure.*)

couple and cross-couple into the polymer following single-electron oxidation analogously to the polymerization of the monolignols themselves (Kim et al. 2003). This possibility comes only with a conceptual problem for researchers wanting to assert the fidelity of lignification as resulting solely from monolignols, as it comes with an implicit recognition that monomer-substitution (by something other than a monolignol, and in fact not an "ol" at all) is already occurring during "all" lignification. Third, it is possible that monolignol biosynthesis in the cytoplasm is not fully complete, even during "normal"

Figure 1.2 (Continued) Proceedings of the National Academy of Sciences. (e) A mulberry variety (Nezumigaeshi) with normal-colored wood (inset) and in which the NMR aldehyde region shows the typical hydroxycinnamaldehyde **X** and benzaldehyde **SA** and **V**) endgroups along with, as we often detect now (Van Acker et al. 2017; Yan et al. 2019), low levels of the product in which sinapaldehyde has 8–*O*–4-cross-coupled with a G-unit in the growing polymer to become part of the internal backbone of the lignin, structure **S′G**, not just an endgroup. *Source:* Yamamoto et al. (2020)/Oxford University Press. (f) The same aldehyde region from the NMR of the lignin isolated from a wild mutant mulberry (Sekizaisou) discovered to have a red-colored wood and leaves that are particularly suited for silkworm feeding that was recently revealed to be a CAD-deficient mutant, the first reported in a hardwood. *Source:* Yamamoto et al. (2020)/Oxford University Press. Units are derived from the 8–*O*–4-cross-coupling of both sinapaldehyde and coniferaldehyde into the polymer (**S′G** and **G′S**, but not **G′G**); by separate experiments, it is possible to determine that sinapaldehyde cross-couples with either guaiacyl or syringyl units, whereas coniferaldehyde cross-couples with only syringyl units (as is also the case *in vitro*) (Kim et al. 2003). The green peak labeled "**Z** (new)" has been observed previously in spectra from our labs from aldehyde-rich lignins and synthetic lignins incorporating coniferaldehyde but had not previously been structurally assigned. (g) The new product has recently been determined to be an unusual benzofuran **Z** that is clearly not a primary coupling product but arises from 8–5-dimerization of coniferaldehyde followed by further single-electron oxidation and then radical disproportionation to a quinone methide intermediate followed by rearomatization (K. Yoshioka, in preparation). Note: Here and in Figure 1.3, we use the standard convention by which the 3-carbon sidechains on monolignol and lignin units are labeled α, β, and γ (from the aromatic C1 attachment), whereas the benzaldehydes and cinnamaldehydes, as well as other "oxidized units" such as ferulate and *p*-coumarate, use 7, 8, and 9. (*See insert for color representation of the figure.*)

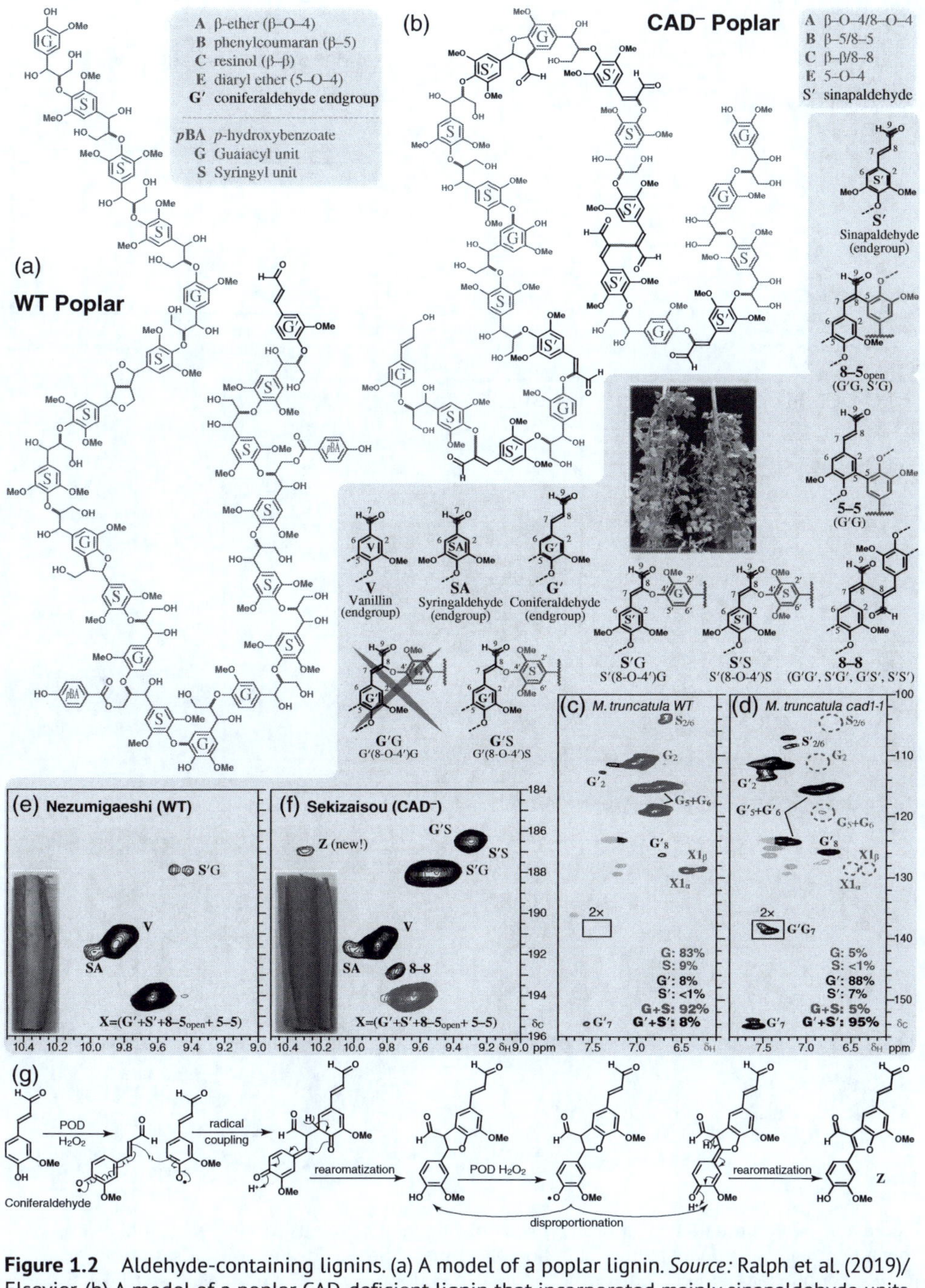

Figure 1.2 Aldehyde-containing lignins. (a) A model of a poplar lignin. *Source:* Ralph et al. (2019)/Elsevier. (b) A model of a poplar CAD-deficient lignin that incorporated mainly sinapaldehyde units. *Source:* Ralph et al. (2019)/Elsevier. (c) The aromatic regions of HSQC NMR spectra from lignins isolated from WT *Medicago truncatula* and (d) a CAD mutant in the same background showing that the lignin is heavily (~95%) hydroxycinnamaldehyde-derived, i.e. that this viable plant has very little of its lignin derived from any of the canonical monolignols. *Source:* Zhao et al. (2013)/with permission of

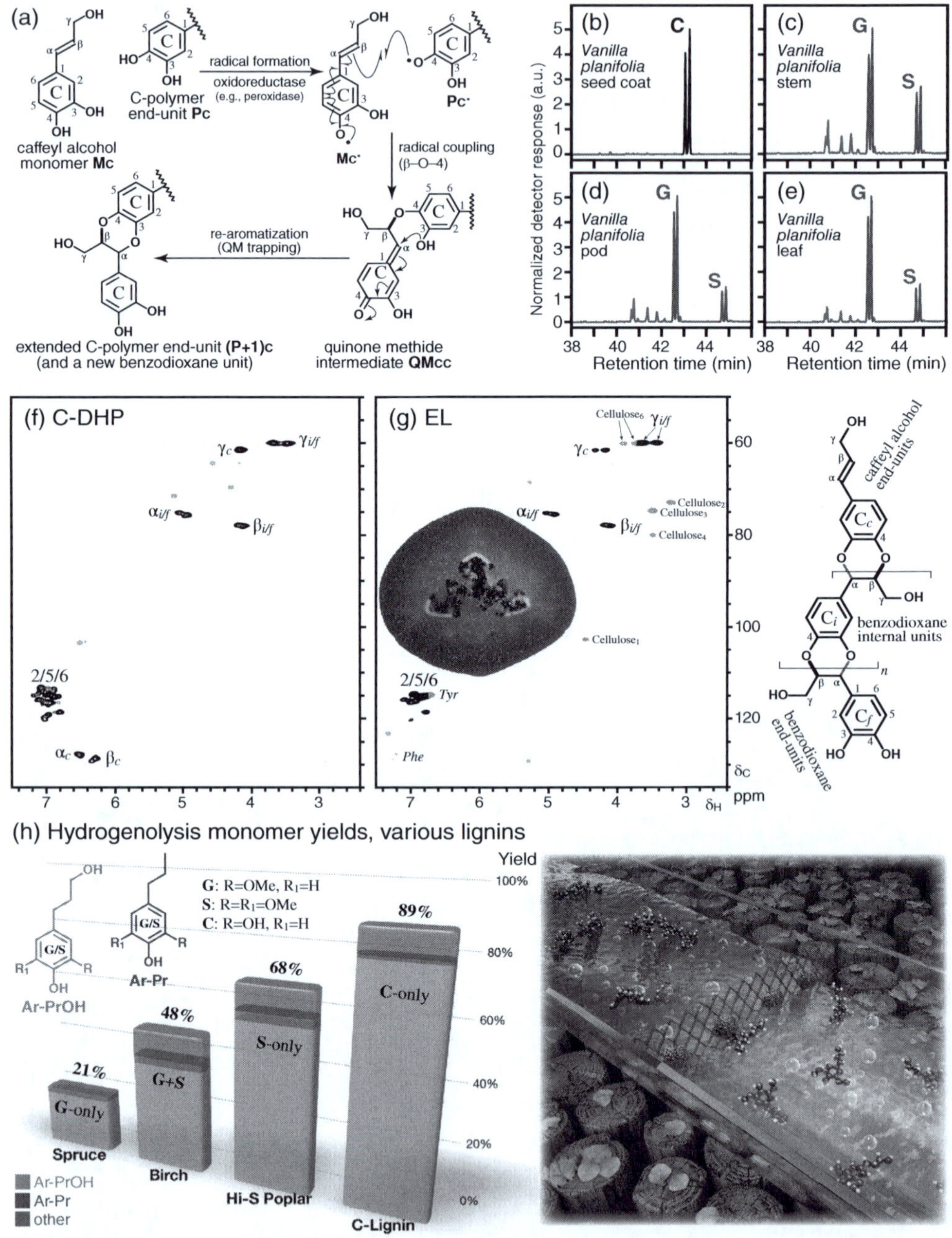

Figure 1.3 C-lignin, derived from caffeyl alcohol. (a) Lignification using caffeyl alcohol as the sole monomer results in an almost pure homopolymer of benzodioxane units arising from β–*O*–4-coupling; although C-units can, in principle, couple with all the options open to G-units, the β–*O*–4-coupling propensity is simply overwhelming such that only low levels of other structures can be found in the polymer. *Source:* Chen et al. (2012)/National Academy of Sciences. (b) Thioacidolysis of vanilla seedcoats releases the typical isomer pair of a sole C-monomer, with no contributions from S- or G-units derived from the canonical monolignols. *Source:* Chen et al. (2012)/with permission of Proceedings of the National Academy of Sciences. (c–e) The stem, pod,

lignification, and that such products of incomplete monolignol biosynthesis simply make it to the wall along with the monolignols, to polymerize purely via their ability to be single-electron-oxidized, and subject to their subsequent chemical coupling and cross-coupling compatibility with the lignification reactions going on around them. As long as transporters required to transport these monomers across the plasma membrane are present, or if transporters are not required as recent *in silico* studies suggest (Vermaas et al. 2019), getting hydroxycinnamaldehydes into the lignin by this mechanism also suffers from no conceptual difficulties in a purely chemical, combinatorial process. We can conclude that a major mechanism is via incorporation of hydroxycinnamaldehyde monomers themselves, the latter two methods, into the polymerization process because, in "natural" wild-type plants, NMR sensitivity is such that we can usually quite easily now detect hydroxycinnamaldehydes that have been incorporated into a growing lignin chain (and are therefore internal units) in addition to the endgroups noted above (Yamamoto et al. 2020), as seen by the S′G structure in Figure 1.2e. At least in the case of CAD-deficient plants, the third mechanism is essentially required, as production of monolignols can be severely curtailed yet, as described below, the lignin is heavily derived from polymerization from hydroxycinnamaldehydes reaching the wall, obviously having originated in that form from the cytoplasm.

CAD-deficient plants, including natural mutants such as brown-midrib varieties of maize (*bmr1*) or sorghum (*bmr6*), have been long known and are now well characterized (Barrière et al. 2004; Jun et al. 2017; Marita et al. 2003). One of the earliest tree mutants was a pine CAD-mutant (Ralph et al. 1997; Sederoff et al. 1999). Across gymnosperms and angiosperms, markers for the incorporation of hydroxycinnamaldehydes, primarily coniferaldehyde and sinapaldehyde, have been identified from the common degradative lignin analytical methods, thioacidolysis, and the derivatization followed by reductive

Figure 1.3 (Continued) and leaf tissues show S/G profiles typical of an angiosperm, with no evidence of C-units. *Source:* Chen et al. (2012)/Proceedings of the National Academy of Sciences. (f) A caffeyl alcohol synthetic lignin (dehydrogenation polymer, DHP) is rather simple, with notable end-units in its HSQC NMR spectrum because of its low molecular weight. The main units are benzodioxanes. *Source:* Adapted from Li et al. (2018). (g) An isolated enzyme lignin (EL) from vanilla seedcoats has a particularly simple HSQC NMR spectrum, showing essentially only benzodioxane units from β–*O*–4-coupling (along with peaks from residual cellulose, C); from the original data used in figure 1a and S1C in Li et al. (2018). The inset is a picture of a sliced-though vanilla seedpod at 10 days post-pollination, showing the black seedcoats containing the C-lignin. *Source:* Chen et al. (2012)/with permission of National Academy of Science. (h) Yield and selectivity of Pd/C-catalyzed hydrogenolysis to produce arylpropanol, arylpropane, and other monomers, for the vanilla seedcoat C-lignin here (in which the arylpropanol monomer is catechylpropanol and the minor arylpropane monomer is catechylpropane) *vs* spruce, birch, and high-S poplar. The C-lignin delivers almost 90% yields of monomers in a product that is 90% one compound, dihydrocaffeyl alcohol (= catechylpropanol); the prior record monomer yield had been from the high-S lignin (Lan et al. 2018a; Shuai et al. 2016). Conditions: 5% Pd/C, 473 K, 15 h, 4 MPa H_2. *Source:* Li et al. (2019)/Royal Society of Chemistry. Picture in the bottom right is part of the picture prepared for a cover issue (Li et al. 2018). Note: Here and in Figure 1.2, we use the standard convention by which the 3-carbon sidechains on monolignol and lignin units are labeled α, β, and γ (from the aromatic C1 attachment), whereas the benzaldehydes and cinnamaldehydes, as well as other "oxidized units" such as ferulate and *p*-coumarate, use 7, 8, and 9. (*See insert for color representation of the figure.*)

cleavage (DFRC) method, and may therefore be used as markers for CAD-deficiency (Kim et al. 2002; Lapierre et al. 2004; Lu and Ralph 1999b). A particularly striking mutant of *Medicago truncatula*, a close relative of alfalfa or lucerne (*Medicago sativa* L.), was identified in Dixon's group and characterized by NMR (Zhao et al. 2013). Despite growing normally in the greenhouse (but admittedly not under high-temperature regimes), the lignin in this plant was derived from hydroxycinnamaldehydes at an astounding 95% level, i.e. its lignin has scarcely even a contribution from normal monolignols (Figure 1.2c, d). A double-CAD knock-out mutant in Arabidopsis produced a lignin that was even more impressively monolignol-deficient, having some >99% coniferaldehyde/sinapaldehyde derivation (Anderson et al. 2015). Traits were also stacked into an F5H mutant (that would normally produce G-only lignin), resulting in Arabidopsis lines derived 99% from just coniferaldehyde (Anderson et al. 2015). Any notion that lignins cannot be derived from monomers other than the three canonical monolignols becomes frankly incongruous in the face of such evidence.

The first hardwood CAD mutant has also been recently revealed (Ikeda et al. 2021; Yamamoto et al. 2020). An unusual red-wood mulberry, Sekizaisou (= "red-colored mulberry wood" in Japanese, see the inset in Figure 1.2f), has been long known and has considerable cultural and historical significance as its leaves constituted a superior silkworm feed for the early silk industry in Japan. Characterization of this plant showed that it was indeed a CAD mutant (Yamamoto et al. 2020). In addition to discovering the usual aldehydic signatures of CAD-deficiency, this sample revealed another minor peak, labeled **Z** in Figure 1.2f, that had been observed in our labs before but had never been identified or reported. Clearly it was time to elucidate this new component. As it was noted to be present in softwoods and appeared to be in a coniferaldehyde-derived synthetic lignin (Hoon Kim, unpublished), we decided to simply use limited-extent radical coupling (dimerization+) to attempt to isolate any component of the mixture with that signature and thus determine its structure unambiguously by the usual array of NMR experiments, supported by hi-res MS analysis. The compound turned out to be derived from 8–5-dehydrodimerization of coniferaldehyde, Figure 1.2g (K. Yoshioka, unpublished). The presence of the double bond in the phenylcoumaran ring is not immediately coherent, however, and requires explanation. As seen in the proposed mechanism in Figure 1.2g, the structure most logically arises from disproportionation of the radical formed from the 8–5-dimer, by disproportionation to the quinone methide followed by rearomatization. Similar disproportionation reactions have been noted for various α-CH_2 compounds such as eugenol (Zanarotti 1983) and dihydroconiferyl alcohol (Ralph et al. 2004b, 2008). One of the implications of finding such structures, intriguingly, is that it suggests that coniferaldehyde is present at sufficient levels to radical-dimerize and that the 8–5-dimer is then present at a sufficient level to not only undergo its own 1-electron oxidation but also to be able to encounter another such radical for disproportionation, i.e. that the concentrations and radical lifetimes must be sufficient to allow this conversion. As such, the structural details regarding coniferaldehyde-derived components in the polymer, revealed by NMR, enhance our appreciation of the environment in these CAD-deficient plants for realizing hydroxycinnamaldehyde-based lignification.

1.2.1.2 OMT Deficiency

Caffeyl alcohol, Figure 1.3a, and 5-hydroxyconiferyl alcohol, two catecholic hydroxycinnamyl alcohols, also shown in Figure 1.1, turn out to be rather good lignin monomers. That is not to say that lignins made from these are ideal for performing the functions required of the polymer, but their incorporation into lignins when canonical monolignol biosynthesis is thwarted again allows plants to retain viability.

Lignins significantly derived from caffeyl alcohol have resisted successful engineering to date; low levels of caffeyl alcohol were incorporated into the lignin in a pine that had the expression of its caffeoyl-CoA: 3-*O*-methyltransferase (CCoAOMT) suppressed (Wagner et al. 2011). However, lignins in the seedcoats from vanilla and various cacti (Chen et al. 2012, 2013; Tobimatsu et al. 2013; Wang et al. 2020b; Zhuo et al. 2019), and the endocarps from jatropha and candlenut (Barta et al. 2014; Li et al. 2018; Rinaldi et al. 2016; Tobimatsu et al. 2013; Yamamura et al. 2012), all derive from high levels of caffeyl alcohol. As has now been well established, the lignins are strikingly linear and homogeneous, Figure 1.3, despite having the same options for coupling, at least on paper, as coniferyl alcohol; coupling is almost entirely β–*O*–4, giving rise, after rearomatization of the quinone methide intermediate by trapping internally with the 3-OH, to benzodioxane units that form long chains (Li et al. 2018). Another feature of these lignins, the high yield of derived monomers shown in Figure 1.3h, will be discussed in Section 1.2.3.

Lignins partially derived from 5-hydroxyconiferyl alcohol have been long known, since their initial discovery in COMT-deficient brown-midrib mutants of maize (*bmr3*) and sorghum (*bmr12*) (Barrière et al. 2004; Lapierre et al. 1988; Marita et al. 2003). Researchers have been considerably more successful in deriving lignins from particularly high levels of 5-hydroxyconiferyl alcohol. In an interesting approach taken simultaneously but independently by two groups, F5H was upregulated in Arabidopsis to help accelerate flux past guaiacyl units and COMT was concomitantly knocked down to prevent the pathway leading fully to sinapyl alcohol (Vanholme et al. 2010b; Weng et al. 2010). The resulting plants contained lignins derived from ~70% 5-hydroxyconiferyl alcohol. Obviously, these lignins are also quite linear and quite regular, also dominated by benzodioxane units in the chain. Similarly high levels of 5H units have not yet been successfully engineered into hardwood models such as poplar, but significant levels (~20%) have been noted, and the characteristic thioacidolysis markers, benzodioxane metabolites, or benzodioxane peaks in the NMR of the lignin or whole cell wall material are easily observed (Jouanin et al. 2000, 2004; Lu et al. 2010; Marita et al. 2001; Morreel et al. 2004b; Ralph et al. 2001a; Ralph et al. 2001b). A cactus, *Escobaria dasyacantha*, appears to have its seedcoat lignin derived solely from 5-hydroxyconiferyl alcohol (Chen et al. 2013).

1.2.2 Phenolics from Beyond the Monolignol Biosynthetic Pathway

When the first monolignol-replacement monomers for lignification were being encountered, there was already speculation about whether non-lignin-pathway phenolics could be used in, conscripted into, or even designed into lignins (Ralph 2006; Ralph et al. 2004b; Sederoff et al. 1999). We and others were certainly wary of the notion, particularly because one of the most logical sources of phenolics could be the lignans, extractive components used for defense that, in some cases, shared the same structure as a dilignol (and in fact originally were by definition – lignans originally referred to dimers of monolignols formed by β–β-coupling, including pinoresinol, for example) (Umezawa 2003). However, lignans were reasonably well accepted as being separated in time and space from lignins (although in-print citations are difficult to unearth) (Ralph et al. 1999). Lignans are optically active (Umezawa 2003), yet all attempts to discover optically active components in the lignin polymer have proved futile, indicating that lignins truly are racemic polymers that do not incorporate lignans (Akiyama et al. 2015; Ralph et al. 1999). In an unpublished study with Joe Karchesy (Oregon State University), we confirmed that Alaskan Cedar contains the syringyl lignan syringaresinol, yet the lignin polymer is, as is typical in softwoods, devoid of detectable syringyl aromatics. If phenolic compounds otherwise identical to lignin dimers that might be produced during lignification are not finding their way to the wall and polymerizing into the lignin polymer, it seemed a little risky to suggest that phenolics from other pathways might.

Phenolics from the flavonoid and stilbenoid pathways are now well-established components of various lignins (del Río et al. 2020, 2021). The first to be discovered was the flavone tricin in wheat straw and, later, in other monocots such as maize, Figure 1.4a (del Río et al. 2012, 2015; Lan et al. 2015, 2016a, 2016b, 2019; Rencoret et al. 2013). The phenolic tricin radical is limited to coupling only at its 4–O-position, and can therefore only start a lignin chain, but it nevertheless cross-couples with monolignols to do just that – a significant fraction of lignin chains in maize have tricin on the starting end, Figure 1.4b, such that the lignin can contain >3% tricin by weight (Lan et al. 2016b). As was later discovered, tricin's incorporation is delightfully combinatorial, Figure 1.4c (Lan et al. 2016a). As various monolignol conjugates, including monolignol acetates and monolignol *p*-coumarates, are also prevalent in maize, metabolic profiling found evidence for many of these dimer and trimer combinations, including T-H, T-G, T-G*p*CA, T-GAc, T-S, T-S*p*CA, T-G-G, T-S-G, T-G-GAc, T-GAc-G, T-G*p*CA-G, T-G-G*p*CA, and T-G-S*p*CA, Figure 1.4b, in which the shorthand is hopefully obvious (Lan et al. 2016a). A phylogenetic study showed that tricin-lignins were a feature of "all" commelinid monocots (Lan et al. 2016b). It is not obvious what advantages the incorporation of tricin gave to monocot evolution, but this is the same plant clade that also has various other cell-wall-defining features including: polysaccharide-polysaccharide cross-linking via hydroxycinnamates (*p*-coumarate, *p*CA and ferulate, FA) acylating arabinoxylans; polysaccharide-lignin crosslinking via (mainly) those same ferulates on arabinoxylans that might act as nucleation sites for lignification; the biosynthesis and incorporation into lignification of monolignol *p*-coumarate (ML-*p*CA) conjugates and, as more recently discovered, monolignol ferulate (ML-FA) conjugates as well; and rather characteristic polysaccharides (Henry and Harris 1997; Karlen et al. 2018; Lan et al. 2016b; Ralph 2020). Tricin-lignins appear in a few lineages beyond the commelinid monocots, including in alfalfa, especially in the leaves (Lan et al. 2016b). The tricin pathway appears to have arisen independently, i.e. via convergent evolution, in this line (Lui et al. 2020; Ralph 2020). As will be noted in Section 1.2.3.3, various other phenolics from flavonoid pathways are being discovered in lignins (Lam et al. 2017; Mahon et al. 2022; Rencoret et al. 2022).

More recently, various hydroxystilbenes have been implicated in lignification, especially in palm endocarp tissues (del Río et al. 2017, 2020, 2021; Elder et al. 2019, 2020, 2021; Rencoret et al. 2019). Although a great deal more needs to be done to provide authentic

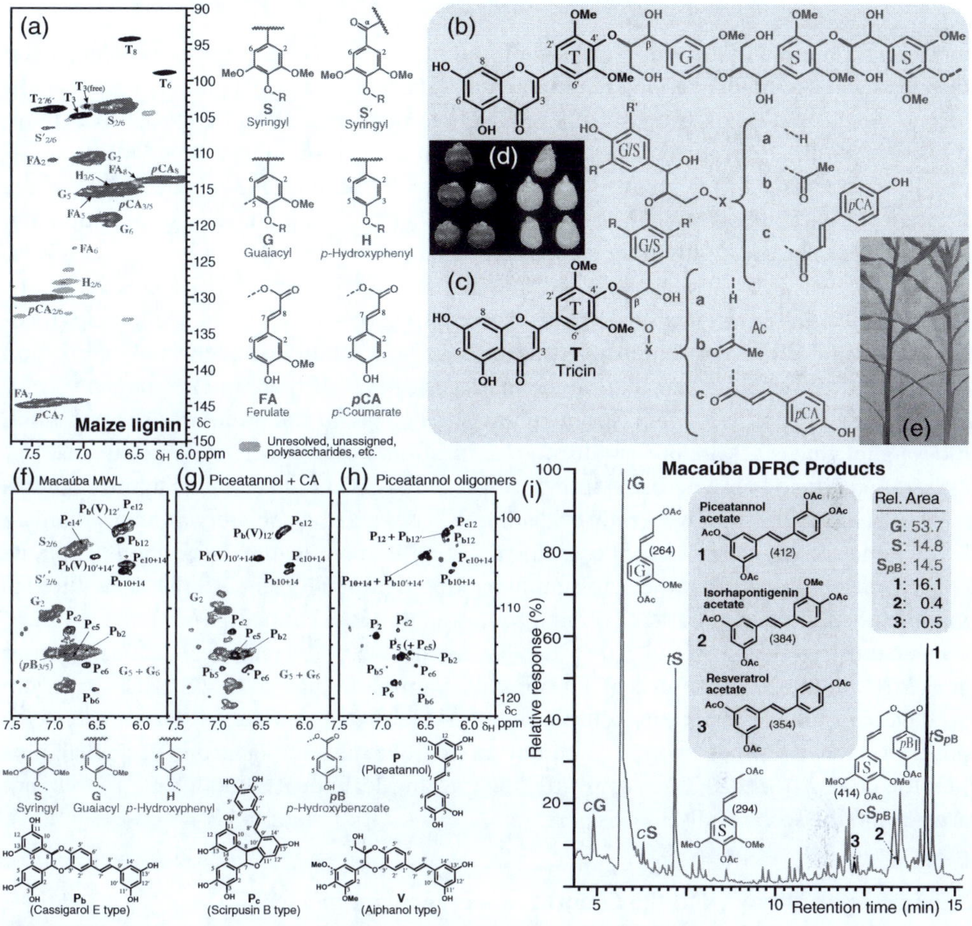

Figure 1.4 Tricin and hydroxystilbenes in lignins. Tricin was first discovered in a wheat straw lignin (del Río et al. 2012). (a) The tricin is clearly evident in "all" grasses, as in the aromatic region of the HSQC-NMR spectrum from maize stem lignin here. *Source:* Lan et al. (2015)/Oxford University Press. (b) Tricin can only start a lignin chain, as shown in this 4-unit model, and is predominantly found coupled to a guaiacyl unit, i.e. from the coupling of coniferyl alcohol with tricin in early lignification, as shown here. (c) Tricin-oligolignol metabolites provided a rich source of information about the combinatorial coupling of tricin with not just coniferyl and sinapyl alcohol, but also with the various acetate and *p*-coumarate conjugates from them, and for the next unit as well. Not all but a great many of the possible combinations indicated have been identified by metabolic profiling. *Source:* Adapted from Lan et al. (2016a). (d, e) Tricin biosynthesis can be completely blocked by knocking out the first committed enzyme (chalcone synthase, CHS, see Figure 1.1) from *p*-coumaroyl-CoA into the flavonoid pathway (Eloy et al. 2017). In a CHS maize mutant, not only is the lignin devoid of tricin (not shown), but the grain and the corn stems lack the dark pigmentation resulting from flavonoids. *Source:* Eloy et al. (2017)/Oxford University Press. (f) Hydroxystilbenes are found in the lignins from various palm fruit endocarp tissues, such as Macaúba, as shown in the aromatic region of HSQC NMR spectra here. *Source:* del Río et al. (2017)/Oxford University Press. (g, h) Synthetic polymers derived from either copolymerizing piceatannol with coniferyl alcohol, or allowing piceatannol alone to polymerize, help identify several of the new peaks in the lignin spectrum in (f). *Source:* del Río et al. (2017)/Oxford University Press. (i) Cleavage of lignin β-ethers using DFRC releases, in addition to the usual G and S monolignols (as their acetates), the acetates of the three hydroxystilbenes piceatannol, isorhapontigenin, and resveratrol, indicating their presence in the lignin as ethers. *Source:* del Río et al. (2017)/Oxford University Press. (*See insert for color representation of the figure.*)

models and NMR data for the various coupling modes, the NMR is quite completely and reasonably assigned for the major components, especially for piceatannol (Figure 1.4f–h) (del Río et al. 2017). As hydroxystilbenes such as piceatannol (and flavones like tricin) possess phenolic OH groups that are not involved in radical coupling, it is possible to incorporate glucosylated piceatannol, such as astringin (piceatannol-3-*O*-glucoside), in the normal way without violating the need for a free-phenolic group for radical formation and coupling (del Río et al. 2021; Elder et al. 2021; Rencoret et al. 2019). Such components were also implicated in lignins, as reviewed in Volume 7 of this series (del Río et al. 2021). We suggested that such glucosylated monomers were a way to explain how phenolic glucosides could be present in lignin, as it seemed to be a conundrum that polymerization required a free-phenolic and that phenolic glucosides could therefore not possibly enter lignification. Recent research has shown how such phenolic glucosides can re-form when monolignol glucosides are present during lignification, however (Miyagawa et al. 2020).

Debate can rage over whether hybrid-pathway polymers should be called lignins, but it is hard to imagine how they can really be distinguished when the polymerization mechanism is the same combinatorial radical coupling that typifies lignification, and when the plants appear to be telling us that they are simply making lignin polymers. All that is required is that the phenolic appear in the lignifying zone, and that it be compatible with the lignification chemistry (Boerjan et al. 2003; del Río et al. 2020; Morreel et al. 2004a; Mottiar et al. 2016; Ralph 2010; Ralph et al. 2004b, 2006; Vanholme et al. 2008, 2012). Clearly tricin and hydroxystilbenes, including glucosylated variants, satisfy both conditions. Other components, such as various hydroxycinnamoyl amides, are also implicated in various lignins (del Río et al. 2018, 2020; Kaur et al. 2012; Negrel and Jeandet 1987; Ralph et al. 1998) but are beyond the scope of discussion here.

1.2.3 Lignin Design, and the Concept of an Ideal Lignin

Once it is realized that non-monolignol phenolics may be incorporated into the lignin polymer, and that such avenues have already been explored in natural evolution, it becomes evident that researchers can at the very least attempt to alter lignins and introduce various components of their own choosing into the polymers *in planta*. This became obvious from the very first perturbation studies in which novel components arose in lignins, such as the hydroxycinnamaldehydes and the two catecholic monomers noted in Section 1.2.1. Broader opportunities were apparent the instant the phenolics tricin and hydroxystilbenes were revealed to be authentic lignin monomers after being discovered as lignin components in various plants (del Río et al. 2012, 2017). We could accept the products that are generated in "natural plants" and in response to gene perturbations, and follow the lead for new phenolics, but would we have the audacity to assume that we might actually be able to dictate new pathways to plants, and create lignins of our own design? And would plants then tolerate such lignins? As we illustrate, we are guilty of underestimating the reach of natural plant development in at least one case, and of arrogantly assuming that we could teach plants a new trick. We always acceded and often stated, however, that some natural plant(s) might already be utilizing any of the plant-compatible ideas that researchers were envisioning.

What would be the purpose of lignin design? At least two major goals are logical (Grabber et al. 2008, 2010; Mansfield et al. 2012; Mottiar et al. 2016; Ralph et al. 2019; Vanholme et al. 2012). The first is to produce a lignin that is easier to depolymerize in existing chemical or biological pretreatments. Lignin depolymerization and solubilization of the released lower-MW components improve the access to the valuable cellulose polymer and enhance polysaccharide accessibility to saccharification enzymes. The second is to add value to the lignin. Technoeconomic analysis and lifecycle assessment demonstrate that lignin valorization is desirable or needed for a successful biorefinery (Corona et al. 2018; Davis et al. 2013, 2018; Huang et al. 2020). In addition to the emergence of "lignin-first" methods that aim to preserve lignin quality for integrated or subsequent conversion to useful products (Abu-Omar et al. 2021), the notion of adding value into the lignin by actually incorporating valuable components right into the biomass feedstock is proving interesting. Obviously, such components need to be (readily) releasable from the polymer to realize their value, and this is far from guaranteed from a polymer that is formed by combinatorial radical coupling chemistry, but such difficulties may be somewhat offset by appreciating the incredible scales, even at just a minor percentage of the lignin, that can be realized by generating valuable compounds in the abundant lignin polymer.

The complexity of lignins often results in complex mixtures of products. Processes like pyrolysis are notorious for producing huge mixtures, whereas oxidation and reductive catalytic fractionation (RCF, or hydrogenolysis) can produce relatively simple mixtures with relatively few, albeit still difficult to separate, components. Another way of enhancing value of lignin is by simply enhancing its homogeneity such that fewer products result from various depolymerization schemes. Researchers began speculating about what kinds of lignins could be most suitable for biorefinery operations, and what kinds of polymer redesign might be pursued. One type of natural lignin proved to have some strikingly ideal properties, as noted below in Section 1.2.3.2.

1.2.3.1 Zip-Lignins

There may have been other ideas envisioned and perhaps even pursued in the same timeframe, but we contend here that the first idea to engineer a new lignin to improve biomass processing came from our labs many years before even the initial publication of the idea in 2008 (Grabber et al. 2008) and its eventual realization in 2014 (Wilkerson et al. 2014). This was the idea to design "zip-lignins" by engineering the production, *in planta*, of new monolignol conjugates that, following polymerization, would result in lignins that had readily chemically cleavable bonds designed into the backbone of the polymer. As summarized in Figure 1.5, the method consisted of inducing plants to biosynthesize ML-FAs to enter lignification. Once genetic methods had developed sufficiently, and thanks to skilled transformation by Mansfield's group, the basic premise succeeded, producing plant materials, including poplar trees (Wilkerson et al. 2014), with unaffected growth and development that indeed had superior processing under a variety of conditions, including pulping (Bhalla et al. 2018; Kim et al. 2017; Zhou et al. 2017). Improvements are still being pursued today as the processing improvements have real economic potential, especially if the levels of esters in the lignin backbone can be further elevated. Better transferases genes/enzymes have been discovered (Bartley et al. 2013; Karlen et al. 2016; Smith et al. 2022a).

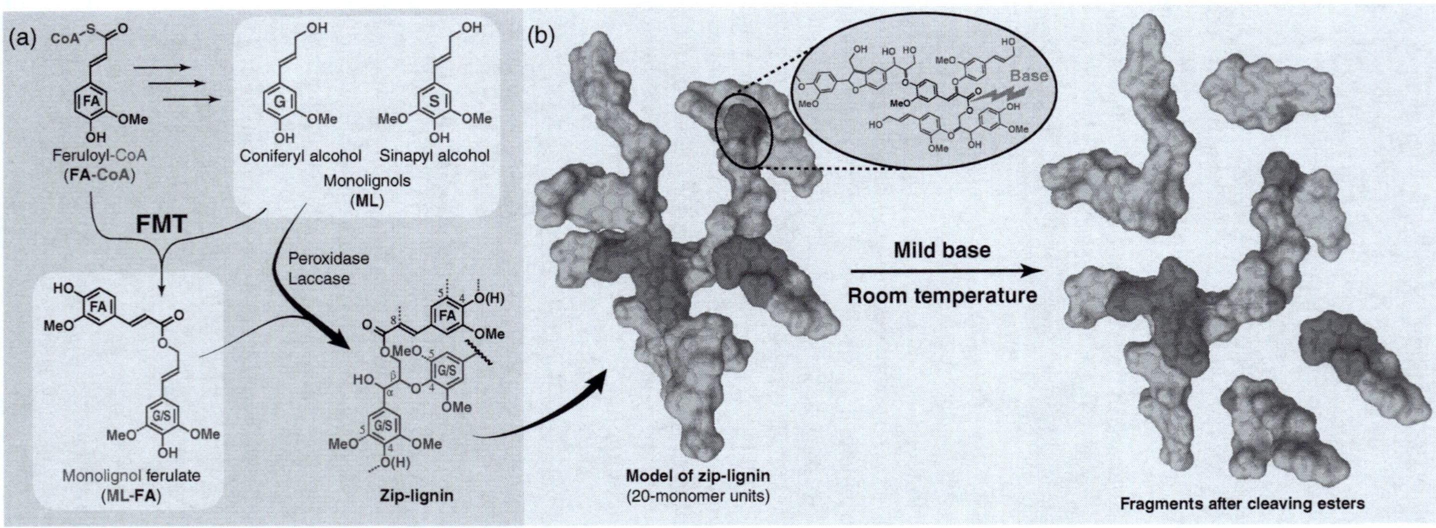

Figure 1.5 Zip-Lignins from lignification with monolignol ferulate (ML-FA) conjugates. (a) A feruloyl-CoA: monolignol transferase (FMT) can be used to produce ML-FA conjugates that can serve as monomers for lignification (Grabber et al. 2008; Ralph 2010; Wilkerson et al. 2014). The resultant lignin has ferulate endgroups but also ferulates that are incorporated into the lignin backbone. *Source:* Adapted from Wilkerson et al. (2014). (b) Such units (dark highlighting in the 3D model) therefore create ester bonds in the polymer backbone, bonds that can be readily chemically cleaved, as shown under a variety of conditions (Bhalla et al. 2018; Kim et al. 2017; Zhou et al. 2017), to enhance delignification and improve access to polysaccharides. *Source:* Adapted from figure 1a, b in Karlen et al. (2016).

We devised new diagnostic and sensitive methods, based on the DFRC method that was established to cleave ethers while retaining esters (Lu and Ralph 1999a, 2008, 2014; Lu et al. 2015; Petrik et al. 2014; Karlen et al. 2016; Regner et al. 2018; Wilkerson et al. 2014), to determine that such conjugates were indeed used in lignification and were creating zip-lignins in our transgenics (Wilkerson et al. 2014). It was not until the emergence of this tool that we discovered that such lignins were already being made naturally, at least with low-level zips (Karlen et al. 2016). Nature is not, of course, driven by the same aims as we have – to think that plants were making themselves easier to chemically degrade at 170 °C in caustic soda for a single annoying branch of evolving hominids is obviously absurd – but merely for competitive, survival, and fitness purposes, and spurred by the random mutation coupled with genetic selection that characterizes evolution. Thus, several times during evolution, various "new" lignin traits have appeared and been sustained to this day. Among them are the zip-lignins, lignins derived in part from ML-FA conjugates (Karlen et al. 2016).

We do not know how zip-lignins evolved, nor what advantage might have been conferred on the plant lines imbued with such a feature. The trait is not found in softwoods but evolved with all grasses that we have surveyed and, in fact, "all" commelinid monocots (Karlen et al. 2016). Lignification using ML-FA conjugates can therefore now be added to the set of special (but not exclusive) monocot plant cell wall features noted above. Sinapyl ferulate conjugates were found to be prevalent in the monocot lignins, whereas coniferyl ferulate was proportionally more heavily featured in the hardwoods (Karlen et al. 2016); zip-lignins appear to have evolved separately in various hardwoods, about half of the small range we have examined. It is tempting to suggest that humans, with their desire to pulp trees for cellulosic fiber, have unwittingly selected for ML-FA incorporation as the favored pulping hardwoods all seem to have the zip-lignin trait. This was intriguingly revealed in a conversation with Prof. Troy Runge (University Wisconsin-Madison) in which he suggested that we should look at maple to support or refute the emerging hypothesis. We had to report to him the disappointing news that the maple we chose appeared to not possess zip-lignin. On discovering that we had selected a hard maple, he commented that hard maple is terrible for pulping and that we needed to look at red maple. As it turned out, red maple did indeed have the zip-lignin signatures, as reported (Karlen et al. 2016). The broad evidence remains insufficient to suggest the veracity of the hypothesis that we have inadvertently selected hardwoods for their zip-lignin trait, but the notion remains intriguing, nevertheless.

Grasses already possess zip-lignins and pretreat more efficiently than hardwoods or softwoods (Karlen et al. 2016). Grass lignins are quite soluble in mild base, a feature that is likely due in part to their zip-lignins and the cross-linking of the cell wall via ferulates on arabinoxylans (Ralph 2010). Realizing an incremental improvement by enhancing zip-levels in grasses is harder but it remains worthwhile to determine what factors might be used to enhance the incorporation of ML-FAs. In the model grass brachypodium, it was demonstrated that removal of competition for monolignol and CoA substrates could enhance levels significantly (Smith et al. 2021). Stacking the expression of an exogenous *FMT*, *AsFMT* from *Angelica sinensis*, in a mutant deficient in PMT (*p*-coumaroyl-CoA: monolignol transferase), the transferase enzyme responsible for producing monolignol *p*-coumarate conjugates, was successful at elevating ML-FA levels by as much as 32-fold over wild-type levels (Smith et al. 2021). Attempts at reducing the flux through feruloyl-CoA to monolignols, through a mutation in maize targeting cinnamoyl-CoA reductase

(CCR), was also successful at elevating zip-levels at the expense of lignin level, but not without deleterious impacts on plant growth and development (Smith et al. 2017a).

Further new discoveries regarding monolignol conjugates are covered in Section 1.3.

1.2.3.2 The Concept of an "Ideal Lignin"

Presented with a completely new and surprising type of lignin, the C-lignin derived entirely from caffeyl alcohol, discovered initially in the seedcoat of the vanilla bean (Chen et al. 2012), led astute young researchers to contemplate the notion of an "ideal lignin" for bioprocessing operations (Li et al. 2018). Such a lignin might benefit from the following properties, for example: "First, if acidic pretreatment is used, then it should be stable under acidic conditions to prevent condensation and the generation of undesired new C—C bonds. Second, it should contain only ether (C—O) interunit linkages in its backbone so that it can be fully depolymerized. Finally, it should be generated *in planta* from a single phenylpropanoid monomer to allow the production of the simplest array of compounds" (Li et al. 2018). High-syringyl-lignin biomass comes close to meeting these goals; as has been long touted, poplar (and Arabidopsis) with ~98% S-lignin (Ciesielski et al. 2014; Franke et al. 2000; Huntley et al. 2003; Marita et al. 1999; Meyer et al. 1998; Skyba et al. 2013; Stewart et al. 2009), and its consequent linearity and high β-ether content, allows hydrogenolytic reactions to yield ~78% monomers with some 90% of those monomers being a single compound, for example (Shuai et al. 2016). This lignin does not satisfy the first condition, stability in acid, although condensation-protection methods have been successfully developed and applied (Abu-Omar et al. 2021; Behaghel de Bueren et al. 2020; Lan and Luterbacher 2019; Lan et al. 2018a; Questell-Santiago et al. 2020; Shuai et al. 2016; Talebi Amiri et al. 2019).

It turns out that the C-lignin derived from, solely, caffeyl alcohol rather admirably meets these three criteria for the above concept of an ideal lignin, including the third that it is from a single aromatic monomer (Li et al. 2018), Figure 1.3. Despite having many conceivable coupling pathways, analogously to those for coniferyl alcohol, potentially leading to β–*O*–4, β–5, β–β, β–1, 4–*O*–5, and 5–5 interunit linkages, caffeyl alcohol's strong propensity is for β–*O*–4 coupling, leading to an essentially linear structure dominated by chains of benzodioxane units (Chen et al. 2012, 2013; Tobimatsu et al. 2013), allows it to meet the second criterion – a polymer with (essentially) only C—O bonds between its units. As a result, hydrogenolysis, one of the most efficient methods to cleave lignin ethers, including those in benzodioxanes, liberates extremely high yields (~90%) of monomers from C-lignin biomass, Figure 1.3h (Li et al. 2018; Stone et al. 2018; Wang et al. 2020a). When coupled with the 90% selectivity achievable for a single product, the catechylpropanol using Pd/C, or the catechylpropane from Rh/C (not shown), for example, these lignins allow an unprecedented 90% yield of monomers that are 90% one single product (Li et al. 2018). Finally, because the quinone methide intermediate resulting from the preferred β–*O*–4-coupling has an internal trapping pathway, via the 3-OH, for rearomatization, the β-ether units in C-lignins differ from those in normal lignins (derived from the canonical monolignols) by being cyclic benzodioxanes, and by not possessing a benzylic-OH (Chen et al. 2012, 2013; Tobimatsu et al. 2013). The primary source of condensation products from lignins under acidic conditions is via attack of the readily formed and quite stable benzylium ions (benzylic carbocations), formed by protonation of the benzylic-OH and E1 elimination of water, on the electron-rich aromatic rings to produce stable new C—C bonds (Li et al. 2018). The benzodioxane units offer protection against such reactions. In fact, C-lignins have been shown to be quite stable even under the harsh 72% sulfuric acid conditions of Klason lignin measurement (Li et al. 2018). The first condition of an ideal lignin is therefore also met, with C-lignins being all but impervious to the acid hydrolysis conditions that may be used to liberate cellulose or the polysaccharides for saccharification and fermentation.

C-lignins are essentially linear homopolymers that are useful in their own right (Berstis et al. 2016; Nar et al. 2016) and are acid-stable yet can be converted to useful catecholic monomers in high yield (Li et al. 2018; Stone et al. 2018; Wang et al. 2020a). They are therefore potentially so useful in a biorefinery context that there is significant impetus to attempt to induce plants to produce (higher levels of) C-lignins, including in their structural tissues. Downregulating the *O*-methyltransferases in softwoods had not been particularly successful – other methylation activities appear to be available (Wagner et al. 2011). Research groups are actively trying to shed light on C-lignin biochemistry (Zhuo et al. 2019). Recently, it was revealed that a specific laccase in *Cleome hassleriana* favors caffeyl alcohol (and sinapyl alcohol) oxidation over coniferyl alcohol and may be involved in biosynthesizing the pure seed-coat polymer after the switch from G- to C-lignin occurs at ~12 days after pollination (Wang et al. 2020b). We do not yet know whether plants will tolerate C-lignin as a replacement for G/S-lignin, but applying tissue-specific methods that are rapidly advancing (Cao et al. 2020; Fanelli et al. 2021; Grimming et al. 2008; Hoffmann et al. 2020; Joo et al. 2021; Meyer et al. 1998; Pyo et al. 2007; Schuetz et al. 2013; Somssich 2020; Takenaka et al. 2018; Vander Mijnsbrugge et al. 1996; Yu et al. 2021) may aid in its tolerance by plants.

1.2.3.3 Introducing New Pathways into Lignification, New Units into Lignins

As noted above, the zip-lignin idea was originally intended to introduce a new trait to plants, but it turns out that such lignins were already being naturally produced at low levels in certain plants. Of the other example approaches described below, we concede that plant evolution may have already explored (some of) those avenues as well.

With the discovery of the flavone tricin in monocot (and a few other) lignins, it is logical to perturb genes in the tricin pathway to determine if intermediates along that pathway might also be incorporated into lignins. Knocking out the pathway completely was first used to illustrate that the component could be manipulated. Thus, a maize mutant deficient in the very first enzyme, chalcone synthase (CHS), Figure 1.1, in the committed flavonoid pathway from the common intermediate *p*-coumaroyl-CoA, was indeed found to produce no tricin in its lignin (Eloy et al. 2017). The plants grew similarly (at least in the greenhouse) and were a striking green color, without the reddish hues in, particularly, the stem and the grain, presumably caused by flavonoids and other components in the tissues, Figure 1.4d, e. As is reasonable for the loss of the enzyme at the beginning of the pathway, no intermediates or other unusual structures were noted in the lignins, but the lignin level was higher presumably as the *p*-coumaroyl-CoA in the pathway was not diverted toward these other components, and the enzyme digestibility of the wall was lower, presumably because of the higher lignin level and because, without the tricin initiation sites, lignin chains were longer on average. Additionally, the lignin structure showed signs that lignification now lacked the chain initiation afforded by tricin, i.e. by having more units (phenylcoumarans and resinols) derived from monolignol dimerization. Tricin levels could also be drastically reduced in rice by downregulating an OMT involved in the last methylation step in tricin biosynthesis (Lam et al. 2019b). Disruption in rice of the P450 flavonoid 3′-hydroxylase (F3′H), the enzyme that hydroxylates the B-ring of apigenin to produce luteolin in the pathway to tricin (Figure 1.1), resulted in a lignin that incorporated apigenin instead of tricin (Lam et al. 2019a). Disruption of FNSII, the enzyme that catalyzes the direct conversion of the flavanone naringenin to the flavone apigenin (Figure 1.1), resulted in a lignin incorporating the intermediate naringenin instead of tricin (Lam et al. 2017).

Because of the multistep pathway and naringenin's and apigenin's incorporation suggesting that intermediates along the tricin pathway could be targeted toward lignification, a fertile new set of modifications to lignin composition and structure could be envisioned. Unlike for tricin itself that can only initiate lignin chains, some of the intermediates have the potential to incorporate into the growing chain because they are not limited to solely 4–O-coupling.

Various other phenolics from flavonoid pathways are beginning to be discovered in natural lignins. For example, naringenin chalcone, naringenin, and dihydrotricin (in addition to tricin), from different flavonoid families, have been discovered and authenticated in papyrus (Rencoret et al. 2022; Rosado et al. 2021), paving the way to consider other classes of flavonoids as lignin monomers. There is evidence that naringenin, for example, has already been incorporated into the lignins in poplar and rice transgenics (Lam et al. 2017; Mahon et al. 2022).

Curcumin was targeted by Boerjan's group as having the potential to improve lignin degradability because it might produce more readily base-cleavable conjugated β-ether units in the polymer. Although that goal was not fully realized, curcumin was successfully incorporated into Arabidopsis lignins, as determined by its characteristic fluorescence and by metabolic profiling but was not authenticated in the polymer by NMR (Oyarce et al. 2019). Scopoletin has been incorporated into lignins in Arabidopsis plants by introducing a 6-hydroxylase (F6H1 or F6′H1) (Vanholme et al. 2019; Sakamoto et al. 2020; Hoengenaert et al. 2022).

1.3 Lignin Conjugates, "Clip-Offs' – New Discoveries, and Enhancing Levels

Various acids have long been known to acylate cell walls in general and lignins in particular (Harris and Hartley 1976, 1980; Ralph 2010; Smith 1955a). Some time ago we provided the evidence that lignin acylation occurred at the monomer level, i.e. that acylated lignins arose from lignification in which acylated monolignols augmented the canonical monolignols as lignin monomers (Ralph 2010; Ralph et al. 1994). Monolignol acetates, *p*-hydroxybenzoates, *p*-coumarates, and, more recently, ferulates have therefore become recognized as authentic lignin precursors, produced by extensions of the monolignol biosynthetic pathway, Figure 1.1. More recently, various other conjugates have been identified, including simple benzoates and vanillates (Karlen et al. 2017; Kim et al. 2020); monolignol acetates, benzoates, *p*-hydroxybenzoates, vanillates, *p*-coumarates, and ferulates have all been found to be present in the lignin of a single plant, Canary Island date palm (*Phoenix canariensis*) (Karlen et al. 2017).

Monolignol conjugates can be used to alter lignins for two purposes. First, as with the zip-lignin technology noted above, alterations to lignin composition and structure can facilitate pretreatment and delignification of plant cell walls. Second, if the components are solely esterified to the wall, they may be easily clipped off to produce relatively clean streams of a small array of simple phenolics (Karlen et al. 2020; Timokhin et al. 2020). Phenolic acids have established markets and significant value. It is tempting to surmise that uses would expand if these compounds were produced sustainably on a much larger scale. It has often been noted that lignins and the phenolic acids represent the largest sustainable source of aromatics available. Although it is difficult to be price-competitive with the petrochemicals that are produced on an incongruously large scale, the planet is already unable to accommodate the ramifications of a reliance on fossil resources. Often, many synthetic steps are required to functionalize simple fossil-derived aromatics such as benzene, toluene, and xylene, into the various phenolic commodity chemicals in use today, yet plants have often provided us with aromatics having the substitution patterns required. It borders on tragic to not be taking advantage of the natural functionality available.

We illustrate the near-absurdity of deriving the common commodity chemical *p*-aminophenol, and the chemical and pharmaceutical acetaminophen (= paracetamol, = Tylenol®) from benzene by comparing the process with that from natural *p*-hydroxybenzoate that can be clipped off various biomass sources and straightforwardly and efficiently converted to these compounds, Figure 1.6 (Mobley et al. 2017). The substitution of the aromatic in *p*-hydroxybenzoate is already in place, allowing ready access to the amide that is then easily converted to the amine by the Hoffman rearrangement (Hofmann 1881), and trivially acetylated in the final step as in the industrial process. By comparison, just getting to phenol from benzene is a major three-step process, albeit benefitting from the clever

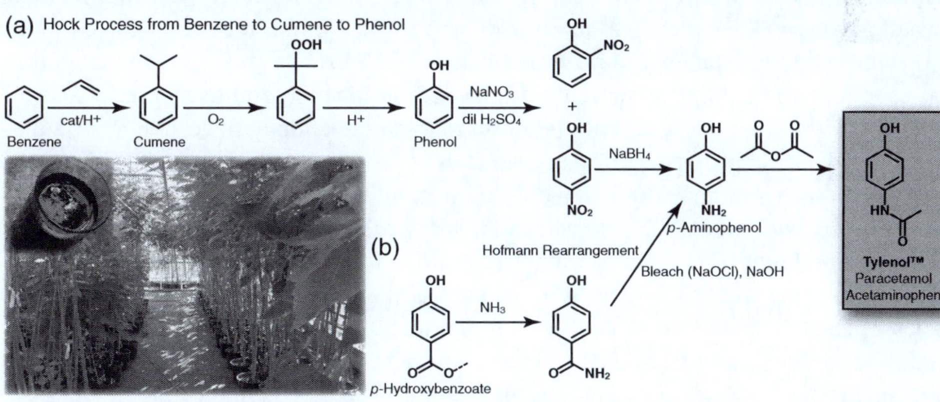

Figure 1.6 Synthesis of acetaminophen (paracetamol, Tylenol®), from fossil *vs* biomass sources. (a) Synthesis from fossil-based benzene requires, for example, a several-step hydroxylation to produce phenol via the Hock process (Hock and Lang 1944), nitration followed by separation to produce *p*-nitrophenol, reduction to *p*-aminophenol, and final acetylation to produce acetaminophen. (b) Because the *p*-hydroxybenzoate on poplar, willow, or palm lignins (Lu et al. 2015; Ralph 2010) naturally has the correct substitution pattern, it is simply a matter of producing the amide from the ester followed by a classical Hofmann rearrangement (Hofmann 1881) to produce the *p*-aminophenol followed by the same acetylation step, a significantly shorter and more rational scheme (Mobley et al. 2017). The picture inset is of greenhouse-grown poplar (courtesy of Shawn Mansfield, UBC) *vs* one fossil source, coal tar (inset, https://www.himadri.com/images/coal_tarpitch_cover_2.jpg), for sourcing the starting materials.

Hock reaction (Hock and Lang 1944). Then nitration is not regiospecific, and only one of the isomers is used for reduction to the amine. In the absence of cheap benzene, sourcing *p*-aminophenol and acetaminophen from it would be imprudent.

1.3.1 Clip-Offs and Their Elevation

Although *p*-coumarate and *p*-hydroxybenzoate esters, being phenolic, can form radicals and undergo radical coupling under *in vitro* polymerization conditions, under radical-limited conditions their primary fate is radical transfer (to other more stable phenolics) rather than radical coupling (Hatfield et al. 2008; Ralph 2010; Ralph et al. 2004a). They are therefore present entirely or largely as free-phenolic appendages acylating lignin side-chains (Ralph and Landucci 2010; Ralph et al. 1994). As esters are easily cleaved, such components have come to be termed "clip-offs" in our own group and beyond (Karlen et al. 2020; Rinaldi et al. 2016; Timokhin et al. 2020), and constitute a viable way of gaining value from the aromatic component of the plants in which they are found – grasses for the *p*-coumarates, and poplar/aspen/willow and palms for the *p*-hydroxybenzoates. The long-anticipated (Ralph 2010; Ralph et al. 2019) transferase enzyme/gene required to produce monolignol *p*-hydroxybenzoates has now been identified by two groups (de Vries et al. 2022; Zhao et al. 2021), and bacterial genes have recently been used to elevate levels in poplar (Mottiar et al 2022). Introducing clip-offs into plants that do not have them is also garnering attention. Monolignol *p*-coumarates and ferulates have been successfully engineered into Arabidopsis, a useful model in the sense that it is normally devoid of the involvement of such conjugates in its lignification (Smith et al. 2015, 2017b; Wilkerson et al. 2014).

As noted, it is easy to clip off the acids *p*-coumaric and *p*-hydroxybenzoic acid from their cell wall polymers (Karlen et al. 2020; Rinaldi et al. 2016; Timokhin et al. 2020); they may fall off under the pulping conditions used to isolate cellulose, or the pretreatment conditions used to prepare plant polysaccharides for saccharification and fermentation. Given the value of these phenolic acids and their uses in the production of novel polymers, adhesives and coatings, pharmaceuticals, cosmetics, and nutritional products (Timokhin et al. 2020), it seems almost criminal to not clip them off to provide value to the biorefinery. Nothing comes without a cost, however, and it is difficult to do so economically even with seemingly simple processes such as mild base extraction. Technoeconomic analyses are beginning to show what factors need to be addressed to enhance the economic viability (Karlen et al. 2020; Liao et al. 2020; Timokhin et al. 2020). A major driver is to "simply" elevate the level of these clip-offs on lignins. As such, various studies have been aimed at enhancing the levels of *p*-coumarates in grasses. In brachypodium, upregulating the endogenous *BdPMT* elevated *p*-coumarate levels by as much as 32-fold (Petrik et al. 2014; Smith et al. 2021). In maize, upregulating the endogenous *ZmPMT* or introducing the *BdPMT* from brachypodium has had only limited success to date (Karlen et al. 2020).

Although this chapter deals mainly with lignins, it is worth remembering that the clip-offs *p*-coumarate and ferulate are also associated with polysaccharides in grasses (see later in Figure 1.11). In the context of a biorefinery operation, the polymer source of useful phenolic acid clip-offs is of only academic relevance, so enhancing the levels of clip-offs in biomass in general is a worthwhile goal. Introduction of a sugarcane BAHD transferase (ScAT10) under the direction of the strong constitutive maize ubiquitin promoter resulted in up to 75% increase in total *p*-coumarate content (Fanelli et al. 2021). Mild hydrolysis and derivatization followed by reductive cleavage (DFRC) analysis (see below in Section 1.6) showed that the *p*-coumarate increase was restricted to the hemicellulosic portion of the cell wall. As the total ferulate content was reduced up to 88%, resulting in a 10-fold increase in the *p*-coumarate/ferulate ratio, this approach is promising for reducing the costs associated with purifying individual hydroxycinnamates and valorizing *p*-coumarate production in biorefineries.

It is worthwhile here to speculate on the levels that might be obtained. Strictly we do not have this information at hand but note the following with respect to conjugates and clip-offs. First, if all available units in lignin were to be acylated (by lignification using 100% monolignol conjugates) for a softwood like *Picea abies*, lignin molecular formula $C_9H_{7.92}O_{2.40}(OMe)_{0.92}$ (average MW per unit of 182.8), and for a hardwood like *Fagus silvatica*, lignin molecular formula $C_9H_{7.49}O_{2.53}(OMe)_{1.39}$ (average MW per unit of 199.1) (Pettersen 1984), there would be literally equimolar clip-offs and lignin units, translating to ~27, 46, and 50 wt% acetate, *p*-hydroxybenzoate, and *p*-coumarate for the softwood and ~25, 43, and 48 wt% for the hardwood, for example. This may be overoptimistic, but there are plants in which monolignol acetate conjugates comprise some 80 mol% of the lignin monomers (del Río et al. 2007, 2008; Martinez et al. 2008). Mature maize stem material studied decades ago as one of the first samples showing the *p*-coumarate to be regiospecifically on the γ-OH of lignin sidechains, thus implicating monolignol conjugates as the source of lignin *p*-coumaroylation, had been noted to be ~17 wt% *p*-coumarate on lignin (Ralph et al. 1994). Attempts to produce higher levels have not yet been particularly successful. *p*-Hydroxybenzoates, primarily on S-units, on poplar are typically only at about the 5–10% level (de Vries et al. 2022; Goacher et al. 2021; Smith 1955b; Zhao et al. 2021), but the level is higher on oil palm empty fruit bunches (Lu et al. 2015), and coconut coir (Rencoret et al. 2013). In a seagrass, *Posidonia oceanica*, particularly high acylation (~60 mol%) by *p*-hydroxybenzoate occurs on both G- and S-units (Rencoret et al. 2020). All-told there seems to be some upside potential in crop grasses if not trees before hitting a level that has deleterious agronomic consequences.

1.3.2 Exploring Monolignol Conjugates in Compositionally Extreme Lignins

The model dicot Arabidopsis, with its simple genome, easy transformation, the availability of a vast collection of single-gene mutants, and the complete absence of monolignol conjugates in its lignins, provides an excellent background in which to tease out details relating to lignification with monolignol conjugates in plants in which only single monolignols comprise the lignin monomer pool. For example, what happens when transferase genes for making conjugates are introduced into the F5H knockout (*fah1*) mutant that normally produces a solely-G lignin (Marita et al. 1999; Meyer et al. 1996, 1998)? Is it possible to even produce *p*-coumarate and ferulate esters when the hydroxylase activity is so upregulated in a transgenic overexpressing the *F5H* gene under the control of the *C4H* promoter (*pC4H::F5H*) that normally produces an almost solely-S lignin (Marita et al. 1999; Meyer et al. 1998; Shuai et al. 2016; Stewart et al. 2009), i.e. are the CoA esters sufficiently available to allow the monolignol conjugates to form before being shunted to S-products, and would high-S plants produce sinapyl sinapate instead of sinapyl ferulate, for example?

Similarly, what happens with *p*-coumarate and ferulate conjugates when genes are knocked out such that the lignin nucleus cannot proceed beyond the H-unit level in the *C3H* null (*ref8, reduced epidermal fluorescence*) mutant background, growth-restored via also mutating *mediator5a* and *mediator5b* subunits (Bonawitz et al. 2014; Franke et al. 2002)? Also, as we know that *p*-coumarate units remain largely free-phenolic, engaging in radical transfer reactions with S and G units rather than radical coupling to incorporate them integrally into the polymer chain, would *p*CA units be more compatible with lignification involving just *p*-coumaryl alcohol, i.e. would HA-*p*CA conjugates constitute a new type of zip-lignin in high-H plants? This idea is intriguing as it takes the opposite tack of the original proposal, i.e. instead of designing ML-FAs so that the conjugate is compatible with the lignification, this approach utilizes the native *p*-coumarate conjugates in grasses and alters the lignin to be radical-coupling-compatible with those *p*-coumarates.

With these and other questions in mind, Prof. Clint Chapple's group (Purdue University) collaborated to produce a set of Arabidopsis mutants/transgenics in which PMT and FMT were separately introduced into H-only, G-only, and S-only backgrounds, Figure 1.7 (Smith et al. 2022b). In answer to the questions posed above: PMT or FMT introductions into the G-only background of the *fah1* mutant do indeed produce the corresponding coniferyl *p*-coumarates or ferulates (with little overlap); PMT or FMT introduction into the high-S *pC4H::F5H* transgenic still allows the production of coniferyl and, primarily, sinapyl *p*-coumarate, but essentially only sinapyl ferulate, and no monolignol sinapates were detected implying that the *p*-coumaroyl-CoA and feruloyl-CoA intermediates that are rather quickly shunted toward S-units in the pathway are nevertheless sufficiently available to form these conjugates; and attempted PMT introduction into the H-only *ref8/mediator*-based line was unfortunately seedling-lethal so we could not directly answer the question (below) of whether *p*-coumarate was more radical-coupling-compatible with H-lignins *in vivo*, but we do have some evidence now for etherified *p*-coumarates in the G-only (*fah1 PMT*) and the high-S (*C4H::F5H*) lines, indicating that these do appear to incorporate their *p*-coumarate moieties into the polymer backbone producing the new brand of zip-lignins (Smith et al. 2022b).

Plant biomass in which lignins are *p*-coumaroylated appears to pretreat better and saccharify more completely after pretreatments; it is not entirely clear why, particularly given that *p*-coumarate units remain pendent free-phenolic units and are, for the most part, not involved in the polymer backbone. Part of this study was to delineate whether *p*-coumarate esters might be more compatible with H-lignins in which case they might be better incorporated into the polymer backbone producing another form of zip-lignin. In synthetic lignins resulting from *in vitro* copolymerization of *p*-coumaryl alcohol with synthetic *p*-coumaryl *p*-coumarate, Figure 1.7d, the HSQC NMR spectrum in Figure 1.7e revealed various products in which *p*-coumarate had cross-coupled with *p*-coumaryl alcohol or with H-lignin, demonstrating the compatibility of *p*-coumarate with radical coupling reactions in an H-lignin environment. Evidence was also obtained from DFRC that the *p*-coumarate was partially etherified and had therefore polymerized into the chain. As it turns out, however, we also discovered that *p*-coumarate could incorporate quite nicely into the high-S lignins; it was basically just in the G/S lignins in normal grasses that it did not incorporate well.

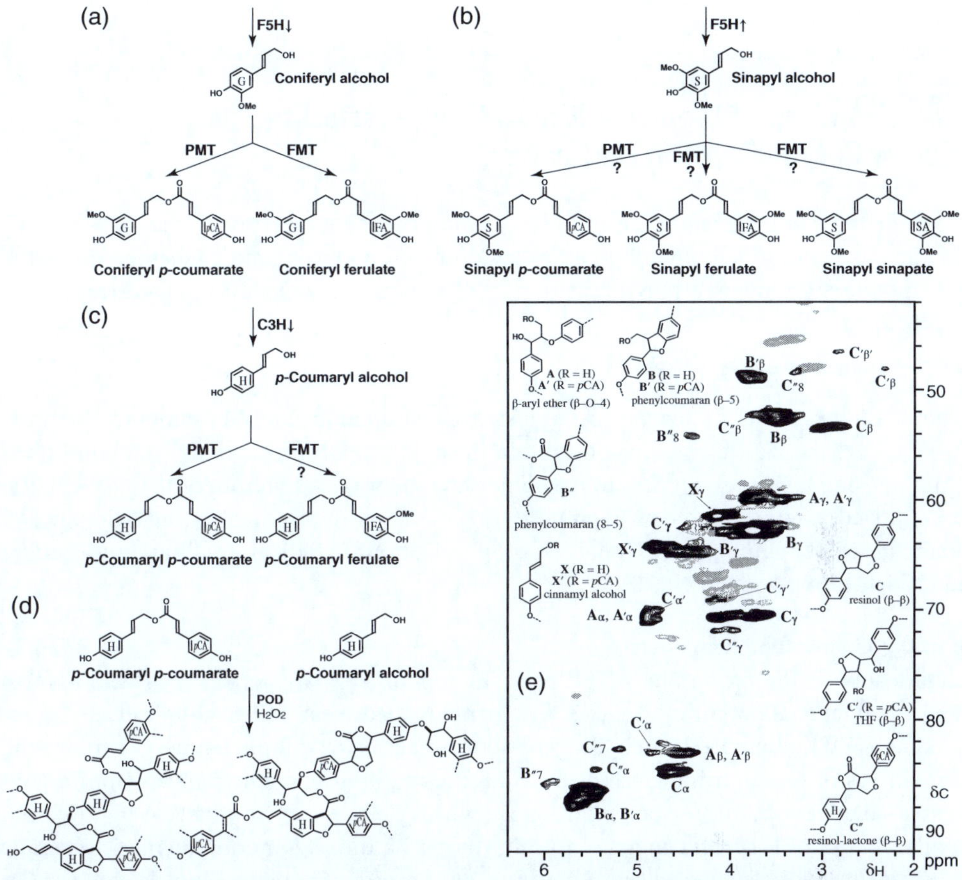

Figure 1.7 Composition-extreme Arabidopsis mutants with introduced hydroxycinnamoyl-CoA: monolignol transferases, PMT and FMT. Arabidopsis is ideal as it normally does not produce monolignol conjugates for lignification (Smith et al. 2015, 2022; Wilkerson et al. 2014). (a) Introducing PMT or FMT into the F5H knockout (*fah1*) mutant, that normally produces solely-G lignin (Marita et al. 1999; Meyer et al. 1996, 1998), allows the production of coniferyl *p*-coumarate and coniferyl ferulate to be examined, as well as the incorporation of those conjugates into guaiacyl lignin. (b) Introducing PMT or FMT into a transgenic overexpressing the *F5H* gene under the control of the *C4H* promoter (*pC4H::F5H*) that normally produces an almost solely-S lignin (Marita et al. 1999; Meyer et al. 1998; Stewart et al. 2009), allows the production of sinapyl *p*-coumarate and sinapyl ferulate to be examined, as well as the incorporation of those conjugates into syringyl lignin (Smith et al. 2022b). It also permits a determination of whether sufficient *p*-coumaroyl-CoA and feruloyl-CoA are available from the monolignol biosynthetic pathway for trapping by the transferase to produce the *p*-coumarate and ferulate conjugates when the pathway flux is so streamlined toward syringyl-level intermediates and products. It also allows an assessment of whether such S-upregulation would result in sinapate conjugates, specifically sinapyl sinapate, a particularly sought zip-conjugate – it does not. (c) Introducing PMT or FMT into the C3H null (*ref8*) mutant, growth-restored via also mutating *mediator5a* and *mediator5b* subunits, that normally produces a solely-H lignin (Bonawitz et al. 2014; Franke et al. 2002), allows the production of *p*-coumaryl *p*-coumarate and *p*-coumaryl ferulate to be examined, as well as the incorporation of those conjugates into H-lignin. It permits a determination of whether such mutants are capable of producing feruloyl-CoA to produce ferulate conjugates when the pathway is truncated at the level of H-aromatics. It also allows an assessment of whether such *p*-coumarate units can incorporate into the backbone of such H-lignins, allowing the *p*-coumaryl *p*-coumarate conjugate to produce a new kind of zip-lignin. *Source:* Adapted from Smith et al. (2022b). (d) That latter point is also examined by elucidating the structures resulting from *in vitro* copolymerization of *p*-coumaryl alcohol with synthetic *p*-coumaryl *p*-coumarate. *Source:* Adapted from Smith et al. (2022b). (e) As seen in the interpreted HSQC NMR spectrum, various products showing that *p*-coumarate has cross-coupled with *p*-coumaryl alcohol or with H-lignin units are identifiable, demonstrating the compatibility of *p*-coumarate with radical coupling reactions in an H-lignin environment.

1.4 Features of Lignification and the Possibility of New Polymerization Pathways

A few features of lignification need to be clarified now that new researchers are entering the field. It is also time to highlight a few conundrums and to examine whether any of the new monomers allow new polymerization mechanisms to operate during lignification.

1.4.1 Features of Lignification

Lignification remains somewhat mysterious. Skilled organic chemists sometimes express surprise over the radical coupling nature and the high proportion of ether- *vs* carbon-linked units, causing lignin researchers to reexamine the theory's status from time to time. Little has changed, however, and new hypotheses or paradigms appear to fall to the rationale and predictive power of the original theory of polymerization via radical coupling (Freudenberg and Neish 1968).

1.4.1.1 Lignification Reminders

Lignification is the process by which phenolic monomers polymerize, *in planta*, to form lignin polymers (Boerjan et al. 2003; Freudenberg and Neish 1968; Ralph et al. 2004b; Sarkanen and Ludwig 1971). Dehydrogenation (1-electron oxidation) using H_2O_2-requiring peroxidases or O_2-requiring laccases provides the required phenolic radicals from monolignols, oligolignols, and/or polymers, as shown in Figure 1.8a for guaiacyl lignins from coniferyl alcohol **1**. Direct contact of the enzymes with the oligomeric/polymeric phenolic substrate may not be required as radical-transfer may occur via radical shuttles such as Mn (III) oxalate (Önnerud et al. 2002), via intermediaries that are more easily oxidized such as *p*-coumarate (Hatfield et al. 2008; Ralph 2010; Ralph et al. 2004a; Takahama et al. 1996), or via monomer radicals themselves (Boerjan et al. 2003; Ralph 2010; Ralph et al. 2004b, 2019; Sasaki et al. 2004; Takahama 1995; Vanholme et al. 2012).

Once the phenolic radical is formed, the subsequent coupling mechanism has features that may surprise organic chemists. Unlike the radical polymerization of styrene to polystyrene, chain-extension is a quenching reaction between a monomer radical and a radical on the phenolic end of the polymer (Figure 1.8a). The newly produced phenolic endgroup is required to be re-oxidized for further polymerization. A likely advantage of such an apparently energy-inefficient process is that it avoids the proliferation of reactive radicals that wreak havoc with living systems.

The initial reaction in the "pure lignification system," as illustrated in Figure 1.8a for coniferyl alcohol, is the (dehydro)dimerization of two monolignol radicals (via β–β-, β–5-, or β–*O*–4-coupling) to produce dehydrodimers **3**. Alternatively, other phenolics in the cell wall may operate as nucleation sites (not shown); examples include ferulate on arabinoxylans in monocots (Ralph 2010), and tricin (del Río et al. 2012; Lan et al. 2016b). Next, the process is combinatorial and entirely chemical, not enzyme- or protein-directed (Freudenberg and Neish 1968; Ralph et al. 1999, 2008). The resulting polymer gains two new optical centers with each coupling of a monomer with another phenolic, resulting in polymers that are structurally and stereochemically diverse. Even a simple β-ether 20-mer homopolymer derived from solely β–*O*–4-coupling has 38 optical centers and consequently

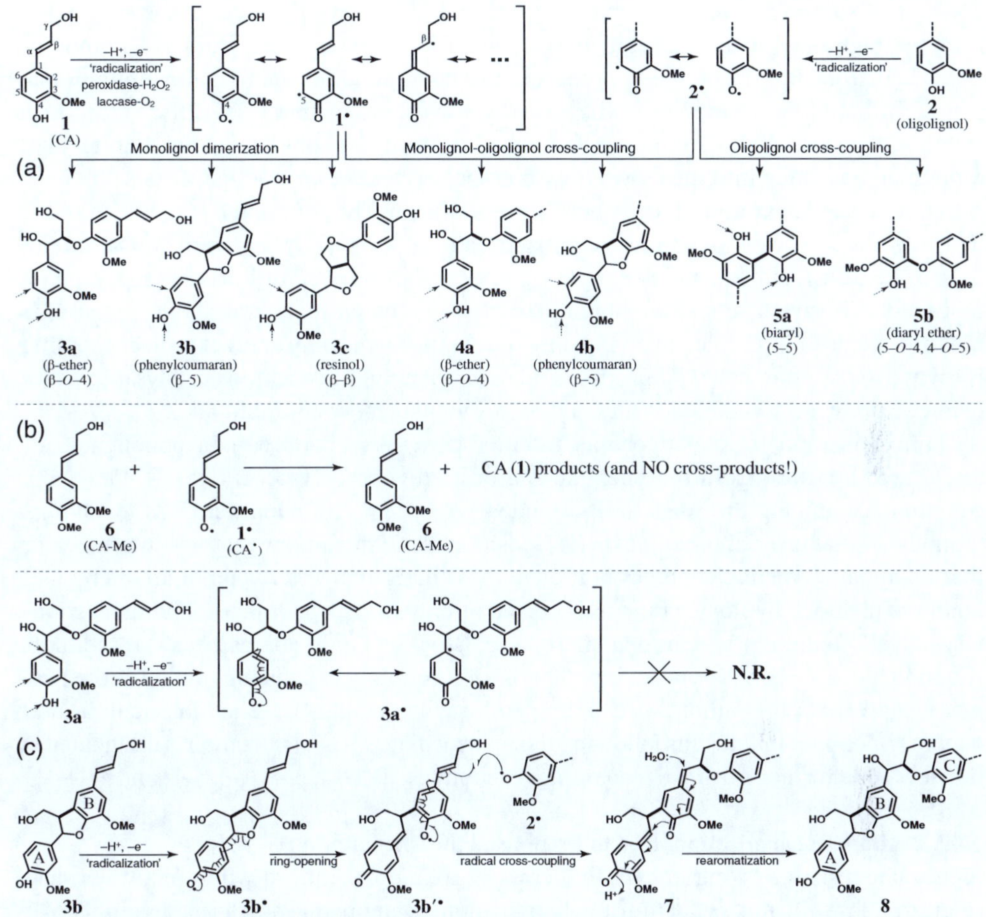

Figure 1.8 Aspects of radical generation, radical coupling, and lignification mechanisms. (a) Generation of the radical **1**• from coniferyl alcohol **1**, a monolignol, and the radical **2**• from an oligolignol or lignin chain **2**. Dimerization of **1** (via its radical **1**•) produces, following rearomatization reactions, one of 3 primary dimers **3a–c**; Cross-coupling of a monolignol **1** with an oligomer **2** (endwise coupling), the main reaction in lignification, (again via the radicals **1**• and **2**• of each) lengthens the chain by one unit and produces new end-units **4a, b**, introducing two new optical centers in the polymer; Cross-coupling can also join two growing chains **2** to produce 5–5- and 4–*O*–5- or 5–*O*–4-coupled units **5a, b**. The bonds formed during the radical coupling step are highlighted by bolding. Additional coupling to continue the lignification can occur at the positions noted by the small dashed arrows; obviously, these new products also have the general structure represented by **2**. (b) An attempt to cross-couple a monolignol **1** (via its radical **1**•) with an etherified (phenol-protected) monolignol such as 4-*O*-methylconiferyl alcohol (3,4-dimethoxycinnamyl alcohol) **6** results in quantitative recovery of starting material **6** along with the usual array of products from the monolignol **1**. (c) A special case in which it appears that lignification on an etherified unit is occurring (Matsushita et al. 2019). In the phenylcoumaran dimer **3b**, the generated radical **3b**• can ring-open, producing the B-ring phenolic radical **3b′**•, which can then undergo radical coupling at its β-position with, e.g. a phenolic unit radical **2**• to produce the intermediate **7**. Following normal rearomatization, the product is **8**, a compound that appears as though it derives from coupling of a monolignol with the etherified B-unit in **3b**, but it is not actually derived that way. *Source:* Ando et al. (2021)/Royal Society of Chemistry/CC BY 3.0. Note: Compound numbers relate only to this figure. N.R.: no reaction.

2^{38} optical isomers, or half that many, 2^{37} (over 137 billion), possible physically distinct isomers. Lignin, as a racemic polymer, is therefore an isomeric nightmare (Ralph and Landucci 2010). But the coupling process is also combinatorial such that a monomer may couple with another monomer or the growing polymer to form several different products even if polymerization is from a single monomer (such as coniferyl alcohol, Figure 1.8a). Monomers assemble into the polymer with no prescribed order, and the various units produced have various coupling regiochemistries available. The coupling is not, however, statistically random (as sometimes suggested) but reflects coupling and cross-coupling propensities in a statistically weighted manner. More possibilities arise if the supplied monomer can vary. The resulting lignin structure is, therefore, complex but the lignification process itself is simple, involving only the single chemical process of radical coupling. Biosynthesis of other cell wall polymers, for example, the hemicellulosic polysaccharides, involves many more genes/enzymes and is enormously more complex.

Chain elongation, the major polymerization pathway, is via coupling of a monolignol radical **1•** with a radical from the phenolic end of a growing polymer **2** (via 4–*O*–β- or 5–β-coupling), producing the single-unit-extended polymer **4**. Coupling reactions involving a monolignol that invariably couples at its β-position result in quinone methide intermediates that rearomatize via nucleophilic addition at the α-position by, e.g. water or an internal aliphatic or phenolic hydroxyl group as observed after β–*O*–4, β–β, or β–5 coupling reactions, respectively, to dimers **3a–c** or oligomers **4a**, **b**. Growing lignin chains **2** may cross-link by coupling at their phenolic ends, via 5–5- or 4–*O*–5-coupling, producing units **5** in the polymer (Figure 1.8a; the minor β–1 coupling pathway is not shown). The dimeric unit in an oligomer/polymer that results following the intermediate's rearomatization is designated by the new bond formed during the coupling reaction (e.g. β–*O*–4, β–5, β–β, 5–5, and 4–*O*–5).

1.4.1.2 Does Polymerization Have to Occur from the Phenolic End?

Lignin chemists have long known that cross-coupling of a radical with a neutral species (Figure 1.8b) will not occur during lignification. Demonstrating that a free-phenol is required and lignification is via radical coupling and not chain propagation, mixing coniferyl alcohol **1** and 4-*O*-methylconiferyl alcohol **6** in a peroxidase/H_2O_2 *in vitro* system (Figure 1.8b) will result only in products from reactions of coniferyl alcohol **1** alone; the etherified monolignol **6** will be quantitatively recovered intact (Ando et al. 2021).

Is it possible to extend the lignin from an etherified hydroxycinnamyl alcohol at the non-phenolic end of the chain? It is well established in the lignin literature that such extensions are not possible as a phenolic radical generated in a structure containing an etherified hydroxycinnamyl alcohol unit is not capable of propagating single-electron density to that sidechain double-bond. For example, Figure 1.8c, in dimer **3b**, the ring B hydroxycinnamyl alcohol is etherified, and therefore cannot enter into any coupling reactions even after generating the ring-A phenolic radical **3b•** – it is not possible to transfer single-electron density to the β-carbon in that B-ring double-bond. An apparent exception was recently revealed in the rather special case of this phenylcoumaran dimer, Figure 1.8c (Matsushita et al. 2019). Because of an actual chemical reaction pathway, the phenylcoumaran radical **3b•** can open the ring to produce a B-ring phenolic radical **3b′•** that now has a resonance form in which single-electron density is at the B-ring β-carbon. Radical coupling can then take place in the B-ring moiety by the usual coupling reactions available to that radical. Following rearomatization and production of the phenylcoumaran ring, comparison of the structure of the product such as **8** (Figure 1.8c) and the starting material **3b** appears to reveal an extension of the chain via the double-bond of the phenol-etherified unit. This clever discovery (Matsushita et al. 2019) provides an apparent exception, but it must be stressed that this process involves true chemical reactions and, as such, does not represent a violation of the strict basic contention that phenol-etherified units cannot inherently generate radicals that, via electron delocalization, could undergo radical coupling.

1.4.1.3 Do New Monomers Propound Possibilities for New Polymerization Mechanisms?

Do the new types of units identified allow other possibilities by which etherified hydroxycinnamyl alcohol end-units might extend the chain in lignins? The catechyl and 5-hydroxyguaiacyl units (C- and 5H-lignin units) noted above in Section 1.2, as well as the catecholic stilbene piceatannol recently discovered in various fruit endocarp tissues and bark (del Río et al. 2017, 2020; Rencoret et al. 2018, 2019), open up the possibility of a new lignification mechanism.

Catecholic monomers could produce *ortho*-quinones under the dehydrogenative (oxidative) conditions of lignification, Figure 1.9; quinones, including *o*-quinones are known plant metabolites (Eyong et al. 2013). As such quinones could conceivably act as the diene component of classical Diels–Alder reactions, with cinnamyl alcohol units, even

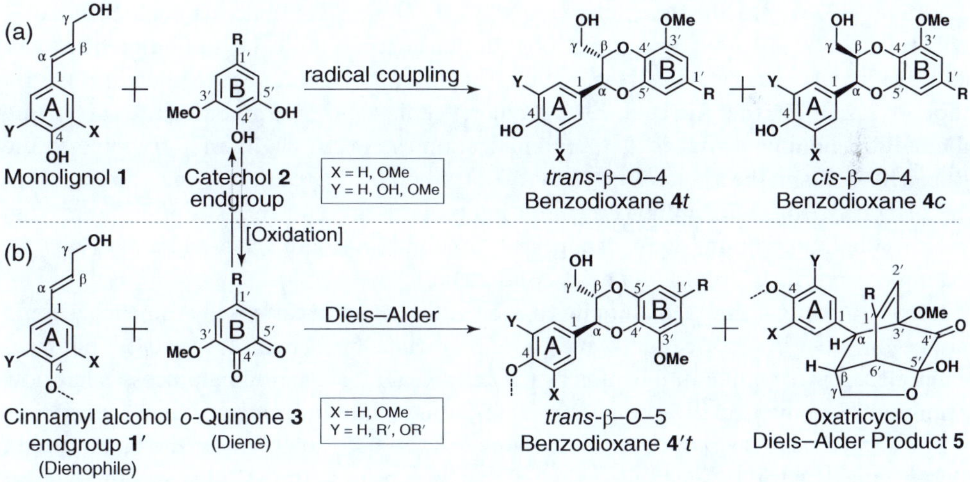

Figure 1.9 Radical coupling of catechols and catechol monomers *vs* Diels–Alder pathways and products determined from *in vitro* reactions on model compounds. (a) Radical coupling of a monolignol **1** with a general catechol lignin end-unit **2** produces *cis*- and *trans*-benzodioxanes **4** from β–*O*–4 coupling (Ando et al. 2021). (b) Diels–Alder reactions of a general cinnamyl alcohol unit **1′** (usually on the starting end of a polymer chain) with an *o*-quinone **3** derived for catechol **2**. A hetero-Diels–Alder reaction produces the *trans*-benzodioxane **4′** that would be described as β–*O*–5-coupled were it derived from radical coupling. The traditional Diels–Alder reaction produces a more traditional oxatricyclo product **5** (following internal trapping of a ketone) (Ando et al. 2021). Bicyclo[2.2.2]octanes (including 5 here) from Diels–Alder reactions have HMBC correlations between a proton and carbons 2- or 3-bonds separated that initially seem to be too numerous to be physically possible, but are in fact a beautiful signature of such structures. For example, proton α correlates with 9 carbons: 6′, β, γ, 3′, 2, 6, 1, 2′, and 4′ (in chemical shift order); proton β correlates with 6 carbons: 6′, α, γ, 3′, 5′, and 1; the 2′ proton correlates with 5 carbons: 6′, 5′, 1′, 7′, and 4′; and proton 6′ correlates with 7 carbons β, α, 5′, 1′, 2′, 7′, and 4′. *Source:* Ando et al. (2021)/Royal Society of Chemistry/CC BY 3.0. Compound numbers relate only to this figure.

those present in etherified structures, operating as the dienophile (Ando et al. 2021). In the special case of the catecholic monomer model, methyl 5-hydroxyvanillate (R = COOMe, Figure 1.9), reacting with non-phenolic cinnamyl alcohols, two types of Diels–Alder reactions did indeed occur *in vitro* (Ando et al. 2021). As summarized in Figure 1.9b, a hetero-Diels–Alder reaction could produce *trans*-benzodioxanes **4′** but with different regiochemistry than the *cis*- and *trans*-benzodioxanes **4** produced via radical coupling, Figure 1.9a. The normal Diels–Alder reaction produced oxatricyclo products **5**, Figure 1.9b, that have the distinction of having an extraordinary number of long-range ^{1}H–^{13}C correlations in HMBC spectra – see the Figure 1.9 caption. We have not been able to find any evidence for the diagnostic products in the lignins examined to date, however (Ando et al. 2021). What is clear is the presence of the *cis*-benzodioxane rings for both C-, 5H-lignins, and hydroxystilbenes (Chen et al. 2012, 2013; del Río et al. 2017; Li et al. 2018; Tobimatsu et al. 2013; Wang et al. 2020b) is consistent with radical coupling but not Diels–Alder reactions, but the latter cannot strictly be ruled out.

Without being able to identify diagnostic Diels–Alder coupling products in lignins, radical coupling represents the only verifiable chain-extension reaction occurring during lignification. It should also be noted that, to date, the only products from 3,4-dihydroxy monomers such as caffeyl alcohol, 5-hydroxyconiferyl alcohol, and piceatannol are those in which the radical derives from the 4-OH and not the other phenols (Chen et al. 2012, 2013; del Río et al. 2017, 2020, 2021; Elder et al. 2019; Li et al. 2018; Rencoret et al. 2018, 2019; Tobimatsu et al. 2013; Wang et al. 2020b). Similarly, tricin radical coupling products can all be rationalized as deriving from the 4′-OH (Figure 1.4B) and not the phenolic OHs on the 5 or 7 positions (del Río et al. 2020; Rencoret et al. 2019). The other feature supporting the radical coupling course is that rearomatization can occur via internal trapping by the catechyl 3-OH and the 5-hydroxyguaiacyl 5-OH, whereby the resultant benzodioxane rings are primarily *trans* with minor *cis* ring-isomers – the presence of *cis* ring-isomers from radical coupling reactions does not appear to be compatible with Diels–Alder reactions, for example (Ando et al. 2021; Elder et al. 2019, 2021).

Conversion of catechols to quinones has been touted as a reason why catechols should not be tolerated in lignification (Anterola and Lewis 2002; Davin et al. 2008). That catechols clearly incorporate into lignins by the canonical radical coupling processes has now been well documented in a range of viable natural, mutant, and transgenic plants, as we have reviewed (Boerjan et al. 2003; Ralph et al. 2004b, 2019; Vanholme et al. 2010a). Despite the *in vitro* demonstration that quinones can form from 5-hydroxyvanillate, for example and produce dimers and higher oligomers by Diels–Alder reactions, the lack of evidence for such reactions even from plants utilizing high levels of catechol monomers, suggests that radical coupling remains the major, if not the sole, pathway operating in lignification.

1.5 The Case for Model Studies and Synthesis

A great deal can be learned from structural studies on *in situ* or isolated lignins from across the diverse array of plant species. Even for lignins that are compositionally normal, sets of spectra can prove extraordinarily useful for assignment and authentication, as recently highlighted in a study of nutshell lignins, the raw datasets of which have been made available (Landucci et al. 2020). This short section is to serve as a reminder of the value of model reactions and biomimetic coupling studies, along with good old organic synthesis for compound authentication and to provide enzyme substrates and products for pathway elucidation (that is not covered here).

1.5.1 The Value of Proper Low-Molecular-Mass Model Compounds

Model compound studies come in and out of vogue. Popular in the early days to understand lignin structure and chemistry, they became criticized for their distraction of researchers from working on real lignins. Model reactions continue to have extreme value, however. Although it may be true that some chemistry that works well on a model compound may well not translate to useful polymer reactions, a corollary is almost certainly valid, i.e. that any chemistry that does not work well on a model compound has little-to-no chance of being of any use for the polymer. Now the value of models is solidly recognized as they allow homogeneous chemistry practices to provide insights into processes of interest.

That said, choosing good model compounds that properly mimic the requisite aspects of the structure of interest is paramount. In the old days, it was common practice to use a 2-carbon-sidechain β-ether model such as guaiacylglycol-β-guaiacyl ether as a model for the main structures in a softwood lignin, for example (Landucci and Ralph 1982). The compound was straightforward to synthesize and was a single isomer, easily crystallized. The more representative model, guaiacylglycerol-β-guaiacyl ether, with the appropriate 3-carbon-sidechain to model real lignin units, was annoying because it required more synthetic steps and came as a mixture of two isomers from which neither was readily obtained in pure form, but using the simpler model could lead to erroneous conclusions (Obst 1983). Of course, those issues have long been solved (Adler et al. 1952; Nakatsubo et al. 1975; Ralph and Helm 1991), and no one would/should think of using anything but the 3-carbon model today. Nevertheless, as other disciplines come into lignin research and in groups that do not have synthetic organic chemistry expertise, it is common practice to use only models that can be purchased. Without meaning to cast dispersions, nowhere is this more evident than in catalytic lignin degradation studies in which the simplest of supposed β-ether models (such as benzyl phenyl ether) are used, and models for other units in lignins are equally poorly chosen. Although something may well be learned about catalytic cleavage of biphenyl ether bonds by using biphenyl ether, for example, the notion that this somehow pertains to lignin is shortsighted.

Just how crucial all of the functionality on a model is to the understanding of what happens with lignin was evident recently with an attempt to elucidate the fate of lignin biphenyl ethers under hydrogenolytic conditions, Figure 1.10a (Li et al. 2020). The cleavage degree, products, and pathway for hydrogenolysis of proper lignin models, Figure 1.10b, were scarcely reminiscent of those from the simpler model, Figure 1.10c. Another benefit is that the model work produced unanticipated marker compounds for reactions of such 4–*O*–5-ethers, unprecedented 1,3,5-*meta*-substituted rearranged monomers, that provide new evidence for the very existence of biphenyl ethers in lignin while simultaneously providing evidence of the reactions occurring with the lignin polymer, Figure 1.10d.

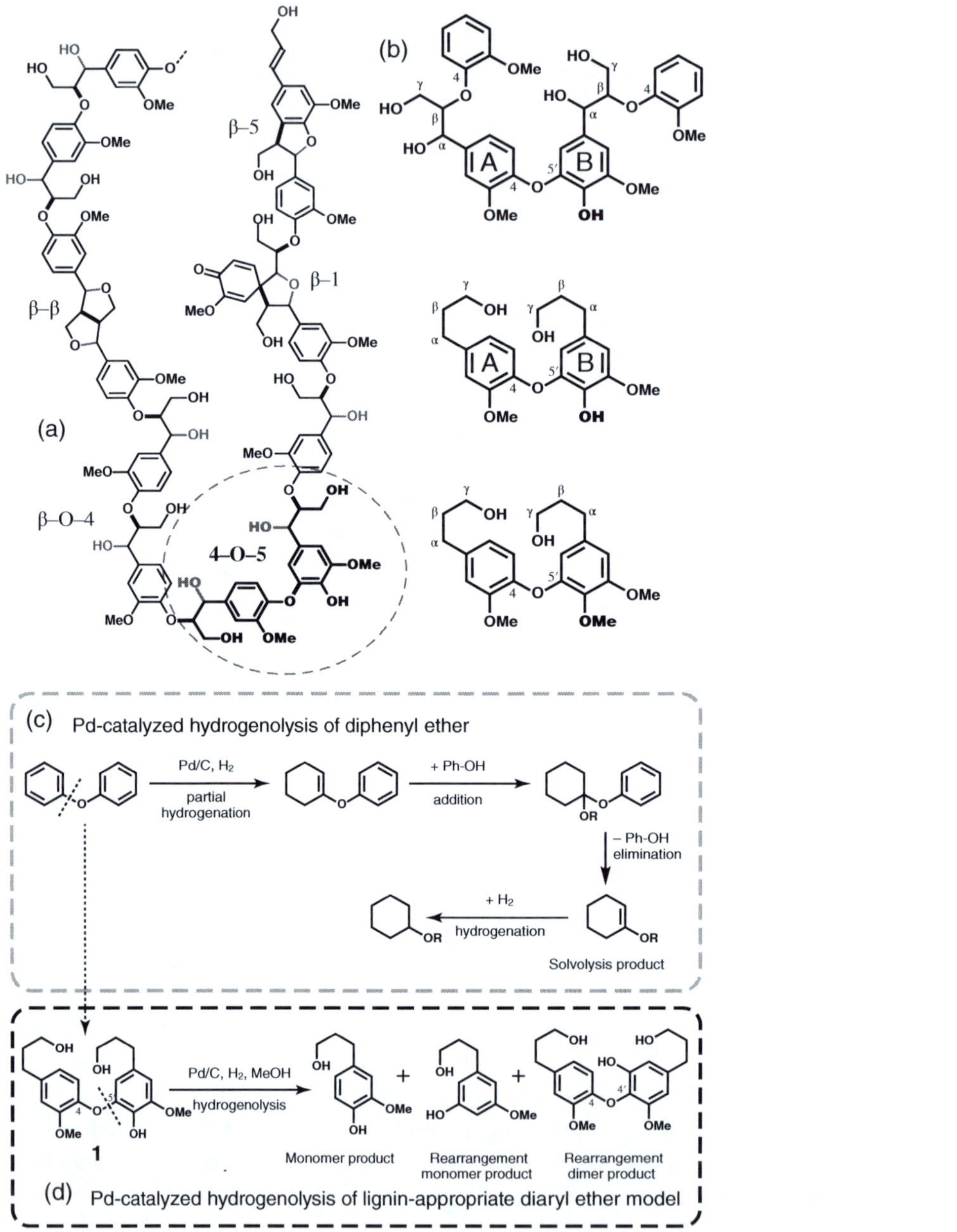

Figure 1.10 Lignin 4–*O*–5 units, appropriate model compounds, and their hydrogenolysis. (a) A model of part of a G-lignin showing the presence of a 4–*O*–5 unit (circled). (b) Model compounds used to study the hydrogenolytic pathways. The top di-β-aryl-ether model is ideal but, given that hydrogenolysis of the β-ether bonds is rapid, the arylpropanol model (middle) is a perfectly good model (and is in fact a likely intermediate product) for studying the hydrogenolytic course of 4–*O*–5 units in lignin; the 4-*O*-methylated analog (bottom) can be used to determine if the phenolic-OH is required by the mechanism. *Source:* Adapted from Li et al. (2020). (c) The products and pathways for the simple diphenyl ether "model". *Source:* Adapted from Cao et al. (2018), Wang et al. (2017, 2018) and Zeng et al. (2018). (d) The contrasting products and pathways from representative lignin models illustrate not only the significant effects of the substituents in determining the course of the reaction but also highlight the newly discovered compounds arising from rearrangement that may be used as markers for 4–*O*–5-coupled units in lignins, compounds that would never have been surmised or determined from models devoid of lignin-relevant features. *Source:* Adapted from Li et al. (2020).

1.5.2 Synthetic Lignin Polymers, Dehydrogenation Polymers (DHPs)

The value of preparing synthetic lignins for understanding the incorporation of new monomers has, if anything, escalated. Examples are too numerous to even tabulate, but the following recent examples (some on studies already alluded to above) illustrate the value and scope of making both polymeric models and using limited polymerization to produce small oligomers that can be fully characterized to help authenticate structures in lignin polymers.

- A DHP from *p*-coumaryl alcohol plus *p*-coumaryl *p*-coumarate has demonstrated that *p*-coumarates can incorporate into the lignin backbone in such lignins, and therefore establishes a variation on the zip-lignin concept (Smith et al. 2022b) (Figure 1.7d, e).
- A new aldehyde component in CAD-deficient lignins has been identified by noting its production in coniferaldehyde DHPs (K. Yoshioka, in preparation). The structural determination revealed, however, that this was not a result of simple dehydrodimerization but required the further disproportionation and rearomatization of a radical derived from such a dimer (Figure 1.2g). Information is therefore revealed about the concentration, lifetime, and reactivity of such a coniferaldehyde dimer (and perhaps cross-coupling products) during lignification in CAD-deficient plants.
- Synthetic lignins significantly aided the understanding of the coupling of piceatannol (and, by analogy, other hydroxystilbenes) and their glucosides into lignins (del Río et al. 2017, 2020, 2021; Rencoret et al. 2018, 2019) (Figure 1.4g, h). Incidentally, *in silico* modeling studies, particularly those using density functional theory (DFT) calculations, are also useful in this context (Elder et al. 2019, 2021).
- Synthetic lignins, along with models and multiply-labeled standards, were crucial to identifying the original incorporation of the flavone tricin into lignins (del Río et al. 2012, 2015; Lan et al. 2015, 2016a, 2016b, 2018b, 2019; Rencoret et al. 2013). Subsequently the incorporation of naringenin and other components was authenticated after early tricin pathway genes were misregulated (Lam et al. 2017, 2019a, 2019b; Lui et al. 2020). Most recently, several flavonoids (naringenin chalcone, naringenin, and dihydrotricin) in addition to tricin, from different flavonoid families, have been discovered and authenticated in papyrus (Rencoret et al. 2022; Rosado et al. 2021), paving way to consider other classes of flavonoids as lignin monomers.
- Lignification using monolignol benzoates (not *p*-hydroxybenzoates) in poplars down-regulated in genes involved in 3- and 4-hydroxylation was authenticated using synthetic lignins incorporating coniferyl benzoate (Kim et al. 2020).

1.6 New or Improved Analytics

With the uptick in interest in phenolic acids and monolignol conjugates, it is opportune to note some enhancements to a pair of orthogonal methods that are diagnostic for two different cell wall processes. When originally developed, the DFRC (derivatization followed by reductive cleavage) method (Lu and Ralph 1997a, 1997b) aspired to compete with analytical thioacidolysis (Lapierre 1993; Rolando et al. 1992), but thioacidolysis remains premier for cleaving β-ether units in lignin to release quantifiable monomers today. DFRC does

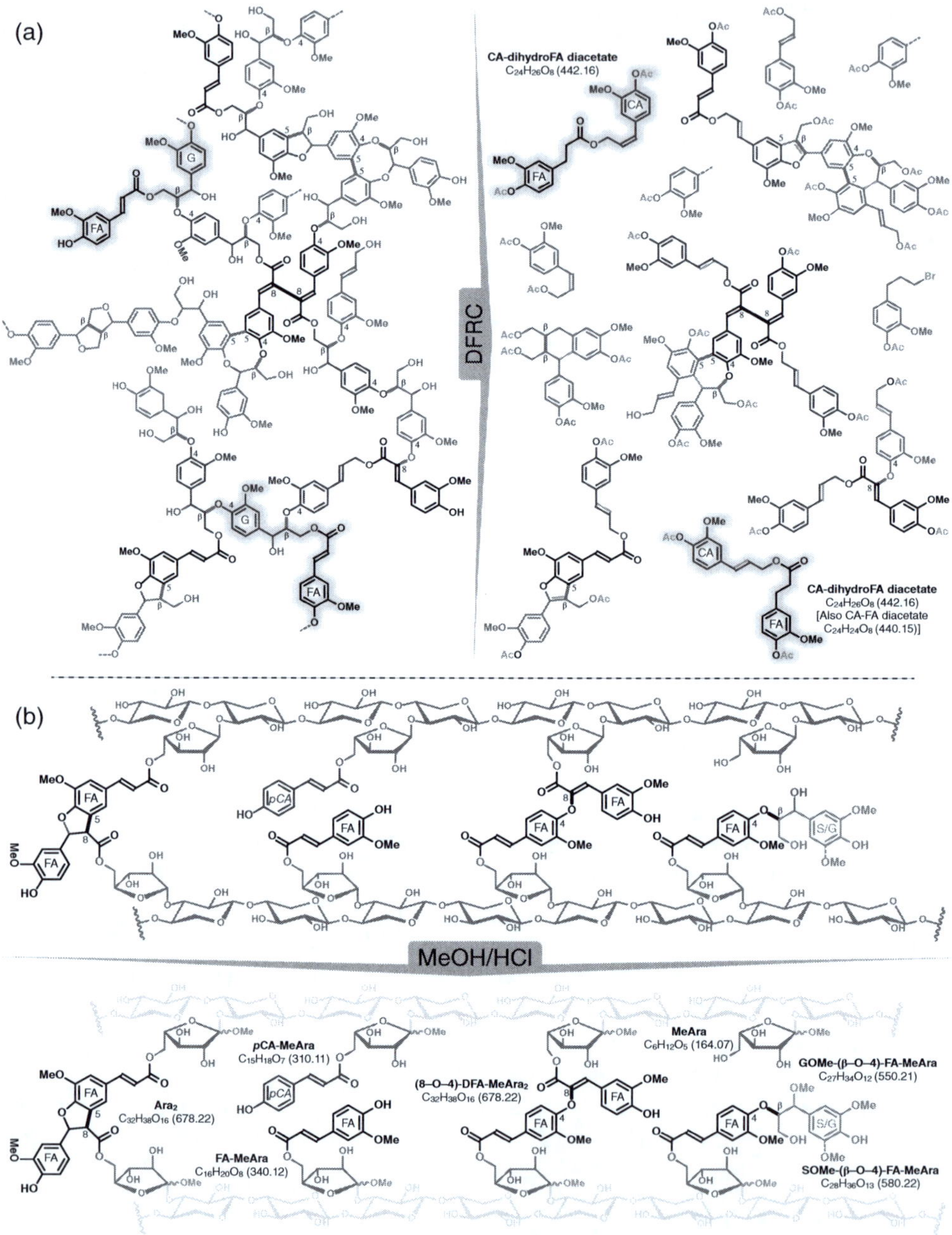

Figure 1.11 DFRC and mild acidolysis, orthogonal methods for determining hydroxycinnamates on lignins *vs* on polysaccharides. (a) A model for a G-lignin derived from coniferyl alcohol (CA) and coniferyl ferulate (CA-FA), i.e. a zip-lignin (*Source:* Adapted from Ralph et al. (2019)) in which a plausible structure derived from the coupling of 7 CA-FA conjugates (green-red in the color figure) and 13 further monolignol moieties (black) is shown. The bonds produced by radical coupling are shown in bold (red in the color figure), and the bonds formed by post-coupling rearomatization reactions (along with the OH from water addition) are in gray in the color figure. Following DFRC, hydroxyls (including those freshly liberated) are acetylated and β-ether bonds are cleaved leaving a cinnamyl double-bond signature, but esters remain intact (Wilkerson et al. 2014); some of the products from the depolymerization expected to occur are shown in the figure to the right with new bonds produced during DFRC and new acetate groups shown in cyan (in the colored version). [A great deal of information on DFRC mechanisms and products is attempted to be conveyed here

have the interesting attribute that it can convert lignin back, partially, to its component monomers, the hydroxycinnamyl alcohols (although, truthfully, as their peracetylated derivatives), Figure 1.11a. One attribute of DFRC – its ability to cleave lignin ethers while leaving esters intact – has rendered it particularly valuable for conjugates research (Grabber et al. 2008; Karlen et al. 2016; Lu and Ralph 1999a, 2014; Lu et al. 2015; Ralph and Lu 1998; Regner et al. 2018). Lignins derived from monolignol acetate, *p*-hydroxybenzoate, *p*-coumarate, ferulate, or other conjugates end up with those acids acylating the γ-OH of lignin sidechains. When subjected to DFRC, the esters, even simple acetates, remain intact such that, instead of monolignols, the monolignol esters are released and may be quantified; a minor adjustment allows the natural acetates to be authenticated (Lu and Ralph 2014; Martone et al. 2009; Ralph and Lu 1998). The DFRC method provided unambiguous evidence for the existence of such conjugates in lignin, and provided a way to assess their (relative) levels in the polymer; DFRC releases conjugates that are involved in β-ether units in lignin that can be cleaved by the method (Bunzel et al. 2008; Karlen et al. 2016, 2017, 2018; Kim et al. 2020; Lu and Ralph 1999a, 2014; Lu et al. 2015; Petrik et al. 2014; Ralph and Lu 1998; Regner et al. 2018; Smith et al. 2015; 2017a; Wilkerson et al. 2014), Figure 1.11. In recent years, DFRC has become an indispensable tool in confirming the incorporation

Figure 1.11 (Continued) in the products on the right from the various structures found in the lignins. Monomer units linked solely by β-ether bonds will release (acetylated) monolignols (Lu and Ralph 1997b); two are shown here along with two possible units for which the sidechains are not specified. A coniferyl alcohol endgroup unit (halfway down on the right) will release one of a variety of monomers (Lu and Ralph 1999b); the acetylated guaiacylbromopropane is shown. Phenylcoumarans typically dehydrate to benzofurans; two are shown, one derived from CA (lower left), and one from CA-FA (top-middle). Resinols undergo acid rearrangement to produce aryltetralins (Ralph et al. 2004b); one is shown (middle-left). As far as we know, even though they are a kind of β-ether, ferulate-8–*O*–4-ethers may not cleave in DFRC; for similar reasons, they do not cleave even upon high-temperature base treatment either (Grabber et al. 1995). Finally, the 8-membered-ring dibenzodioxocins resulting initially from 5–5-coupling of guaiacyl units appear to undergo a rearrangement to a 7-membered ring structure, at least in model systems (Argyropoulos et al. 2002); two examples are shown here, one from normal lignin guaiacyl units and the other from 5–5-diferulate.] In the model on the left, there are two ferulates deriving from the incorporation of CA-FA conjugates (highlighted in gray or yellow shading on the left) by means that will release quantifiable CA-FA and/or coniferyl dihydroferulate conjugates (similarly highlighted on the right) upon DFRC (Regner et al. 2018; Wilkerson et al. 2014). Other incorporated CA-FA units will not be able to release the simple conjugates; other lower molecular mass fragments will also be released by the β-ether cleavage but as more complex oligomers for which authentic standards and quantitative methods are not yet available. (b) Methanolysis of hydroxycinnamoylated arabinoxylan (Eugene et al. 2020; Lapierre et al. 2018, 2019). Top: Schematic arabinoxylan chains shown with the arabinosyl substitutions on every third unit as revealed recently (Busse-Wicher et al. 2016), along with *p*CA and FA substitutions on arabinosyl units, two representative diferulates (8–*O*–4 and 8–5) formed by radical coupling, and an ML-FA conjugate, all known to be present in grass cell walls; acetate substitutions, also known, are not shown. The bonds produced by radical coupling are shown in bold-black, and the bonds formed by post-coupling rearomatization reactions (along with the OH from water addition in the case of the ML-FA coupling) are in gray. Bottom: The derived products released for structural analysis and quantification upon methanolysis of the arabinosyl unit from the xylan backbone leave the arabinosyl esters largely intact. The original xylan chain is shown with low opacity for easier visualization of the source of the products, and the elemental formulae and exact masses (for MS analysis) are shown. *Source:* Figures and data are adapted from Eugene et al. (2020) and Lapierre et al. (2019). (*See insert for color representation of the figure.*)

of monolignol acetate (Bunzel et al. 2008; del Río et al. 2007; Ralph and Lu 1998), *p*-coumarate (Petrik et al. 2014; Smith et al. 2015), benzoate (Karlen et al. 2017; Kim et al. 2020), *p*-hydroxybenzoate (Karlen et al. 2017; Lu et al. 2015; Rencoret et al. 2020), vanillate (Karlen et al. 2017), and ferulate (Karlen et al. 2016; Smith et al. 2017a; Wilkerson et al. 2014) conjugates into the lignin polymer. The original methods have now been updated to MS-based methods using multi-labeled internal standards for improved quantification (Regner et al. 2018).

The availability of the DFRC method was key to determining whether we had successfully introduced ML-FAs into lignification to produce zip-lignins (Figure 1.5) – there simply was no other diagnostic method at the time (Mottiar et al. 2016; Ralph et al. 2019; Wilkerson et al. 2014). It was also the availability of that method that revealed that natural lignins from various plant lineages already utilized ML-FAs in their lignification (Karlen et al. 2016), a natural trait that had not previously been discovered because there was no diagnostic method for the determination (and/or the question had never been posed).

Given that hydroxycinnamates acylate polysaccharides as well as monolignols in grasses, it is particularly useful to have orthogonal analytical methods to distinguish between the sources of these clipped-off hydroxycinnamates. To this end, we collaborated on introducing an improved method for clipping arabinosyl units from arabinoxylan polysaccharides in the presence of methanol, releasing methyl 5-*O*-feruloyl arabinofuranoside (FA-MeAra) and methyl 5-*O*-*p*-coumaroyl arabinofuranoside (*p*CA-MeAra) for analysis (Lapierre et al. 2019), Figure 1.11b. The resultant 1-*O*-methyl arabinosides allow for sharper HPLC chromatograms relative to the 1-OH arabinosides produced by the original aqueous HCl method (Lapierre et al. 2018). The improved peak shapes are attributed to the elimination of interconverting α- and β-anomers of the parent 1-OH arabinose through methylation. The method can also be used to characterize crosslinked hydroxycinnamates, predominantly various dehydrodiferulates, as well as ferulate-monolignol cross-products in which the ferulate moieties are esterified to 1-*O*-methyl arabinosides (Lapierre et al. 2018, 2019), Figure 1.11b, bottom panel. An "anhydrous" method that limits hydrolytic reactions produced a cleaner product distribution (Eugene et al. 2020).

1.7 Conclusions and Opportunities

If lignin research were not already interesting enough, it has certainly become enthralling lately. Various factors are at play including: a coupling of the ability to poke and prod the circuitry of biosynthetic pathways, a plant's often but not always predictable response, and our ability to figure out what is going on in a structural and mechanistic sense; the evolution of analytical and instrumental methods (including the scarcely mentioned herein MS-based metabolomics and solution- and solid-state NMR-based polymer analysis tools); and more of the secrets of natural plant evolution are becoming open to researchers' investigations ranging from inquisitive "look-and-see" studies to mechanistically driven critical inquiry. Revelations about the polymer and its diversity can already inform attempts to utilize sustainable resources more efficiently to produce the products we need. Natural evolution can inspire new approaches to adding value to biomass, particularly when it is appreciated that an expanding array of components produced in lignins may become available on a previously incomprehensible scale in the lignin that has traditionally been regarded as a waste product. Coupled with an enhanced appreciation of the need to consider "lignin-first" principles (Abu-Omar et al. 2021) to preserve lignin structure to optimize biomass value and utility, some of the developments coming online now will hopefully help current generations succeed in lowering the human impact on the planet.

Acknowledgments

All authors except JCdR and JRe were funded by the DOE Great Lakes Bioenergy Research Center (DOE Office of Science BER DE-SC0018409). JR, CS, AE, YL, and MC were partially funded for hydrogenolysis work by Swiss National Science Foundation (Synergia) grant # CRS115_180258. JCdR and JRe were funded MCIN/AEI/ 10.13039/501100011033 (projects PID2020-118968RB-I00 and AGL2017-83036-R, the later also financed by ERDF A way of making Europe), and the Junta de Andalucía (project P20-00017).

We are also grateful to an enormous number of wonderful group members, colleagues, and collaborators over many years. They cannot all be noted, but the ones most involved in work covered by this chapter include the groups of: Wout Boerjan and Ruben Vanholme (VIB and Gent University, Belgium); Shawn Mansfield (UBC, Canada); Rick Dixon and Fang Chen (University North Texas); Clint Chapple (Purdue University); Catherine Lapierre (INRA, France); Ron Hatfield, Jane Marita, and John Grabber (USDA-ARS); Shinya Kajita and Yoshi Ohya (Tokyo University, Japan); Ana Gutiérrez (IRNAS-CSIC, Spain); Antje Potthast and Thomas Rosenau (Boku, Vienna, Austria); Tom Elder (USDA-FS); Yuki Tobimatsu (Kyoto University, Japan); John Sedbrook (Illinois State University); Vincent Chiang and Ron Sederoff (North Carolina State University); Curtis Wilkerson (Michigan State University); Gregg Beckham (NREL); Roberto Rinaldi (Imperial College, UK); Jeremy Luterbacher (EPFL, Switzerland); Jim Dumesic, Shannon Stahl, Tim Donohue, Dan Noguera, Shawn Kaeppler, Jeff Piotrowski, and Brian Fox (University of Wisconsin); CJ Tsai (University of Georgia); Xueming Zhang (SCUT, China); and many, many more. Various researchers in or visiting our own group have also added significantly to the stories told here, including especially Stéphane Quideau, Richard Helm, Takuya Akiyama, Ali Azarpira, Ruili Gao, Justin Mobley, Purba Mukerjee, Mathish Nambiar-Veetil, Martina Opietnik, Dharshana Padmakshan, Matt Regner, Li Shuai, Yukiko Tsuji, Fengxia Yue, Dan Yelle, and Yimin Zhu – apologies to all you others who were also important contributors to the laboratory's endeavors!

References

Abu-Omar, M.M., Barta, K., Beckham, G.T., et al. (2021). Guidelines for performing lignin-first biorefining. *Energy & Environmental Science*, **14**: 262–292.

Adler, E., Björkquist, K.J., and Häggroth, S. (1948). Über die Ursache der Farbreaktionen des Holzes. *Acta Chemica Scandinavica*, **2**: 93–94.

Adler, E., Lindgren, B.O., and Saedén, U. (1952). The α-guaiacyl ether of α-veratrylglycerol as a lignin model. *Svensk Papperstidning*, **15**: 245–254.

Akiyama, T., Magara, K., Meshitsuka, G., et al. (2015). Absolute configuration of β- and α-asymmetric carbons within β-O-4-structures in hardwood lignin. *Journal of Wood Chemistry and Technology*, **35**: 8–16.

Anderson, N.A., Tobimatsu, Y., Ciesielski, P.N., et al. (2015). Manipulation of guaiacyl and syringyl monomer synthesis in an Arabidopsis cinnamyl alcohol dehydrogenase mutant results in atypical lignin biosynthesis and modified cell wall structure. *Plant Cell*, **27**: 2195–2209.

Ando, D., Lu, F., Kim, H., et al. (2021). Incorporation of catechyl monomers into lignins: lignification from the non-phenolic end via Diels-Alder cycloaddition? *Green Chemistry*, **23**: 8995–9013.

Anterola, A.M. and Lewis, N.G. (2002). Trends in lignin modification: a comprehensive analysis of the effects of genetic manipulations/mutations on lignification and vascular integrity. *Phytochemistry*, **61**: 221–294.

Argyropoulos, D.S., Jurasek, L., Kristofova, L., et al. (2002). Abundance and reactivity of dibenzodioxocins in softwood lignin. *Journal of Agricultural and Food Chemistry*, **50**: 658–666.

Barrière, Y., Ralph, J., Méchin, V., et al. (2004). Genetic and molecular basis of grass cell wall biosynthesis and degradability. II. Lessons from brown-midrib mutants. *Comptes Rendus Biologies*, **327**: 847–860.

Barta, K., Warner, G.R., Beach, E.S., and Anastas, P.T. (2014). Depolymerization of organosolv lignin to aromatic compounds over Cu-doped porous metal oxides. *Green Chemistry*, **16**: 191–196.

Bartley, L.E., Peck, M.L., Kim, S.R., Ebert, B., Manisseri, C., Chiniquy, D.M., et al. (2013). Overexpression of a BAHD acyltransferase, OsAt10, alters rice cell wall hydroxycinnamic acid content and saccharification. *Plant Physiology*, **161**: 1615–1633.

Behaghel de Bueren, J., Héroguel, F., Wegmann, C., et al. (2020). Aldehyde-assisted fractionation enhances lignin valorization in endocarp waste biomass. *ACS Sustainable Chemistry & Engineering*, **8**: 16737–16745.

Berstis, L., Elder, T., Crowley, M., and Beckham, G.T. (2016). Radical nature of C-lignin. *ACS Sustainable Chemistry & Engineering*, **4**: 5327–5335.

Bhalla, A., Bansal, N., Pattathil, S., et al. (2018). Engineered lignin in poplar biomass facilitates Cu-AHP pretreatment. *ACS Sustainable Chemistry & Engineering*, **6**: 2932–2941.

Boerjan, W., Ralph, J., and Baucher, M. (2003). Lignin biosynthesis. *Annual Reviews in Plant Biology*, **54**: 519–546.

Bonawitz, N.D. and Chapple, C. (2010). The genetics of lignin biosynthesis: connecting genotype to phenotype. *Annual Review of Genetics*, **44**: 337–363.

Bonawitz, N.D., Kim, J.I., Tobimatsu, Y., et al. (2014). Disruption of Mediator rescues the stunted growth of a lignin-deficient Arabidopsis mutant. *Nature*, **509**: 376–380.

Bunzel, M., Heuermann, B., Kim, H., and Ralph, J. (2008). Peroxidase-catalyzed oligomerization of ferulic acid esters. *Journal of Agricultural and Food Chemistry*, **56**: 10368–10375.

Busse-Wicher, M., Li, A., Silveira, R.L., et al. (2016). Evolution of xylan substitution patterns in gymnosperms and angiosperms: implications for xylan interaction with cellulose. *Plant Physiology*, **171**: 2418–2431.

Cao, D., Zeng, H., and Li, C.-J. (2018). Formal cross-coupling of diaryl ethers with ammonia by dual C (Ar)–O bond cleavages. *ACS Catalysis*, **8**: 8873–8878.

Cao, S., Huang, C., Luo, L., et al. (2020). Cell-specific suppression of 4-coumarate-CoA ligase gene reveals differential effect of lignin on cell physiological function in *Populus*. *Frontiers in Plant Science*, **11** (589729): 1–11.

Chen, F., Tobimatsu, Y., Havkin-Frenkel, D., et al. (2012). A polymer of caffeyl alcohol in plant seeds. *Proceedings of the National Academy of Sciences, USA*, **109**: 1772–1777.

Chen, F., Tobimatsu, Y., Jackson, L., et al. (2013). Novel seed coat lignins in the Cactaceae: structure, distribution and implications for the evolution of lignin diversity. *The Plant Journal*, **73**: 201–211.

Chrichton, M. (1990). Jurassic Park: Alfred A. Knopf.

Ciesielski, P.N., Resch, M.G., Hewetson, B., et al. (2014). Engineering plant cell walls: tuning lignin monomer composition for deconstructable biofuel feedstocks or resilient biomaterials. *Green Chemistry*, **16**: 2627–2635.

Corona, A., Biddy, M.J., Vardon, D.R., et al. (2018). Life cycle assessment of adipic acid production from lignin. *Green Chemistry*, **20**: 3857–3866.

Davin, L.B., Jourdes, M., Patten, A.M., et al. (2008). Dissection of lignin macromolecular configuration and assembly: comparison to related biochemical processes in allyl/propenyl phenol and lignan biosynthesis. *Natural Product Reports*, **25**: 1015–1090.

Davis, R., Tao, L., Tan, E.C.D., et al. (2013). Process Design and Economics for the Conversion of Lignocellulosic Biomass to Hydrocarbons: Dilute-acid and Enzymatic Deconstruction of Biomass to Sugars and Biological Conversion of Sugars to Hydrocarbons. *Laboratory Technical Report NREL/TP-5100-60223*, National Renewable Energy Lab (NREL), Golden, CO, USA. https://www.nrel.gov/docs/fy14osti/60223.pdf

Davis, R.E., Grundl, N.J., Tao, L., et al. (2018). Process Design and Economics for the Conversion of Lignocellulosic Biomass to Hydrocarbon Fuels and Coproducts: 2018 Biochemical Design Case Update; Biochemical Deconstruction and Conversion of Biomass to Fuels and Products Via Integrated Biorefinery Pathways. *Laboratory Technical Report NREL/TP-5100-71949*, National Renewable Energy Lab (NREL), Golden, CO, USA. https://www.nrel.gov/docs/fy19osti/71949.pdf

Dixon, R.A., Chen, F., Guo, D.J., and Parvathi, K. (2001). The biosynthesis of monolignols: a "metabolic grid", or independent pathways to guaiacyl and syringyl units? *Phytochemistry*, **57**: 1069–1084.

Elder, T., del Río, J.C., Ralph, J., et al. (2019). Radical coupling reactions of piceatannol and monolignols: a density functional theory study. *Phytochemistry*, **164**: 12–23.

Elder, T., del Río, J.C., Ralph, J., et al. (2020). Coupling and reactions of lignols and new lignin monomers: a density functional theory study. *ACS Sustainable Chemistry & Engineering*, **8**: 11033–11045.

Elder, T., Rencoret, J., del Río, J.C., et al. (2021). Radical coupling reactions of hydroxystilbene glucosides and coniferyl alcohol: a density functional theory study. *Frontiers in Plant Science*, **12** (642848): 1–13.

Eloy, N.B., Voorend, W., Lan, W., et al. (2017). Silencing CHALCONE SYNTHASE in maize impedes the incorporation of tricin into lignin and increases lignin content. *Plant Physiology*, **173**: 998–1016.

Eugene, A., Lapierre, C., and Ralph, J. (2020). Improved analysis of arabinoxylan-bound hydroxycinnamate conjugates in grass cell walls. *Biotechnology for Biofuels*, **13** (202): 1–7.

Eyong, K.O., Kuete, V., and Efferth, T. (2013). Quinones and benzophenones from the medicinal plants of Africa. In: Medicinal Plant Research in Africa, (ed. Kuete, V.), 351–391. Amsterdam, Netherlands: Elsevier.

Fanelli, A., Rancour, D.M., Sullivan, M., et al. (2021). Overexpression of a sugarcane BAHD acyltransferase alters hydroxycinnamate content in maize cell wall. *Frontiers in Plant Science*, **12**, 626168: 1–14.

Franke, R., McMichael, C.M., Meyer, K., et al. (2000). Modified lignin in tobacco and poplar plants over-expressing the Arabidopsis gene encoding ferulate 5-hydroxylase. *Plant Journal*, **22**: 223–234.

Franke, R., Humphreys, J.M., Hemm, M.R., et al. (2002). The Arabidopsis REF8 gene encodes the 3-hydroxylase of phenylpropanoid metabolism. *The Plant Journal*, **30**: 33–45.

Freudenberg, K. and Neish, A.C. (1968). Constitution and Biosynthesis of Lignin. Berlin-Heidelberg-New York: Springer-Verlag.

Goacher, R.E., Mottiar, Y., and Mansfield, S.D. (2021). ToF-SIMS imaging reveals that *p*-hydroxybenzoate groups specifically decorate the lignin of fibres in the xylem of poplar and willow. *Holzforschung*, **75**: 452–462.

Grabber, J.H., Hatfield, R.D., Ralph, J., et al. (1995). Ferulate cross-linking in cell walls isolated from maize cell suspensions. *Phytochemistry*, **40**: 1077–1082.

Grabber, J.H., Hatfield, R.D., Lu, F., and Ralph, J. (2008). Coniferyl ferulate incorporation into lignin enhances the alkaline delignification and enzymatic degradation of maize cell walls. *Biomacromolecules*, **9**: 2510–2516.

Grabber, J.H., Schatz, P.F., Kim, H., et al. (2010). Identifying new lignin bioengineering targets: 1. Monolignol-substitute impacts on lignin formation and cell wall fermentability. *BMC Plant Biology*, **10**: 1–13.

Grimming, B., Kneusel, R.E., Junghanns, K.T., and Matern, U. (2008). Expression of bifunctionaI caffeoyl-CoA 3-*O*-methyltransferase in stress compensation and lignification. *Plant Biology*, **1**: 299–310.

Harris, P.J. and Hartley, R.D. (1976). Detection of bound ferulic acid in cell walls of the Gramineae by ultraviolet fluorescence microscopy. *Nature*, **259**: 508–510.

Harris, P.J. and Hartley, R.D. (1980). Phenolic constituents of the cell walls of monocotyledons. *Biochemical Systematics and Ecology*, **8**: 153–160.

Hatfield, R.D., Ralph, J., and Grabber, J.H. (2008). A potential role of sinapyl *p*-coumarate as a radical transfer mechanism in grass lignin formation. *Planta*, **228**: 919–928.

Henry, R.J. and Harris, P.J. (1997). Molecular distinction between monocotyledons and dicotyledons: more than a simple dichotomy. *Plant Molecular Biology Reporter*, **15**: 216–218.

Higuchi, T. (2006). Look back over the studies of lignin biochemistry. *Journal of Wood Science*, **52**: 2–8.

Hock, H. and Lang, S. (1944). Autoxydation von Kohlenwasserstoffen, IX. Mitteil.: Über Peroxyde von Benzol-Derivaten. *Berichte der Deutschen Chemischen Gesellschaft*, **77**: 257–264.

Hoengenaert, L., Wouters, M., Kim, H., De Meester, B., Morreel, K., Vandersyppe, S., Pollier, J., Desmet, S., Goeminne, G., Ralph, J., Boerjan, W. and Vanholme, R. (2022) Overexpression of the scopoletin biosynthetic pathway enhances lignocellulosic biomass processing. *Science Advances* **8**: eabo5738.

Hoffmann, N., Benske, A., Betz, H., et al. (2020). Laccases and peroxidases co-localize in lignified secondary cell walls throughout stem development. *Plant Physiology*, **184**: 806–822.

Hofmann, A.W. (1881). Über die Einwirkung des Broms in alkalischer Lösung auf Amide [On the action of bromine in alkaline solution on amides]. *Berichte der Deutschen Chemischen Gesellschaft*, **14**: 2725–2736.

Huang, K., Fasahati, P., and Maravelias, C.T. (2020). System-level analysis of lignin valorization in lignocellulosic biorefineries. *Iscience*, **23**: 100751.

Humphreys, J.M., Hemm, M.R., and Chapple, C. (1999a). Ferulate 5-hydroxylase from Arabidopsis is a multifunctional cytochrome P450-dependent monooxygenase catalyzing parallel hydroxylations in phenylpropanoid metabolism. *Proceedings of the National Academy of Sciences, USA*, **96**: 10045–10050.

Humphreys, J.M., Hemm, M.R., and Chapple, C. (1999b). New routes for lignin biosynthesis defined by biochemical characterization of recombinant ferulate 5-hydroxylase, a multifunctional cytochrome P450-dependent monooxygenase. *Proceedings of the National Academy of Sciences, USA*, **96**: 10045–10050.

Huntley, S.K., Ellis, D., Gilbert, M., et al. (2003). Significant increases in pulping efficiency in C4H-F5H-transformed poplars: improved chemical savings and reduced environmental toxins. *Journal of Agricultural and Food Chemistry*, **51**: 6178–6183.

Ikeda, T., Takata, N., Sakamoto, S., et al. (2021). Improved chemical pulping and saccharification of a natural mulberry mutant deficient in cinnamyl alcohol dehydrogenase. *Holzforschung*, **75**: 968–977.

Joo, Y., Kim, H., Kang, M., et al. (2021). Pith specific lignification in *Nicotiana attenuata* as a defense against a stem-boring herbivore. *New Phytologist*, **232**: 332–344.

Jouanin, L., Goujon, T., de Nadaï, V., et al. (2000). Lignification in transgenic poplars with extremely reduced caffeic acid *O*-methyltransferase activity. *Plant Physiology*, **123**: 1363–1373.

Jouanin, L., Gujon, T., Sibout, R., et al. (2004). Comparison of the consequences on lignin content and structure of COMT and CAD downregulation in poplar and *Arabidopsis thaliana*. In: Plantation Forest Biotechnology in the 21st Century, (eds. Walter, C. and Carson, M.), 219–229. Kerala, India: Research Signpost.

Jun, S.-Y., Walker, A., Kim, H., et al. (2017). The enzyme activity and substrate specificity of two major cinnamyl alcohol dehydrogenases in sorghum (*Sorghum bicolor*), SbCAD2 and SbCAD4. *Plant Physiology*, **174**: 2128–2145.

Karlen, S.D., Zhang, C., Peck, M.L., et al. (2016). Monolignol ferulate conjugates are naturally incorporated into plant lignins. *Science Advances*, **2**, e1600393: 1–9.

Karlen, S.D., Smith, R.A., Kim, H., et al. (2017). Highly decorated lignins occur in leaf base cell walls of the Canary Island date palm *Phoenix canariensis*. *Plant Physiology*, **175**: 1058–1067.

Karlen, S.D., Free, H.C.A., Padmakshan, D., et al. (2018). Commelinid monocotyledon lignins are acylated by *p*-coumarate. *Plant Physiology*, **177**: 513–521.

Karlen, S.D., Fasahati, P., Mazaheri, M., et al. (2020). Assessing the viability of recovering hydroxycinnamic acids from lignocellulosic biorefinery alkaline pretreatment waste streams. *ChemSusChem*, **13**: 2012–2024.

Kaur, H., Shaker, K., Heinzel, N., et al. (2012). Environmental stresses of field growth allow cinnamyl alcohol dehydrogenase-deficient *Nicotiana attenuata* plants to compensate for their structural deficiencies. *Plant Physiology*, **159**: 1545–1570.

Kim, H., Ralph, J., Lu, F., et al. (2002). Identification of the structure and origin of thioacidolysis marker compounds for cinnamyl alcohol dehydrogenase deficiency in angiosperms. *Journal of Biological Chemistry*, **277**: 47412–47419.

Kim, H., Ralph, J., Lu, F., et al. (2003). NMR analysis of lignins in CAD-deficient Plants. Part 1. Incorporation of hydroxycinnamaldehydes and hydroxybenzaldehydes into lignins. *Organic and Biomolecular Chemistry*, **1**: 268–281.

Kim, K.H., Dutta, T., Ralph, J., et al. (2017). Impact of lignin polymer backbone esters on ionic liquid pretreatment of poplar. *Biotechnology for Biofuels*, **10** (101): 1–10.

Kim, H., Li, Q., Karlen, S.D., et al. (2020). Monolignol benzoates incorporate into the lignin of transgenic *Populus trichocarpa* depleted in C3H and C4H. *ACS Sustainable Chemistry & Engineering*, **8**: 3644–3654.

Lam, P.Y., Tobimatsu, Y., Takeda, Y., et al. (2017). Disrupting flavone synthase II alters lignin and improves biomass digestibility. *Plant Physiology*, **174**: 972–985.

Lam, P.Y., Lui, A.C., Yamamura, M., et al. (2019a). Recruitment of specific flavonoid B-ring hydroxylases for two independent biosynthesis pathways of flavone-derived metabolites in grasses. *New Phytologist*, **223**: 204–219.

Lam, P.Y., Tobimatsu, Y., Matsumoto, N., et al. (2019b). OsCAldOMT1 is a bifunctional *O*-methyltransferase involved in the biosynthesis of tricin-lignins in rice cell walls. *Nature Scientific Reports*, **9**, 11597: 1–13.

Lan, W. and Luterbacher, J.S. (2019). Preventing lignin condensation to facilitate aromatic monomer production. *Chimia*, **73**: 591–598.

Lan, W., Lu, F., Regner, M., et al. (2015). Tricin, a flavonoid monomer in monocot lignification. *Plant Physiology*, **167**: 1284–1295.

Lan, W., Morreel, K., Lu, F., et al. (2016a). Maize tricin-oligolignol metabolites and their implications for monocot lignification. *Plant Physiology*, **171**: 810–820.

Lan, W., Rencoret, J., Lu, F., et al. (2016b). Tricin-lignins: occurrence and quantitation of tricin in relation to phylogeny. *The Plant Journal*, **88**: 1046–1057.

Lan, W., Talebi Amiri, M., Hunston, C.M., and Luterbacher, J.S. (2018a). Protection group effects during α,γ-diol lignin stabilization promote high-selectivity monomer production. *Angewandte Chemie International Edition*, **57**: 1356–1360.

Lan, W., Yue, F., Rencoret, J., et al. (2018b). Elucidating tricin-lignin structures: assigning correlations in HSQC spectra of monocot lignins. *Polymers*, **10**: 916 1-6.

Lan, W., Rencoret, J., del Río, J.C., and Ralph, J. (2019). Tricin in grass lignin: biosynthesis, characterization, and quantitation. In: Lignin: Biosynthesis, Functions, and Economic Significance, (eds. Lu, F. and Yue, F.), 51–78. Hauppauge, NY, USA: Nova Science Publisher, Inc.

Landucci, L.L. and Ralph, J. (1982). Adducts of anthrahydroquinone and anthranol with lignin model quinone methides. 1. Synthesis and characterization. *Journal of Organic Chemistry*, **47**: 3486–3495.

Landucci, L., Smith, R.A., Liu, S., et al. (2020). Eudicot nutshells: cell-wall composition and biofuel feedstock potential. *Energy & Fuels*, **34**: 16274–16283.

Lapierre, C. (1993). Application of new methods for the investigation of lignin structure. In: Forage Cell Wall Structure and Digestibility, (eds. Jung, H.G., Buxton, D.R., Hatfield, R.D., and Ralph, J.), 133–166. Madison, WI, USA: American Society of Agronomy, Crop Science Society of America, Soil Science Society of America.

Lapierre, C., Tollier, M.T., and Monties, B. (1988). A new type of constitutive unit in lignins from the corn *bm3* mutant. *Comptes Rendus de l'Académie des Sciences. Série III*, **307**: 723–728.

Lapierre, C., Pilate, G., Pollet, B., et al. (2004). Signatures of cinnamyl alcohol dehydrogenase deficiency in poplar lignins. *Phytochemistry*, **65**: 313–321.

Lapierre, C., Voxeur, A., Karlen, S.D., et al. (2018). Evaluation of feruloylated and *p*-coumaroylated arabinosyl units in grass arabinoxylans by acidolysis in dioxane/methanol. *Journal of Agricultural and Food Chemistry*, **66**: 5418–5424.

Lapierre, C., Voxeur, A., Boutet, S., and Ralph, J. (2019). Arabinose conjugates diagnostic of ferulate-ferulate and ferulate-monolignol cross-coupling are released by mild acidolysis of grass cell walls. *Journal of Agricultural and Food Chemistry*, **67**: 12962–12971.

Li, L., Popko, J.L., Umezawa, T., and Chiang, V.L. (2000). 5-Hydroxyconiferyl aldehyde modulates enzymatic methylation for syringyl monolignol formation, a new view of monolignol biosynthesis in angiosperms. *Journal of Biological Chemistry*, **275**: 6537–6545.

Li, Y., Shuai, L., Kim, H., et al. (2018). An “ideal lignin” facilitates full biomass utilization. *Science Advances*, **4**: eaau2968 1-10.

Li, Y., Demir, B., Vázquez Ramos, L.M., et al. (2019). Kinetic and mechanistic insights into hydrogenolysis of lignin to monomers in a continuous flow reactor. *Green Chemistry*, **21**: 3561–3572.

Li, Y., Karlen, S.D., Demir, B., et al. (2020). Mechanistic study of diaryl ether bond cleavage during palladium-catalyzed lignin hydrogenolysis. *ChemSusChem*, **13**: 4487–4494.

Liao, Y.H., Koelewijn, S.F., Van den Bossche, G., et al. (2020). A sustainable wood biorefinery for low-carbon footprint chemicals production. *Science*, **367**: 1385–1390.

Lu, F. and Ralph, J. (1997a). Derivatization followed by reductive cleavage (DFRC method), a new method for lignin analysis: protocol for analysis of DFRC monomers. *Journal of Agricultural and Food Chemistry*, **45**: 2590–2592.

Lu, F. and Ralph, J. (1997b). The DFRC method for lignin analysis. Part 1. A new method for β-aryl ether cleavage: lignin model studies. *Journal of Agricultural and Food Chemistry*, **45**: 4655–4660.

Lu, F. and Ralph, J. (1999a). Detection and determination of *p*-coumaroylated units in lignins. *Journal of Agricultural and Food Chemistry*, **47**: 1988–1992.

Lu, F. and Ralph, J. (1999b). The DFRC method for lignin analysis. 7. The behavior of cinnamyl end-groups. *Journal of Agricultural and Food Chemistry*, **47**: 1981–1987.

Lu, F. and Ralph, J. (2008). Novel tetrahydrofuran structures derived from β–β-coupling reactions involving sinapyl acetate in Kenaf lignins. *Organic & Biomolecular Chemistry*, **6**: 3681–3694.

Lu, F. and Ralph, J. (2014). The DFRC (derivatization followed by reductive cleavage) method and its applications for lignin characterization. In: Lignin: Structural Analysis, Applications in Biomaterials, and Ecological Significance, (ed. Lu, F.), 27–65. Hauppauge, New York, USA: Nova Science Publishers, Inc.

Lu, F., Marita, J.M., Lapierre, C., et al. (2010). Sequencing around 5-hydroxyconiferyl alcohol-derived units in caffeic acid *O*-methyltransferase-deficient poplar lignins. *Plant Physiology*, **153**: 569–579.

Lu, F., Karlen, S.D., Regner, M., et al. (2015). Naturally *p*-hydroxybenzoylated lignins in palms. *Bioenergy Research*, **8**: 934–952.

Lui, A.C.W., Lam, P.Y., Chan, K.H., et al. (2020). Convergent recruitment of 5′-hydroxylase activities by CYP75B flavonoid B-ring hydroxylases for tricin biosynthesis in Medicago legumes. *New Phytologist*, **228**: 269–284.

Mahon, E.L., de Vries, L., Jang, S.-K., et al. (2022). Exogenous chalcone synthase expression in developing poplar xylem incorporates naringenin into lignins. *Plant Physiology*, **188** (2): 984–996. doi: 10.1093/plphys/kiab499.

Mansfield, S.D., Kang, K.Y., and Chapple, C. (2012). Designed for deconstruction – poplar trees altered in cell wall lignification improve the efficacy of bioethanol production. *New Phytologist*, **194**: 91–101.

Marita, J.M., Ralph, J., Hatfield, R.D., and Chapple, C. (1999). NMR characterization of lignins in *Arabidopsis* altered in the activity of ferulate-5-hydroxylase. *Proceedings of the National Academy of Sciences, USA*, **96**: 12328–12332.

Marita, J.M., Ralph, J., Lapierre, C., et al. (2001). NMR characterization of lignins from transgenic poplars with suppressed caffeic acid *O*-methyltransferase activity. *Journal of the Chemical Society*, Perkin Transactions 1, **22**: 2939–2945.

Marita, J.M., Vermerris, W., Ralph, J., and Hatfield, R.D. (2003). Variations in the cell wall composition of maize *brown midrib* mutants. *Journal of Agricultural and Food Chemistry*, **51**: 1313–1321.

Martinez, A.T., Rencoret, J., Marques, G., et al. (2008). Monolignol acylation and lignin structure in some nonwoody plants: a 2D NMR study. *Phytochemistry*, **69**: 2831–2843.

Martone, P.T., Estevez, J.M., Lu, F., et al. (2009). Discovery of lignin in seaweed reveals convergent evolution of cell-wall architecture. *Current Biology*, **19**: 169–175.

Matsui, N., Chen, F., Yasuda, S., and Fukushima, K. (2000). Conversion of guaiacyl to syringyl moieties on the cinnamyl alcohol pathway during the biosynthesis of lignin in angiosperms. *Planta*, **210**: 831–835.

Matsushita, Y., Oyabu, Y., Aoki, D., and Fukushima, K. (2019). Unexpected polymerization mechanism of dilignol in the lignin growing. *Royal Society Open Science*, **6**: 190445 1-8.

Meyer, K., Cusumano, J.C., Somerville, C., and Chapple, C.C.S. (1996). Ferulate-5-hydroxylase from *Arabidopsis thaliana* defines a new family of cytochrome P450-dependent monooxygenases. *Proceedings of the National Academy of Sciences, USA*, **93**: 6869–6874.

Meyer, K., Shirley, A.M., Cusumano, J.C., et al. (1998). Lignin monomer composition is determined by the expression of a cytochrome P450-dependent monooxygenase in *Arabidopsis*. *Proceedings of the National Academy of Sciences, USA*, **95**: 6619–6623.

Miyagawa, Y., Tobimatsu, Y., Lam, P.Y., et al. (2020). Possible mechanisms for the generation of phenyl glycoside-type lignin-carbohydrate linkages in lignification with monolignol glucosides. *Plant Journal*, **104**: 156–170.

Mobley, J., Karlen, S.D., and Ralph, J. (2017) Synthesis of paracetamol from *p*-hydroxybenzamide. US Patent Awarded/issued 3/7/2019, U.S. Patent No. 10,286,504 (WARF P170066US02, Filed 12/12/2018; Original WARF P170066US01, Filed 3/16/2017).

Morreel, K., Ralph, J., Kim, H., et al. (2004a). Profiling of oligolignols reveals monolignol coupling conditions in lignifying poplar xylem. *Plant Physiology*, **136**: 3537–3549.

Morreel, K., Ralph, J., Lu, F., et al. (2004b). Phenolic profiling of caffeic acid *O*-methyltransferase-deficient poplar reveals novel benzodioxane oligolignols. *Plant Physiology*, **136**: 4023–4036.

Mottiar, Y., Vanholme, R., Boerjan, W., et al. (2016). Designer lignins: harnessing the plasticity of lignification. *Current Opinion in Biotechnology*, **37**: 190–200.

Mottiar, Y., Karlen, S.D., Goacher, R.E., Ralph, J. and Mansfield, S.D. (2022) Metabolic engineering of *p*-hydroxybenzoate in poplar lignin. *Plant Biotechnology Journal*: EarlyAccess (DOI: 10.1111/pbi.13935).

Nakatsubo, F., Sato, K., and Higuchi, T. (1975). Synthesis of guaiacylglycerol-β-guaiacyl ether. *Holzforschung*, **29**: 165–168.

Nar, M., Rizvi, H.R., Dixon, R.A., et al. (2016). Superior plant based carbon fibers from electrospun poly-(caffeyl alcohol) lignin. *Carbon*, **103**: 372–383.

Negrel, J. and Jeandet, P. (1987). Metabolism of tyramine and feruloyltyramine in TMV inoculated leaves of *Nicotiana tabacum*. *Phytochemistry*, **26**: 2185–2190.

Obst, J.R. (1983). Kinetics of alkaline cleavage of β-aryl ether bonds in lignin models: significance to delignification. *Holzforschung*, **37**: 23–28.

Önnerud, H., Zhang, L., Gellerstedt, G., and Henriksson, G. (2002). Polymerization of monolignols by redox shuttle-mediated enzymatic oxidation: a new model in lignin biosynthesis I. *Plant Cell*, **14**: 1953–1962.

Osakabe, K., Tsao, C.C., Li, L., et al. (1999). Coniferyl aldehyde 5-hydroxylation and methylation direct syringyl lignin biosynthesis in angiosperms. *Proceedings of the National Academy of Sciences, USA*, **96**: 8955–8960.

Oyarce, P., De Meester, B., Fonseca, F., et al. (2019). Introducing curcumin biosynthesis in Arabidopsis enhances lignocellulosic biomass processing. *Nature Plants*, **5**: 225–237.

Petrik, D.L., Karlen, S.D., Cass, C.L., et al. (2014). *p*-Coumaroyl-CoA: monolignol transferase (PMT) acts specifically in the lignin biosynthetic pathway in *Brachypodium distachyon*. *The Plant Journal*, **77**: 713–726.

Pettersen, R.C. (1984). The chemical composition of wood. In: The Chemistry of Solid Wood. *Vol.* 207, (ed. Rowell, R.M.), 57–126. Washington, DC, USA: American Chemical Society.

Pomar, F., Merino, F., and Ros Barceló, A. (2002). *O*-4-linked coniferyl and sinapyl aldehydes in lignifying cell walls are the main targets of the Wiesner (phloroglucinol-HCl) reaction. *Protoplasma*, **220**: 17–28.

Pyo, H.J., Demura, T., and Fukuda, H. (2007). TERE; a novel cis-element responsible for a coordinated expression of genes related to programmed cell death and secondary wall formation during differentiation of tracheary elements. *Plant Journal*, **51**: 955–965.

Questell-Santiago, Y.M., Galkin, M.V., Barta, K., and Luterbacher, J.S. (2020). Stabilization strategies in biomass depolymerization using chemical functionalization. *Nature Reviews Chemistry*, **4**: 311–330.

Ralph, J. (2006). What makes a good monolignol substitute? In: The Science and Lore of the Plant Cell Wall Biosynthesis, Structure and Function, (ed. Hayashi, T.), 285–293. Boca Raton, FL, USA: Universal Publishers (BrownWalker Press).

Ralph, J. (2010). Hydroxycinnamates in lignification. *Phytochemistry Reviews*, **9**: 65–83.

Ralph, J. (2020). Tricin and tricin-lignins in Medicago versus in monocots. *New Phytologist*, **228**: 11–14.

Ralph, J. and Helm, R.F. (1991). Rapid proton NMR method for determination of *threo:erythro* ratios in lignin model compounds and examination of reduction stereochemistry. *Journal of Agricultural and Food Chemistry*, **39**: 705–709.

Ralph, J. and Landucci, L.L. (2010). NMR of lignins. In: Lignin and Lignans; Advances in Chemistry, (eds. Heitner, C., Dimmel, D.R., and Schmidt, J.A.), 137–234. Boca Raton, FL, USA: CRC Press (Taylor & Francis Group).

Ralph, J. and Lu, F. (1998). The DFRC method for lignin analysis. Part 6. A modified method to determine acetate regiochemistry on native and isolated lignins. *Journal of Agricultural and Food Chemistry*, **46**: 4616–4619.

Ralph, J., Hatfield, R.D., Quideau, S., et al. (1994). Pathway of *p*-coumaric acid incorporation into maize lignin as revealed by NMR. *Journal of the American Chemical Society*, **116**: 9448–9456.

Ralph, J., MacKay, J.J., Hatfield, R.D., et al. (1997). Abnormal lignin in a loblolly pine mutant. *Science*, **277**: 235–239.

Ralph, J., Hatfield, R.D., Piquemal, J., et al. (1998). NMR characterization of altered lignins extracted from tobacco plants down-regulated for lignification enzymes cinnamyl-alcohol dehydrogenase and cinnamoyl-CoA reductase. *Proceedings of the National Academy of Sciences, USA*, **95**: 12803–12808.

Ralph, J., Peng, J., Lu, F., et al. (1999). Are lignins optically active? *Journal of Agricultural and Food Chemistry*, **47**: 2991–2996.

Ralph, J., Lapierre, C., Lu, F., et al. (2001a). NMR evidence for benzodioxane structures resulting from incorporation of 5-hydroxyconiferyl alcohol into lignins of *O*-methyl-transferase-deficient poplars. *Journal of Agricultural and Food Chemistry*, **49**: 86–91.

Ralph, J., Lapierre, C., Marita, J., et al. (2001b). Elucidation of new structures in lignins of CAD- and COMT-deficient plants by NMR. *Phytochemistry*, **57**: 993–1003.

Ralph, J., Bunzel, M., Marita, J.M., et al. (2004a). Peroxidase-dependent cross-linking reactions of *p*-hydroxycinnamates in plant cell walls. *Phytochemistry Reviews*, **3**: 79–96.

Ralph, J., Lundquist, K., Brunow, G., et al. (2004b). Lignins: natural polymers from oxidative coupling of 4-hydroxyphenylpropanoids. *Phytochemistry Reviews*, **3**: 29–60.

Ralph, J., Akiyama, T., Kim, H., et al. (2006). Effects of coumarate-3-hydroxylase downregulation on lignin structure. *Journal of Biological Chemistry*, **281**: 8843–8853.

Ralph, J., Brunow, G., Harris, P.J., et al. (2008). Lignification: are lignins biosynthesized via simple combinatorial chemistry or via proteinaceous control and template replication? In: Recent Advances in Polyphenol Research. *Vol.* **1**, (eds. Daayf, F., El Hadrami, A., Adam, L., and Ballance, G.M.), 36–66. Oxford, UK: Wiley-Blackwell Publishing.

Ralph, J., Lapierre, C., and Boerjan, W. (2019). Lignin structure and its engineering. *Current Opinion in Biotechnology*, **56**: 240–249.

Regner, M., Bartuce, A., Padmakshan, D., et al. (2018). Reductive cleavage method for quantitation of monolignols and low-abundance monolignol conjugates. *ChemSusChem*, **11**: 1600–1605.

Rencoret, J., Ralph, J., Marques, G., et al. (2013). Structural characterization of the lignin from coconut (*Cocos nucifera*) coir fibers. *Journal of Agricultural and Food Chemistry*, **61**: 2434–2445.

Rencoret, J., Kim, H., Evaristo, A.B., et al. (2018). Variability in lignin composition and structure in cell walls of different parts of macaúba (*Acrocomia aculeata*) palm fruit. *Journal of Agricultural and Food Chemistry*, **66**: 138–153.

Rencoret, J., Neiva, D., Marques, G., et al. (2019). Hydroxystilbene glucosides are incorporated into Norway spruce bark lignin. *Plant Physiology*, **180**: 1310–1321.

Rencoret, J., Marques, G., Serrano, O., et al. (2020). Deciphering the unique structure and acylation pattern of *Posidonia oceanica* lignin. *ACS Sustainable Chemistry & Engineering*, **8**: 12521–12533.

Rencoret, J., Rosado, M.J., Kim, H., et al. (2022). The flavonoids naringenin chalcone, naringenin, dihydrotricin, and tricin are lignin monomers incorporated into the rind lignin of papyrus (*Cyperus papyrus* L.). *Plant Physiology*, **188** (1): 208–219. doi: 10.1093/plphys/kiab469.

Rinaldi, R., Jastrzebshi, R., Clough, M.T., et al. (2016). Paving the way for lignin valorisation: recent advances in bioengineering, biorefining and catalysis. *Angewandte Chemie International Edition*, **55**: 8164–8215.

del Río, J.C., Marques, G., Rencoret, J., et al. (2007). Occurrence of naturally acetylated lignin units. *Journal of Agricultural and Food Chemistry*, **55**: 5461–5468.

del Río, J.C., Rencoret, J., Marques, G., et al. (2008). Highly acylated (acetylated and/or *p*-coumaroylated) native lignins from diverse herbaceous plants. *Journal of Agricultural and Food Chemistry*, **56**: 9525–9534.

del Río, J.C., Rencoret, J., Prinsen, P., et al. (2012). Structural characterization of wheat straw lignin as revealed by analytical pyrolysis, 2D-NMR, and reductive cleavage methods. *Journal of Agricultural and Food Chemistry*, **60**: 5922–5935.

del Río, J.C., Lino, A.G., Colodette, J.L., et al. (2015). Differences in the chemical structure of the lignins from sugarcane bagasse and straw. *Biomass & Bioenergy*, **81**: 322–328.

del Río, J.C., Rencoret, J., Gutiérrez, A., et al. (2017). Hydroxystilbenes are monomers in palm fruit endocarp lignins. *Plant Physiology*, **174**: 2072–2082.

del Río, J.C., Rencoret, J., Gutierrez, A., et al. (2018). Structural characterization of lignin from maize (*Zea mays* L.) fibers: evidence for diferuloylputrescine incorporated into the lignin polymer in maize kernels. *Journal of Agricultural and Food Chemistry*, **66**: 4402–4413.

del Río, J.C., Rencoret, J., Gutiérrez, A., et al. (2020). Lignin monomers from beyond the canonical monolignol biosynthetic pathway: another brick in the wall. *ACS Sustainable Chemistry & Engineering*, **8**: 4997–5012.

del Río, J.C., Rencoret, J., Gutiérrez, A., et al. (2021). Lignin monomers derived from the flavonoid and hydroxystilbene biosynthetic pathways. In: Recent Advances in Polyphenol Research. *Vol.* **7**, (eds. Reed, J., de Freitas, V., and Quideau, S.), 177–206. Oxford, UK: Wiley-Blackwell.

Rolando, C., Monties, B., and Lapierre, C. (1992). Thioacidolysis. In: Methods in Lignin Chemistry, (eds. Dence, C.W. and Lin, S.Y.), 334–349. Berlin-Heidelberg, Germany: Springer-Verlag.

Rosado, M.J., Bausch, F., Rencoret, J., et al. (2021). Differences in the content, composition and structure of the lignins from rind and pith of papyrus (*Cyperus papyrus* L.) culms. *Industrial Crops and Products*, **174**, 114226: 1–11.

Sakamoto, S., Kamimura, N., Tokue, Y., et al. (2020). Identification of enzymatic genes with the potential to reduce biomass recalcitrance through lignin manipulation in Arabidopsis. *Biotechnology for Biofuels*, **13** (97): 1–16.

Sarkanen, K.V. and Ludwig, C.H. (1971). Lignins: Occurrence, Formation, Structure and Reactions. New York: Wiley-Interscience.

Sasaki, S., Nishida, T., Tsutsumi, Y., and Kondo, R. (2004). Lignin dehydrogenative polymerization mechanism: a poplar cell wall peroxidase directly oxidizes polymer lignin and produces in vitro dehydrogenative polymer rich in β-*O*-4 linkage. *FEBS Letters*, **562**: 197–201.

Schuetz, M., Smith, R., and Ellis, B. (2013). Xylem tissue specification, patterning, and differentiation mechanisms. *Journal of Experimental Botany*, **64**: 11–31.

Sederoff, R.R., MacKay, J.J., Ralph, J., and Hatfield, R.D. (1999). Unexpected variation in lignin. *Current Opinion in Plant Biology*, **2**: 145–152.

Shi, R., Sun, Y.-H., Li, Q., et al. (2010). Towards a systems approach for lignin biosynthesis in *Populus trichocarpa*: transcript abundance and specificity of the monolignol biosynthetic genes. *Plant and Cell Physiology*, **51**: 144–163.

Shuai, L., Amiri, M.T., Questell-Santiago, Y.M., et al. (2016). Formaldehyde stabilization facilitates lignin monomer production during biomass depolymerization. *Science*, **354**: 329–333.

Skyba, O., Douglas, C.J., and Mansfield, S.D. (2013). Syringyl-rich lignin renders poplars more resistant to degradation by wood decay fungi. *Applied and Environmental Microbiology*, **79**: 2560–2571.

Smith, D.C.C. (1955a). Ester groups in lignin. *Nature*, **176**: 267–268.

Smith, D.C.C. (1955b). *p*-Hydroxybenzoates groups in the lignin of aspen (*Populus tremula*). *Journal of the Chemical Society*: 2347–2351.

Smith, R.A., Gonzales-Vigil, E., Karlen, S.D., et al. (2015). Engineering monolignol *p*-coumarate conjugates into Poplar and Arabidopsis lignins. *Plant Physiology*, **169**: 2992–3001.

Smith, R.A., Cass, C.L., Mazaheri, M., et al. (2017a). Suppression of CINNAMOYL-CoA REDUCTASE increases the level of monolignol ferulates incorporated into maize lignins. *Biotechnology for Biofuels*, **10** (109): 1–10.

Smith, R.A., Scheutz, M., Karlen, S.D., et al. (2017b). Defining the diverse cell populations contributing to lignification in Arabidopsis stems. *Plant Physiology*, **174**: 1028–1036.

Smith, R.A., Cass, C.L., Petrik, D.L., et al. (2021). Stacking *AsFMT* over-expression with a *Bdpmt* mutation enhances monolignol ferulate production in *Brachypodium distachyon*. *Plant Biotechnology Journal*, **19**: 1878–1886.

Smith, R.A., Beebe, E.T., Bingman, C.A., Vander Meulen, K.A., Eugene, A., Steiner, A., et al. (2022a). Identification and characterization of a set of monocot BAHD monolignol transferases. *Plant Physiology*, **189**: 37–48.

Smith, R.A., Lu, F., Cusumano, J.C., et al. (2022b). Manipulation of lignin monomer composition and introduction of monolignol conjugate biosynthesis leads to synergistic changes in lignin structure. *Plant Cell Physiology*, **63**: 744–754.

Somssich, M. (2020). Lignification and oxidative enzymes: localization, localization, localization! *Plant Physiology*, **184**: 554–555.

Stewart, J.J., Akiyama, T., Chapple, C.C., et al. (2009). The effects on lignin structure of overexpression of ferulate 5-hydroxylase in hybrid poplar. *Plant Physiology*, **150**: 621–635.

Stone, M.L., Anderson, E.M., Meek, K.M., et al. (2018). Reductive catalytic fractionation of C-lignin. *ACS Sustainable Chemistry & Engineering*, **6**: 11211–11218.

Takahama, U. (1995). Oxidation of hydroxycinnamic acid and hydroxycinnamyl alcohol derivatives by laccase and peroxidase. Interactions among *p*-hydroxyphenyl, guaiacyl and syringyl groups during the oxidation reactions. *Physiologia Plantarum*, **93**: 61–68.

Takahama, U., Oniki, T., and Shimokawa, H. (1996). A possible mechanism for the oxidation of sinapyl alcohol by peroxidase-dependent reactions in the apoplast: enhancement of the oxidation by hydroxycinnamic acids and components of the apoplast. *Plant and Cell Physiology*, **37**: 499–504.

Takenaka, Y., Watanabe, Y., Schuetz, M., et al. (2018). Patterned deposition of xylan and lignin is independent from that of the secondary wall cellulose of Arabidopsis xylem vessels. *Plant Cell*, **30**: 2663–2676.

Talebi Amiri, M., Dick, G.R., Questell-Santiago, Y.M., and Luterbacher, J.S. (2019). Fractionation of lignocellulosic biomass to produce uncondensed aldehyde-stabilized lignin. *Nature Protocols*, **14**: 921–954.

Timokhin, V.I., Regner, M., Motagamwala, A.H., et al. (2020). Production of *p*-coumaric acid from corn GVL-lignin. *ACS Sustainable Chemistry & Engineering*, **8**: 17427–17438.

Tobimatsu, Y., Chen, F., Nakashima, J., et al. (2013). Coexistence but independent biosynthesis of catechyl and guaiacyl/syringyl lignins in plant seeds. *The Plant Cell*, **25**: 2587–2600.

Umezawa, T. (2003). Diversity in lignan biosynthesis. *Phytochemistry Reviews*, **2**: 371–390.

Van Acker, R., Déjardin, A., Desmet, S., et al. (2017). Different routes for conifer- and sinapaldehyde and higher saccharification upon deficiency in the dehydrogenase CAD1. *Plant Physiology*, **175**: 1018–1039.

Vander Mijnsbrugge, K., Van Montagu, M., Inze, D., and Boerjan, W. (1996). Tissue-specific expression conferred by the S-adenosyl-L-methionine synthetase promoter of *Arabidopsis thaliana* in transgenic poplar. *Plant and Cell Physiology*, **37**: 1108–1115.

Vanholme, R., Morreel, K., Ralph, J., and Boerjan, W. (2008). Lignin engineering. *Current Opinion in Plant Biology*, **11**: 278–285.

Vanholme, R., Demedts, B., Morreel, K., et al. (2010a). Lignin biosynthesis and structure. *Plant Physiology*, **153**: 895–905.

Vanholme, R., Ralph, J., Akiyama, T., et al. (2010b). Engineering traditional monolignols out of lignins by concomitant up-regulation *F5H1* and down-regulation of *COMT* in Arabidopsis. *The Plant Journal*, **64**: 885–897.

Vanholme, R., Morreel, K., Darrah, C., et al. (2012). Metabolic engineering of novel lignin in biomass crops. *New Phytologist*, **196**: 978–1000.

Vanholme, R., De Meester, B., Ralph, J., and Boerjan, W. (2019a). Lignin biosynthesis and its integration into metabolism. *Current Opinion in Biotechnology*, **56**: 230–239.

Vanholme, R., Sundin, L., Seetso, K.C., Kim, H., Liu, X., De Meester, B., Hoengenaert, L., Goeminne, G., Morreel, K., Haustraete, J., Tsai, H.-H., Schmidt, W., Vanholme, B., Ralph, J. and Boerjan, W. (2019b) COSY catalyzes trans-cis isomerization and lactonization in the biosynthesis of coumarins. *Nature Plants* **5**: 1066-1075.

Vermaas, J.V., Dixon, R.A., Chen, F., et al. (2019). Passive membrane transport of lignin-related compounds. *Proceedings of the National Academies of Sciences, USA*, **116**: 23117–23123.

de Vries, L., MacKay, H.A., Smith, R.A., et al. (2022). *p*HBMT1, a BAHD-family monolignol acyltransferase, mediates lignin acylation in poplar. *Plant Physiology*, **188**: 1014–1027.

Wagner, A., Tobimatsu, Y., Phillips, L., et al. (2011). *CCoAOMT* suppression modifies lignin composition in *Pinus radiata*. *The Plant Journal*, **67**: 119–129.

Wang, M., Shi, H., Camaioni, D.M., and Lercher, J.A. (2017). Palladium-catalyzed hydrolytic cleavage of aromatic C—O bonds. *Angewandte Chemie International Edition*, **56**: 2110–2114.

Wang, M., Gutiérrez, O.Y., Camaioni, D.M., and Lercher, J.A. (2018). Palladium-catalyzed reductive insertion of alcohols into aryl ether bonds. *Angewandte Chemie International Edition*, **130**: 3809–3813.

Wang, S.Z., Su, S.H., Xiao, L.P., et al. (2020a). Catechyl lignin extracted from Castor seed coats using deep eutectic solvents: Characterization and depolymerization. *ACS Sustainable Chemistry & Engineering*, **8**: 7031–7038.

Wang, X., Zhuo, C., Xiao, X., et al. (2020b). Substrate-specificity of LACCASE 8 facilitates polymerization of caffeyl alcohol for C-lignin biosynthesis in the seed coat of *Cleome hassleriana*. *Plant Cell*, **32**: 3825–3845.

Weng, J.K. and Chapple, C. (2010). The origin and evolution of lignin biosynthesis. *New Phytologist*, **187**: 273–285.

Weng, J.-K., Mo, H., and Chapple, C. (2010). Over-expression of F5H in COMT-deficient Arabidopsis leads to enrichment of an unusual lignin and disruption of pollen wall formation. *The Plant Journal*, **64**: 898–911.

Wilkerson, C.G., Mansfield, S.D., Lu, F., et al. (2014). Monolignol ferulate transferase introduces chemically labile linkages into the lignin backbone. *Science*, **344**: 90–93.

Yamamoto, M., Tomiyama, H., Koyama, A., et al. (2020). A century-old mystery unveiled: sekizaisou is a natural lignin mutant. *Plant Physiology*, **182**: 1821–1828.

Yamamura, M., Akashi, K., Yokota, A., et al. (2012). Characterization of *Jatropha curcas* lignins. *Plant Biotechnology*, **29**: 179–183.

Yan, X., Liu, J., Kim, H., et al. (2019). CAD1 and CCR2 protein complex formation in monolignol biosynthesis in *Populus trichocarpa*. *New Phytologist*, **222**: 244–260.

Yoo, H., Widhalm, J.R., Qian, Y., et al. (2013). An alternative pathway contributes to phenylalanine biosynthesis in plants via a cytosolic tyrosine:phenylpyruvate aminotransferase. *Nature Communications*, **4** (2833): 1–11.

Yu, H., Liu, C., and Dixon, R.A. (2021). A gene-editing/complementation strategy for tissue-specific lignin reduction while preserving biomass yield. *Biotechnology for Biofuels*, **14** (175): 1–13.

Zanarotti, A. (1983). Synthesis and reactivity of lignin model quinone methides. Biomimetic synthesis of 8.0.4' neolignans. *Journal of Chemical Research, Synopses*: 306–307.

Zeng, H., Cao, D., Qiu, Z., and Li, C.J. (2018). Palladium-catalyzed formal cross-coupling of diaryl ethers with amines: slicing the 4-*O*-5 linkage in lignin models. *Angewandte Chemie International Edition*, **57**: 3752–3757.

Zhao, Q., Tobimatsu, Y., Zhou, R., et al. (2013). Loss of function of *Cinnamyl Alcohol Dehydrogenase 1* causes accumulation of an unconventional lignin and a temperature-sensitive growth defect in *Medicago truncatula*. *Proceedings of the National Academy of Sciences, USA*, **110**: 13660–13665.

Zhao, Y., Yu, X., Lam, P.Y., et al. (2021). Monolignol acyltransferase for lignin *p*-hydroxybenzoylation in *Populus*. *Nature Plants*, **7**: 1288–1300.

Zhou, S., Runge, T., Karlen, S.D., et al. (2017). Chemical pulping advantages of Zip-lignin hybrid poplar. *ChemSusChem*, **10**: 3565–3573.

Zhuo, C.L., Rao, X.L., Azad, R., et al. (2019). Enzymatic basis for C-lignin monomer biosynthesis in the seed coat of *Cleome hassleriana*. *Plant Journal*, **99**: 506–520.

2

Synthesis of Epigallocatechin Gallate, Nobiletin, and Their Derivatives for Chemical-Biological Studies

Tomohiro Asakawa[1], Makoto Inai[2], and Toshiyuki Kan[2]

[1] *School of Marine Science and Technology, Department of Fisheries-Food Science, Tokai University, Shizuoka, Japan*
[2] *School of Pharmaceutical Sciences, University of Shizuoka, Shizuoka, Japan*

2.1 Synthetic Investigations of Catechin Derivatives

(−)-Epigallocatechin gallate (**1**, EGCg) (Figure 2.1), a major constituent of green tea extract (Wheeler and Wheeler 2004; Nagle et al. 2006; Zaveri 2006; Friedman 2007), exhibits anti-tumor (Yang et al. 2009; Yiannakopoulou 2014), antiviral (Song and Seong 2007), and other important bioactivities (Cabrera et al. 2006; Thielecke and Boschmann 2009). Consequently, EGCg and its derivatives are considered to be promising lead compounds for drug development (Dell'Agli et al. 2005; Wan et al. 2005; Moon et al. 2006). Although various activities of catechins have been reported, the mechanisms of these activities are not sufficiently clear. Thus, there is a requirement for imaging probes and other analytical tools to enable biochemical studies of **1**. There have been several reports on the synthesis of catechin probes (Fukuhara et al. 2002, 2009a, 2009b; Hakamata et al. 2006). However, direct modification of **1** to introduce probe moieties without the isomerization of the stereochemistry of chroman structure or the decomposition of galloyl ester is difficult due to the structural instability of **1** and the lack of appropriate tethering functional groups. In our previous synthetic investigations, we found that the deoxy derivative **2**, which was constructed by 6-endo cyclization of epoxy phenol, exhibits similar anti-influenza infection activity to natural **1** (Furuta et al. 2007). Inspired by this finding, we aimed to synthesize 6-(5-aminopentyl)-5,7-deoxyepigallocatechin gallate (APDOEGCg, **3**), which contains a linker and a reactive nitrogen atom (Yoshida et al. 2011), as an EGCg probe precursor (Figure 2.1). Since direct incorporation of a linker unit into DOEGCg (**2**), seemed difficult, we decided to employ a cross-coupling reaction. Incorporation of a nitrogen atom at the terminal of the linker was found to be compatible with our Ns nitrobenzenesulfonyl (Ns) strategy (Fukuyama et al. 1999; Kan and Fukuyama 2001, 2004; Kan et al. 2003, 2007).

This paper is dedicated to Professor Hidetoshi Yamada, who passed away on 23 November 2019.

Recent Advances in Polyphenol Research, Volume 8, First Edition. Edited by Juha-Pekka Salminen, Kristiina Wähälä, Victor de Freitas, and Stéphane Quideau.

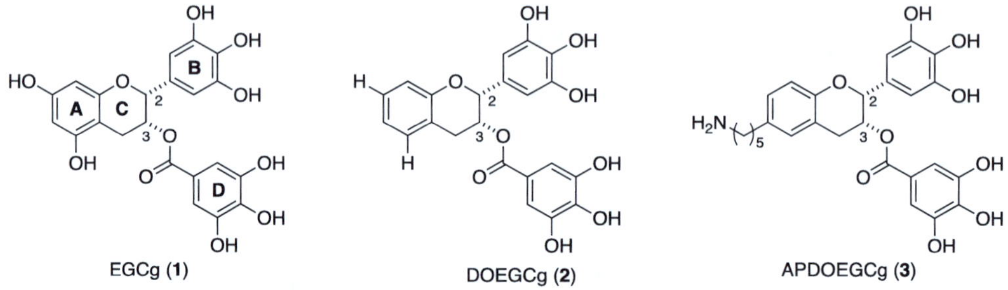

Figure 2.1 Structures of EGCg (**1**), DOEGCg (**2**), and APDOEGCg (**3**).

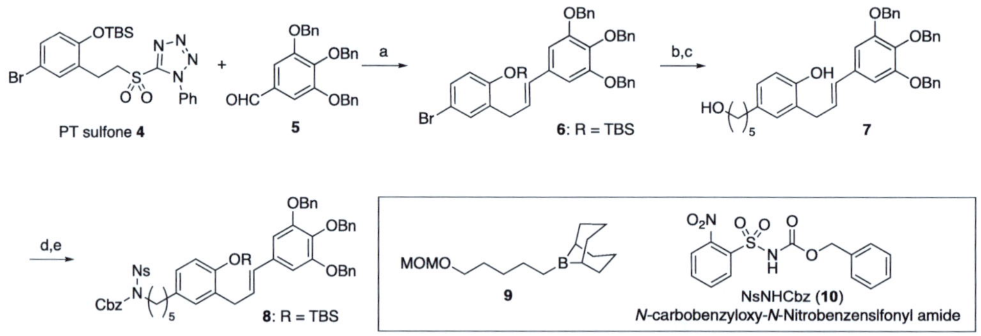

Scheme 2.1 Condensation of A and B rings and incorporation of amino linker moiety. Reagents and conditions: (a) Lithium hexamethyldisilazide (LiHMDS), THF, −78 °C, 96%; (b) **9**, $PdCl_2(dppf){\cdot}CH_2Cl_2$, 3 M aq. NaOH, THF, reflux; (c) conc. HCl, MeOH, 60 °C, 95% (2 steps), (d) NsNHCbz (**10**), di-2-methoxyethyl Azodicarboxylate (DMEAD), PPh_3, toluene, 89%; (e) TBSCl, imidazole, dimethylformamide (DMF), 95%.

As shown in Scheme 2.1, coupling reaction of the A- and B-rings was accomplished by Julia–Kocieński reaction (Blakemore et al. 1998; Blakemore 2002; Lebrun et al. 2006) between phenyltetrazole (PT)-sulfone **4** and aldehyde **5** to provide **6** in 96% yield with *E*-selectivity. Next, a linker group was incorporated by Suzuki-Miyaura reaction (Miyaura and Suzuki 1995). The desired coupling reaction of **6** and borate **9** in the presence of $PdCl_2$(dppf) catalyst and 3 M aqueous NaOH proceeded smoothly to give **7**. After acidic hydrolysis of the methoxymethyl (MOM) group, incorporation of the amino group was accomplished by Mitsunobu reaction (Mitsunobu 1981; Hughes 1992; Sugimura and Hagiya 2007) with our *N*-carbobenzyloxy-*N*-2-nitrobenzensulfonyl amide (**10**) (Ns strategy) to give **8**.

After enantioselective conversion of **8** to diol **11** by Sharpless asymmetric dihydroxylation (Kolb et al. 2014), we found that mono-selective acylation of **11** with acyl chloride **17** proceeded in the presence of *n*-Bu_2SnO (Nagashima and Ohno 1987). Although the ester was obtained as a 1:1 regioisomer, treatment of this mixture with 1,8-diazabicyclo[5.4.0]undec-7-ene (DBU) provided stable ester **12** through an interesting migration reaction. This migration of acyl group to less hindered hydroxyl group would be accelerated by the steric hindrance of the neighboring aryl group. Oxidation of the secondary alcohol **12** was performed with 1-Me-AZADO (Shibuya et al. 2006) and $PhI(OAc)_2$ to afford ketone **13**. After deprotection of the tributylsilyl (TBS) group, reductive cyclization of **14** with Et_3SiH

and $BF_3{\cdot}OEt_2$ provided *cis*-dihydrobenzopyran **16**. Because this reaction proceeds through cationic intermediate **15**, the attack of hydride occurs predominantly from the β-face (Kitada 2006; Tanaka et al. 2007) (Scheme 2.2). Finally, removal of the Ns (Fukuyama et al. 1995; Kurosawa et al. 2002) and benzyl groups of **16** provided the desired APDOEGCg (**3**), as shown in Scheme 2.3.

With the desired derivative **3** in hand, we evaluated the inhibitory activity against influenza virus infection in a cellular model. As shown in Table 2.1, APDOEGCg (**3**) potently inhibited infection of Madin–Darby canine kidney (MDCK) cells with the influenza virus A/Memphis/1/71 (H3N2), with the IC_{50} value of 4.40 μM, being more potent than natural **1** and synthetic derivative **2** (Mori et al. 2008). Thus, the introduction of the linker moiety did not affect the activity, and so incorporation of probe moieties at this site might be feasible.

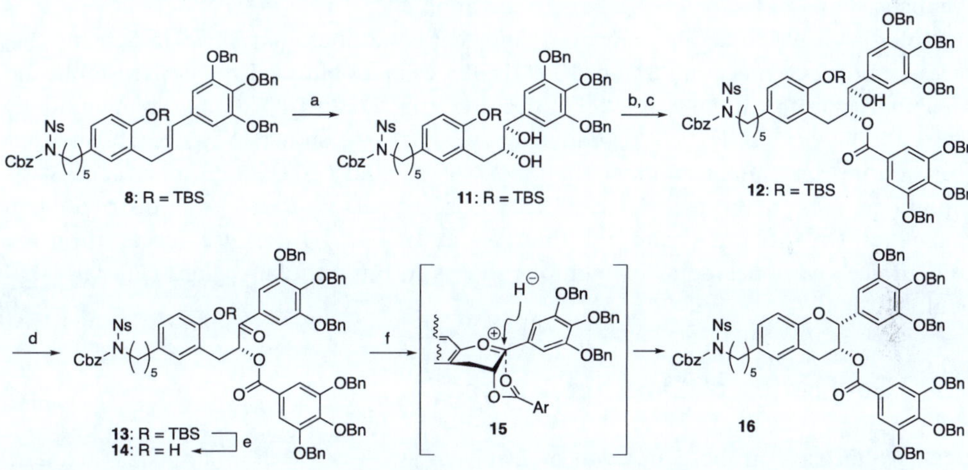

Scheme 2.2 Stereoselective construction of the *cis*-benzopyran ring. Reagents and conditions: (a) AD-mix β, $MeSO_2NH_2$, *t*-BuOH–H_2O, rt, 76%; (b) Ar–COCl (**17**), *n*-Bu_2SnO, Et_3N, CH_2Cl_2, rt; (c) DBU, toluene, −20 °C, 89% (2 steps); (d) 1-Me-AZADO, $PhI(OAc)_2$, CH_2Cl_2, rt; (e) TBAF, AcOH, THF, 0 °C, 96%; (f) Et_3SiH, $BF_3{\cdot}OEt_2$, CH_2Cl_2, 84%.

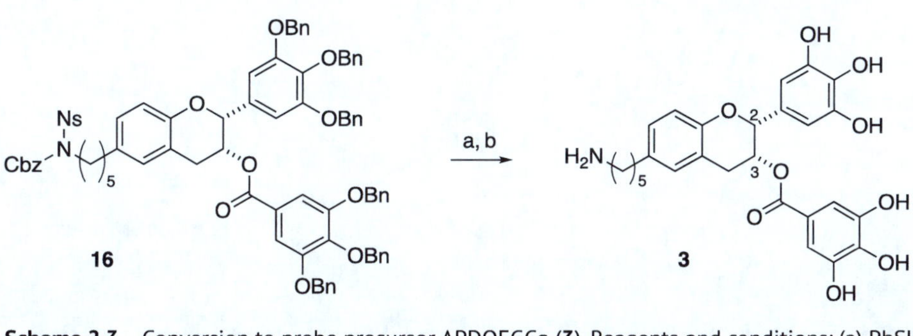

Scheme 2.3 Conversion to probe precursor APDOEGCg (**3**). Reagents and conditions: (a) PhSH, Cs_2CO_3, MeCN, rt; (b) H_2, $Pd(OH)_2$, THF-MeOH, rt.

Table 2.1 Inhibition of influenza A viral infectivity toward MDCK cells.

Compound	Complement inhibition IC_{50} [a] (μM)
EGCg (**1**)	66.3 (± 9.21)
DOEGCg (**2**)	9.05 (± 2.26)
APDOGCg (**3**)	4.40 (± 2.36)

[a] Values are reported as the mean of three experiments, with the standard deviation in parentheses.

2.2 Synthesis and Application of Fluorescent Catechin Probes

Encouraged by these results, we set out to prepare a fluorescent probe molecule from **3**. We focused on TokyoGreen (TG: **20**) (Urano et al. 2005), a fluorescein derivative that is suitable for *in vivo* imaging under physiological conditions. As shown in Scheme 2.4, reaction of probe precursor **3** and TG-activated ester **18** afforded the desired probe **19**. The biodistribution of the synthetic fluorescent probe EGCg-TG (**19**) in human umbilical vein endothelial cells (HUVECs) (Yamakawa et al. 2004) is shown in Figure 2.2. The confocal fluorescence images show that most of the EGCg-TG (**19**) fluorescence was present as dots in the cytoplasm of HUVECs, suggesting that it was distributed in organelles. On the other hand, the fluorescence of free TG (**20**) was localized on the cell surface and appeared as filamentous shapes in intracellular regions (Piyaviriyakul et al. 2011).

2.3 Generation of Catechin Antibody

Next, we focused on the generation of EGCg antibodies (Kawai et al. 2008; Kuzuhara et al. 2008), which are expected to be useful for immunohistology, as well as for developing enzyme-linked immunosorbent assays (ELISA) with color or fluorescence endpoints for quantitating low levels of EGCg in serum. We synthesized the immunogen **23** by conjugation of **4** to carrier protein **21** (HSA: human serum albumin) via glutaraldehyde (**22**) as a

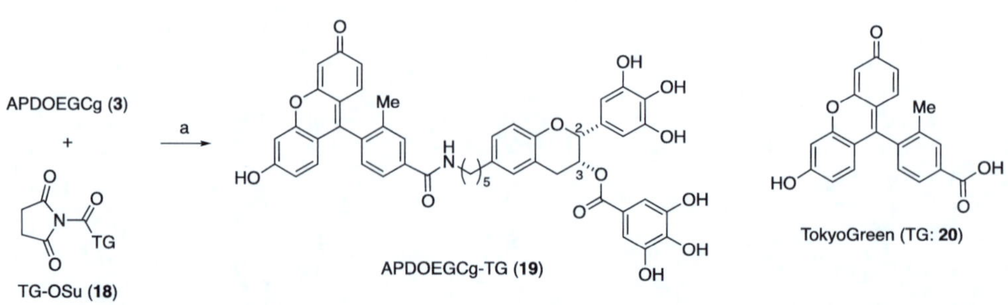

Scheme 2.4 Synthesis of the fluorescein probe **19** from APDOEGCg (**3**). Reagents and conditions: (a) DMF, rt.

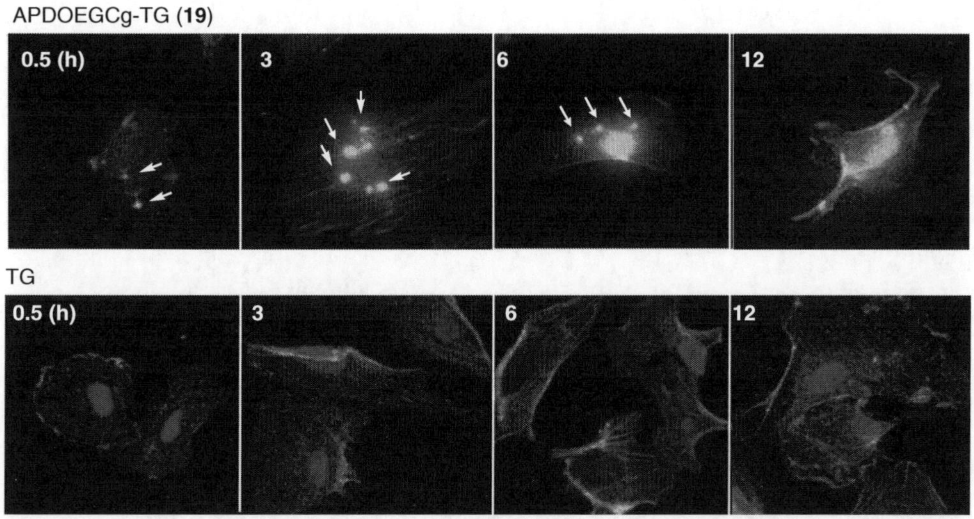

Figure 2.2 Intracellular localization of APDOEGCg-TG (**19**) in HUVEC. Condition: HUVECs, 37 °C, 1 μM of APDOEGCg-TG (**19**) or TG (**20**) with 5 U/mL of rhodamine/pholloidin and 10 μg/mL of DAPI.

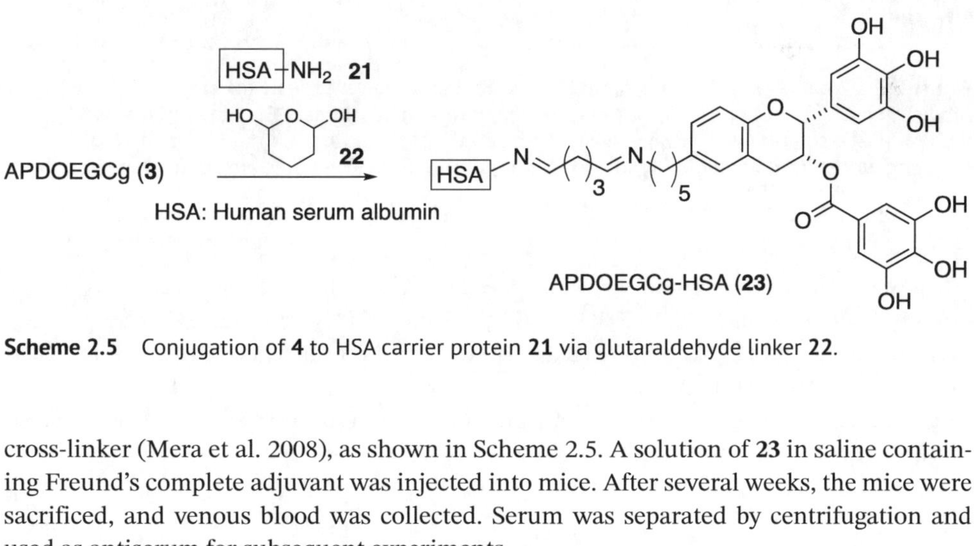

Scheme 2.5 Conjugation of **4** to HSA carrier protein **21** via glutaraldehyde linker **22**.

cross-linker (Mera et al. 2008), as shown in Scheme 2.5. A solution of **23** in saline containing Freund's complete adjuvant was injected into mice. After several weeks, the mice were sacrificed, and venous blood was collected. Serum was separated by centrifugation and used as antiserum for subsequent experiments.

2.4 PET Imaging of Biodistribution of Catechin

Next, we turned our attention to positron emission tomography (PET) analysis of EGCg derivatives. PET is a noninvasive imaging technique that is used for clinical diagnosis in many fields, including oncology and neurology. PET can provide real-time three-dimensional images of the biodistribution of a positron-labeled agent in the whole body

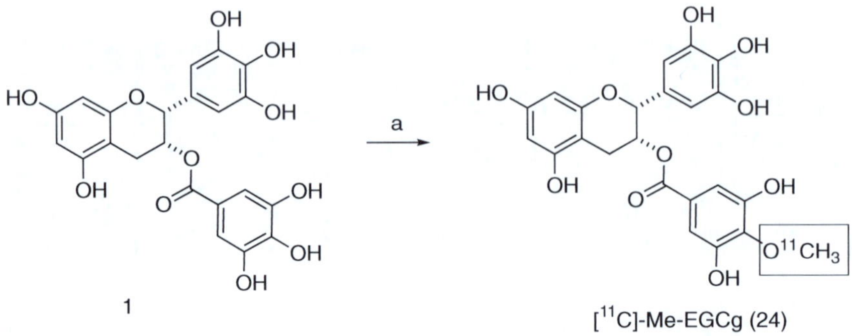

Scheme 2.6 Rapid synthesis of PET probe **24** from EGCg (**1**). Reagents and conditions: (a) $^{11}CH_3I$, aq. *n*-Bu_4NOH, DMSO, 75 °C, three minutes; HPLC separation.

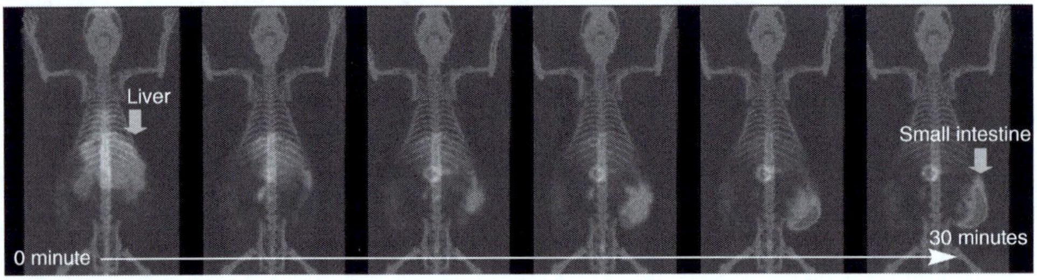

Figure 2.3 Whole-body PET imaging of the distribution of intravenously injected [^{11}C] Me-EGCg. Biodistribution of [^{11}C] Me-EGCg was determined with a small-animal PET system (Clairvivo PET). [^{11}C] Me-EGCg solution was intravenously injected into a Wistar rat via a tail vein, and PET scanning was started immediately. Images acquired from 0 to 30 minutes are shown. (*See insert for color representation of the figure.*)

after the injection of microgram-level doses. As shown in Scheme 2.6, we accomplished the rapid incorporation of ^{11}C into EGCg (**1**) in the form of CH_3 on the most reactive 4”-OH (Shimizu et al. 2014). Positron-labeled 4”-[^{11}C]methyl-epigallocatechin gallate (**24**) was intravenously injected into rats, and the biodistribution was imaged for 60 minutes with a small-animal PET system. Typical real-time images of absorption and biodistribution of Me-EGCg (**24**) are illustrated in Figure 2.3.

2.5 Practical Synthesis of Nobiletin

Nobiletin (**25**), isolated from citrus fruits (Tseng 1938) (Figure 2.4), has various biological activities, including modulation of the extracellular signal-related protein kinase (ERK)/ cAMP response element-binding protein (CREB) pathway in cell culture systems, and is a potential lead compound for novel anti-dementia agents (Murakami et al. 2000; Rooprai et al. 2001). However, commercially available natural nobiletin (**25**) is expensive, and so a practical total synthesis of **25** is required (Asakawa et al. 2019). During the course of our flavone investigations, we have developed an efficient synthetic method for the flavone

Nobiletin (**25**) Sudachitin (**26**)

Figure 2.4 Structures of nobiletin (**25**) and sudachitin (**26**).

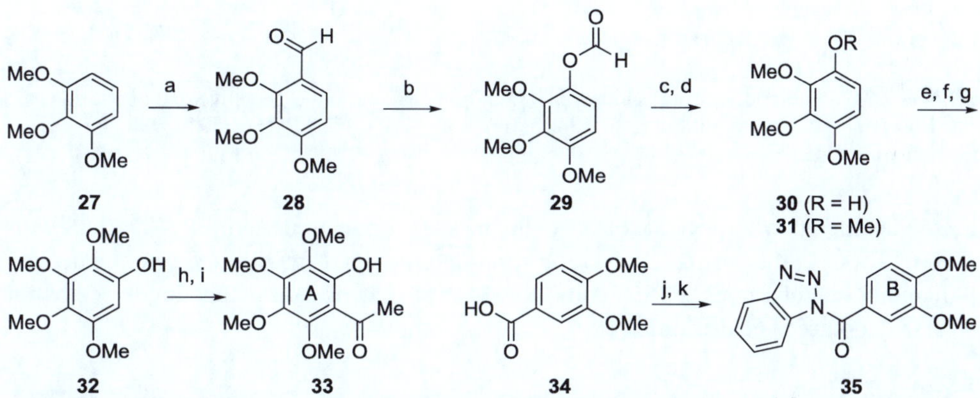

Scheme 2.7 Preparation of A-ring acetophenone **33** and B-ring acyl donor. Reagents and conditions: (a) $TiCl_4$, Cl_2CHOMe, CH_2Cl_2; (b) SeO_2 (cat.), 30% H_2O_2, *t*-BuOH, 50 °C; (c) Et_3N, MeOH; (d) MeI, K_2CO_3, acetone, 84% (4 steps); (e) PhN(Me)CHO, $POCl_3$, 60 °C; (f) SeO_2 (cat.), 30% H_2O_2, *t*-BuOH, 50 °C; (g) Et_3N, MeOH; (h) MeI, K_2CO_3, acetone, 94% (4 steps); (i) AcCl, $AlCl_3$, CH_2Cl_2, 70%; (j) $SOCl_2$, DMF; (k) benzotriazole, Et_3N, 98% (2 steps).

skeleton via a β-diketone intermediate, enabling large-scale synthesis of **25** and its demethyl derivative sudachitin (**26**) (Horie et al. 1961) (Figure 2.4).

As shown in Scheme 2.7, incorporation of a hydroxyl group into electron-rich trimethoxybenzene (**27**) was achieved by sequential formylation and Baeyer–Villiger-type oxidation reactions. Upon treatment of **27** with Cl_2CHOMe in the presence of $TiCl_4$, the formylation reaction proceeded to afford **28**. Subsequent Baeyer–Villiger-type reaction was performed with the aid of a catalytic amount of selenium dioxide (Fukuyama and Yang 1986) and hydrogen peroxide to afford formyl ester **29**, which, without purification, was transformed to 1,2,3,4-tetramethoxybenzene (**31**). Formylation of **31** readily proceeded under modified Vilsmeier–Haack conditions (Jones and Stanforth 2000) due to the more electron-rich aromatic rings, compared with **27**. Baeyer–Villiger oxidation and solvolysis of the generated formyl group under the same conditions used for **28** afforded the phenol **32**. Methylation of **32** and Friedel–Crafts acylation with acetyl chloride and aluminum chloride with concomitant demethylation afforded acetophenone **33**. On the other hand, acyl benzotriazole derivative **35** (Katritzky et al. 1992; Katritzky and Pastor 2000) was readily obtained from **34** by conversion to the corresponding acid chloride and treatment with benzotriazole. Total synthesis of nobiletin was accomplished by acylation reaction of

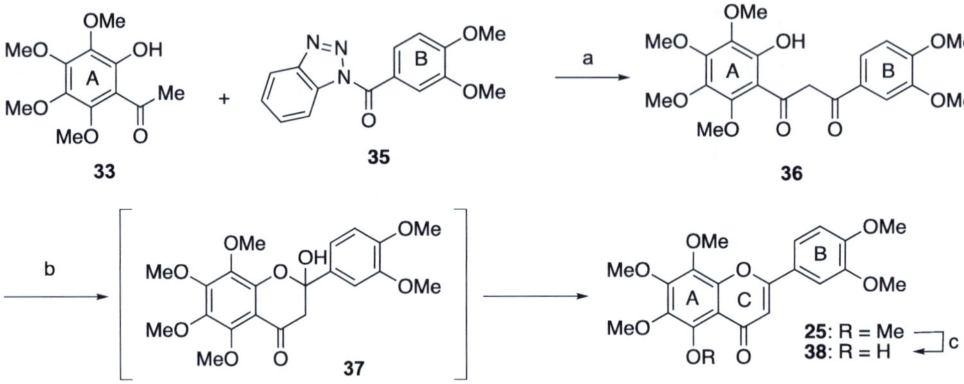

Scheme 2.8 Completion of total synthesis of nobiletin (**25**) and regio-selective demethylation reaction. Reagents and conditions: (a) LiHMDS, THF, 0 °C, 91%; (b) TFA, MeOH, 50 °C, 98%; (c) $AlCl_3$, EtSH, CH_2Cl_2, 75%.

acetophenone **33** with acyl donor **35** (Scheme 2.8). Upon treatment of **33** and **35** with LiHMDS at 0 °C, the desired *C*-acylation reaction proceeded smoothly to give the desired β-diketone **36**. Subsequent cyclization (**36** → **37**) and dehydration by heating in methanol with trifluoroacetic acid afforded **25**.

2.6 Derivatization of Desmethyl Nobiletins

Several desmethyl nobiletin derivatives have been reported to show biological activity. Our synthetic strategy of nobiletin was also applicable for the preparation of desmethyl derivatives **26** and **39–44** in Figure 2.5.

Sudachitin (**26**), a 5,7,4'-tridesmethyl nobiletin derivative isolated from citrus sudachi, also possesses interesting biological activity (Horie et al. 1961). As shown in Scheme 2.9, synthesis of sudachitin (**26**) was commenced with acetophenone **33**, which is readily available on a large scale as a synthetic intermediate of nobiletin. Upon treatment of **33** with LiCl, regioselective demethylation reaction occurred at the 4-methoxy group to provide **45** (Sagara et al. 2018). Further treatment with benzyl bromide and K_2CO_3 resulted in incorporation of a benzyl group to afford **45**. Acylation reaction of acetophenone **11** and acylbenzotriazole **47** (Hiza et al. 2014) in the presence of LiHMDS afforded **48**.

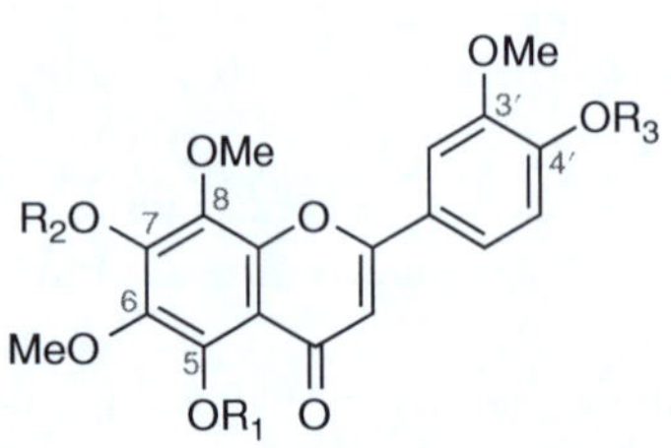

Nobiletin **25** ($R_1 = R_2 = R_3 = Me$)
Sudachitin **26** ($R_1 = R_2 = R_3 = H$)
39 ($R_1 = Me, R_2 = R_3 = H$)
40 ($R_1 = R_2 = Me, R_3 = H$)
41 ($R_1 = R_3 = H, R_2 = Me$)
42 ($R_1 = H, R_3 = R_2 = Me$)
43 ($R_1 = R_3 = Me, R_2 = H$)
44 ($R_1 = R_2 = H, R_3 = Me$)

Figure 2.5 Structure of nobiletin (**25**) and its desmethyl derivatives (**26, 39–44**).

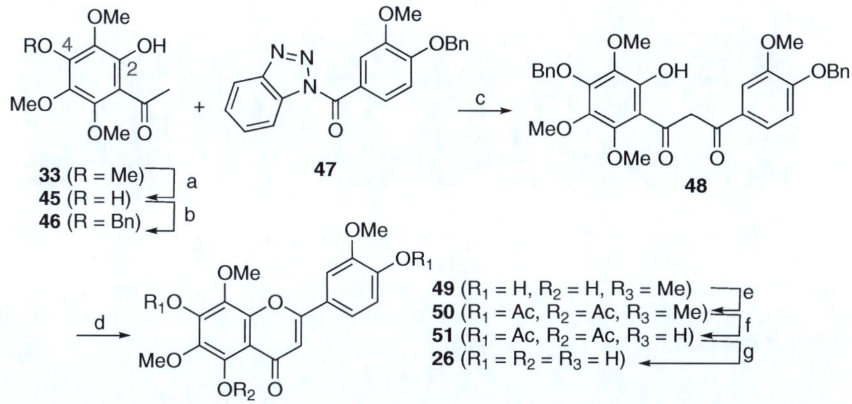

Scheme 2.9 Synthetic route to sudachitin (**26**). Reagents and conditions: (a) LiCl, DMF, 120 °C, 59% (brsm 70%); (b) benzyl bromide (BnBr), K_2CO_3, DMF, 79%; (c) LiHMDS, THF, −78 °C to 0 °C, 87%; (d) TsOH, toluene, 80 °C, 93%; (e) Ac_2O, *N*,*N*-dimethylaminopyridine (DMAP), pyridine; (f) $AlCl_3$, EtSH, CH_2Cl_2, 35% (two steps); (g) K_2CO_3, MeOH, 31%.

Acidic treatment of **48** resulted in simultaneous cyclization and dehydration with concomitant cleavage of the benzyl ether to afford **49**. After acetylation of the phenol groups of **49**, regioselective removal of the methyl group was carried out by treatment with EtSH in the presence of $AlCl_3$ (Node et al. 1980) to afford **51**. Finally, ammonia-mediated ammonolysis of the acetate provided **25**. Other desmethyl derivatives of nobiletin (**39**–**44**) were obtained via a similar approach. These desmethyl nobiletins are useful for Structure–Activity Relationship study (Takii et al. 2017) and biosynthetic study (Seoka et al. 2020).

2.7 PET Imaging of Biodistribution of Nobiletin

Inspired by our success with EGCg, we next examined the PET analysis of nobiletin (Asakawa et al. 2011). To obtain a suitable probe, we focused on regioselective demethylation. 5-Desmethyl nobiletin (**38**) was obtained by regioselective demethylation at the 5-position of **25**, as shown in Scheme 2.10. Methylation with [^{11}C]methyl iodide and tetrabutylammonium hydroxide furnished ^{11}C-labeled nobiletin (**25a**) within one minute.

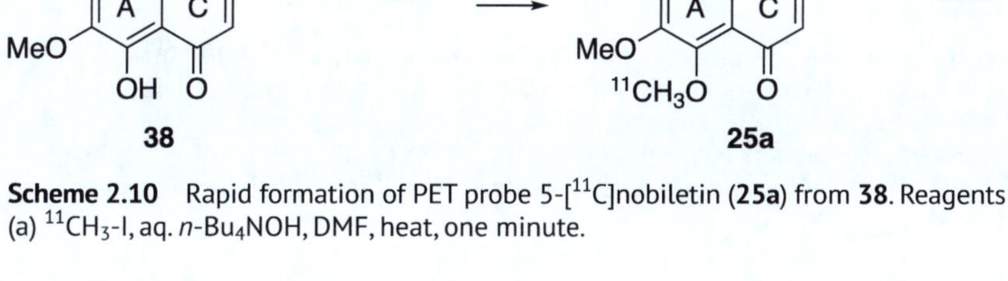

Scheme 2.10 Rapid formation of PET probe 5-[^{11}C]nobiletin (**25a**) from **38**. Reagents and conditions: (a) $^{11}CH_3$-I, aq. *n*-Bu_4NOH, DMF, heat, one minute.

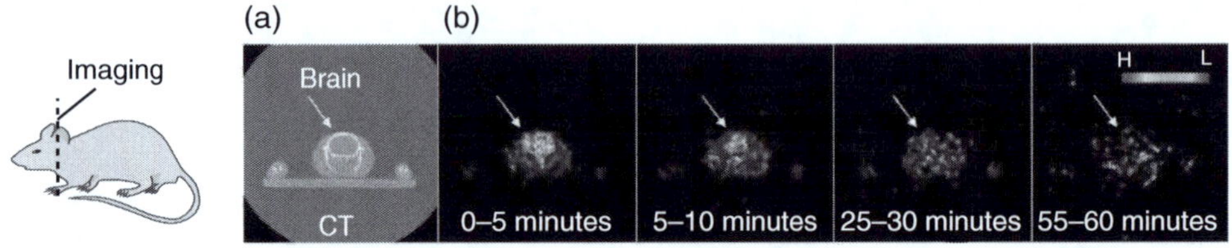

Figure 2.6 PET imaging of the accumulation of 5-[^{11}C]nobiletin (**25a**) in rat brain. **25a** was intravenously injected via the tail vein of an eight-month-old male SD rat, and its distribution was imaged using a Clairvivo PET system (Shimadzu Corporation). The data were acquired for 60 minutes and integrated every five minutes. The figure shows a CT image (a) and representative PET images (b). White arrows indicate the brain region.

This PET probe was used to investigate the biodistribution of [^{11}C]**25a** in a rat model, focusing on accumulation in the brain (Figure 2.6), since nobiletin (**25**) was recently reported to be effective for Alzheimer's disease.

2.8 Synthesis and Application of Fluorescent Nobiletin Probe

Next, we examined the preparation of probe molecules for chemical-biological studies of nobiletin (**25**) from probe precursor **38**, focusing on incorporation of a probe moiety via the Hüisgen reaction (Huisgen 1963; Kolb et al. 2001), which should be readily applicable to a wide range of linkers and probes. Thus, introduction of a terminal alkyne unit was accomplished by condensation of reactive amine with a linker unit at the 5-position of **38** (Scheme 2.11). Alkylation of **38** with *tert*-butoxycarbonyl (Boc)-5-aminopentyl-1-iodide and successive deprotection of the Boc group provided the reactive amino alkyl nobiletin

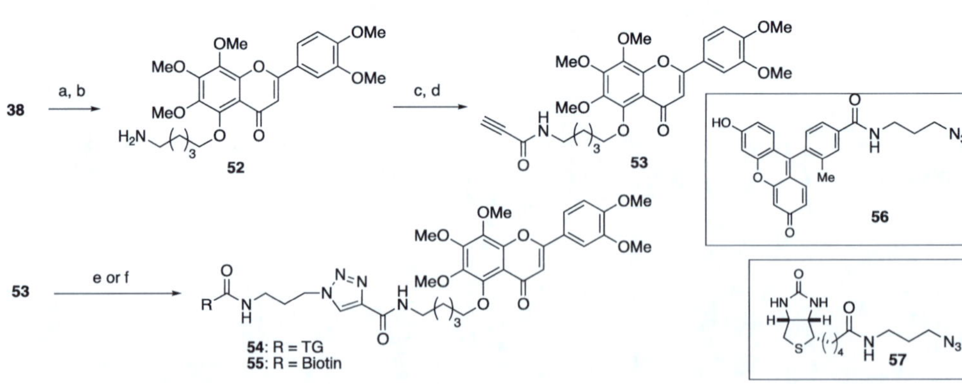

Scheme 2.11 Synthesis of nobiletin probes **54** and **55** from the versatile precursor **38** via Hüisgen reaction. Reagents and conditions: (a) *N*-Boc-5-aminopentyl-1-iodide K_2CO_3, acetone, rt; (b) 2 M HCl, MeOH, CH_2Cl_2, rt, 71% (2 steps); (c) 3-(trimethylsilyl)propiolic acid, 1-Ethyl-3-(3-dimethylaminopropyl)carbodiimide Hydrochloride (EDCI), HOBt, CH_2Cl_2, rt; (d) TBAF, CH_2Cl_2, rt, 68% (2 steps); € $CuSO_4{\cdot}5H_2O$, sodium ascorbate, CH_2Cl_2, MeOH, rt, 71%; (f) $CuSO_4{\cdot}5H_2O$, sodium ascorbate, CH_2Cl_2, MeOH, rt, 48%.

52. After the incorporation of 3-trimethylsilyl (TMS) propiolic acid, treatment w tetrabutyl ammonium fluoride (TBAF) for the removal of the TMS group was carried out to give the desired nobiletin probe precursor **53**. Using this probe precursor, we prepared the fluorescein- and biotin-based probes **54** and **55**. As shown in Scheme 2.11, Hüisgen reaction of nobiletin probe precursor **53** and TG-conjugated alkyl azide **56** proceeded smoothly to provide the fluorescent probe **54**. Incorporation of a biotin moiety into the nobiletin precursor **53** was carried out by cyclization reaction with the biotin-alkyl azide **57**. Our group is currently undertaking fluorescence imaging studies with **54** and target-protein detection with **55** (Yasuda et al. 2014).

2.9 Conclusions

We accomplished an asymmetric synthesis of APDOEGCg (**3**) as a chemical-probe precursor via cationic cyclization utilizing neighboring group participation of the gallate carbonyl group as a key step. APDOEGCg (**3**) was efficiently converted to fluorescent probe **19** and immunogen **22** by utilizing the high reactivity of the amine functional group. We confirmed the usefulness of these probes for imaging studies and generation of antibodies, respectively. We also efficiently synthesized PET probe **24** by incorporation of ^{11}C into EGCg (**1**).

Our strategy was also applied to develop a practical synthesis of nobiletin (**25**) and its demethylated analogs, such as sudachitin (**26**), on a 100-g scale. 5-Desmethyl nobiletin was readily converted to a series of probes by incorporation of fluorescein, biotin, or ^{11}C. Among them, the PET probe 5-[^{11}C]nobiletin (**25a**) was confirmed to be useful for imaging the biodistribution in rats. Further applications are under study in our laboratory and with collaborators.

References

Asakawa, T., Hiza, A., Nakayama, M., et al. (2011). PET imaging of nobiletin based on a practical total synthesis. *Chemical Communications*, **47**: 2868–2870.

Asakawa, T., Sagara, H., Kanakogi, M., et al. (2019). Practical synthesis of polymethylated flavones: nobiletin and its desmethyl derivatives. *Organic Process Research & Development*, **23**: 595–602.

Blakemore, P.R. (2002). The modified Julia olefination: alkene synthesis via the condensation of metallated heteroarylalkylsulfones with carbonyl compounds. *Journal of the Chemical Society, Perkin Transactions* 1: 2563–2585 https://pubs.rsc.org/en/content/articlelanding/2002/p1/b208078h.

Blakemore, P.R., Cole, W.J., and Kocienski, P.J. (1998). A stereoselective synthesis of *trans*-1,2-disubstituted alkenes based on the condensation of aldehydes with metallated 1-phenyl-1*H*-tetrazol-5-yl sulfones. *Synlett*: 26–28.

Cabrera, C., Artacho, R., and Giménez, R. (2006). Beneficial effects of green tea – a review. *Journal of the American College of Nutrition*, **25**: 79–99.

Dell'Agli, M., Bellosta, S., Rizzi, L., et al. (2005). A structure-activity study for the inhibition of metalloproteinase-9 activity and gene expression by analogues of gallocatechin-3-gallate. *Cellular and Molecular Life Sciences*, **62**: 2896–2903.

Friedman, M. (2007). Overview of antibacterial, antitoxin, antiviral, and antifungal activities of tea flavonoids and teas. *Molecular Nutrition & Food Research*, **51**: 116–134.

Fukuhara, K., Nakanishi, I., Kansui, H., et al. (2002). Enhanced radical-scavenging activity of a planar catechin analogue. *Journal of the American Chemical Society*, **124**: 5952–5953.

Fukuhara, K., Nakanishi, I., Ohkubo, K., *et al.* (2009a) A planar catechin derivative incorporating a lysine moiety was synthesized and showed ~400-fold increased radical-scavenging activity relative to naturally-occurring (+)-catechin. *Chemical Communications*, 6180–6182. https://pubs.rsc.org/en/content/articlelanding/2009/CC/b913714a.

Fukuhara, K., Ohno, A., Nakanishi, I., et al. (2009b). Novel ninhydrin adduct of catechin with potent antioxidative activity. *Tetrahedron Letters*, **50**: 6989–6992.

Fukuyama, T. and Yang, L. (1986). Synthetic approaches toward mitomycins. I. Stereoselective synthesis of a tetracyclic intermediate. *Tetrahedron Letters*, **27**: 6299–6300.

Fukuyama, T., Jow, C.-K., and Cheung, M. (1995). 2- and 4-Nitrobenzenesulfonamides: exceptionally versatile means for preparation of secondary amines and protection of amines. *Tetrahedron Letters*, **36**: 6373–6374.

Fukuyama, T., Cheung, M., and Kan, T. (1999). *N*-Carboalkoxy-2-Nitrobenzene sulfonamides: a practical preparation of *N*-Boc-, *N*-Alloc-, and *N*-Cbz-protected primary amines. *Synlett*: 1301–1303 https://www.thieme-connect.com/products/ejournals/abstract/10.1055/s-1999-2827.

Furuta, T., Hirooka, Y., Abe, A., et al. (2007). Concise synthesis of dideoxy-epigallocatechin gallate (DO-EGCG) and evaluation of its anti-influenza virus activity. *Bioorganic and Medicinal Chemistry Letters*, **17**: 3095–3098.

Hakamata, W., Nakanishi, I., Masuda, Y., et al. (2006). Planar catechin analogues with alkyl side chains, as a potent antioxidant and an α-glucosidase inhibitor. *Journal of the American Chemical Society*, **128**: 6524–6525.

Hiza, A., Tsukaguchi, Y., Ogawa, T., et al. (2014). Synthetic studies of fisetin, myricetin and nobiletin analogs and related probe molecules. *Heterocycles*, **88**: 1371–1396.

Horie, T., Masumura, M., and Okumura, F.S. (1961). Sudachitin, a new flavone pigment of sudachi. *Bulletin of the Chemical Society of Japan*, **34**: 1547–1580.

Hughes, D.L. (1992). The Mitsunobu reaction. *Organic Reactions*, **42**: 335–656.

Huisgen, R. (1963). 1,3-dipolar cycloadditions. Past and future. *Angewandte Chemie International Edition*, **2**: 565–598.

Jones, G. and Stanforth, S.P. (2000). The Vilsmeier reaction of non-aromatic compounds. *Organic Reactions*, **56**: 355–686.

Kan, T. and Fukuyama, T. (2001). Highly versatile synthesis of nitrogen-containing compounds by means of nitrobenzenesulfonamides. *Journal of Synthetic Organic Chemistry, Japan*, **59**: 779–789.

Kan, T. and Fukuyama, T. (2004). Ns strategies: a highly versatile synthetic method for amines. *Chemical Communications*: 353–359 https://pubs.rsc.org/en/content/articlelanding/2004/cc/b311203a.

Kan, T., Tominari, Y., Morohashi, Y., et al. (2003). Solid-phase synthesis of photoaffinity probes: highly efficient incorporation of biotin-tag and cross-linking groups. *Chemical*

Communications: 2244–2245 https://pubs.rsc.org/en/content/articlelanding/2003/cc/b306970b.

Kan, T., Kita, Y., Morohashi, Y., et al. (2007). Convenient synthesis of photoaffinity probes and evaluation of their labeling abilities. *Organic Letters*, **9**: 2055–2058.

Katritzky, A.R. and Pastor, A. (2000). Synthesis of beta-dicarbonyl compounds using 1-acylbenzotriazoles as regioselective C-acylating reagents. *Journal of Organic Chemistry*, **65**: 3679–3682.

Katritzky, A.R., Shobana, N., Pernak, J., et al. (1992). Sulfonyl derivatives of benzotriazole: part 1. A novel approach to the activation of carboxylic acids. *Tetrahedron*, **48**: 7817–7822.

Kawai, Y., Tanaka, H., Murota, K., et al. (2008). (−)-Epicatechin gallate accumulates in foamy macrophages in human atherosclerotic aorta: implication in the anti-atherosclerotic actions of tea catechins. *Biochemical and Biophysical Research Communications*, **374**: 527–532.

Kitada, M., Ohno, Y., Tanaka, H., and Takahashi, T. (2006). An efficient synthesis of (±)-epigallocatechin gallate by reductiveintramolecular etherification. *Synlett*: 2827–2829 https://www.thieme-connect.com/products/ejournals/abstract/10.1055/s-2006-950283.

Kolb, H.C., Finn, M.G., and Sharpless, K.B. (2001). Click chemistry: diverse chemical function from a few good reactions. *Angewandte Chemie International Edition*, **40**: 2004–2021.

Kolb, H., VanNiewenhze, M.S., and Sharpless, K.B. (2014). Catalytic asymmetric dihydroxylation. *Chemical Reviews*, **94**: 2483–2547.

Kurosawa, W., Kan, T., Fukuyama, T., and T. (2002). Preparation of secondary amines from primary amines via 2-nitrobenzenesulfonamides: *N*-(4-methoxybenzyl)-3-phenylpropylamine. *Organic Synthesis*, **79**: 186–192.

Kuzuhara, T., Kise, D., Shirakawa, Y., et al. (2008). Generation of mouse monoclonal antibody against (−)-epigallocatechin gallate. *Biological and Pharmaceutical Bulletin*, **31**: 816–819.

Lebrun, M.-E., Marquand, P.L., and Berthelette, C. (2006). Stereoselective synthesis of *Z* alkenyl halides via Julia olefination. *Journal of Organic Chemistry*, **71**: 2009.

Mera, K., Nagai, M., Brock, J.W., et al. (2008). Glutaraldehyde is an effective cross-linker for production of antibodies against advanced glycation end-products. *Journal of Immunological Methods*, **334**: 82–90.

Mitsunobu, O. (1981). The use of diethyl azodicarboxylate and triphenylphosphine in synthesis and transformation of natural products. *Synthesis*: 1–28.

Miyaura, N. and Suzuki, A. (1995). Palladium-catalyzed cross-coupling reactions of organoboron compounds. *Chemical Reviews*, **95**: 2457–2483.

Moon, Y.-H., Lee, J.-H., Ahn, J.-S., et al. (2006). Synthesis, structure analyses, and characterization of novel epigallocatechin gallate (EGCG) glycosides using the glucansucrase from leuconostoc mesenteroides B-1299CB. *Journal of Agricultural and Food Chemistry*, **54**: 1230–1237.

Mori, S., Miyake, S., Kobe, T., et al. (2008). Enhanced anti-influenza A virus activity of (−)-epigallocatechin-3-O-gallate fatty acid monoester derivatives: effect of alkyl chain length. *Bioorganic and Medicinal Chemistry Letters*, **18**: 4249–4252.

Murakami, A., Nakamura, Y., Torikai, K., et al. (2000). Inhibitory effect of citrus nobiletin on phorbol ester-induced skin inflammation, oxidative stress, and tumor promotion in mice. *Cancer Research*, **60**: 5059–5066.

Nagashima, M. and Ohno, M. (1987). An efficient *O*-monoalkylation of dimethyl L-tartrate via *O*-stannylene acetal with alkyl halides in the presence of cesium fluoride. *Chemistry Letters*: 141–144 https://www.journal.csj.jp/doi/10.1246/cl.1987.141.

Nagle, D.G., Ferreira, D., and Zhou, Y.-D. (2006). Epigallocatechin-3-gallate (EGCG): chemical and biomedical perspectives. *Phytochemistry*, **67**: 1849–1855.

Node, M., Nishida, K., Fuji, K., and Fujita, E. (1980). Hard acid and soft nucleophile system. 2. Demethylation of methyl ethers of alcohol and phenol with an aluminum halide-thiol system. *Journal of Organic Chemistry*, **45**: 4275–4277.

Piyaviriyakul, S., Shimizu, K., Asakawa, T., et al. (2011). Anti-angiogenic activity and intracellular distribution of epigallocatechin-3-gallate analogs. *Biological and Pharmaceutical Bulletin*, **34**: 396–400.

Rooprai, H.K., Kandanearatchi, A., Maidment, S.L., et al. (2001). Evaluation of the effects of swainsonine, captopril, tangeretin and nobiletin on the biological behaviour of brain tumour cells in vitro. *Neuropathology and Applied Neurobiology*, **27**: 29–39.

Sagara, H., Kanakogi, M., Tara, Y., et al. (2018). Concise synthesis of polymethoxyflavone sudachitin and its derivatives, and biological evaluations. *Tetrahedron Letters*, **59**: 1816–1818.

Seoka, M., Ma, G., Zhang, L., et al. (2020). Expression and functional analysis of the nobiletin biosynthesis-related gene CitOMT in citrus fruit. *Scientific Reports*, **10**: 15288.

Shibuya, M., Tomizawa, M., Suzuki, I., and Iwabuchi, Y. (2006). 2-Azaadamantane *N*-Oxyl (AZADO) and 1-Me-AZADO: highly efficient organocatalysts for oxidation of alcohols. *Journal of the American Chemical Society*, **128**: 8412–8413.

Shimizu, K., Asakawa, T., Harada, N., et al. (2014). Use of positron emission tomography for real-time imaging of biodistribution of green tea catechin. *PLoS One*, **9**: e85520.

Song, J.M. and Seong, B.L. (2007). Tea catechins as a potential alternative anti-infectious agent. *Expert Review of Anti-Infective Therapy*, **5**: 497–506.

Sugimura, T. and Hagiya, K. (2007). Di-2-methoxyethyl azodicarboxylate (DMEAD): an inexpensive and separation-friendly alternative reagent for the Mitsunobu reaction. *Chemistry Letters*, **36**: 566–567.

Takii, M., Kaneko, Y.K., Akiyama, K., et al. (2017). Insulinotropic and anti-apoptotic effects of nobiletin in INS-1D β-cells. *Journal of Functional Foods*, **30**: 8–15.

Tanaka, H., Miyoshi, H., Chuang, Y.-C., et al. (2007). Solid-phase synthesis of epigallocatechin gallate derivatives. *Angewandte Chemie International Edition*, **46**: 5934–5937.

Thielecke, F. and Boschmann, M. (2009). The potential role of green tea catechins in the prevention of the metabolic syndrome – a review. *Phytochemistry*, **70**: 11–24.

Tseng, K. (1938). Nobiletin. Part I. *Journal of the Chemical Society*: 1003–1004 https://pubs.rsc.org/en/content/articlelanding/1938/JR/JR9380001003.

Urano, Y., Kamiya, M., Kanda, K., et al. (2005). Evolution of fluorescein as a platform for finely tunable fluorescence probes. *Journal of the American Chemical Society*, **127**: 4888–4894.

Wan, S.B., Landis-Piwowar, K.R., Kuhn, D.J., et al. (2005). Structure-activity study of *epi*-gallocatechin gallate (EGCG) analogs as proteasome inhibitors. *Bioorganic & Medicinal Chemistry*, **13**: 2177–2185.

Wheeler, D.S. and Wheeler, W.J. (2004). The medicinal chemistry of tea. *Drug Development Research*, **61**: 45–65.

Yamakawa, S., Asai, T., Uchida, T., et al. (2004). (−)-Epigallocatechin gallate inhibits membrane-type 1 matrix metalloproteinase, MT1-MMP, and tumor angiogenesis. *Cancer Letters*, **210**: 47–55.

Yang, C.S., Wang, X., Lu, G., and Picinich, S.C. (2009). Cancer prevention by tea: animal studies, molecular mechanisms and human relevance. *Nature Reviews Cancer*, **9**: 429–439.

Yasuda, N., Ishii, T., Oyama, D., et al. (2014). Neuroprotective effect of nobiletin on cerebral ischemia-reperfusion injury in transient middle cerebral artery-occluded rat. *Brain Research*, **1559**: 46–54.

Yiannakopoulou, E.C. (2014). Interaction of green tea catechins with breast cancer endocrine treatment: a systematic review. *Pharmacology*, **94**: 245–248.

Yoshida, A., Hirooka, Y., Sugata, Y., et al. (2011). Concise synthesis of catechin probes enabling analysis and imaging of EGCg. *Chemical Communications*, **47**: 1794–1796.

Zaveri, N.T. (2006). Green tea and its polyphenolic catechins: medicinal uses in cancer and noncancer applications. *Life Sciences*, **78**: 2073–2080.

3

Procyanidins in the Onset and Progression of Colorectal Cancer

Recent Advances and Open Questions

Wei Zhu, Gerardo G. Mackenzie, and Patricia I. Oteiza

Department of Nutrition, University of California-Davis, Davis, CA, USA

3.1 Introduction

Colorectal cancer (CRC) has the third highest cancer incidence in the United States (Siegel et al. 2020), with over 140,000 new CRC cases diagnosed every year. Over the last decades, there has been an improvement in CRC survival, largely attributed to increased screening, early detection, and recent advances in systemic and local treatment modalities. Despite these advances, CRC remains a leading cause of cancer-related mortality in the United States, with over 50,000 deaths from CRC in 2019 (Siegel et al. 2020). Genetics only accounts for 15–20% of CRC cases, while the rest is associated with environmental factors. In terms of genetic factors, patients with ulcerative colitis have an increased cumulative risk to develop CRC, being 5–10% after 20 years, 15–20% after 30 years, and 25–40% after 40 years in patients without prophylactic colectomy (Ekbom et al. 1992, Harpaz and Talbot 1996). In order to continue improving CRC patient survival, current clinical and epidemiological data stress the critical need for new strategies in CRC prevention.

Given that over 80% of CRC cases are caused by environmental factors and lifestyle, this provides great opportunities for cancer prevention. For instance, practicing a healthy lifestyle, including maintaining healthy body weight, being physically active, refraining from smoking, limiting alcohol intake, and eating a healthy diet, have been associated with a substantial reduction in CRC incidence and mortality (Huxley et al. 2009; Erdrich et al. 2015; Song and Giovannucci 2016; Wei et al. 2017). It is noteworthy that adherence to a healthy lifestyle is associated with a substantial reduction in CRC incidence and mortality independent of endoscopic screening, highlighting the importance of a healthy lifestyle in the prevention of CRC (Wang et al. 2021). Among lifestyle factors, nutrition has emerged as a major preventive approach (Platz et al. 2000; Kim and Milner 2007). Diets rich in fat and beef and low in fiber, fruit, and vegetables are a risk factor for CRC (Platz et al. 2000; Kim and Milner 2007). On the other hand, the consumption of diets rich in fruits, vegetables, spices, and grains possesses beneficial effects on the intestine,

Recent Advances in Polyphenol Research, Volume 8, First Edition. Edited by Juha-Pekka Salminen, Kristiina Wähälä, Victor de Freitas, and Stéphane Quideau.

particularly the colon (Divisi et al. 2006; Vieira et al. 2017; Schwingshackl et al. 2018) with an inversed correlation observed between the consumption of diets rich in fruit and vegetables and the risk of developing CRC (Franceschi et al. 1997; Terry et al. 2001; Michels et al. 2006; Theodoratou et al. 2007; Rossi et al. 2010; Fernández-Villa et al. 2020; Hidaka et al. 2020).

Unlike other tumor cells, CRC cells are exposed directly to specific dietary components. Thus, there is a growing interest to identify bioactives or groups of compounds present in fruit and vegetables that may have chemopreventive properties. For example, among hundreds of compounds, fiber and flavonoids, present in high amounts in fruit and vegetables, could, in part, account for their CRC-protective effects (Franceschi et al. 1997; Terry et al. 2001; Michels et al. 2006; Theodoratou et al. 2007; Rossi et al. 2010; Hidaka et al. 2020). Among flavonoids, procyanidin (PC) consumption was found to be inversely correlated with the risk of developing CRC (Rossi et al. 2010). Importantly, given the very limited absorption of parent PC at the gastrointestinal (GI) tract, they can reach significant concentrations in the colon/rectum lumen where, as parent compounds or as their microbial metabolites, they could exert anti-CRC actions.

There is an increasing body of evidence in humans and rodents suggesting that dietary PC could reduce CRC risk. This review summarizes current knowledge on the chemopreventive capacity of PC in CRC. We discuss current findings in humans and rodents that support such effects and describe the signaling pathways that could mediate PC's potential CRC-preventive actions.

3.2 Procyanidins: Chemistry and Metabolism

PC, oligomers and polymers of flavan-3-ols, are among the most abundant polyphenols present in the human diet (Gu et al. 2004). Figure 3.1 shows the chemical structures of (−)-epicatechin (EC), dimeric and oligomeric PC, as well as examples of the different unions among subunits leading to different PC types (Haslam 1998). PC are present in large amounts in many edible plants, e.g. grapes, cocoa, tea, apples, and derived foods (Prior and Gu 2005), being large oligomers/polymers the most abundant in nature (Haslam 1998). Generally, there are mainly two types of linkages, with some plants presenting a single type. For example, in cocoa, grapes, and apples, the monomeric units are linked through 4β→8 C—C bonds forming mostly B-type PC, while in peanut skins and cranberries, the monomeric units are linked by 4β→8 C—C and 2β→7 C—O—C bonds generating A-type PC. A-type PC are more hydrophobic due to a steric hindrance effect, which causes them to have a higher affinity for cell membranes (Mackenzie et al. 2009). PC are poorly absorbed in the GI tract (Manach et al. 2005) due to their complexity and high molecular weight. Only PC dimers seem to be absorbed intact (Baba et al. 2002; Sano et al. 2003; Tsang et al. 2005; Appeldoorn et al. 2009b; Stoupi et al. 2010b), but at much lower efficiency than monomers (Manach et al. 2005). Given their limited absorption PC are present throughout the GI tract and found in feces upon consumption of PC-rich foods (Choy et al. 2013; Choy et al. 2014). PC are not cleaved to render monomeric subunits. In this regard, PC consumed as a cocoa drink do not contribute to the circulating flavanol pool (Ottaviani et al. 2012).

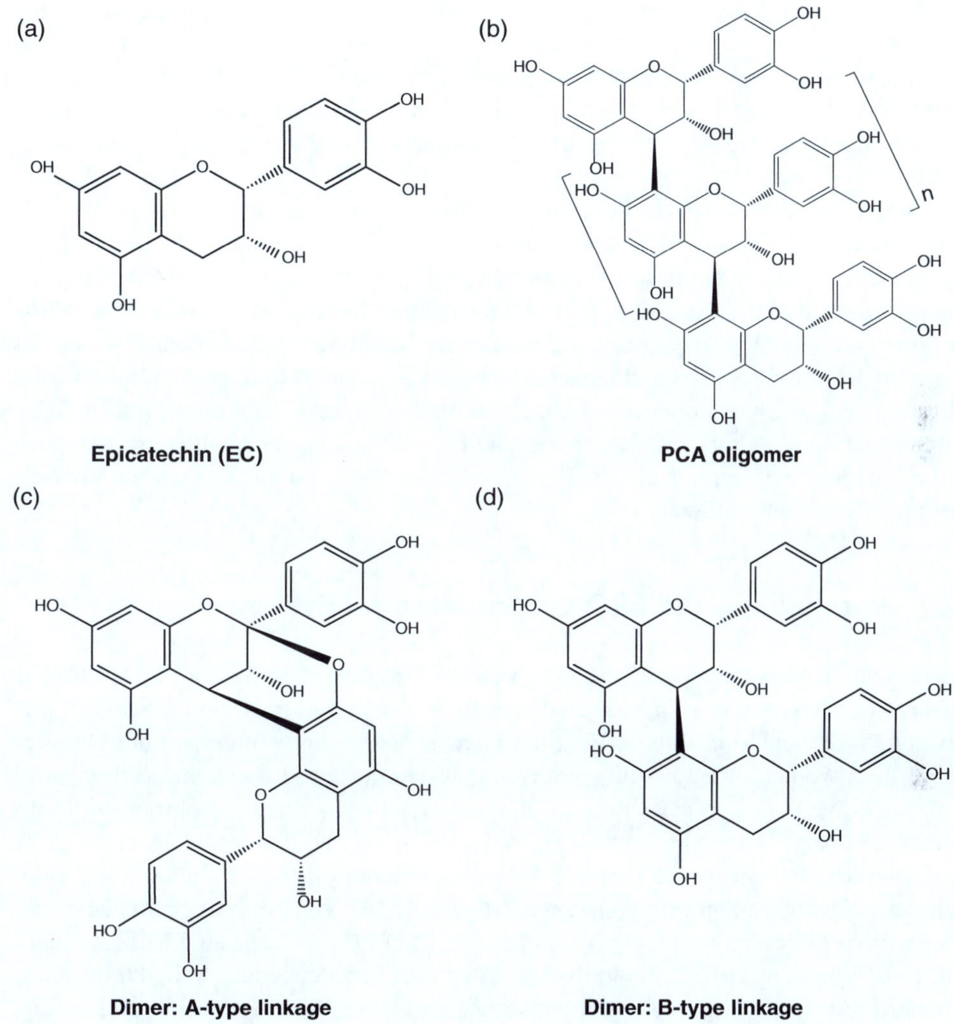

Figure 3.1 Chemical structure of PC: (a) monomeric flavan-3-ol (−)-epicatechin (EC), (b) EC B-type oligomers. The number of subunits (n) depends on the food source. (c, d) dimeric PC bound through (c) A-type linkage: PC A2 (EC-(C4β→C8, 2β→O7)-EC) and (d) B-type linkage: PC B2 (EC-(C4β→C8)-EC).

In the colon, PC are metabolized by the microbiota to various aromatic acid metabolites that are more readily absorbed (Deprez et al. 2000; Düweler and Rohdewald, 2000; Stoupi et al. 2010a). Initial evidence showed that PC polymers are fully metabolized *in vitro* by human gut microbiota within 48 hours incubation, being the major metabolites detected in phenylacetic, phenylpropionic, and phenylvaleric acids (Deprez et al. 2000). Also *in vitro*, human colonic microbiota-catabolized dimeric PC into 2-(3,4-dihydroxyphenyl) acetic acid and 5-(3,4-dihydroxyphenyl)-γ-valerolactone (Appeldoorn et al. 2009a). Delta-(3,4-dihydroxy-phenyl)-γ-valerolactone metabolites were also found in urine upon consumption of a PC-rich cocoa drink by healthy human males (Ottaviani et al. 2012). A limited number of metabolites, including phenylacetic acid, 3-phenylpropionic acid, and 3-(4-hydroxyphenyl) propionic acid, were detected after 72 hours incubation of black tea and red wine/grape juice with human intestinal microbiota (Gross et al. 2010). Grape seed extract incubated with human intestinal microbiota for 36 hours resulted in the generation of mainly acetic and propionic acid (Zhou et al. 2016). Consumption of a PC-rich pine bark extract (Jerez et al. 2007), commercially called "Pycnogenol," by healthy individuals led to the detection of urinary delta-(3,4-dihydroxy-phenyl)-γ-valerolactone and delta-(3-methoxy-4-hydroxyphenyl)-γ-valerolactone conjugated with glucuronic acid/sulfate (Düweler and Rohdewald 2000). Importantly, metabolite profiles varied among individuals, which might imply that an individual's health benefits depend in part on microbiota bioconversion (Gross et al. 2010). Furthermore, given the variation in PC-type composition in different foods, metabolites present in the GI tract and in the circulation/tissues will depend on the type of PC food source.

In summary, the characteristics of PC metabolism and absorption in the GI tract indicate that large amounts of parent PC reach the colon and several PC metabolites are generated by the microbiota. Thus, it is plausible that both parent PC and metabolites can underlie the anti-CRC actions of dietary PC.

3.3 Procyanidins and CRC: Epidemiological Evidence

Multiple studies have evaluated the link between diet, flavonoid content, and CRC risk (Table 3.1). However, the current evidence from epidemiological studies on the link between PC consumption and CRC is still limited. Among cohort studies, a study based on the US Iowa Women's Health Study observed that flavanol consumption protected against upper digestive tract [(+)-catechin and EC] and rectal cancer (tea catechins), while the impact of PC was not analyzed (Arts et al. 2002). However, the monomers (flavan-3-ols) are associated with beneficial effects on CRC. And a case-control study including 1,953 cases with the incident, histologically confirmed CRC and 4,154 controls admitted for acute non-neoplastic conditions, found that high dietary intake of PC was associated with a reduced CRC risk (Rossi et al. 2010). This study found that the effect depended on the PC degree of polymerization, with larger PC having a higher protective effect. Two other case-control studies found a significant inverse association between CRC and PC intake, but they did not differentiate on the action of particular types of PC (Theodoratou et al. 2007; Zamora-Ros et al. 2013). In contrast, two studies based on data from the European Prospective Investigation into Cancer and Nutrition (EPIC) study found no association between PC consumption and CRC risk (Zamora-Ros et al. 2017; Zamora-Ros et al. 2018).

Green tea is rich in monomeric catechins, particularly in (−)-epigallocatechin (EGC) and (−)-epigallocatechin 3-*O*-gallate (EGCG). Green tea PC include mostly dimers, trimers and tetramers with different monomer combinations (Ricci et al. 2017). Although not identifying particular phenolic compounds, a prospective cohort study showed that consumption of 10 cups/day of green tea, which is rich in PC, significantly reduced cancer incidence in women and nonsmoking men compared to those consuming 3 cups/day (Nakachi et al. 2000). High green tea intake was also associated with a delay in cancer onset.

Supporting findings from cohort studies, a randomized, placebo-controlled, multicenter trial study showed that three years consumption of 300 mg EGCG per day, calculated to

Table 3.1 Epidemiological studies evaluating the preventive effects of PC or PC-rich diet on CRC.

PC sources	Samples	Intervention	Study design	Effects	References
Diet	1,953 cases (1,225 colon cancers, 728 rectal cancers), 4,154 controls[a]		Case-control study	Decreased risk of CRC (except monomers)	Rossi et al. (2010)
Diet	424 cases with CRC and 401 hospital-based controls	past 12 months	Case-control study	Decreased risk of CRC (include monomers)	Zamora-Ros et al. (2013)
Diet	1,635 cases with CRC and 815 hospital-based controls		Case-control study	Decreased risk of CRC (fruits and vegetables, not tea consumption)	Xu et al. (2016)
Diet	477,312 adult men and women	11 years' follow-up	Prospective cohort study	No association with CRC risk	Zamora-Ros et al. (2017)
Diet	476,160 men and women	14 years' follow-up	Prospective cohort study	No association with CRC risk	Zamora-Ros et al. (2018)
Green tea (drinking)	8,552 men and women	≤3 cups/day 3–9 cups/day ≥10 cups/day	Prospective cohort study	Decreased risk of cancer residence, delayed cancer onset	Nakachi et al. (2000)
Green tea	38,540 participants (14,783 men, 23,667 women)	13–15 years follow-up	Prospective cohort study	No association	Jun et al. (2001)
Catechins	34,651 postmenopausal cancer-free women aged 55–69 years	13 years follow-up	Prospective cohort study	(+)-catechin and (−)-epicatechin from fruits protect against upper digestive tract cancer and tea catechins protect against rectal cancer	Arts et al. (2002)
Green tea extract (GTE) tablets	136 patients, 65 in the control group and 71 in the GTE group	1.5 g GTE per day for 12 months	Randomized trial	Reduced incidence of metachronous adenomas and size of relapsed adenomas	Shimizu et al. (2008)

(*Continued*)

Table 3.1 (Continued)

PC sources	Samples	Intervention	Study design	Effects	References
GTE tablets	143 patients (71 in the control group and 72 in the GTE group)	0.9 g GTE per day for 12 months	Randomized control trials	Reduced incidence of metachronous adenomas and number of relapsed adenomas	Shin et al. (2018)
Apple	1,953 patients[a]	<1 apple/day ≥1 apple/day	Hospital-based case-control	≥1 apple/day reduced risk of CRC	Gallus et al. (2005)
Apple	186 cases, 211 controls[a]	<1 apple/day ≥1 apple/day	Hospital-based case-control	≥1 apple/day reduced risk of CRC	Jedrychowski and Umberto (2009)
Chocolate	171 cases with CRC, 309 population controls		Case-control study	Chocolate is a risk factor for CRC	Boutron-Ruault et al. (1999)
Chocolate	234 cases with adenomatous polyps, 407 controls		Case-control study	No increase in adenoma prevalence	
Chocolate	465 cases (hyperplastic polyps (n = 103), adenomas<10 mm, (n = 154) or larger adenomas (n = 208)) versus 426 polyp-free controls		Case-control study	No correlation between high chocolate intake and colorectal polyps, adenoma or CRC	Rouillier et al. (2005)

[a] Admitted for acute, non-neoplastic condition. GTE: Green tea extracts.

correspond to 4–7 cups of green tea per day reduced the risk of metachronous adenomas and CRC (Stingl et al. 2011). Two separate randomized trials found that consumption of 0.9 and 1.5 g green tea extract/day (equivalent to 4 and 6 cups of green tea/day, respectively) for 12 months significantly decreased the incidence of metachronous adenomas and the number and size of relapsed adenomas (Shimizu et al. 2008; Shin et al. 2018). On the other hand, other clinical trials have failed to show a beneficial impact of tea intake on the incidence of CRC. For example, a hospital-based case-control study found that consumption of all subclasses of flavonoids including monomers and PC from vegetables and fruits, but not those from tea, were inversely associated with CRC risk (Xu et al. 2016). A prospective cohort study of 34,651 postmenopausal women (including 635 colon cancer and 132 rectal cancer) found that intake of tea catechins was inversely associated with the incidence of rectal cancer, but not of colon cancer (Arts et al. 2002). Another large-scale prospective cohort study (38,540 people) found no relationship between regular green tea intake and lower CRC risk through 15 years of follow-up (Nagano et al. 2001).

Apples and cocoa are abundant in B-type PC (Hammerstone et al. 2000). Consumption of apples has shown protective effects against CRC in hospital-based case-control studies (Gallus et al. 2005; Jedrychowski and Umberto 2009). Among the very few human studies addressing the influence of cocoa and chocolate on CRC risk, evidence suggests a lack of association (M-C Boutron-Ruault et al. 1999; Rouillier et al. 2005). However, a critical factor to consider is that apples and cocoa show major variations in PC content depending, among other factors, on varieties, climate, and cultivation. Additionally, the consumption of chocolates with high relative content of sugar and fat and low PC content can be a major confounding factor when associating chocolate consumption with CRC.

As noted earlier, multiple human studies support the capacity of PC to mitigate CRC onset and/or progression, although such findings are not always consistent. Such inconsistencies may be in part due to the type of epidemiological study. Prospective cohort studies provide clarity of the temporal sequence and allow for a calculation of incidence, but they may have less detailed information on specific aspects of diet on each subject, compared to case-control studies. Case-control studies are easy to be conducted but have limitations of bias that relies on memory or past records and cannot distinguish between cause and associated factors. In this regard, a comprehensive meta-analysis of 11 prospective and 9 case-control studies found that subclasses of flavanols, proanthocyanidins, and catechins lowered the risk of CRC among case-control studies while no beneficial effects were found when analyzing results from prospective studies (Grosso et al. 2017). Another limitation of these studies is that PC intake is estimated through different strategies of self-reported questionnaires which, although widely used and accepted, have significant margins of error at the time of analyzing dietary PC amounts and types. Besides systematic bias, food composition data do not account for the fact that each fruit or vegetable can have different content of PC depending on seasonal variations, fruit varieties, etc. Current efforts to identify biomarkers of individual polyphenol intake are critical to establishing a link between their intake and human health. In this regard, 5-(3′,4′-dihydroxyphenyl)-γ-valerolactone has been recently validated as a biomarker of flavan-3-ol intake suitable to be used in large epidemiological studies (Ottaviani et al. 2018). While this metabolite also originates from PC microbiota metabolism (Ottaviani et al. 2012), there is still no specific biomarker available to evaluate the consumption of PC. Thus, additional research on this area is currently needed.

Overall, our understanding of the effects of PC consumption on CRC risk needs more epidemiological studies that take into account, among other factors, lifestyle, dietary patterns, environment, and genetics, and use reliable biomarkers to assess PC intake.

3.4 Procyanidins and CRC: Rodent Studies

Studies in rodents support the anti-CRC activity of dietary PC and PC-rich extracts. Studies have particularly focused on PC present in grape seeds, apples, cocoa, cranberries, and teas (Table 3.2). Grape seed extracts (GSE) are popular dietary supplements rich in PC, which constitute up to 90% of all GSE phenolic components (Prieur et al. 1994). GSE exerted anti-CRC actions in a variety of *in vivo* rodent models (Ravindranathan et al. 2018b). For example, consumption of 200 mg/kg body weight (BW) of a GSE containing 89.3% (w/w) PC and 6.6% monomeric flavonols, decreased tumor volume in an HT-29 tumor xenograft in athymic nude mice (Kaur et al. 2006). Dimeric PC B2 isolated from grape seed dose-dependently (25–100 mg B2/kg BW) reduced tumor volume and weight in a LoVo cells xenograft in nude mice (Zhang et al. 2019). A grape seed-oligomeric PC extract (OPCs) inhibited: (i) the formation and growth of tumor organoids derived from $Apc^{Min/+}$ mice and patients, (ii) the *in vitro* growth of CRC stem cells, and (iii) tumor growth in an HCT-116 xenograft mouse model (Toden et al. 2018). Importantly, at 100 mg B2/kg BW, the inhibitory potency was comparable to that of the positive control, 5-fluorouracil, a commonly used chemotherapeutic drug (Zhang et al. 2019). Mechanistically, B2 promoted tumor cell apoptosis via downregulation of the phosphoinositide 3-kinase (PI3K)/protein kinase B (Akt) pathway, and enhanced autophagy, evidenced by increased levels of LC3II, beclin1 and Atg5, and decreased LC3I levels (Zhang et al. 2019).

PC could also potentiate the efficiency of other bioactives and drugs that have anti-CRC effects. For example, grape seed OPCs (100 mg/kg BW)-inhibited tumor growth in patient-derived organoids, and their combination with curcumin increased their capacity to inhibit tumor growth in HCT-116 xenografts in mice (Ravindranathan et al. 2018a). Grape seed OPCs acted cooperatively with resveratrol to suppress azoxymethane (AOM)-induced colon tumor incidence in mice through the suppression of the Wnt/β-catenin pathway and promotion of apoptosis (Reddivari et al. 2016). Co-administration of grape seed OPCs (100 mg/kg BW) with the chemotherapeutic drugs 5-fluorouracil and oxaliplatin significantly decreased chemo-resistant xenograft tumor growth in mice (Ravindranathan et al. 2019). This was in part due to their effect decreasing the expression of key drug transporters, i.e. MRP2, MDR1, ABCG1, ABCC3, and ABCA5 (Ravindranathan et al. 2019). The above results suggest that PC not only could exhert anti-CRC bioactivities per se but also increase the efficacy of anticancer drugs and of other bioactives.

With regard to apple PC, Barth et al. used a well-established 1,2-dimethylhydrazine (DMH)-induced rat CRC model to examine the protective effects of PC-enriched apple juice with low or high pectin content (Barth et al. 2005; Barth et al. 2007). Pectin content in the juice enhanced the anticancer effect of apple phenols. When comparing the effects of apple monomers and PC, only the PC-enriched fraction had growth-inhibiting properties in SW620 cells while monomers had no inhibitory effects at comparable concentrations

Table 3.2 Rodent studies evaluating the preventive effects of PC or PC-rich diet on CRC.

Source	PC composition	Dose	Duration	Model	Effect mechanisms	References
Grape seed extract	89% (w/w) PC	200 mg/kg BW	40 days	HT29 tumor xenograft, mice	Induced apoptosis and inhibited tumor growth	Kaur et al. (2006)
Grape seed extract	Oligomers (DP = 1–3)	0.12% (w/w diet) combined with 0.03%–(w/w) resveratrol	7 days	AOM-induced CRC, mice	Suppressed tumor incidence	Reddivari et al. (2016)
Grape seed extract	Oligomers (DP = 1–3)	50, 100 mg/kg BW	14 days	HCT116-xenograft, mice	Inhibited tumor growth by hippo pathway and oncogenes SOX4 and JAG1	Toden et al. (2018)
Grape seed extract	Oligomers (DP = 1–3)	50, 100 mg/kg BW	15 days	HCT116-xenograft, mice	Decreased tumor growth	
Grape seed extract	PC B2	25, 50, 100 mg/kg BW	28 days	LoVo tumor-xenograft athymic nude mice	Induced apoptosis and autophagy	Zhang et al. (2019)
Grape seed extract	Oligomers (DP = 1–3)	100 mg/kg BW	42 days	HCT116-xenograft, mice	Attenuated tumor growth, decreased drug transporters gene expression	Ravindranathan et al. (2019)
Apple extract	78% of PC in mixture	6 mg PC/kg BW	6 weeks	AOM-induced CRC, rat	Reduced preneoplastic lesions	Gossé et al. (2005)
Apple juice	PC B1/B2	2–3 mg PC/kg BW	7 weeks	DMH-induced CRC, rat	Reduced DNA damage and hyperproliferation, lowered large aberrant crypt foci number	Barth et al. (2005, 2007)
Cranberry extract	NC	100 mg/kg BW	20 days (10 injections)	HT29 nude mice	Inhibited explant tumor growth	Ferguson et al. (2006)

(*Continued*)

Table 3.2 (Continued)

Source	PC composition	Dose	Duration	Model	Effect mechanisms	References
Cranberry juice	NC	20% (v/v) in drinking water	13 weeks	AOM-induced, rats	Reduced aberrant crypt foci number	Boateng et al. (2007)
Cranberry extract	11% phenolics of powder	0.1, 1% extract (w/w diet)	6 weeks	DSS-induced colitis, mice	Lowered inflammation and disease activity index	Xiao et al. (2015)
Cocoa extract	26% PC of extract	500 mg extract/kg BW	8 days	DSS-induced ulcerative colitis, mice	Reduced acute ulcerative colitis, Inhibited STAT3 and NF-κB	Andújar et al. (2011)
Cocoa extract	33% PC of total phenols	12% extract (w/w diet)	8 weeks	AOM-induced CRC, rats	Reduced cell proliferation, induced apoptosis, prevented oxidative stress	Rodríguez-Ramiro et al. (2013)
Cocoa extract	NC	5 and 10% extract (w/w diet)	9 weeks	AOM/DSS-induced colitis, mice	Decreased the tumor incidence and size. Inhibited IL-6/STAT3 and activated Nrf2/Keap1 pathway	Pandurangan et al. (2015) and Saadatdoust et al. (2015)
Dark chocolate	41–44 mg PC/g chocolate	10% (w/w diet)	6 weeks	AOM-induced rats	Decreased colon aberrant crypt foci number	Hong et al. (2013)
GTE and EGCG	39 and 75% catechins w/w powder	0.005% EGCG or 0.01–0.1% GTE (w/w diet)	12 weeks	ENNG and AOM-induced CRC, rats	Decreased colon tumor incidence	Yamane et al. (1996)
Green tea extract	27% catechins of solids	0.002–0.05% in drinking water	35 weeks	ENNG-induced CRC, rats	Decreased colon tumor incidence	Narisawa and Fukaura (1993)

Green tea and tea pigments	NC	2 and 0.1% (w/w diet)	16 and 32 weeks	DMH-induced CRC, rats	Inhibited aberrant crypt foci and colonic tumors formation	Jia and Han (2000)
EGCG or green tea catechins mix	60% EGCG, 12% EGC, 10% other catechin	0.01 or 0.1% in drinking water	17 weeks	AOM/DSS-induced CRC, mice	Suppressed the multiplicity and volume of colonic neoplasms	Shirakami et al. (2008)
GTE	70% catechins	0.01–1% (w/w diet)	20 weeks	DMH/DSS-induced CRC, mice	0.5 and 1% dose GTE increased colon tumors	Kim et al. (2010)
EGCG	NC	50–200 mg/kg BW	8 weeks	DMH-induced CRC, rats	Reduced tumors number and volume, and ACF count	
Green tea and white tea	NC	1.5% (w/v drinking water)	12 weeks	$Apc^{Min/+}$ mice	Inhibited intestinal polyp formation. Inhibited β-catenin/APC pathway	Orner et al. (2003)
Black tea	NC	40 mg/kg BW	10 days	DMH-induced CRC rats	Inhibited oxidative DNA damage	

NC: Not characterized. AOM: azoxymethane; APC: Adenomatous polyposis coli; DMH: 1,2-dimethylhydrazine; DP: Degree of polymerization; DSS: Dextran sodium sulfated; EGCG: (-)-Epigallocatechin 3-O-gallate; ENNG: N-ethyl-N'-nitro-N-nitrosoguanidine; GTE: Green tea extracts; IL-6: Interleukin-6; JAG1:Jagged1; NF-κB: nuclear factor kappa-light-chain-enhancer of activated B cells; Nrf2: Nuclear factor erythroid 2-related factor 2; SOX4: Sex-determining region Y-related high-mobility group box 4; STAT3: Signal transducer and activator of transcription.

(Gossé et al. 2005). In AOM-induced colon carcinogenesis in rats, supplementation with an apple PC mixture (approximately 6 mg PC/kg BW) for six weeks significantly reduced pre-neoplastic lesions, including 50% fewer aberrant crypts compared to controls (Gossé et al. 2005). Considering PC amounts in dessert apples (Guyot et al. 2002), the amount of PC consumed by rats (4–10 mg/kg BW) in this study would be close to that provided by two apples to a normal-weight human.

Cranberries are rich in A-type PC, and preclinical studies support the concept that they could also exert anti-CRC actions (Weh et al. 2016). In rodents, cranberry PC (100 mg/kg BW) inhibited the growth of explant tumors of HT-29 cells in nude mice (Ferguson et al. 2006). Cranberry extracts lowered inflammation and the disease activity index in dextran sodium sulfate (DSS)-induced colitis in mice, a model of inflammatory bowel disease (Xiao et al. 2015). This is relevant considering the link between inflammatory bowel diseases and an increased risk for CRC. Additionally, supplementation of rats with 20% (v/v) cranberry juice in drinking water for 15 weeks reduced the number of AOM-induced aberrant crypt foci in rats by 77% (Boateng et al. 2007).

Cocoa has been shown to have anti-CRC effects in various rodent studies (Martin et al. 2013). For example, in AOM-induced colon cancer in rats, supplementation with a cocoa-rich diet (12% w/w diet, 33% PC to total phenols) supplementation for 8 weeks inhibited colon carcinogenesis, preventing oxidative stress, reducing cancer cell proliferation and inducing apoptosis (Rodríguez-Ramiro et al. 2013). In AOM/DSS-treated mice, dietary cocoa supplementation (5 and 10% w/w diet, no characterization of PC content) for nine weeks protected against colitis-associated cancer, inhibiting tumorigenesis in part by inhibiting the interleukin (IL)-6/STAT3 pathway (Saadatdoust et al. 2015). In this model, dietary cocoa also improved colon antioxidant defenses through the activation of the Nrf2/Keap1 pathway (Pandurangan et al. 2015). A polyphenol-enriched cocoa extract, which included EC, PC B2, catechin, and PC B1 as the major phenolics, protected mice against DSS-induced ulcerative colitis, mainly through the inhibition of STAT3 and NF-κB (Andújar et al. 2011). In addition, a diet including 10% (w/w diet) normal (44 mg PC/g chocolate) or bloomed dark (41 mg/g chocolate) chocolate decreased colon aberrant crypt foci in AOM-induced colon cancer in rats (Hong et al. 2013).

The chemopreventive and therapeutic effects against CRC of tea PC and EGCG in rodents have been recently reviewed (Wang et al. 2020a). Several studies showed that tea PC fractions inhibit the development of AOM-induced colon carcinogenesis in rats (Yamane et al. 1991; Weisburger et al. 1998; Jung and Ellis, 2001; Xiao et al. 2008; Hao et al. 2017). EGCG and green tea extracts mitigated N-ethyl-N′-nitro-N-nitrosoguanidine (ENNG)-induced development of duodenal tumors and AOM-induced colon carcinogenesis (Yamane et al. 1996). Supplemented at 0.002% (w/v) in drinking water, a green tea extract (27% catechins) inhibited ENNG-induced colon carcinogenesis in F344 rats (Narisawa and Fukaura, 1993). EGCG and a green tea catechins mix (containing 60% EGCG, 7% EC, 12% EGC, 1% ECG, and 2% GCG) in drinking water (0.01 or 0.1%) for 17 weeks protected mice against AOM/DSS-induced colitis-related colonic tumors (Shirakami et al. 2008). Supplementation with green tea and tea pigments (2% and 0.1% (w/w diet), respectively, no characterization of catechin/PC content) to DMH-treated rats for 16 or 32 weeks inhibited ACF and colonic tumors formation (Jia and Han 2000). EGCG (50–200 mg/kg BW) supplementation for 8 weeks inhibited DMH-induced CRC cancer in rats by reducing tumor formation rates, the total number of tumors, cancerous and noncancerous tumors, tumor volume, ascites formation, and aberrant crypt foci count (Wang et al. 2020b). In $Apc^{Min/+}$ mice, white tea, in combination with the nonsteroidal anti-inflammatory drug sulindac, provided greater tumor-suppression effects than either tea or sulindac treatment alone (Orner et al. 2003). Importantly, the dose might determine the beneficial or harmful effects of green tea extracts on CRC. Supplementation with 0.5 and 1% (w/w diet) green tea extracts containing 70% total catechins tends to favor DMH- and DSS-induced colon carcinogenesis, while at a lower concentration (0.1% (w/w diet), the effect was beneficial (Kim et al. 2010).

Overall, the extensive evidence in rodents supports an anti-CRC action for PC. It is important to consider that many studies used either whole fruit/food or poorly characterized extracts. Thus, it is often impossible to calculate the amount of PC given to the rodents, nor to evaluate the potential contribution of other components of the food/extracts to the overall anti-CRC effects. This limits the possibility of extrapolating data to an action that can be specifically attributed to PC and to particular PC characteristics, e.g. degree of polymerization, type of bonds, isomerization, and subunits.

3.5 Procyanidins and CRC: Mechanisms of Action

PC and their metabolites could have indirect and direct inhibitory effects in the development/progression of CRC. In terms of direct effects, PC and metabolites could regulate signaling pathways in the transformed cells that limit their proliferation and survival, or modulate other processes that facilitate tumor growth, e.g. angiogenesis.

Additionally, but not discussed in this review, PC-induced modulation of the colonic microbiota toward healthier profiles could be an indirect mechanism involved in the anti-CRC actions by PC (MacDonald and Wagner 2012, Liu et al. 2017). Available evidence suggests that mechanisms involved in the direct anti-CRC actions of PC include interactions with the cell membrane, mitigation of chronic inflammation and oxidative stress, and inhibition of signaling pathways that promote CRC cell proliferation and resistance to apoptosis, i.e. NF-κB, STAT, epidermal growth factor receptor (EGFR)- and insulin-like growth factor 1 receptor (IGF1R)-activated pathways (extracellular signal-regulated kinase 1/2(ERK1/2) and Akt) (Figure 3.2). We discuss these mechanisms next.

3.5.1 Interactions with Membranes

PC interact with membranes and affect their biophysical properties. These interactions are the main mechanisms that explain the capacity of large PC, which do not enter cells, to regulate cell signaling and exert biological actions. The interactions of PC with membranes have been the subject of thorough review (Verstraeten et al. 2015, Zhu et al. 2019). Thus, here we will briefly describe the main aspects of these interactions.

The interactions of PC with membranes depend on membrane lipid composition, PC concentration, degree of polymerization, type and conformation (Verstraeten et al. 2003). Binding of cocoa B-type PC to membranes causes changes in membrane physical properties, including decreased membrane fluidity that extends from the surface to the hydrophobic core of the monolayer (Erlejman et al. 2004). PC-bilayer interactions are relevant to the

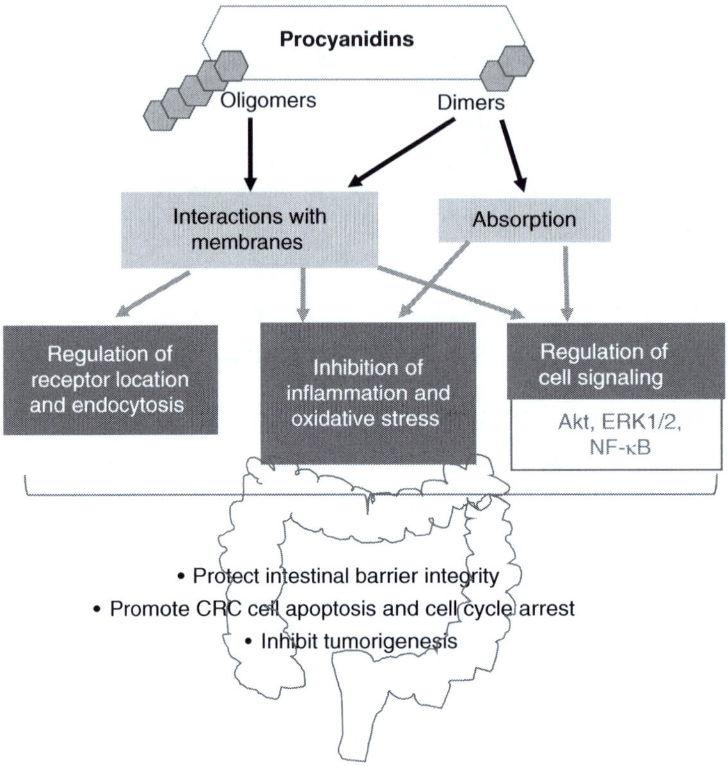

Figure 3.2 Potential mechanisms potentially involved in the anti-CRC actions of PC. While PC dimers can be absorbed by cells, larger oligomers can exert biological actions through their interactions with membranes. PC have a particular affinity for plasma membrane lipid rafts which concentrate numerous proteins involved in cell signaling. Through interactions with the cell membrane, PC have been shown to decrease the EGFR location at lipid rafts and promote its subsequent endocytosis and degradation. PC-membrane interactions also inhibit inflammatory cascades, e.g. triggered by TNFα and bile acids, and inhibit NADPH oxidase and the subsequent increase in oxidants' production and activation of redox-regulated signaling. Through all the mentioned mechanisms, as well as through direct interactions of dimers with intracellular targets, PC regulate signaling cascades involved in CRC cell proliferation and survival.

capacity of PC to act as antioxidants in the membrane environment (Verstraeten et al. 2003; Erlejman et al. 2004; Erlejman et al. 2006). Thus, cocoa B-type PC inhibited liposome lipid oxidation with a larger effect at a higher PC degree of polymerization (Verstraeten et al. 2003). These actions correlated with changes in membrane physical properties, being PC-membrane interactions mostly at the level of the polar head groups (Verstraeten et al. 2003; Erlejman et al. 2004). A grape seed extract containing B-type PC also caused changes in membrane polarization in enteroendocrine cells (Gonzalez-Abuin et al. 2014). In terms of specific membrane domains, cocoa hexameric PC particularly interact with cholesterol both in lipid raft-like liposomes and in Caco-2 lipid rafts (Verstraeten et al. 2013). The latter is very important in terms of biological actions given that lipid rafts concentrate proteins involved, among others, in cell signaling (e.g. EGFR and IGF1R, immune signaling, transport, and cell growth) (Sezgin et al. 2017). Overall, the interactions of B-type PC with membranes seem to be involved in their antioxidant activity (Verstraeten et al. 2005), as well as in their capacity to regulate cell signaling (Da Silva et al. 2012) and exert anti-CRC actions (Choy et al. 2016; Daveri et al. 2021).

Dimeric PC isolated from persimmons are constituted by subunits of either ECG or EGCG. These dimers also interact with membranes in a process that depends on the type of subunit and linkage type, polymerization degree, and the presence of galloyl moieties (Zhu et al. 2016a; Zhu et al. 2018). Comparative studies on the interactions of different types of dimers showed that A-type ECG and EGCG dimers had a high affinity for the lipid bilayer interacting with deep lipid oxygen atoms, while A-type and B-type EC dimers interact mostly with surface oxygen atoms (Zhu et al. 2015; Zhu et al. 2016b). As initially observed for hexameric B-type PC, A-type ECG and EGCG dimers interacted with lipid raft cholesterol and modulate lipid rafts-triggered signaling (Zhu et al. 2017). Molecular dynamics studies in synthetic lipid bilayers showed that these dimers change the lipid bilayer structural properties such as area per lipid, bilayer thickness, volume per lipid, lateral lipid diffusion, and lipid tail order parameters (Zhu et al. 2016a). In CRC (Caco-2) cells, both ECG and EGCG dimers decreased the area of lipid raft domains, which was paralleled by an increase in EGFR internalization and decreased cell growth (Zhu et al. 2020).

Overall, the interaction of PC with membranes can largely explain the capacity of this large and nonabsorbable compound to modulate cell signaling, act as membrane antioxidants, and exert anti-CRC actions.

3.5.2 Inflammation and the NF-κB Pathway

3.5.2.1 Inflammation

Extensive evidence supports the concept that chronic inflammation increases CRC risk. Inflammatory responses play critical roles at multiple stages of tumor development, including initiation, promotion, invasion, and metastasis (Grivennikov et al. 2010). Inflammation also affects immune surveillance and immune response to therapy. Immune cells that infiltrate tumors engage in an extensive and dynamic crosstalk with cancer cells and some of the molecular events that mediate this cell-to-cell communication are now better understood (Aran et al. 2016). These immunological mediators may regulate cancer development by directly or indirectly altering the intracellular signaling landscape in tumor cells. Two transcription factors, NF-κB and STAT3, are essential for inflammation-promoted cancer development and progression (Yu et al. 2007; Grivennikov and Karin 2010; He and Karin 2011; DiDonato et al. 2012).

In support of the central role of inflammation in CRC, a large proportion of patients with the pro-inflammatory disease ulcerative colitis develop CRC (UC-CRC) with a high mortality rate (Ekbom et al. 1992; Harpaz and Talbot 1996; Feagins et al. 2009). Chronic colorectal inflammation is a proposed underlying cause leading to UC-CRC. Colon tumors, either if they originate from UC or not, share alterations in cumulative pro-oncogenic genetic mutations, alterations in several signaling pathways (STAT3 and NF-κB, β-catenin, Wnt, p53, and TGFβ), and stages of cancer development: preneoplastic crypt foci, polyps, adenomas, and carcinoma (Terzic et al. 2010).

Interestingly, EC, catechin, and PC have shown a preventive and therapeutic potential in inflammatory bowel diseases. For example, polyphenol-enriched cocoa extract, which included EC, PC B2, catechin, and PC B1 as the major phenolics, protected mice against DSS-induced ulcerative colitis, mainly through the inhibition of STAT3 and NF-κB (Andújar et al. 2011). This section further discussed studies showing protective effects of PC in rodent models of inflammation-induced CRC.

A normal intestinal barrier function is required to prevent GI inflammation (Groschwitz and Hogan 2009). Disruption of the epithelial barrier is a crucial event that starts a cross-talk between resident inflammatory and endothelial cells leading to increased production of cytokines, activation of NF-κB and mitogen-activated protein kinases (MAPK), causing further barrier disruption and exacerbation of the inflammatory conditions. PC can protect the intestinal barrier from permeabilization. Hexameric PC protected differentiated Caco-2 cell monolayers from bile acid- and reactive oxygen species-mediated disruption of tight junctions, which control intestinal epithelium paracellular transport (Erlejman et al. 2006). The PC monomer EC also protects the intestinal barrier from diet- and inflammation-induced permeabilization through a decrease in oxidative stress and the inhibition of NF-κB and ERK1/2 (Contreras et al. 2015; Cremonini et al. 2018; Sheng et al. 2020).

Therefore, many studies have explored the possibility of mitigating CRC-induced inflammation by modulating transcription factors, e.g. NF-κB, STAT3, MAPK, given their central role in promoting cancer cell survival and growth. Antagonists to some of these targets are being actively investigated as potential therapies for various human cancers. Moreover, multiple efforts are being conducted to find natural agents that can target these signaling pathways.

3.5.2.2 NF-κB

Originally described as a transcription factor present in B cells, NF-κB is now known to be ubiquitously present in different types of cells and to regulate the transcription of multiple genes (Lenardo and Baltimore 1989). Several stimuli can trigger NF-κB activation, including pro-inflammatory cytokines, bacteria, viruses, and environmental pollutants (Karin and Greten 2005; Karin 2006). In the classical pathway, the activation of NF-κB is triggered by the IκB kinase (IKK)-mediated phosphorylation of two conserved serine residues of IκB (Delhase et al. 1999), followed by the ubiquitination and degradation of IκB by the 26S proteasome, unmasking of the nuclear localization signal and NF-κB nuclear translocation (Gorlich and Mattaj 1996). The binding of NF-κB to its consensus sequence leads to the transcription of genes, which, in general, mediate the adaptation of cells to stress. NF-κB signaling promotes cancer development by its activity within both cancer cells and tumor-infiltrating immune cells (Karin and Greten 2005). Thus, NF-κB regulates the transcription of genes involved in inflammation (e.g. tumor necrosis factor alpha (TNFα), cyclooxygenase-2, IL-6, IL-17, and IL-23), cell survival (e.g. Bcl-xl, Bcl-2, XIAP, c-IAP2, and c-FLIP), proliferation (e.g. cyclin D1 and c-myc), tumor promotion (e.g. inducible nitric oxide synthase and metalloproteinase 9) and angiogenesis (e.g. monocyte chemoattractant protein-1 (MCP-1), vascular endothelial growth factor (VEGF), IL-8) (Greten et al. 2004; Karin and Greten 2005; Karin 2006; Wang and Karin 2015). NF-κB is aberrantly expressed in many inflammation-dependent cancers. In CRC, this inflammatory signature contributes to tumor progression and metastasis (Elyada et al. 2011; Pribluda et al. 2013; Schwitalla et al. 2013). Loss of tumor suppressors can also inhibit proper DNA repair and accelerate DNA damage, which can trigger DNA damage-induced inflammatory pathways (Andriani et al. 2016).

PC can regulate NF-κB at multiple steps in the pathway through both redox and non-redox mechanisms (Fraga and Oteiza 2011; Fraga et al. 2018). In terms of redox mechanisms, hexameric PC inhibited TNFα-induced NADPH oxidase activation, the subsequent increased oxidants production, and NF-κB activation. In terms of non-redox mechanisms, dimeric PC B2 could also directly interact with DNA κB sites in the promoter of genes preventing NF-κB-driven gene transcription. Molecular modeling showed that B2 inhibited NF-κB-DNA binding given its chemical/conformational similarities to the NF-κB proteins p65 and p50 (Mackenzie et al. 2004). Dimeric and hexameric PC inhibited NF-κB in intestinal cells and lymphocytes under proinflammatory conditions (Mackenzie et al. 2004; Erlejman et al. 2008; Mackenzie et al. 2008; Mackenzie et al. 2009). Dimeric PC B1 and B2, but not A1 and A2, inhibited NF-κB-dependent IL-2 production in Jurkat T cells (Mackenzie et al. 2004; Mackenzie et al. 2009). On the other hand, an A1 dimer inhibited NF-κB activation in lipopolysaccharide-treated RAW264.7 macrophages (Han et al. 2019). B2 inhibited constitutively active NF-κB in Hodgkin's lymphoma cells, decreasing cell viability, and the production of IL-6, TNFα, and Rantes (Mackenzie et al. 2008). In a JB6 P+ mouse epidermal cell model of neoplastic transformation, a fraction of cocoa PC and pure B2 exerted inhibitory actions in part by inhibiting NF-κB and ERK1/2 (Kang et al. 2008). Importantly, this study also showed a lack of effect for theobromine, another major component of cocoa, which helps support the concept that the observed findings could be largely attributed to PC.

3.5.3 EGFR and IGF1R Pathways

Tyrosine kinase receptors EGFR and IGF1R have a central role in transformation and tumorigenicity (Pollak 2008; Arteaga and Engelman 2014). The expression and activation of both EGFR and IGF1R are frequently altered in CRC (Markowitz et al. 1990; Malecka-Panas et al. 1997; Hakam et al. 1999). EGFR, as well as other receptor tyrosine kinase receptors including the IGF1R, are localized at lipid rafts (Mineo et al. 1996; Pike and Casey 2002). Importantly, cholesterol in these domains modulates EGFR activation and internalization (Pike and Casey 2002). Thus, considering the previously described interactions of PC with lipid rafts, and their particular affinity for cholesterol, those receptors emerge as potential targets of PC anti-CRC activities.

PC isolated from GSE inhibited the EGFR tyrosine kinase activity, with effects that depended on PC polymerization degree and isomerization in purified EGFR assays (Fridrich et al. 2007). The inhibitory capacity of PC on EGFR tyrosine kinase activity decreased with polymerization degree in HT29 human CRC cells (Fridrich et al. 2007). On the other hand, in Caco-2 cells, hexameric PC (containing EC and catechin subunits) isolated from cocoa were more efficient than dimeric PC (containing ECG or EGCG) in inhibiting EGFR phosphorylation at the Tyr 1,068 residue (Zhu et al. 2020; Daveri et al. 2021). Such differences in the dependence of the effects on PC polymerization degree may be related to differences in PC subunits and on the particular alteration in the EGFR pathways present in different CRC cells. Downstream the EGFR, two central pathways are activated, i.e. PI3K/Akt and MAPK kinase ERK1/2, which promote cell proliferation and resistance to apoptosis favoring malignant transformation and tumor growth (Chang et al. 2003; Larue and Bellacosa 2005; Steelman et al. 2010). We initially observed that hexameric PC inhibited deoxycholic acid-induced ERK1/2 and Akt activation in Caco-2 cells differentiated into intestinal epithelial cells (Da Silva et al. 2012). Hexameric PC inhibited the growth of several human CRC cell lines, also inhibiting the ERK1/2 and Akt pathways in proliferating Caco-2 cells (Choy et al. 2016). This resulted in the induction of apoptosis via the mitochondrial pathway and cell cycle arrest in G_2/M phases. Recent evidence showed that upstream Akt and ERK1/2, hexameric PC targeted the EGFR pathway in Caco-2 cells (Daveri et al. 2021). Hexameric PC acted inhibiting EGF-mediated: (i) EGFR dimerization and downstream activation of ERK1/2 and Akt, (ii) increased oxidants' production that helps prolong the activation of the pathway, and (iii) localization of the receptor at lipid rafts. Hexameric PC also promoted receptor internalization and degradation via the ubiquitin/proteasome pathway (Daveri et al. 2021). The interactions of hexameric PC with lipid rafts can explain part of the observed actions. Similarly, dimeric PC isolated from persimmons and composed of either ECG or EGCG-inhibited human CRC cell growth targeting the EGFR and IGF1R receptors, and downstream ERK1/2 and Akt activation (Zhu et al. 2020). Also, dimeric PC promoted apoptosis via the mitochondrial pathway and arrested the cell cycle at G_2/M phases. Dimeric PC inhibited Caco-2 cell colony formation, migration and invasion, decreasing the activity of metalloproteinases, which are required for CRC metastasis (Zhu et al. 2020). While not affecting EGFR localization at lipid rafts, both ECG and EGCG dimers inhibited EGFR dimerization and activation, promoting its internalization and proteasomal degradation. The latter was mediated through dimers-induced EGFR phosphorylation at the Tyr 1,045 residue, providing a docking site for ubiquitin ligase c-Cbl that results in subsequent EGFR ubiquitylation (Zhu et al. 2020). Overall, evidence supports the concept that the EGFR and IGF1R signals initiated at lipid rafts represent a major target of PC-mediated regulation of oncogenic signaling.

3.6 Conclusions and Open Questions

As the data described earlier indicate, the effects PC have on CRC growth are complex. Although the majority of the current evidence from rodent and human studies suggests that select PC may reduce CRC risk and reduce tumor growth, we would deem the existing data insufficient for a general conclusion regarding the impact of PC on CRC carcinogenesis. Throughout our discussion, we have identified various specific shortfalls with regard to the available animal and human data that represent key opportunity areas for future advancement. For example, PC are complex molecules, with different subunits, linkages, degrees of polymerization and conformation, which can explain conflicting findings not only when studying complex mixtures but also pure compounds (Mackenzie et al. 2009).

To enhance rigor and reproducibility in rodent studies, we recommend following the ARRIVE (Animal Research: Reporting of *In vivo* Experiments) guidelines for animal studies, which have been recently updated: ARRIVE 2.0 (Percie du Sert et al. 2020). Clear reporting of sample size, sex, age, and strain, in the "Essential 10" is fundamental. It is critical to describe food type, exact diet composition, and access as outlined in the ARRIVE 2.0 guidelines. Standardized reporting of diet composition (including detailed PC analysis composition) will definitely strengthen the outcomes of animal studies, and provide a better understanding of the role of PC in the onset and progression of CRC.

Regarding human studies, future research should aim to include a large and diverse population. Studies should consider how lifestyle, dietary patterns, environment, and genetics interact with each other and may affect disease states rather than attributing a single nutrient to disease progression (Salem and Mackenzie 2018). In addition, whenever possible, the use of well-characterized biomarkers of PC intake is encouraged since they do not rely on recall and are a less biased approach to dietary assessment. Finally, we strongly encourage future clinical studies to consider genetic factors, embrace minority populations and both sexes in their cohorts, and include biomarkers of dietary PC intake in order to better understand how intake of dietary PC influence the risk of developing CRC.

Acknowledgments

This work was supported by grants from NIFA-USDA (CA-D*-NTR-7244-H) to PIO, and (CA-D-NTR-2397-H) to GGM. WZ was a recipient of the China International Postdoctoral Exchange Fellowship by China Postdoctoral Council.

Conflict of Interest Disclosure

Authors have no conflict of interest to declare.

References

Andriani, G.A., Almeida, V.P., Faggioli, F., et al. (2016). Whole chromosome instability induces senescence and promotes SASP. *Scientific Reports*, **6**: 35218.

Andújar, I., Recio, M.C., Giner, R.M., et al. (2011). Inhibition of ulcerative colitis in mice after oral administration of a polyphenol-enriched cocoa extract is mediated by the inhibition of STAT1 and STAT3 phosphorylation in colon cells. *Journal of Agricultural and Food Chemistry*, **59**: 6474–6483.

Appeldoorn, M.M., Vincken, J.P., Aura, A.M., et al. (2009a). Procyanidin dimers are metabolized by human microbiota with 2-(3,4-dihydroxyphenyl)acetic acid and 5-(3,4-dihydroxyphenyl)-gamma-valerolactone as the major metabolites. *Journal of Agricultural and Food Chemistry*, **57**: 1084–1092.

Appeldoorn, M.M., Vincken, J.P., Gruppen, H., and Hollman, P.C. (2009b). Procyanidin dimers A1, A2, and B2 are absorbed without conjugation or methylation from the small intestine of rats. *Journal of Nutrition*, **139**: 1469–1473.

Aran, D., Lasry, A., Zinger, A., et al. (2016). Widespread parainflammation in human cancer. *Genome Biology*, **17**: 145.

Arteaga, C.L. and Engelman, J.A. (2014). ERBB receptors: from oncogene discovery to basic science to mechanism-based cancer therapeutics. *Cancer Cell*, **25**: 282–303.

Arts, I.C., Jacobs, D.R., Gross, M., et al. (2002). Dietary catechins and cancer incidence among postmenopausal women: the Iowa Women's Health Study (United States). *Cancer Causes & Control*, **13**: 373–382.

Baba, S., Osakabe, N., Natsume, M., and Terao, J. (2002). Absorption and urinary excretion of procyanidin B2 [Epicatechin-(4b-8)-epicatechin] in rats. *Free Radical Biology and Medicine*, **33**: 142–148.

Barth, S.W., Fahndrich, C., Bub, A., et al. (2005). Cloudy apple juice decreases DNA damage, hyperproliferation and aberrant crypt foci development in the distal colon of DMH-initiated rats. *Carcinogenesis*, **26**: 1414–1421.

Barth, S.W., Fahndrich, C., Bub, A., et al. (2007). Cloudy apple juice is more effective than apple polyphenols and an apple juice derived cloud fraction in a rat model of colon carcinogenesis. *Journal of Agricultural and Food Chemistry*, 55: 1181–1187.

Boateng, J., Verghese, M., Shackelford, L., et al. (2007). Selected fruits reduce azoxymethane (AOM)-induced aberrant crypt foci (ACF) in Fisher 344 male rats. *Food and Chemical Toxicology*, **45**: 725–732.

Boutron-Ruault, M.-C., Senesse, P., Faivre, J., et al. (1999). Foods as risk factors for colorectal cancer: a case-control study in Burgundy (France). *European Journal of Cancer Preventation*, **8**: 229–235.

Chang, F., Lee, J.T., Navolanic, P.M., et al. (2003). Involvement of PI3K/Akt pathway in cell cycle progression, apoptosis, and neoplastic transformation: a target for cancer chemotherapy. *Leukemia*, **17**: 590–603.

Choy, Y.Y., Jaggers, G.K., Oteiza, P.I., and Waterhouse, A.L. (2013). Bioavailability of intact proanthocyanidins in the rat colon after ingestion of grape seed extract. *Journal of Agricultural and Food Chemistry*, **61**: 121–127.

Choy, Y.Y., Quifer-Rada, P., Holstege, D.M., et al. (2014). Phenolic metabolites and substantial microbiome changes in pig feces by ingesting grape seed proanthocyanidins. *Food & Function*, **5**: 2298–2308.

Choy, Y.Y., Fraga, M., Mackenzie, G.G., et al. (2016). The PI3K/Akt pathway is involved in procyanidin-mediated suppression of human colorectal cancer cell growth. *Molecular Carcinogenesis*, **55**: 2196–2209.

Contreras, T.C., Ricciardi, E., Cremonini, E., and Oteiza, P.I. (2015). (−)-Epicatechin in the prevention of tumor necrosis alpha-induced loss of Caco-2 cell barrier integrity. *Archives of Biochemistry and Biophysics*, **573**: 84–91.

Cremonini, E., Wang, Z., Bettaieb, A., et al. (2018). (−)-Epicatechin protects the intestinal barrier from high fat diet-induced permeabilization: implications for steatosis and insulin resistance. *Redox Biology*, **14**: 588–599.

Da Silva, M., Jaggers, G.K., Verstraeten, S.V., et al. (2012). Large procyanidins prevent bile-acid-induced oxidant production and membrane-initiated ERK1/2, p38, and Akt activation in Caco-2 cells. *Free Radical Biology and Medicine*, **52**: 151–159.

Daveri, E., Adamo, A.M., Alfine, E., et al. (2021). Hexameric procyanidins inhibit colorectal cancer cell growth through both redox and non-redox regulation of the epidermal growth factor signaling pathway. *Redox Biology*, **38**: 101830.

Delhase, M., Hayakawa, M., Chen, Y., and Karin, M. (1999). Positive and negative regulation of IkappaB kinase activity through IKKbeta subunit phosphorylation. *Science*, **284**: 309–313.

Deprez, S., Brezillon, C., Rabot, S., et al. (2000). Polymeric proanthocyanidins are catabolized by human colonic microflora into low-molecular-weight phenolic acids. *Journal of Nutrition*, **130**: 2733–2738.

DiDonato, J.A., Mercurio, F., and Karin, M. (2012). NF-kappaB and the link between inflammation and cancer. *Immunological Reviews*, **246**: 379–400.

Divisi, D., Di Tommaso, S., Salvemini, S., et al. (2006). Diet and cancer. *Acta BioMedica*, **77**: 118–123.

Düweler, K.G. and Rohdewald, P. (2000). Urinary metabolites of French maritime pine bark extract in humans. *Pharmazie*, **55**: 364–368.

Ekbom, A., Helmick, C.G., Zack, M., et al. (1992). Survival and causes of death in patients with inflammatory bowel disease: a population-based study. *Gastroenterology*, **103**: 954–960.

Elyada, E., Pribluda, A., Goldstein, R.E., et al. (2011). CKIα ablation highlights a critical role for p53 in invasiveness control. *Nature*, **470**: 409–413.

Erdrich, J., Zhang, X., Giovannucci, E., and Willett, W. (2015). Proportion of colon cancer attributable to lifestyle in a cohort of US women. *Cancer Causes and Control*, **26**: 1271–1279.

Erlejman, A.G., Verstraeten, S.V., Fraga, C.G., and Oteiza, P.I. (2004). The interaction of flavonoids with membranes: potential determinant of flavonoid antioxidant effects. *Free Radical Research*, **38**: 1311–1320.

Erlejman, A.G., Fraga, C.G., and Oteiza, P.I. (2006). Procyanidins protect Caco-2 cells from bile acid- and oxidant-induced damage. *Free Radical Biology and Medicine*, **41**: 1247–1256.

Erlejman, A.G., Jaggers, G., Fraga, C.G., and Oteiza, P.I. (2008). TNFalpha-induced NF-kappaB activation and cell oxidant production are modulated by hexameric procyanidins in Caco-2 cells. *Archives of Biochemistry and Biophysics*, **476** (2): 186–195.

Feagins, L.A., Souza, R.F., and Spechler, S.J. (2009). Carcinogenesis in IBD: potential targets for the prevention of colorectal cancer. *Nature Reviews Gastroenterology & Hepatology*, **6**: 297–305.

Ferguson, P.J., Kurowska, E.M., Freeman, D.J., et al. (2006). In vivo inhibition of growth of human tumor lines by flavonoid fractions from cranberry extract. *Nutrition and Cancer*, **56**: 86–94.

Fernández-Villa, T., Álvarez-Álvarez, L., Rubín-García, M., et al. (2020). The role of dietary patterns in colorectal cancer: a 2019 update. *Expert Review of Gastroenterology & Hepatology*, **14**: 281–290.

Fraga, C.G. and Oteiza, P.I. (2011). Dietary flavonoids: role of (−)-epicatechin and related procyanidins in cell signaling. *Free Radical Biology and Medicine*, **51**: 813–823.

Fraga, C.G., Oteiza, P.I., and Galleano, M. (2018). Plant bioactives and redox signaling: (−)-Epicatechin as a paradigm. *Molecular Aspects of Medicine*, 61: 31–40.

Franceschi, S., Favero, A., La Vecchia, C., et al. (1997). Food groups and risk of colorectal cancer in Italy. *International Journal of Cancer*, **72**: 56–61.

Fridrich, D., Kern, M., Fritz, J., et al. (2007). The epidermal growth factor receptor and human topoisomerases represent potential cellular targets of oligomeric procyanidins. *Molecular Nutrition & Food Research*, **51**: 192–200.

Gallus, S., Talamini, R., Giacosa, A., et al. (2005). Does an apple a day keep the oncologist away? *Annals of Oncology*, **16**: 1841–1844.

Gonzalez-Abuin, N., Martinez-Micaelo, N., Blay, M., et al. (2014). Grape-seed procyanidins modulate cellular membrane potential and nutrient-induced GLP-1 secretion in STC-1 cells. *American Journal of Physiology-Cell Physiology*, **306**: C485–C492.

Gorlich, D. and Mattaj, I.W. (1996). Nucleocytoplasmic transport. *Science*, **271**: 1513–1518.

Gossé, F., Guyot, S., Roussi, S., et al. (2005). Chemopreventive properties of apple procyanidins on human colon cancer-derived metastatic SW620 cells and in a rat model of colon carcinogenesis. *Carcinogenesis*, **26**: 1291–1295.

Greten, F.R., Eckmann, L., Greten, T.F., et al. (2004). IKKbeta links inflammation and tumorigenesis in a mouse model of colitis-associated cancer. *Cell*, **118**: 285–296.

Grivennikov, S.I. and Karin, M. (2010). Dangerous liaisons: STAT3 and NF-kappaB collaboration and crosstalk in cancer. *Cytokine and Growth Factor Reviews*, **21**: 11–19.

Grivennikov, S.I., Greten, F.R., and Karin, M. (2010). Immunity, inflammation, and cancer. *Cell*, **140**: 883–899.

Groschwitz, K.R. and Hogan, S.P. (2009). Intestinal barrier function: molecular regulation and disease pathogenesis. *The Journal of Allergy and Clinical Immunology*, **124**: 3–20 quiz 21-2.

Gross, G., Jacobs, D.M., Peters, S., et al. (2010). In vitro bioconversion of polyphenols from black tea and red wine/grape juice by human intestinal microbiota displays strong interindividual variability. *Journal of Agricultural and Food Chemistry*, **58**: 10236–10246.

Grosso, G., Godos, J., Lamuela-Raventos, R., et al. (2017). A comprehensive meta-analysis on dietary flavonoid and lignan intake and cancer risk: level of evidence and limitations. *Molecular Nutrition & Food Research*, **61**: 1600930.

Gu, L., Kelm, M.A., Hammerstone, J.F., et al. (2004). Concentrations of proanthocyanidins in common foods and estimations of normal consumption. *Journal of Nutrition*, **134**: 613–617.

Guyot, S., Le Bourvellec, C., Marnet, N., and Drilleau, J.-F. (2002). Procyanidins are the most abundant polyphenols in dessert apples at maturity. *LWT- Food Science and Technology*, **35**: 289–291.

Hakam, A., Yeatman, T.J., Lu, L., et al. (1999). Expression of insulin-like growth factor-1 receptor in human colorectal cancer. *Human Pathology*, **30**: 1128–1133.

Hammerstone, J.F., Lazarus, S.A., and Schmitz, H.H. (2000). Procyanidin content and variation in some commonly consumed foods. *Journal of Nutrition*, **130**: 2086S–2092S.

Han, S., Gao, H., Chen, S., et al. (2019). Procyanidin A1 Alleviates Inflammatory Response induced by LPS through NF-κB, MAPK, and Nrf2/HO-1 Pathways in RAW264.7 cells. *Scientific Reports*, **9**: 15087.

Hao, X., Xiao, H., Ju, J., et al. (2017). Green tea polyphenols inhibit colorectal tumorigenesis in azoxymethane-treated F344 rats. *Nutrition and Cancer*, **69**: 623–631.

Harpaz, N. and Talbot, I.C. (1996). Colorectal cancer in idiopathic inflammatory bowel disease. *Seminars in Diagnostic Pathology*, **13**: 339–357.

Haslam, E. (1998). *Practical Polyphenolics: From Structure to Molecular Recognition and Physiological Action*. Cambridge, United Kingdom: Cambridge University Press.

He, G. and Karin, M. (2011). NF-kappaB and STAT3 - key players in liver inflammation and cancer. *Cell Research*, **21**: 159–168.

Hidaka, A., Harrison, T.A., Cao, Y., et al. (2020). Intake of dietary fruit, vegetables, and fiber and risk of colorectal cancer according to molecular subtypes: a pooled analysis of 9 studies. *Cancer Research*, **80**: 4578–4590.

Hong, M.Y., Nulton, E., Shelechi, M., et al. (2013). Effects of dark chocolate on azoxymethane-induced colonic aberrant crypt foci. *Nutrition and Cancer*, **65**: 677–685.

Huxley, R.R., Ansary-Moghaddam, A., Clifton, P., et al. (2009). The impact of dietary and lifestyle risk factors on risk of colorectal cancer: a quantitative overview of the epidemiological evidence. *International Journal of Cancer*, **125**: 171–180.

Jedrychowski, W. and Umberto, M. (2009). An apple a day may hold colorectal cancer at bay: recent evidence from a case-control study. *Review in Environmental Health*, **24**: 59–74.

Jerez, M., Tourino, S., Sineiro, J., et al. (2007). Procyanidins from pine bark: relationships between structure, composition and antiradical activity. *Food Chemistry*, **104**: 518–527.

Jia, X.D. and Han, C. (2000). Chemoprevention of tea on colorectal cancer induced by dimethylhydrazine in Wistar rats. *World Journal of Gastroenterology*, **6**: 699.

Jung, Y.D. and Ellis, L.M. (2001). Inhibition of tumour invasion and angiogenesis by epigallocatechin gallate (EGCG), a major component of green tea. *International Journal of Experimental Pathology*, **82**: 309–316.

Kang, N.J., Lee, K.W., Lee, D.E., et al. (2008). Cocoa procyanidins suppress transformation by inhibiting mitogen-activated protein kinase kinase. *Journal of Biological Chemistry*, **283**: 20664–20673.

Karin, M. (2006). NF-kappaB and cancer: mechanisms and targets. *Molecular Carcinogenesis*, **45**: 355–361.

Karin, M. and Greten, F.R. (2005). NF-kappaB: linking inflammation and immunity to cancer development and progression. *Nature Reviews Immunology*, **5**: 749–759.

Kaur, M., Singh, R.P., Gu, M., et al. (2006). Grape seed extract inhibits in vitro and in vivo growth of human colorectal carcinoma cells. *Clinical Cancer Research*, **12**: 6194–6202.

Kim, Y.S. and Milner, J.A. (2007). Dietary modulation of colon cancer risk. *Journal of Nutrition*, **137**: 2576S–2579S.

Kim, M., Murakami, A., Miyamoto, S., et al. (2010). The modifying effects of green tea polyphenols on acute colitis and inflammation-associated colon carcinogenesis in male ICR mice. *BioFactors*, **36**: 43–51.

Larue, L. and Bellacosa, A. (2005). Epithelial-mesenchymal transition in development and cancer: role of phosphatidylinositol 3' kinase/AKT pathways. *Oncogene*, **24**: 7443–7454.

Lenardo, M.J. and Baltimore, D. (1989). NF-kappa B: a pleiotropic mediator of inducible and tissue-specific gene control. *Cell*, **58**: 227–229.

Liu, W., Zhao, S.Q., Wang, J.Q., et al. (2017). Grape seed proanthocyanidin extract ameliorates inflammation and adiposity by modulating gut microbiota in high-fat diet mice. *Molecular Nutrition & Food Research*, **61**: 1601082.

MacDonald, R.S. and Wagner, K. (2012). Influence of dietary phytochemicals and microbiota on colon cancer risk. *Journal of Agricultural and Food Chemistry*, **60**: 6728–6735.

Mackenzie, G.G., Carrasquedo, F., Delfino, J.M., et al. (2004). Epicatechin, catechin, and dimeric procyanidins inhibit PMA-induced NF-kappaB activation at multiple steps in Jurkat T cells. *FASEB Journal*, **18**: 167–169.

Mackenzie, G.G., Adamo, A.M., Decker, N.P., and Oteiza, P.I. (2008). Dimeric procyanidin B2 inhibits constitutively active NF-kappaB in Hodgkin's lymphoma cells independently of the presence of IkappaB mutations. *Biochemical Pharmacology*, **75**: 1461–1471.

Mackenzie, G.G., Delfino, J.M., Keen, C.L., et al. (2009). Dimeric procyanidins are inhibitors of NF-kappaB-DNA binding. *Biochemical Pharmacology*, **78**: 1252–1262.

Malecka-Panas, E., Kordek, R., Biernat, W., et al. (1997). Differential activation of total and EGF receptor (EGF-R) tyrosine kinase (tyr-k) in the rectal mucosa in patients with adenomatous polyps, ulcerative colitis and colon cancer. *Hepato-Gastroenterology*, **44**: 435–440.

Manach, C., Williamson, G., Morand, C., et al. (2005). Bioavailability and bioefficacy of polyphenols in humans. I. Review of 97 bioavailability studies. *Amercian Journal of Clinical Nutrition*, **81**: 230S–242S.

Markowitz, S.D., Molkentin, K., Gerbic, C., et al. (1990). Growth stimulation by coexpression of transforming growth factor-alpha and epidermal growth factor-receptor in normal and adenomatous human colon epithelium. *Journal of Clinical Investigation*, **86**: 356–362.

Martin, M.A., Goya, L., and Ramos, S. (2013). Potential for preventive effects of cocoa and cocoa polyphenols in cancer. *Food and Chemical Toxicology*, **56**: 336–351.

Michels, K.B., Giovannucci, E., Chan, A.T., et al. (2006). Fruit and vegetable consumption and colorectal adenomas in the Nurses' Health Study. *Cancer Research*, **66**: 3942–3953.

Mineo, C., James, G.L., Smart, E.J., and Anderson, R.G. (1996). Localization of epidermal growth factor-stimulated Ras/Raf-1 interaction to caveolae membrane. *Journal of Biological Chemistry*, **271**: 11930–11935.

Nagano, J., Kono, S., Preston, D.L., and Mabuchi, K. (2001). A prospective study of green tea consumption and cancer incidence, Hiroshima and Nagasaki (Japan). *Cancer Causes and Control*, **12**: 501–508.

Nakachi, K., Matsuyama, S., Miyake, S., et al. (2000). Preventive effects of drinking green tea on cancer and cardiovascular disease: epidemiological evidence for multiple targeting prevention. *BioFactors*, **13**: 49–54.

Narisawa, T. and Fukaura, Y. (1993). A very low dose of green tea polyphenols in drinking water prevents N-methyl-N-nitrosourea-induced colon carcinogenesis in F344 rats. *Japanese Journal of Cancer Research*, **84**: 1007–1009.

Orner, G.A., Dashwood, W.-M., Blum, C.A., et al. (2003). Suppression of tumorigenesis in the Apc min mouse: down-regulation of β-catenin signaling by a combination of tea plus sulindac. *Carcinogenesis*, **24**: 263–267.

Ottaviani, J.I., Kwik-Uribe, C., Keen, C.L., and Schroeter, H. (2012). Intake of dietary procyanidins does not contribute to the pool of circulating flavanols in humans. *Amercian Journal of Clinical Nutrition*, **95**: 851–858.

Ottaviani, J.I., Fong, R., Kimball, J., et al. (2018). Evaluation at scale of microbiome-derived metabolites as biomarker of flavan-3-ol intake in epidemiological studies. *Scientific Reports*, **8**: 9859.

Pandurangan, A.K., Saadatdoust, Z., Mohd. Esa, N., et al. (2015). Dietary cocoa protects against colitis-associated cancer by activating the N rf2/K eap1 pathway. *BioFactors*, **41**: 1–14.

Percie du Sert, N., Hurst, V., Ahluwalia, A., et al. (2020). The ARRIVE guidelines 2.0: updated guidelines for reporting animal research. *PLoS Biology*, **18**: e3000410.

Pike, L.J. and Casey, L. (2002). Cholesterol levels modulate EGF receptor-mediated signaling by altering receptor function and trafficking. *Biochemistry*, **41**: 10315–10322.

Platz, E.A., Willett, W.C., Colditz, G.A., et al. (2000). Proportion of colon cancer risk that might be preventable in a cohort of middle-aged US men. *Cancer Causes and Control*, **11**: 579–588.

Pollak, M. (2008). Insulin and insulin-like growth factor signalling in neoplasia. *Nature Reviews Cancer*, **8**: 915–928.

Pribluda, A., Elyada, E., Wiener, Z., et al. (2013). A senescence-inflammatory switch from cancer-inhibitory to cancer-promoting mechanism. *Cancer Cell*, **24**: 242–256.

Prieur, C., Rigaud, J., Cheynier, V., and Moutounet, M.J.P. (1994). Oligomeric and polymeric procyanidins from grape seeds. *Phytochemistry*, **36**: 781–784.

Prior, R.L. and Gu, L. (2005). Occurrence and biological significance of proanthocyanidins in the American diet. *Phytochemistry*, **66**: 2264–2280.

Ravindranathan, P., Pasham, D., Balaji, U., et al. (2018a). A combination of curcumin and oligomeric proanthocyanidins offer superior anti-tumorigenic properties in colorectal cancer. *Scientific Reports*, **8**: 13869.

Ravindranathan, P., Pasham, D., Balaji, U., et al. (2018b). Mechanistic insights into anticancer properties of oligomeric proanthocyanidins from grape seeds in colorectal cancer. *Carcinogenesis*, **39**: 767–777.

Ravindranathan, P., Pasham, D., and Goel, A. (2019). Oligomeric proanthocyanidins (OPCs) from grape seed extract suppress the activity of ABC transporters in overcoming chemoresistance in colorectal cancer cells. *Carcinogenesis*, **40**: 412–421.

Reddivari, L., Charepalli, V., Radhakrishnan, S., et al. (2016). Grape compounds suppress colon cancer stem cells in vitro and in a rodent model of colon carcinogenesis. *BMC Complementary Medicine and Therapies*, **16**: 278.

Ricci, A., Parpinello, G.P., Palma, A.S., et al. (2017). Analytical profiling of food-grade extracts from grape (Vitis vinifera sp.) seeds and skins, green tea (Camellia sinensis) leaves and Limousin oak (Quercus robur) heartwood using MALDI-TOF-MS, ICP-MS and spectrophotometric methods. *Journal of Food Composition and Analysis*, **59**: 95–104.

Rodríguez-Ramiro, I., Ramos, S., López-Oliva, E., et al. (2013). Cocoa polyphenols prevent inflammation in the colon of azoxymethane-treated rats and in TNF-α-stimulated Caco-2 cells. *British Journal of Nutrition*, **110**: 206–215.

Rossi, M., Negri, E., Parpinel, M., et al. (2010). Proanthocyanidins and the risk of colorectal cancer in Italy. *Cancer Causes and Control*, **21**: 243–250.

Rouillier, P., Senesse, P., Cottet, V., et al. (2005). Dietary patterns and the adenomacarcinoma sequence of colorectal cancer. *European Journal of Nutrition*, **44**: 311–318.

Saadatdoust, Z., Pandurangan, A.K., Sadagopan, S.K.A., et al. (2015). Dietary cocoa inhibits colitis associated cancer: a crucial involvement of the IL-6/STAT3 pathway. *Journal of Nutritional Biochemistry*, **26**: 1547–1558.

Salem, A.A. and Mackenzie, G.G. (2018). Pancreatic cancer: a critical review of dietary risk. *Nutrition Research*, **52**: 1–13.

Sano, A., Yamakoshi, J., Tokutake, S., et al. (2003). Procyanidin B1 is detected in human serum after intake of proanthocyanidin-rich grape seed extract. *Bioscience, Biotechnology, and Biochemistry*, **67**: 1140–1143.

Schwingshackl, L., Schwedhelm, C., Hoffmann, G., et al. (2018). Food groups and risk of colorectal cancer. *International Journal of Cancer*, **142**: 1748–1758.

Schwitalla, S., Ziegler, P.K., Horst, D., et al. (2013). Loss of p53 in enterocytes generates an inflammatory microenvironment enabling invasion and lymph node metastasis of carcinogen-induced colorectal tumors. *Cancer Cell*, **23**: 93–106.

Sezgin, E., Levental, I., Mayor, S., and Eggeling, C. (2017). The mystery of membrane organization: composition, regulation and roles of lipid rafts. *Nature Reviews Molecular Cell Biology*, **18**: 361–374.

Sheng, K., Zhang, G., Sun, M., et al. (2020). Grape seed proanthocyanidin extract ameliorates dextran sulfate sodium-induced colitis through intestinal barrier improvement, oxidative stress reduction, and inflammatory cytokines and gut microbiota modulation. *Food & Function*, **11**: 7817–7829.

Shimizu, M., Fukutomi, Y., Ninomiya, M., et al. (2008). Green tea extracts for the prevention of metachronous colorectal adenomas: a pilot study. *Cancer Epidemiology, Biomarkers & Prevention*, **17**: 3020–3025.

Shin, C.M., Lee, D.H., Seo, A.Y., et al. (2018). Green tea extracts for the prevention of metachronous colorectal polyps among patients who underwent endoscopic removal of colorectal adenomas: a randomized clinical trial. *Clinical Nutrition*, **37**: 452–458.

Shirakami, Y., Shimizu, M., Tsurumi, H., et al. (2008). EGCG and Polyphenon E attenuate inflammation-related mouse colon carcinogenesis induced by AOM plus DDS. *Molecular Medicine Reports*, **1**: 355–361.

Siegel, R.L., Miller, K.D., and Jemal, A. (2020). Cancer statistics, 2020. *CA: A Cancer Journal for Clinicians*, **70**: 7–30.

Song, M. and Giovannucci, E. (2016). Preventable incidence and mortality of carcinoma associated with lifestyle factors among white adults in the United States. *JAMA Oncology*, **2**: 1154–1161.

Steelman, L.S., Abrams, S.L., Shelton, J.G., et al. (2010). Dominant roles of the Raf/MEK/ERK pathway in cell cycle progression, prevention of apoptosis and sensitivity to chemotherapeutic drugs. *Cell Cycle*, **9**: 1629–1638.

Stingl, J.C., Ettrich, T., Muche, R., et al. (2011). Protocol for minimizing the risk of metachronous adenomas of the colorectum with green tea extract (MIRACLE): a randomised controlled trial of green tea extract versus placebo for nutriprevention of metachronous colon adenomas in the elderly population. *BMC Cancer*, **11**: 1–9.

Stoupi, S., Williamson, G., Drynan, J.W., et al. (2010a). Procyanidin B2 catabolism by human fecal microflora: partial characterization of 'dimeric' intermediates. *Archives of Biochemistry and Biophysics*, **501**: 73–78.

Stoupi, S., Williamson, G., Viton, F., et al. (2010b). In vivo bioavailability, absorption, excretion, and pharmacokinetics of [14C]procyanidin B2 in male rats. *Drug Metabolism and Disposition*, **38**: 287–291.

Terry, P., Giovannucci, E., Michels, K.B., et al. (2001). Fruit, vegetables, dietary fiber, and risk of colorectal cancer. *Journal of the National Cancer Institute*, **93**: 525–533.

Terzic, J., Grivennikov, S., Karin, E., and Karin, M. (2010). Inflammation and colon cancer. *Gastroenterology*, **138**: 2101–2114.e5.

Theodoratou, E., Kyle, J., Cetnarskyj, R., et al. (2007). Dietary flavonoids and the risk of colorectal cancer. *Cancer Epidemiology, Biomarkers & Prevention*, **16**: 684–693.

Toden, S., Ravindranathan, P., Gu, J., et al. (2018). Oligomeric proanthocyanidins (OPCs) target cancer stem-like cells and suppress tumor organoid formation in colorectal cancer. *Scientific Reports*, **8**: 3335.

Tsang, C., Auger, C., Mullen, W., et al. (2005). The absorption, metabolism and excretion of flavan-3-ols and procyanidins following the ingestion of a grape seed extract by rats. *British Journal of Nutrition*, **94**: 170–181.

Verstraeten, S.V., Keen, C.L., Schmitz, H.H., et al. (2003). Flavan-3-ols and procyanidins protect liposomes against lipid oxidation and disruption of the bilayer structure. *Free Radical Biology and Medicine*, **34**: 84–92.

Verstraeten, S.V., Hammerstone, J.F., Keen, C.L., et al. (2005). Antioxidant and membrane effects of procyanidin dimers and trimers isolated from peanut and cocoa. *Journal of Agricultural and Food Chemistry*, **53**: 5041–5048.

Verstraeten, S.V., Jaggers, G.K., Fraga, C.G., and Oteiza, P.I. (2013). Procyanidins can interact with Caco-2 cell membrane lipid rafts: involvement of cholesterol. *Archives of Biochemistry and Biophysics*, **1828**: 2646–2653.

Verstraeten, S.V., Fraga, C.G., and Oteiza, P.I. (2015). Interactions of flavan-3-ols and procyanidins with membranes: mechanisms and the physiological relevance. *Food & Function*, **6**: 32–41.

Vieira, A.R., Abar, L., Chan, D.S.M., et al. (2017). Foods and beverages and colorectal cancer risk: a systematic review and meta-analysis of cohort studies, an update of the evidence of the WCRF-AICR Continuous Update Project. *Annals of Oncology*, **28**: 1788–1802.

Wang, K. and Karin, M. (2015). Tumor-elicited inflammation and colorectal cancer. *Advances in Cancer Research*, **128**: 173–196.

Wang, S.-T., Cui, W.-Q., Pan, D., et al. (2020a). Tea polyphenols and their chemopreventive and therapeutic effects on colorectal cancer. *World Journal of Gastroenterology*, **26**: 562.

Wang, Y., Jin, H.-Y., Fang, M.-Z., et al. (2020b). Epigallocatechin gallate inhibits dimethylhydrazine-induced colorectal cancer in rats. *World Journal of Gastroenterology*, **26**: 2064.

Wang, K., Ma, W., Wu, K., et al. (2021). Healthy lifestyle, endoscopic screening, and colorectal cancer incidence and mortality in the United States: a nationwide cohort study. *PLoS Medicine*, **18**: e1003522.

Weh, K.M., Clarke, J., and Kresty, L.A.J.A. (2016). Cranberries and cancer: an update of preclinical studies evaluating the cancer inhibitory potential of cranberry and cranberry derived constituents. *Antioxidants*, **5**: 27.

Wei, E.K., Colditz, G.A., Giovannucci, E.L., et al. (2017). A comprehensive model of colorectal cancer by risk factor status and subsite using data from the Nurses' Health Study. *American Journal of Epidemiology*, **185**: 224–237.

Weisburger, J.H., Rivenson, A., Aliaga, C., et al. (1998). Effect of tea extracts, polyphenols, and epigallocatechin gallate on azoxymethane-induced colon cancer. *Proceedings of the Society for Experimental Biology and Medicine*, **217**: 104–108.

Xiao, H., Hao, X., Simi, B., et al. (2008). Green tea polyphenols inhibit colorectal aberrant crypt foci (ACF) formation and prevent oncogenic changes in dysplastic ACF in azoxymethane-treated F344 rats. *Carcinogenesis*, **29**: 113–119.

Xiao, X., Kim, J., Sun, Q., et al. (2015). Preventive effects of cranberry products on experimental colitis induced by dextran sulphate sodium in mice. *Food Chemistry*, **167**: 438–446.

Xu, M., Chen, Y.-M., Huang, J., et al. (2016). Flavonoid intake from vegetables and fruits is inversely associated with colorectal cancer risk: a case-control study in China. *British Journal of Nutrition*, **116**: 1275–1287.

Yamane, T., Hagiwara, N., Tateishi, M., et al. (1991). Inhibition of azoxymethane-induced colon carcinogenesis in rat by green tea polyphenol fraction. *Japanese Journal of Cancer Research*, **82**: 1336–1339.

Yamane, T., Nakatani, H., Kikuoka, N., et al. (1996). Inhibitory effects and toxicity of green tea polyphenols for gastrointestinal carcinogenesis. *Cancer*, **77**: 1662–1667.

Yu, H., Kortylewski, M., and Pardoll, D. (2007). Crosstalk between cancer and immune cells: role of STAT3 in the tumour microenvironment. *Nature Reviews Immunology*, **7**: 41–51.

Zamora-Ros, R., Not, C., Guinó, E., et al. (2013). Association between habitual dietary flavonoid and lignan intake and colorectal cancer in a Spanish case-control study (the Bellvitge colorectal cancer study). *Cancer Causes and Control*, **24**: 549–557.

Zamora-Ros, R., Barupal, D.K., Rothwell, J.A., et al. (2017). Dietary flavonoid intake and colorectal cancer risk in the European prospective investigation into cancer and nutrition (EPIC) cohort. *International Journal of Cancer*, **140**: 1836–1844.

Zamora-Ros, R., Cayssials, V., Jenab, M., et al. (2018). Dietary intake of total polyphenol and polyphenol classes and the risk of colorectal cancer in the European prospective investigation into cancer and nutrition (EPIC) cohort. *Europen Journal of Epidemiology*, **33**: 1063–1075.

Zhang, R., Yu, Q., Lu, W., et al. (2019). Grape seed procyanidin B2 promotes the autophagy and apoptosis in colorectal cancer cells via regulating PI3K/Akt signaling pathway. *Oncotargets and Therapy*, **12**: 4109–4118.

Zhou, L., Wang, W., Huang, J., et al. (2016). in vitro extraction and fermentation of polyphenols from grape seeds (Vitis vinifera) by human intestinal microbiota. *Food & Function*, **7**: 1959–1967.

Zhu, W., Zou, B., Nie, R., et al. (2015). A-type ECG and EGCG dimers disturb the structure of 3T3-L1 cell membrane and strongly inhibit its differentiation by targeting peroxisome proliferator-activated receptor γ with miR-27 involved mechanism. *Journal of Nutritional Biochemistry*, **26**: 1124–1135.

Zhu, W., Xiong, L., Peng, J., et al. (2016a). Molecular Insight into Affinities of Gallated and Nongallated Proanthocyanidins Dimers to Lipid Bilayers. *Scientific Reports*, **6**: 37680.

Zhu, W., Xiong, L., Peng, J., et al. (2016b). Structure-dependent membrane-perturbing potency of four proanthocyanidin dimers on 3T3-L1 preadipocytes. *Journal of Agricultural and Food Chemistry*, **64**: 7022–7032.

Zhu, W., Deng, X., Peng, J., et al. (2017). A-type ECG and EGCG dimers inhibit 3T3-L1 differentiation by binding to cholesterol in lipid rafts. *Journal of Nutritional Biochemistry*, **48**: 62–73.

Zhu, W., Khalifa, I., Peng, J., and Li, C. (2018). Position and orientation of gallated proanthocyanidins in lipid bilayer membranes: influence of polymerization degree and linkage type. *Journal of Biomolecular Structure and Dynamics*, **36**: 2862–2875.

Zhu, W., Wang, R.-F., Khalifa, I., and Li, C.-M. (2019). Understanding toward the biophysical interaction of polymeric proanthocyanidins (persimmon condensed tannins) with biomembranes: Relevance for biological effects. *Journal of Agricultural and Food Chemistry*, **67**: 11044–11052.

Zhu, W., Li, M.C., Wang, F.R., et al. (2020). The inhibitory effect of ECG and EGCG dimeric procyanidins on colorectal cancer cells growth is associated with their actions at lipid rafts and the inhibition of the epidermal growth factor receptor signaling. *Biochemical Pharmacology*, **175**: 113923.

4

The Potential of Low Molecular Weight (Poly)phenol Metabolites for Attenuating Neuroinflammation and Treatment of Neurodegenerative Diseases

Daniela Marques, Rafael Carecho, Diogo Carregosa, and Cláudia Nunes dos Santos

NOVA Medical School, Faculdade de Ciências Médicas, Universidade NOVA de Lisboa, Lisboa, Portugal

4.1 Introduction: Neurodegenerative Disorders, Dietary (Poly)phenols and Neuroinflammation

Neurodegenerative diseases (NDs) are a heterogeneous group of disorders characterized by chronic and progressive neurological dysfunction that leads to cell degeneration by losing both structure and function of neurons. These disorders are associated with pathophysiological changes, such as age, affecting mental thinking and movement function (Brettschneider et al. 2015). Generally, these diseases arise from hereditary or sporadic conditions and exhibit neuron pathological alterations and dysfunction (Soto 2003).

There are several diseases that are known to affect the nervous systems, such as Alzheimer's disease (AD), Parkinson's disease (PD), Huntington's disease (HD), amyotrophic lateral sclerosis (ALS), multiple sclerosis (MS), brain cancer, degenerative nerve diseases, encephalitis, and epilepsy. Although these diseases show alterations in different brain regions, they share pathological processes: the aggregation of misfolded proteins (Soto 2003); increased oxidative stress; protein carbonylation; lipid peroxidation; DNA damage; mitochondrial dysfunction; endoplasmic reticulum (ER) stress; unfolded protein response (UPR); dysregulation and imbalanced proteostasis, and neuroinflammation. These common ND hallmarks will require multitargeted interventions to tackle disease progression. These pathologies, which lack modern therapies, present a significant projected rise in incidence and are among the most burdensome age-related chronic diseases, with both social and economic consequences. Therefore, novel strategies are urgently needed to tackle their growing burden.

Associated with phenolic consumption, several epidemiological studies have revealed that diets rich in fruits and vegetables may improve human health (Li et al. 2014;

Recent Advances in Polyphenol Research, Volume 8, First Edition. Edited by Juha-Pekka Salminen, Kristiina Wähälä, Victor de Freitas, and Stéphane Quideau.

Rodriguez-Mateos et al. 2014; Medina-Remón et al. 2015) and prevent degenerative diseases and cognitive decline (Macready et al. 2009; Nooyens et al. 2011; Nehlig 2013; Rendeiro et al. 2015; Psaltopoulou et al. 2019). (Poly)phenol-rich diets in animal models can activate neuronal receptors, interact with signaling pathways (e.g. through a mitogen-activated protein kinase (MAPK) signal transduction pathway), and control the expression of specific genes (Stalmach et al. 2012).

Besides that, (poly)phenol compounds are described as having pleiotropic effects, such as the modulation of synaptic plasticity e.g. memory and/or learning improvement in both animals and humans (Spencer 2008; Williams and Spencer 2012; Rodriguez-Mateos et al. 2014). A recent study has also correlated the cognitive benefits of phenolics consumption, with a decrease in several markers of systemic inflammation that can influence the neuroinflammation status of the brain (Mayr et al. 2018). Various studies have shown an association between increased consumption of berries with a slower progression of cognitive decline in older women (Devore et al. 2012) and consumption of (poly)phenol-rich foods with cognitive benefits in humans (Macready et al. 2009; Nehlig 2013; Kennedy 2014). Furthermore, wild blueberry consumption was proven to improve cognitive function in older adults (Krikorian et al. 2010), and chronic blueberry supplementation have shown improvement in brain perfusion, task-related activation, and cognitive function in healthy older adults, suggesting the improvement of brain activation in areas associated with cognitive function (Bowtell et al. 2017). Despite this, in a phase three clinical trial, treatment with that epigallocatechin gallate treatment in patients did not modify the progression of multiple system atrophy (Levin et al. 2019). However, the understanding of their mechanism of action on the central nervous system (CNS) remains in its infancy (Figueira et al. 2017).

This shows that even though a lot of research is conducted on (poly)phenols and NDs, some key aspects of their brain bioavailability and mechanism are still missing. Therapeutic and nutritional strategies are urgently needed to tackle these multifactorial diseases, and (poly)phenols are emerging as effective agents in slowing cognitive decline, preventing, or postponing the onset of NDs and delaying or slowing their progression. Yet, we still need to unlock some key aspects of the journey of (poly)phenols within our bodies until they reach the brain.

4.2 (Poly)phenols: Metabolism and Distribution

(Poly)phenols are a class of chemical compounds characterized by at least two benzene rings with one or more hydroxyl groups. In practice, the term (poly)phenol has been extensively used to designate plant biosynthetic compounds from the shikimate pathways, i.e. flavonoids (Quideau et al. 2011). However, some flavonoid monomeric units like phenolic acids have been widely nominated polyphenols. To avoid confusion, the term polyphenols was recently updated to (poly)phenols in order to engulf a wider range of structurally and metabolically related phenolics (Rodriguez-Mateos et al. 2013; Feliciano et al. 2016).

(Poly)phenols can be found in a vast array of natural dietary sources including fruits (e.g. orange, mango, pomegranate, apple, and berries), vegetables (e.g. cabbage, onion, bell pepper, and aromatic plants), beverages, coffee, and cocoa. All these examples and many others include a variety of different and interesting molecules with interesting capabilities to change health conditions. As such, dietary habits including fruits, plants, and vegetables with an increased presence of (poly)phenols can be a very attractive field for human health applications. Such research could have a huge impact on the prevention and treatment of chronic diseases since this will avoid many setbacks in the administration of synthetic drugs to patients.

For decades, dietary (poly)phenols' research has been focused to understand all the potential health benefits of a vast array of different molecules. Evidences associated with these compounds include anti-inflammatory, anti-tumoral, anti-diabetic, and anti-viral potential.

Although our knowledge of flavonoids in dietary sources is vast and we have some understanding of their potential biological functions, their metabolism in humans has not been fully unveiled. Yet, our knowledge about their bioavailability in humans has shifted in the last one/two decades. Several intervention studies show that flavonoids, the most abundant molecules in dietary sources, are not the most abundant in human circulation. Five main factors are responsible for this fact: their poor availability across the membranes; studies had neglected the enterohepatic circulation; metabolic reactions carried out by gastrointestinal cells and liver cells; catabolism carried by microbiota that degrades the flavonoids into smaller molecules; and finally even upon reaching circulation, flavonoids do not tend to accumulate and spend short limited periods of time in circulation before being excreted (de Ferrars et al. 2014; Williamson et al. 2018).

Upon ingestion, flavonoids present in dietary sources go through the stomach almost untouched (Williamson et al. 2018). Reaching the small intestine, conjugated flavonoids like some quercetin-glucosides may be cleaved by the action of lactase phlorizin hydrolase (LPH) (Day et al. 2000), entering enterocytes as aglycones possibly through passive diffusion (Zhang et al. 2020). Flavonoid conjugates that surpass the action of LPH may still enter enterocytes through transporters like SGTL-1 or GLUT2 and be deconjugated by the action of cytosolic β-glucosidase (CBG) (Zhang et al. 2020). Inside enterocytes and in the liver, flavonoids may undergo phase I and II metabolism, resulting in a variety of flavonoid conjugated metabolites, such as sulfate, methyl, or glucuronide. Such metabolites have a short lifespan in circulation, reaching their maximum concentration in the first hours after ingestion, usually in the first 2–4 hours. Flavonoids not absorbed in the small intestine make their way through the gut where they are subjected to metabolism by intestinal microbiota. The microbiota can cleave the flavonoid backbone into new metabolites and do successive catabolic steps to further cleave these molecules into progressively smaller molecules (Figure 4.1).

(Poly)phenol metabolites resulting from microbiota catabolism and present in the gut can then be absorbed in the gut and are also able to undergo phase I and II metabolism in enterocytes or the liver. Such metabolic reactions include conjugation with methyl, sulfate, glucuronides, or even amino acids, like glycine, generating a vast array of low molecular weight (LMW) (poly)phenol metabolites in human circulation (Del Rio et al. 2013). These metabolites include hydroxybenzenes, benzoic acids, phenylacetic acids, phenylpropionic acids, cinnamic acids, 5-(phenyl)valeric acids, and phenyl-gamma-valerolactones, and their phase II conjugates correspond to the culmination of both microbiota and human

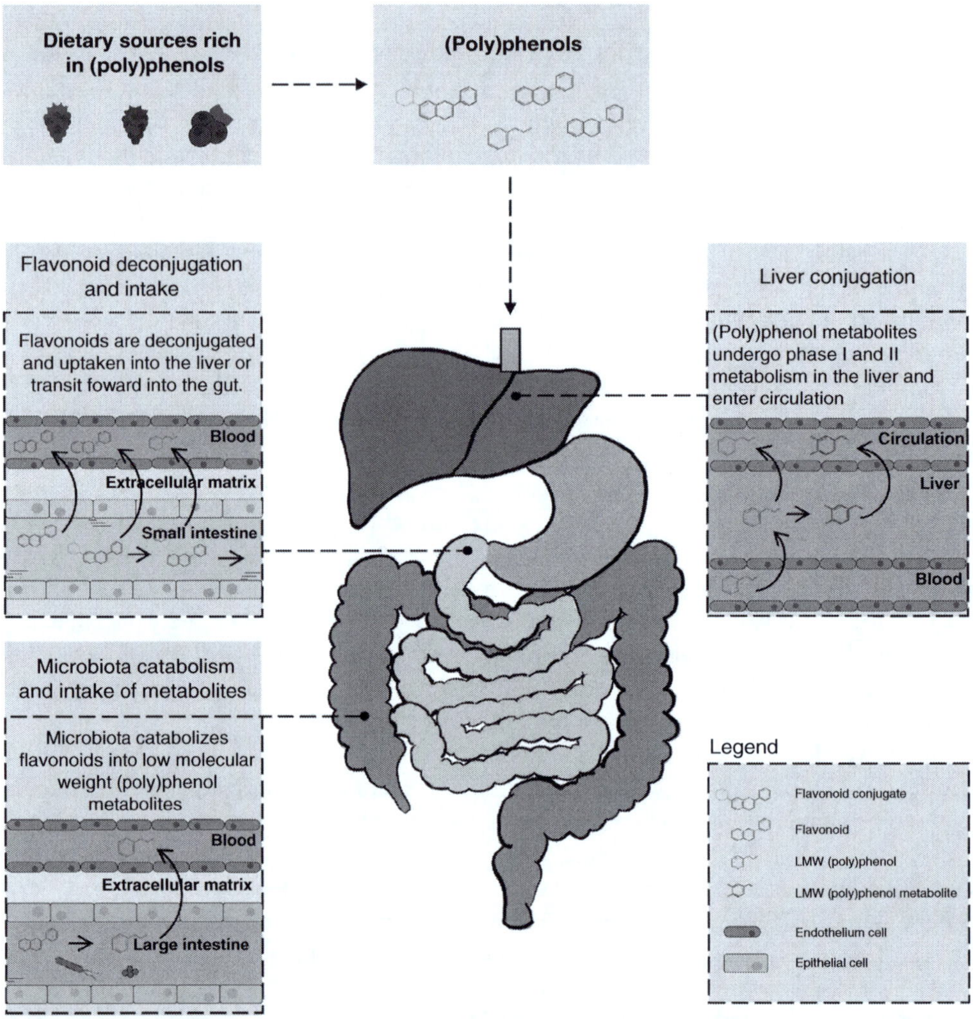

Figure 4.1 (Poly)phenols route for absorption and metabolism in the human gastrointestinal system. (Poly)phenols, represented by flavonoids, enter the gastrointestinal tract from different fruits and vegetables rich in (poly)phenols. (Poly)phenols start by being unconjugated and parts are absorbed in the small intestine. The remaining unconjugated flavonoids are catabolized by the microbiota into low molecular weight (LMW) (poly)phenol metabolites. Both parent compounds and LMW (poly)phenols suffer metabolism in the epithelium and liver and reach circulation. ECM, extracellular matrix; SI, small intestine; LI, large intestine; LMW; low molecular weight. (*See insert for color representation of the figure.*)

reactions. Metabolites like 3′-hydroxycinnamic acid appear minutes after ingestion, while others like hippuric acids tend to appear later, 6–8 hours after ingestion (Figure 4.2). Moreover, contrasting with their parent compounds, metabolites tend to stay in circulation for longer periods of time, reaching the 24 and 48 hours like hippuric acid and some valeric acid and phenyl-gamma-valerolactones (Figure 4.2).

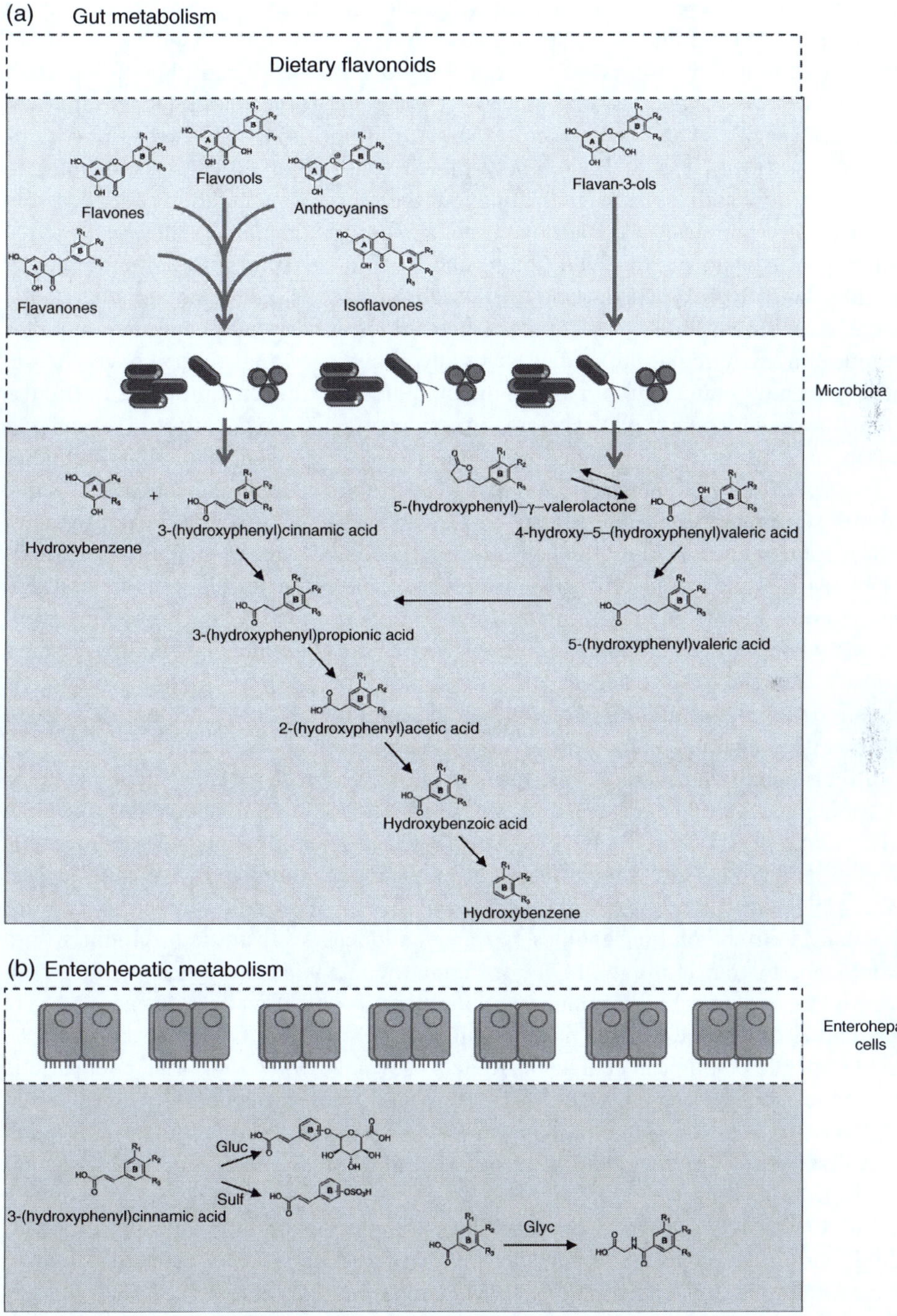

Figure 4.2 (Poly)phenol catabolism into low molecular weight (LMW) metabolites. (a) (Poly) phenol uptaken in the diet suffers catabolism by the microbiota leading to the generation of (poly) phenol metabolites. (b) Any of these molecules are uptaken and may undergo phase I and II metabolism produced by epithelium cells and liver cells in the enterohepatic system. Gluc, glucuronidation; sulf, sulfation; glyc, glycosylation.

The correlation between parent flavonoids and the derived (poly)phenol metabolites is puzzling; many LMW phenolic metabolites can arise through the metabolism of a wide group of structurally diverse flavonoids, e.g. flavonols, flavan-3-ols, anthocyanins, chalcones, isoflavones, flavanones, and flavones, while for some, e.g. ellagitannins and flavan-3-ols, specific metabolites are formed: urolithins and 5-(phenyl)valeric acid/phenyl-gamma-valerolactones, respectively. Moreover, some of these (poly)phenols can be traced back to dietary sources, indicating that the source in circulation could be a mix between dietary sources and microbiota/endogenous metabolism. Examples include 4′-hydroxy-3′methoxycinnamic acid (i.e. ferulic acid), a common metabolite present in coffee, or 3,4,5-trihydroxybenzoic acid (gallic acid) present in significant amounts in tea. Nevertheless, although LMW (poly)phenol metabolites are dependent on microbiota for their appearance in circulation, studies have shown that these metabolites converge from different flavonoid sources culminating in much more relevant concentrations in circulation than flavonoids and can also stay for long. For example, Czank et al. (2013) found that ingestion of 500 mg of ^{13}C-labelled cyanidin-3-glucoside provided a 42-fold higher abundance of ^{13}C-labeled metabolites relative to the original compound at their respective maximum serum concentration. Moreover, Ottaviani et al. (2016) found extensive biotransformation on the flavan-3-ol (−)-epicatechin that culminated in more than 20 metabolites in plasma and urine of human volunteers, while the levels of unmetabolized compound were below the limits of detection. Pimpão et al. (2015), in a human intervention study using 500 mL of a puree of berries, have found that phenolic and hydroxybenzoic phase II metabolites reached concentrations of over 20 μM in some volunteers. Moreover, de Ferrars et al. (2014) discovered the final metabolite, the hippuric acid, in circulation up to 24 hours and reaching concentrations well above 25 μM.

According to the endogenous production or dietary exposure to parent compounds, some LMW (poly)phenol metabolites are already described in the human metabolome database (HMDB) (Table 4.1) (Wishart et al. 2018). Besides that, some of them have an endogenous production and from these similarities with endogenous compounds, we could expect already a biological function and/or an interference in a specific biological pathway. Additionally, Benzene diol metabolites appear mentioned as substrates or products from catechol-*O*-methyltransferase (COMT) involved in the metabolism of catecholamine neurotransmitters and catechol hormones, (Kanehisa et al. 2016) while benzene-1,2,3-triol (pyrogallol) and benzene-1,3,5-triol (phloroglucinol) are inhibitors of the same enzyme (Hattori et al. 1969; Bonifácio et al. 2007; Kanehisa et al. 2016). Besides that, phenylacetic acids are also the class of phenolics metabolites with more evidences described in the HMDB as being detected in the human brain (Wishart et al. 2018), although not much more is described concerning their disposition within human organs and pathways that could be involved (Table 4.1).

Overall, these LMW (poly)phenol metabolites represent a vast array of molecules with favorable absorption, distribution, metabolism, and excretion (ADME) properties, known to be present in relevant concentrations, show no reported signs of toxicity at such concentrations and stay in circulation for long periods of time (Ofosu et al. 2020; Cladis et al. 2020). For these reasons, LMW (poly)phenol metabolites are believed to be of interest to study their potential on chronic diseases and, in particular, NDs. Evidences of (poly)phenol metabolites having anti-inflammatory effects on systemic immune cells support the idea

Table 4.1 LMW (poly)phenol metabolites resulting from dietary sources rich in (poly)phenolic content, mainly flavonoids: Molecular weight, presence in plasma or excreted in urine, disposition within human organs and pathways that are affected after dietary ingestion by those compounds.

Family	Recommended name[a] (common name)	Structure	Detected	MW (g/mol)	References[b]	Organs detected (HMD)[c]	Pathways that could be involved (KEGG)[d]
Benzene diols and triols	Benzene-1,2-diol (catechol)	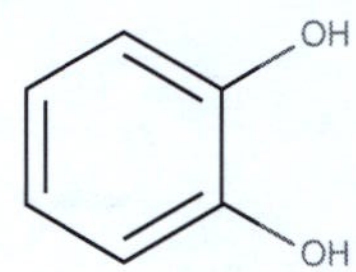	Urine	110.11	Loke et al. (2009)	Intestine/bone marrow/brain/liver/prostate	Predicted to be involved in COMT activity in the metabolism of catecholamine neurotransmitters and catechol hormones
	Phenol-2-sulfate (catechol-sulfate)	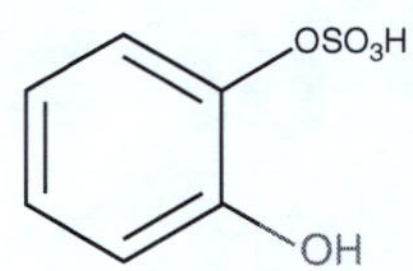	Plasma	190.17	Pimpão et al. (2015)	—	—
	2-methoxyphenol (guaiacol)	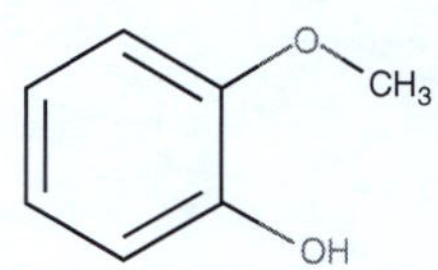	Plasma	124.14	Cialdella-Kam et al. (2013)	—	Predicted to be involved in COMT activity in the metabolism of catecholamine neurotransmitters and catechol hormones
	Methoxybenzene-2-sulfate (guaiacol-sulfate)	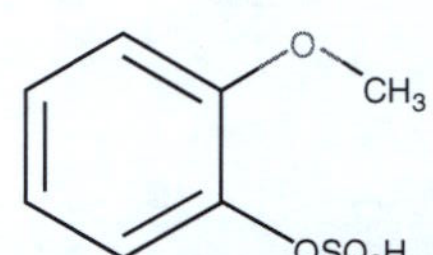	Urine	204.20	van der Hooft et al. (2012)	—	—

(*Continued*)

Table 4.1 (Continued)

Family	Recommended name[a] (common name)	Structure	Detected	MW (g/mol)	References[b]	Organs detected (HMD)[c]	Pathways that could be involved (KEGG)[d]
	Benzene-1,3-diol (resorcinol)	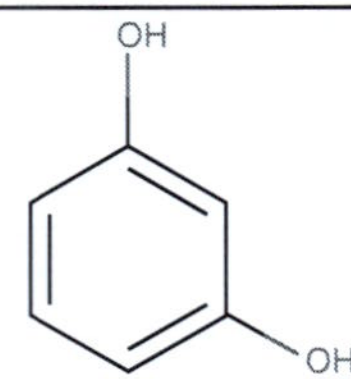	Plasma	110.11	Welsch (2008)	—	—
	Phenol-3-sulfate (resorcinol-sulfate)	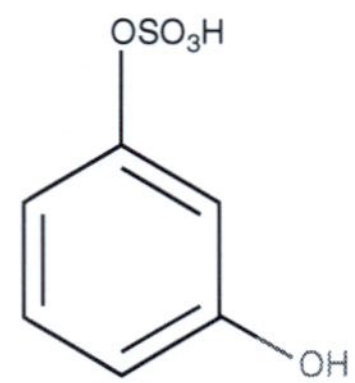	Plasma	190.17	Curzon and Pratt (1964)	—	—
	Benzene-1,2,3-triol (pyrogallol)	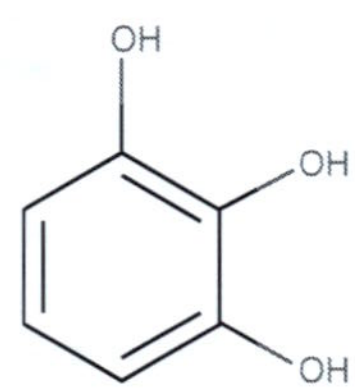	Plasma	126.11	Guo and Li (2009)	—	COMT inhibitor
	Benzene-1,2-diol-3-sulfate (pyrogallol-1-sulfate)	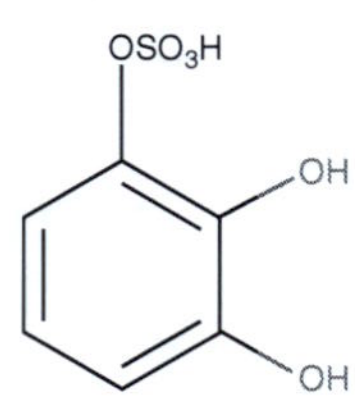	Plasma	206.17	Pimpão et al. (2015)	—	—

Benzene-1,2-diol-3-glucuronide (pyrogallol-1-glucuronide)	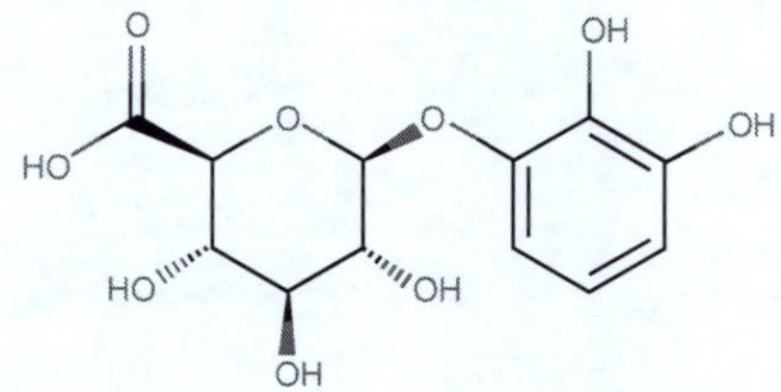	Urine	302.24	van der Hooft et al. (2012)	Liver/Kidney	—
3-methoxybenzene-1,2-diol (1-methylpyrogallol)	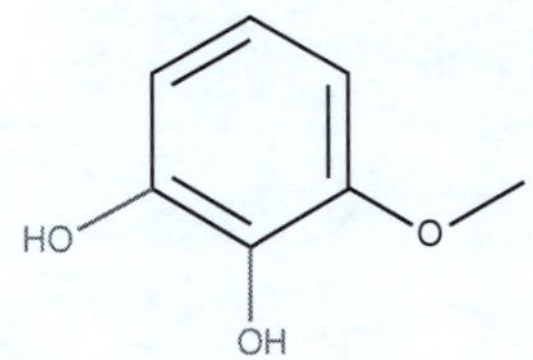	Urine	140.14	Toshimitsu et al. (1981)	—	—
2-hydroxy-1-methoxybenzene-3-sulfate (1-methylpyrogallol-3-sulfate)	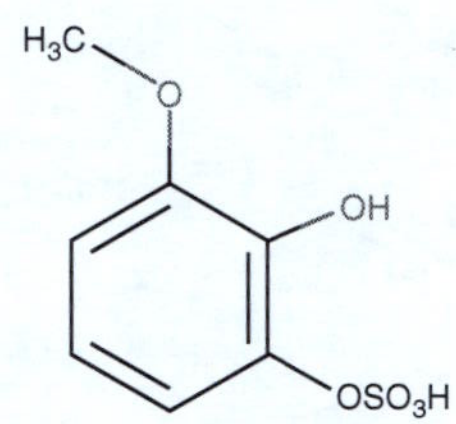	Plasma	220.20	Pimpão et al. (2015)	—	—
2-methoxybenzene-1,3-diol (2-methylpyrogallol)	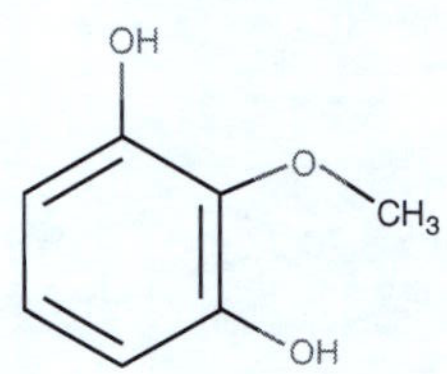	Urine	140.14	Toshimitsu et al. (1981)	—	—
2-methoxyphenol-3-sulfate (2-methylpyrogallol-sulfate)	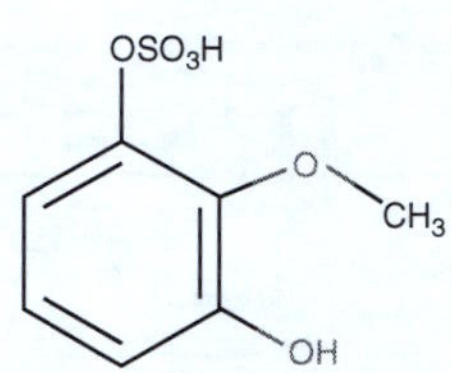	Plasma	220.20	Pimpão et al. (2015)	—	—

(*Continued*)

Table 4.1 (Continued)

Family	Recommended name[a] (common name)	Structure	Detected	MW (g/mol)	References[b]	Organs detected (HMD)[c]	Pathways that could be involved (KEGG)[d]
	Benzene-1,3-diol-5-sulfate (phloroglucinol)	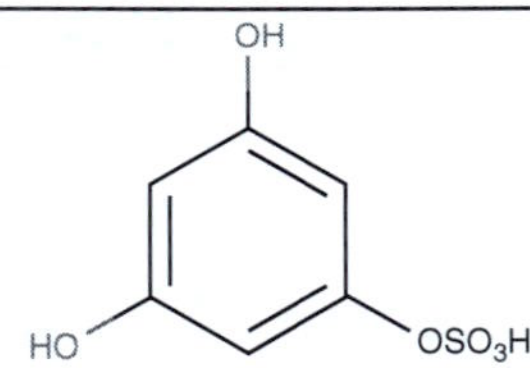	Plasma	206.17	Baldrick et al. (2018)	—	Phloroglucinol is described as COMT inhibitor
	1,3,5-trimethoxybenzene (trimethoxyphloroglucinol)	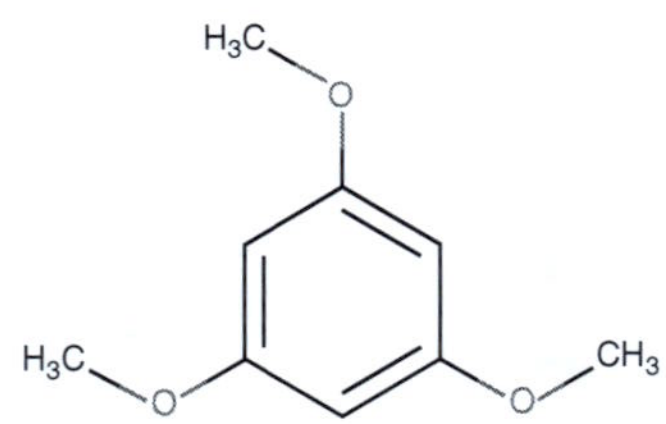	Plasma	168.19	Loke et al. (2009)	—	—
Benzalde-hydes	4-hydroxybenzaldehyde	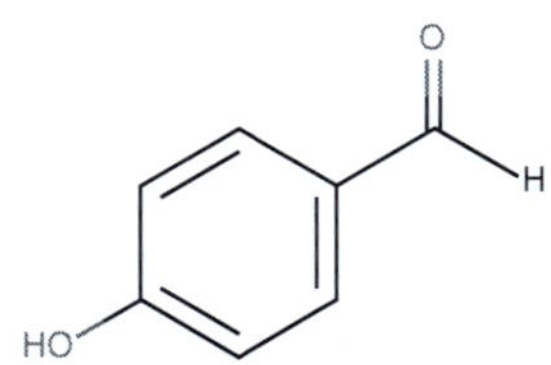	Plasma	122.12	de Ferrars et al. (2014)	—	—
	3,4-dihydroxybenzaldehyde (protocatechualdehyde)	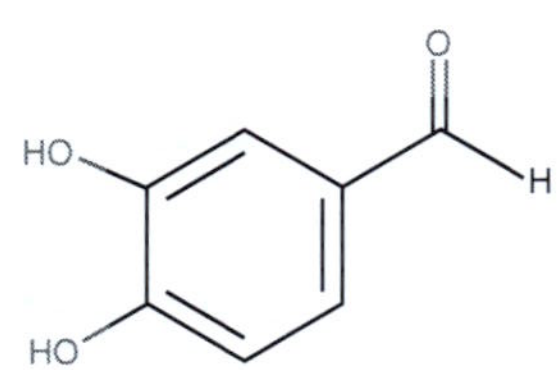	Plasma	138.12	de Ferrars et al. (2014)	—	—

	4-hydroxy-3-methoxybenzaldehyde (vanillin)	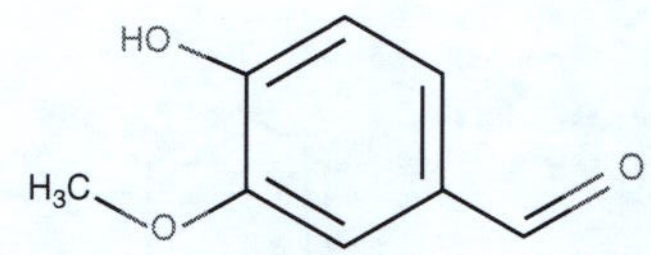	Plasma	152.15	Loke et al. (2009)	—	—
	2,4,6-trihydroxybenzaldehyde (phloroglucinaldehyde)	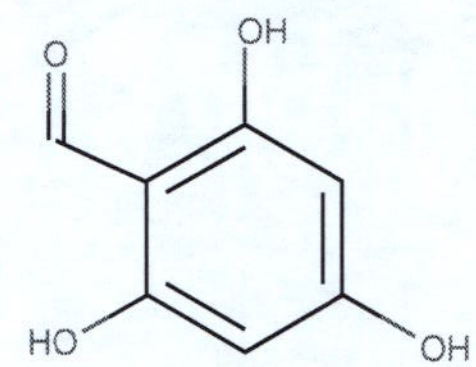	Plasma	154.12	de Ferrars et al. (2014)	—	—
Benzoic acids	Benzoic acid	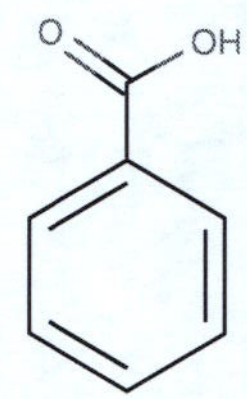	Plasma	122.12	de Ferrars et al. (2014)	Testicle/ Bladder/Liver/ Skin/Kidney	—
	2-hydroxybenzoic acid (salicylic acid)	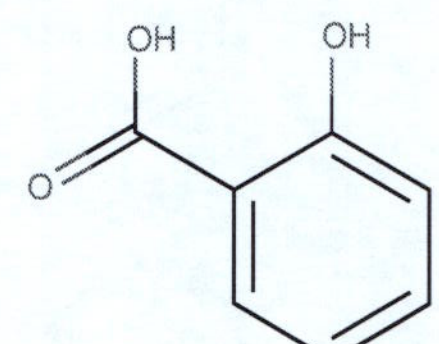	Plasma	138.12	Blacklock et al. (2001)	Liver/Skin	—
	3-hydroxybenzoic acid	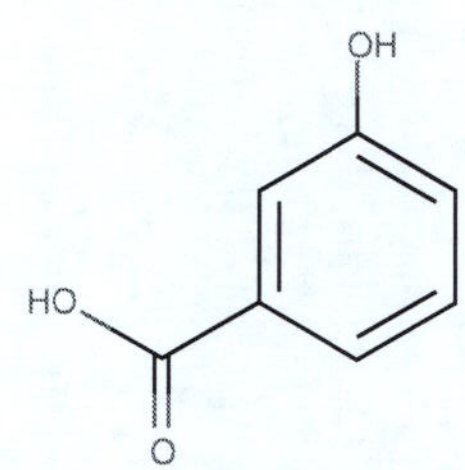	Plasma	138.12	de Ferrars et al. (2014)	—	—

(*Continued*)

Table 4.1 (Continued)

Family	Recommended name[a] (common name)	Structure	Detected	MW (g/mol)	References[b]	Organs detected (HMD)[c]	Pathways that could be involved (KEGG)[d]
	4-hydroxybenzoic acid	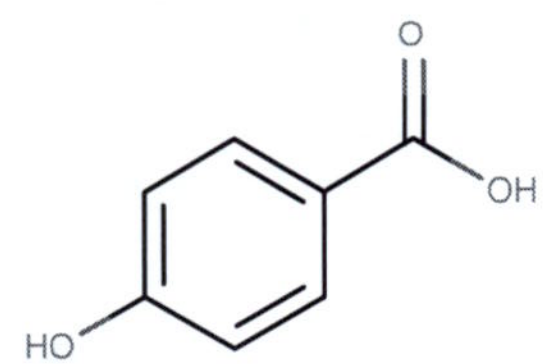	Plasma	138.12	de Ferrars et al. (2014)	Cerebrospinal Fluid	—
	Benzoic acid-4-glucuronide	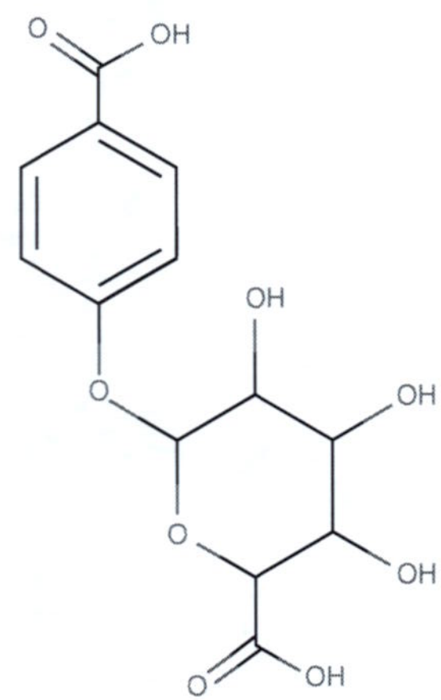	Plasma	314.25	de Ferrars et al. (2014)	Kidney/Liver	—
	3,4-dihydroxybenzoic acid (protocatechuic acid)	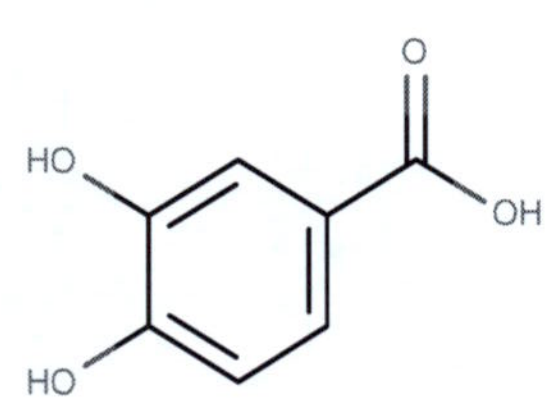	Plasma	154.12	de Ferrars et al. (2014)	Testicle	—

4-hydroxybenzoic acid-3-sulfate (protocatechuic acid-3-sulfate)	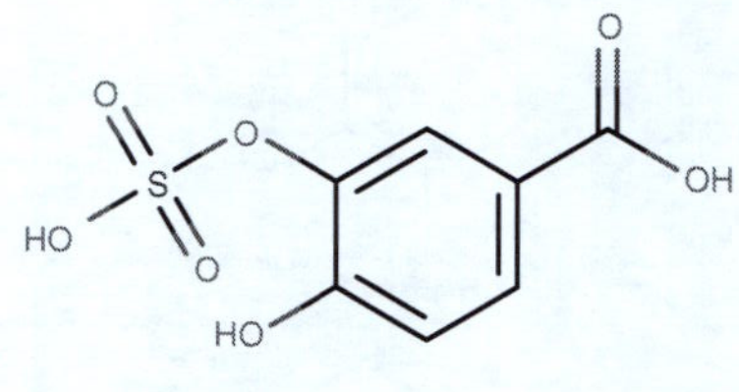	Plasma	234.18	de Ferrars et al. (2014)	—	—
4-hydroxybenzoic-3-glucuronide (protocatechuic acid-3-glucuronide)	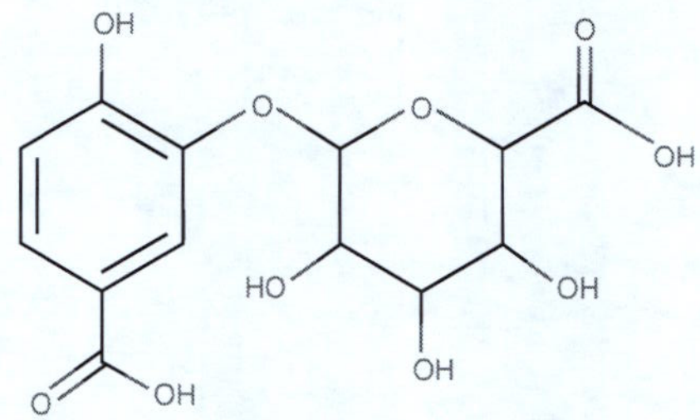	Plasma	330.25	de Ferrars et al. (2014)	—	—
3,4-dimethoxybenzoic acid (veratric acid)	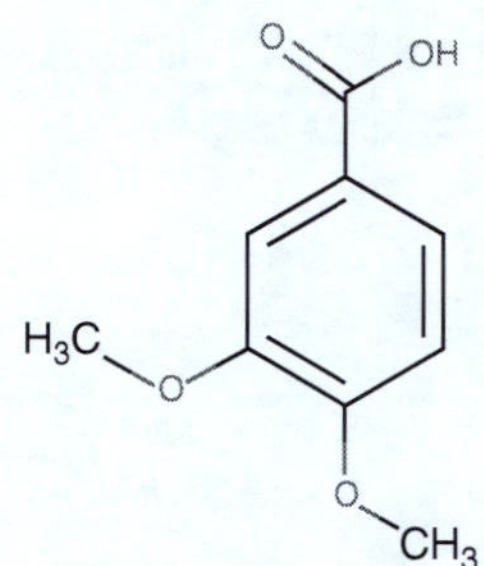	Plasma	182.18	Anson et al. (2010)	—	—
3-hydroxy-4-methoxybenzoic acid (isovanillic acid)	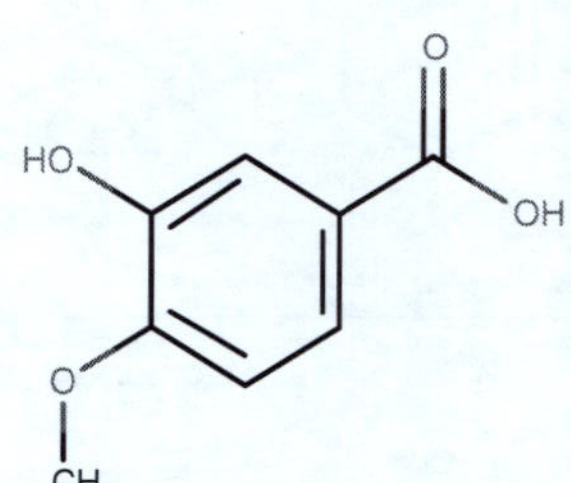	Plasma	168.15	de Ferrars et al. (2014)	—	—

(*Continued*)

Table 4.1 (Continued)

Family	Recommended name[a] (common name)	Structure	Detected	MW (g/mol)	References[b]	Organs detected (HMD)[c]	Pathways that could be involved (KEGG)[d]
	4-methoxybenzoic acid-3-sulfate (isovanillic acid-sulfate)	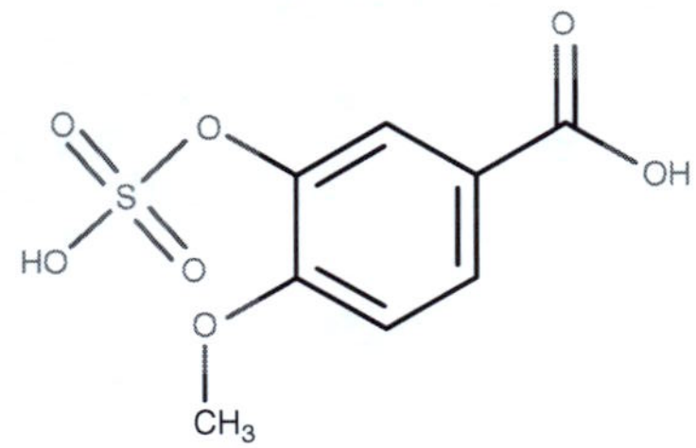	Plasma	248.21	de Ferrars et al. (2014)	—	—
	4-methoxybenzoic acid-3-glucuronide (isovanillic acid-glucuronide)	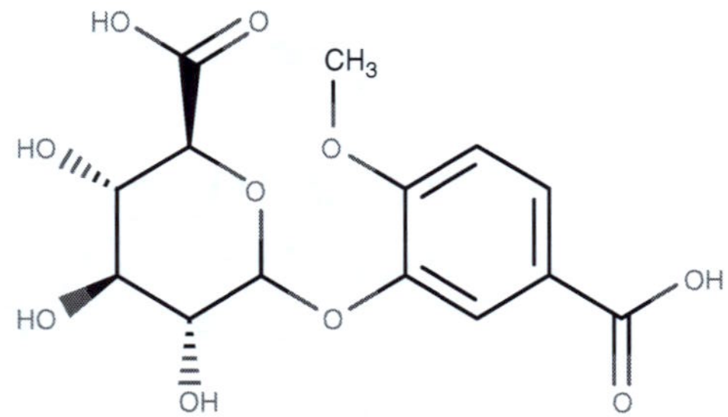	Plasma	344.27	de Ferrars et al. (2014)	—	—
	4-hydroxy-3-methoxybenzoic acid (vanillic acid)	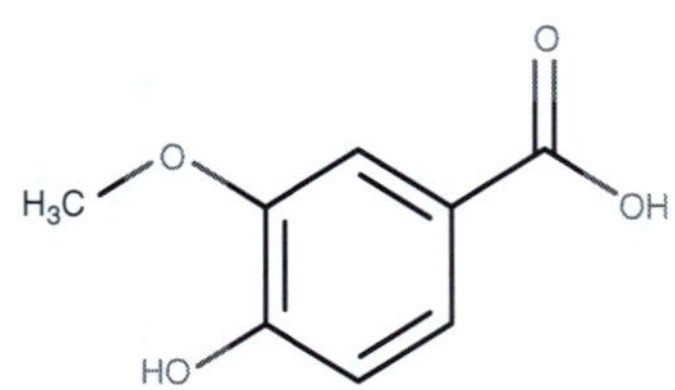	Plasma	168.15	de Ferrars et al. (2014)	—	—
	3-methoxybenzoic acid-4-sulfate (vanillic acid-sulfate)	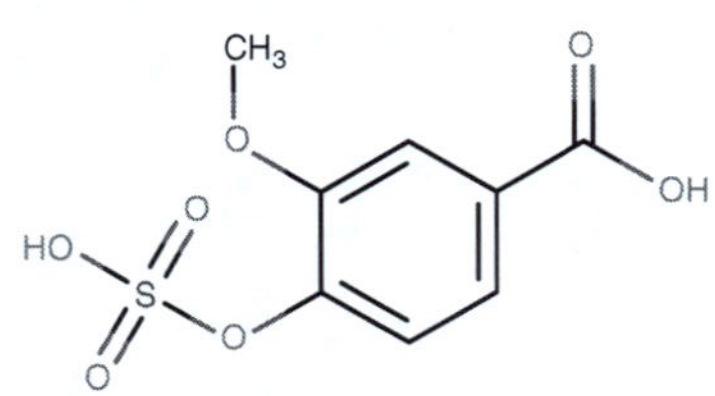	Plasma	248.21	de Ferrars et al. (2014)	—	—

3-methoxybenzoic acid-4-glucuronide (vanillic acid-glucuronide)	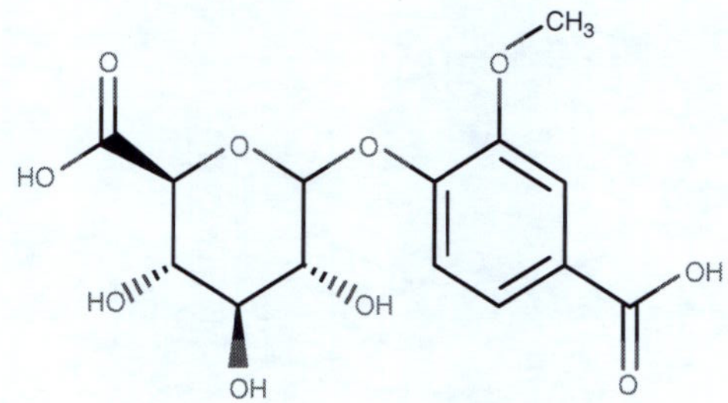	Plasma	344.27	de Ferrars et al. (2014)		—
2,4-dihydroxybenzoic acid	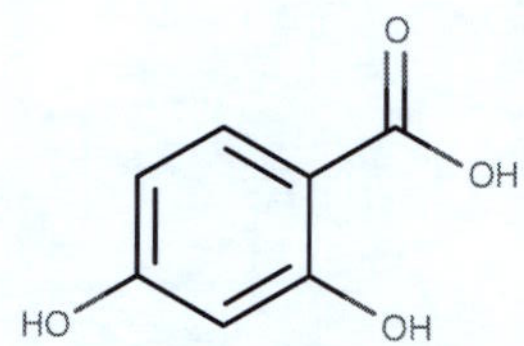	Plasma	154.12	Boto-Ordóñez et al. (2013)	—	—
2,5-dihydroxybenzoic acid (gentisic acid)	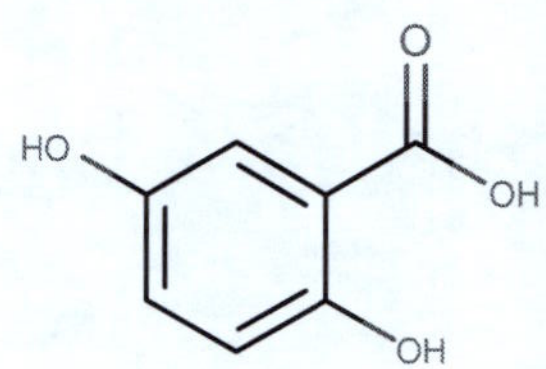	Plasma	154.12	Perez-Ternero et al. (2017)	—	Product of tyrosine metabolism
2,6-dihydroxybenzoic acid	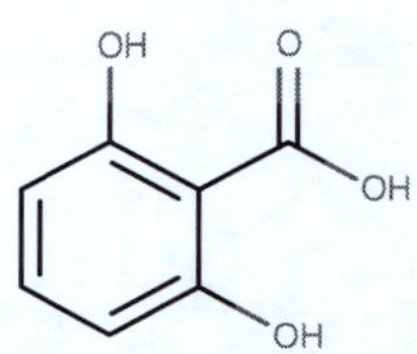	Plasma	154.12	Perez-Ternero et al. (2017)	—	—

(*Continued*)

Table 4.1 (Continued)

Family	Recommended name[a] (common name)	Structure	Detected	MW (g/mol)	References[b]	Organs detected (HMD)[c]	Pathways that could be involved (KEGG)[d]
	3,5-dihydroxybenzoic acid	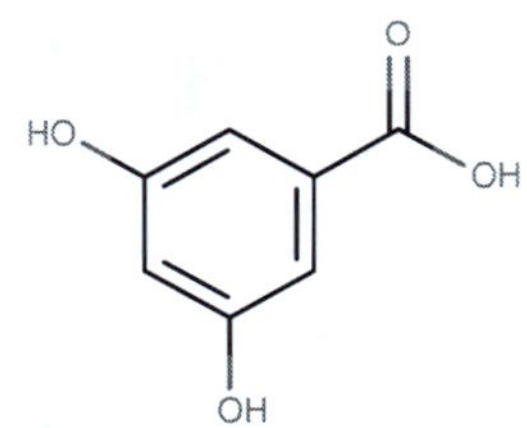	Plasma	154.12	Perez-Ternero et al. (2017)	—	—
	3,4,5-trihydroxybenzoic acid (gallic acid)	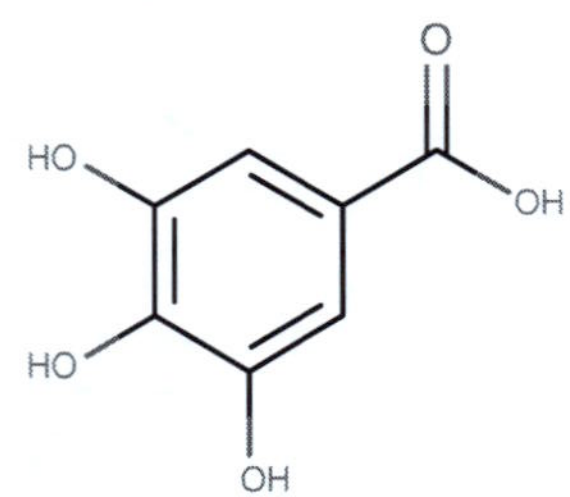	Plasma	170.12	de Ferrars et al. (2014)	—	—
	3,4-dihydroxy-5-methoxybenzoic acid (3-methylgallic acid)	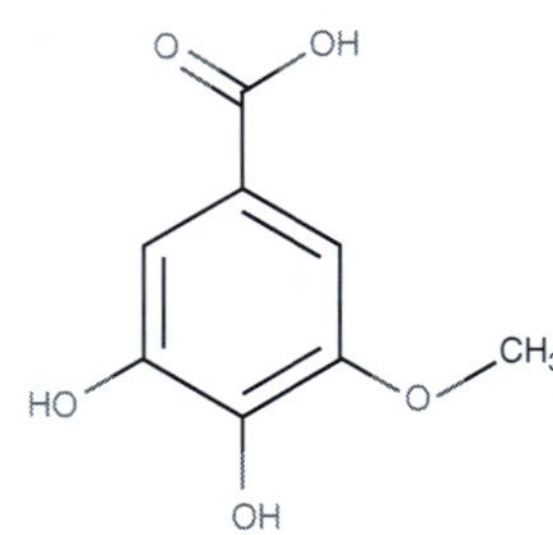	Urine	184.15	Loke et al. (2009)	—	—

3,5-dihydroxy-4-methoxybenzoic acid (4-methylgallic acid)	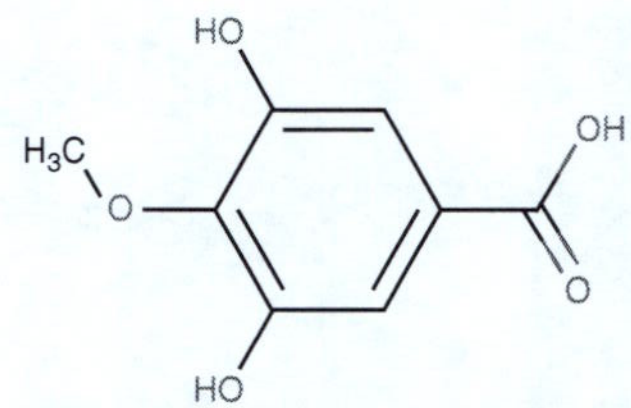	Plasma	184.15	de Ferrars et al. (2014)	—	—
3-hydroxy-4-methoxybenzoic acid-5-sulfate (4-methylgallic acid-sulfate)	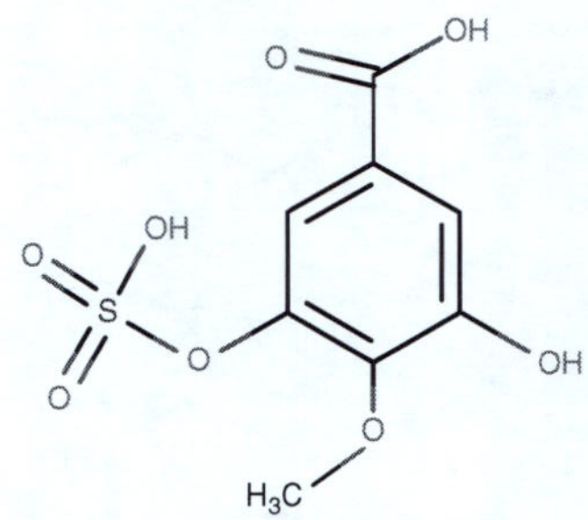	Plasma	360.27	Pimpão et al. (2015)	—	—
3,5-dimethoxy-4-hydroxybenzoic acid (syringic acid)	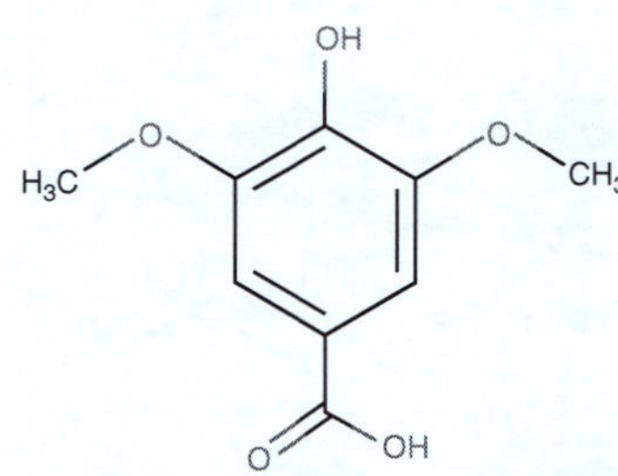	Plasma	198.17	de Ferrars et al. (2014)	—	—

(*Continued*)

Table 4.1 (Continued)

Family	Recommended name[a] (common name)	Structure	Detected	MW (g/mol)	References[b]	Organs detected (HMD)[c]	Pathways that could be involved (KEGG)[d]
	Ethyl 3,4,5-trihydroxybenzoate (ethyl gallate)	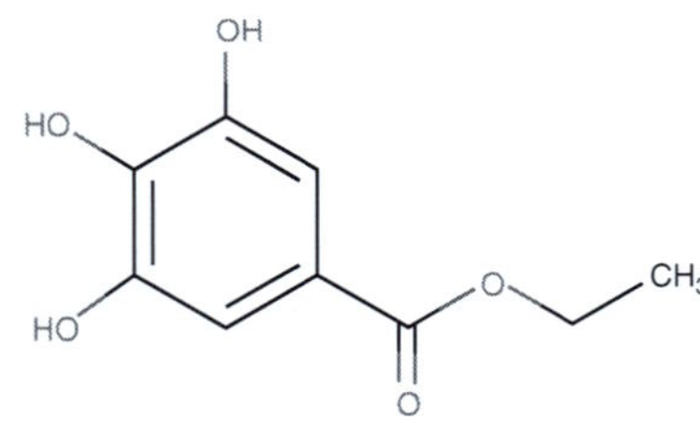	Plasma	198.17	Boto-Ordóñez et al. (2013)	—	—
Phenyl-acetic acids	Phenylacetic acid	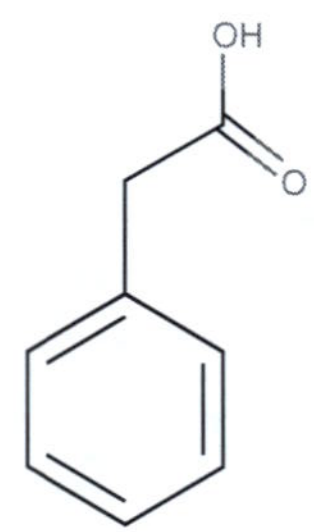	Plasma	136.15	Loke et al. (2009)	Cerebrospinal fluid	Product of phenylalanine metabolism
	2′-methoxyphenylacetic acid	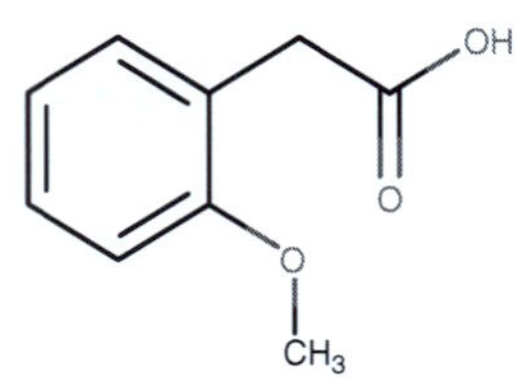	Urine	166.18	Loke et al. (2009)	—	—
	3′-hydroxyphenylacetic acid	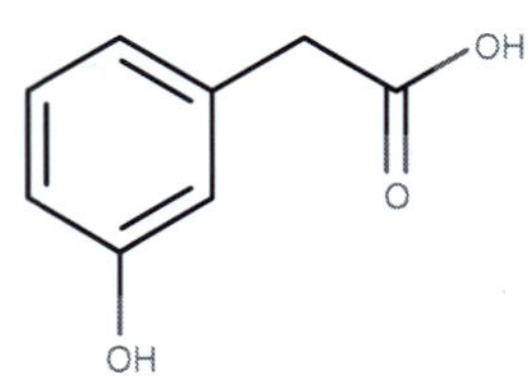	Urine	152.15	van Dorsten et al. (2010)	—	Product of phenylalanine metabolism

	4′-hydroxyphenylacetic acid	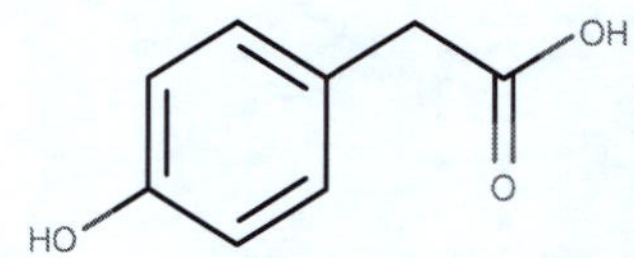	Plasma	152.15	de Ferrars et al. (2014)	Bladder/ Epidermis/ Kidney/Prostate	Product of tyrosine metabolism
	3′,4′-dihydroxyphenylacetic (homoprotocatechuic acid, DOPAC)	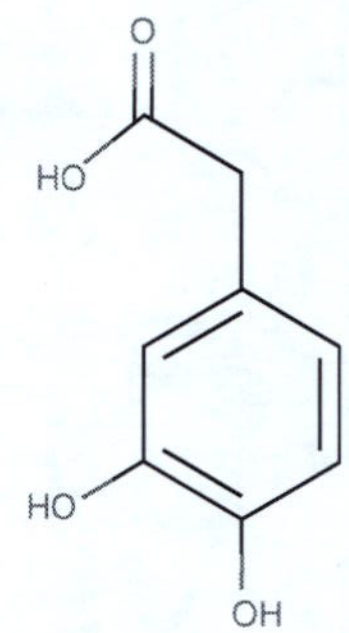	Plasma	168.15	de Ferrars et al. (2014)	Brain	Product of dopamine metabolism by monoamine oxidase (MAO) and further metabolized by COMT
	4′-hydroxy-3′-methoxyphenylacetic acid (homovanillic acid)	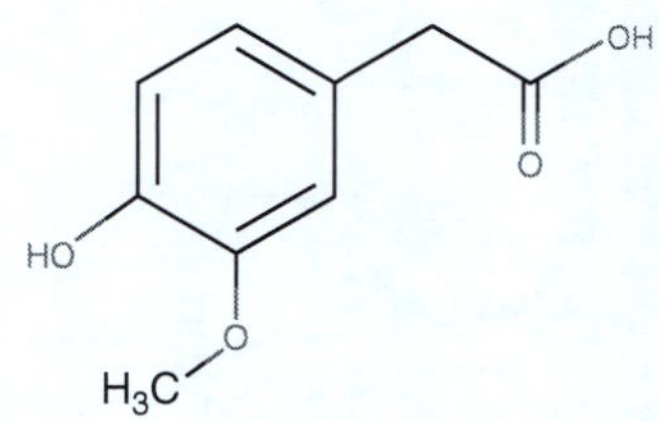	Urine	182.18	van Dorsten et al. (2010)	Brain/Central Nervous System/Kidney/ Spinal Cord	Product of dopamine metabolism after DOPAC metabolism by COMT
Mandelic acids	2-hydroxy-2-(phenyl)acetic acid	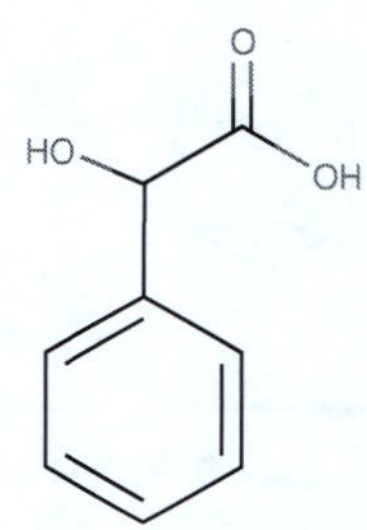	Urine	152.15	Olthof et al. (2003)	—	—

(*Continued*)

Table 4.1 (Continued)

Family	Recommended name[a] (common name)	Structure	Detected	MW (g/mol)	References[b]	Organs detected (HMD)[c]	Pathways that could be involved (KEGG)[d]
	2-hydroxy-2-(4′hydroxyphenyl) acetic acid	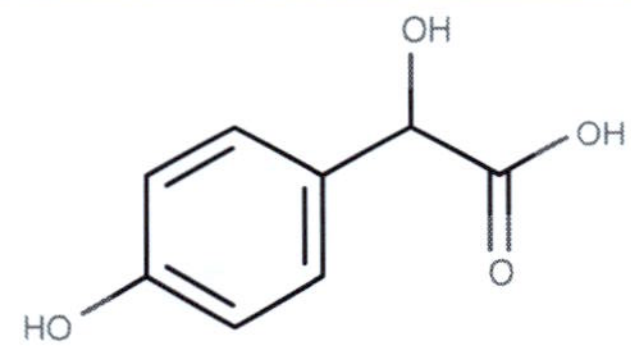	Urine	168.15	van Dorsten et al. (2010)	Retina/Vitreous humor	—
	2-hydroxy-2-(4′hydroxy-3′-methoxyphenyl)acetic acid	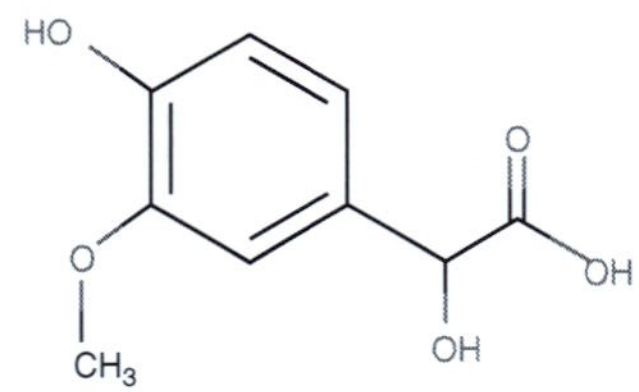	Urine	198.18	van Dorsten et al. (2010)	Adrenal Gland/ Epidermis/ Thyroid Gland	Product of tyrosine metabolism
Hippuric acids	Hippuric acid	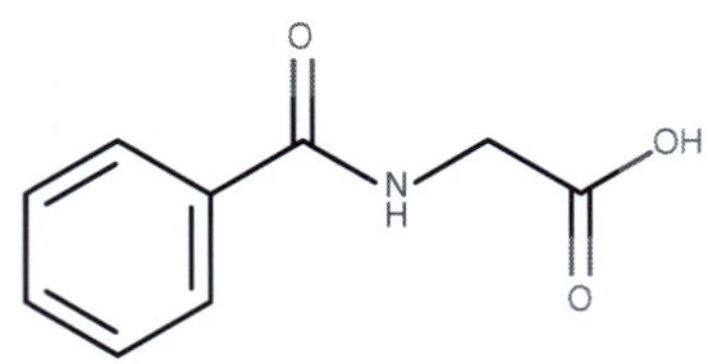	Plasma	179.18	Clifford et al. (2000) and de Ferrars et al. (2014)	Kidney/Liver/ Prostate	Product of phenylalanine metabolism
	3′-hydroxyhippuric acid	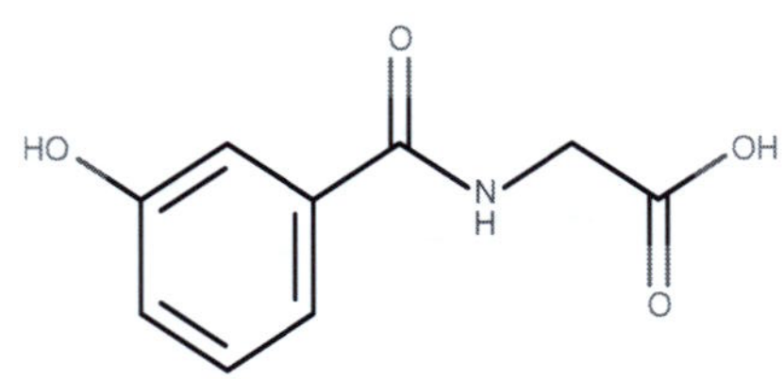	Plasma	195.17	Guertin et al. (2015)	—	—
	4′-hydroxyhippuric acid	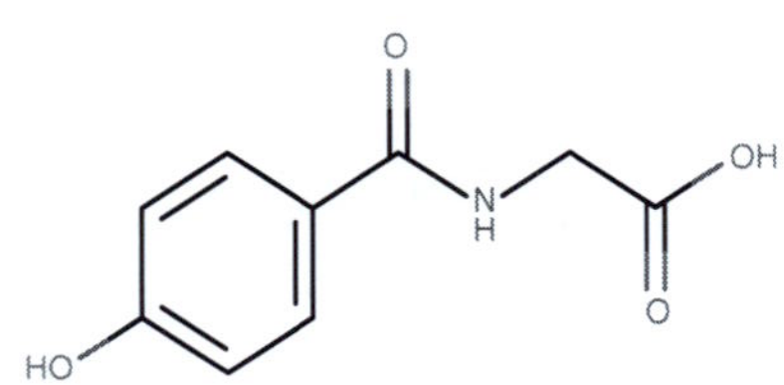	Urine	195.17	van Dorsten et al. (2010)	—	—

Cinnamic acids	Cinnamic acid	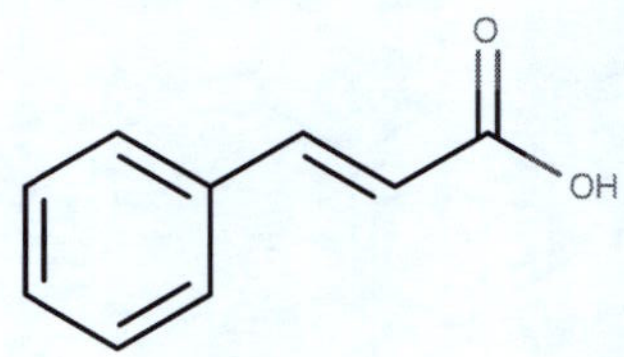	Plasma	148.16	Loke et al. (2009)	Liver	—
	3′-hydroxycinnamic acid (m-coumaric acid)	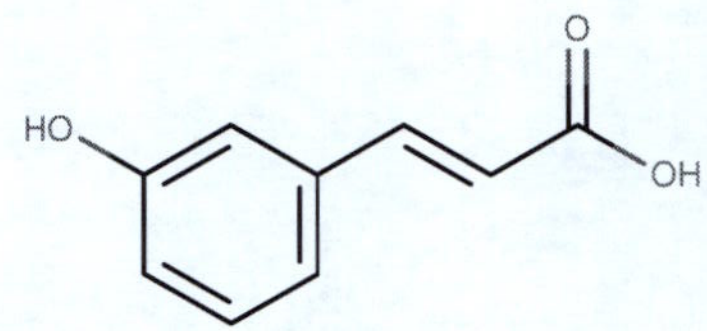	Urine	164.16	Zamora-Ros et al. (2016)	—	—
	4′-hydroxycinnamic acid (p-coumaric acid)	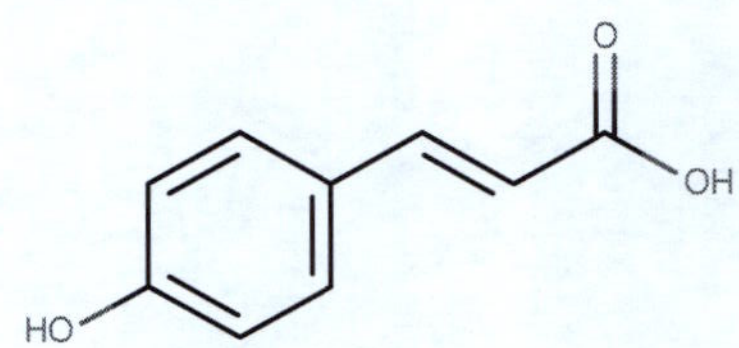	Urine	164.16	Loke et al. (2009)	—	—
	2′-hydroxycinnamic acid (o-coumaric acid)	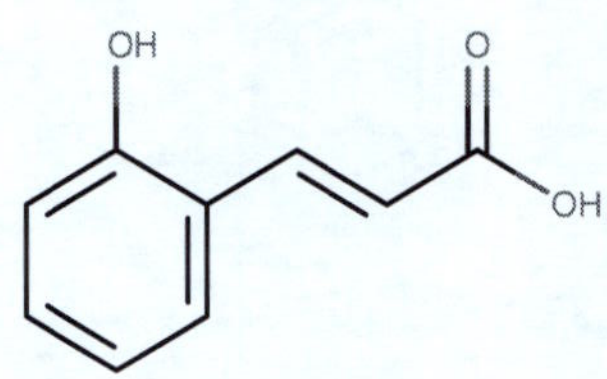	Urine	164.16	Loke et al. (2009)	—	—
	3′4′-dihydroxycinnamic acid (caffeic acid)	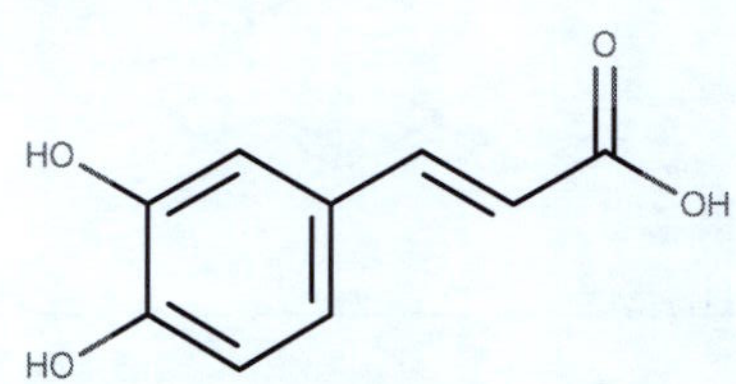	Plasma	180.16	de Ferrars et al. (2014)	Prostate	—

(Continued)

Table 4.1 (Continued)

Family	Recommended name[a] (common name)	Structure	Detected	MW (g/mol)	References[b]	Organs detected (HMD)[c]	Pathways that could be involved (KEGG)[d]
	3′-hydroxy-4′methoxycinnamic acid (ferulic acid)	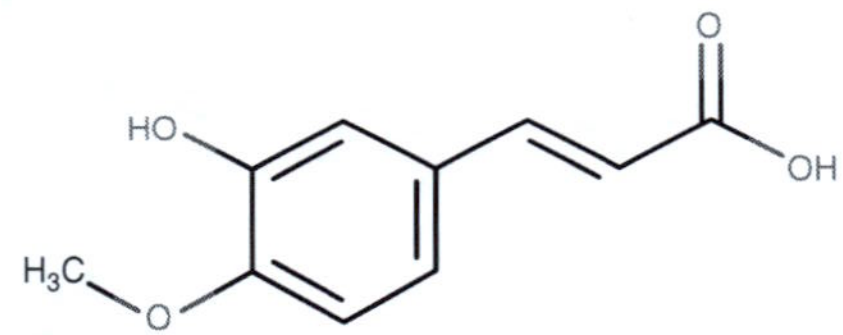	Plasma	194.19	Guy et al. (2009)	—	—
	4′-hydroxy-3′methoxycinnamic acid (isoferulic acid)	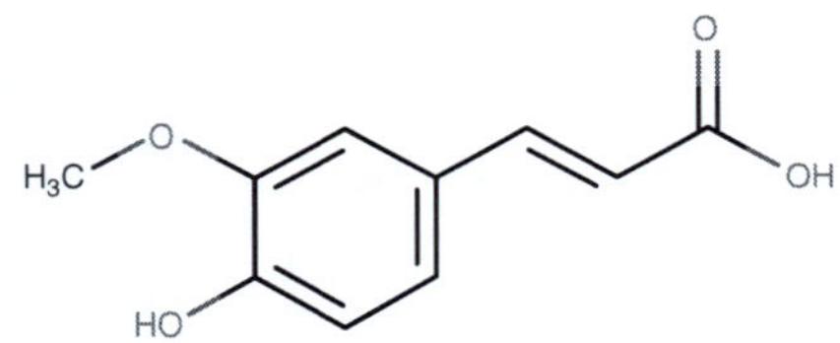	Plasma	194.19	de Ferrars et al. (2014)	—	—
	3′,5′-dimethoxy-4′-hydroxycinnamic acid (sinapic acid)	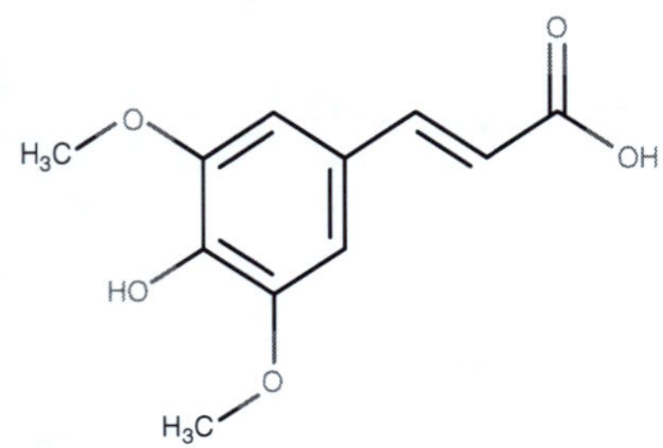	Urine	224.21	Kern et al. (2003)	—	—
Phenyl-propionic acids	3-(3′-hydroxyphenyl)propanoic acid	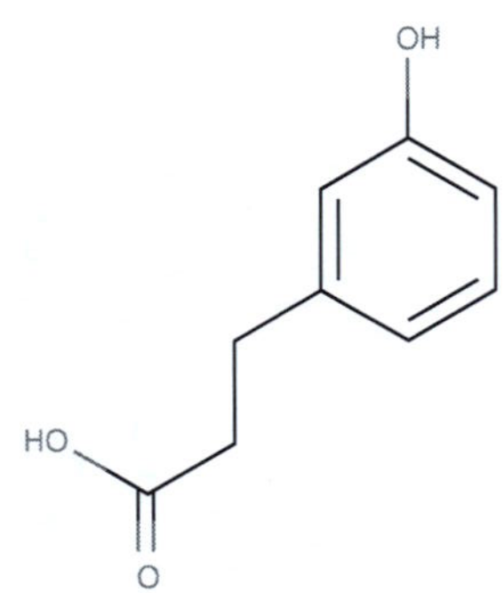	Urine	166.18	Guertin et al. (2015)	—	—

3-(4′-hydroxyphenyl)propanoic acid	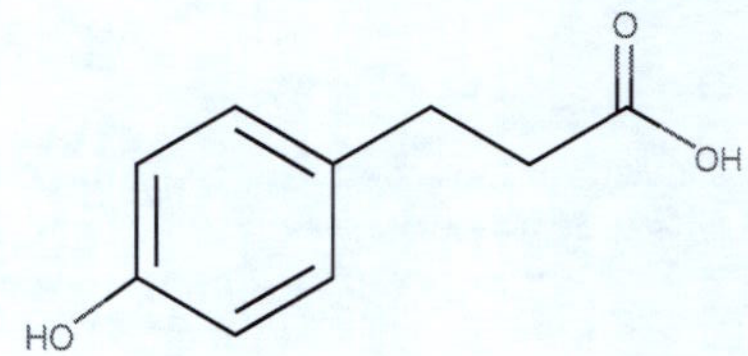	Urine	166.18	Loke et al. (2009)	Kidney/Liver/ Spleen	—
3-(3′,4′-dihydroxyphenyl)propanoic acid (dihydrocaffeic acid)	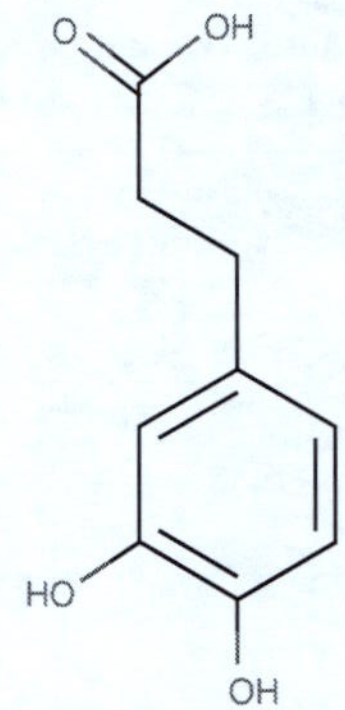	Urine	182.18	van Dorsten et al. (2010)	—	—
3-(4′-hydroxy-3′-methoxyphenyl) propanoic acid (dihydroferulic acid)	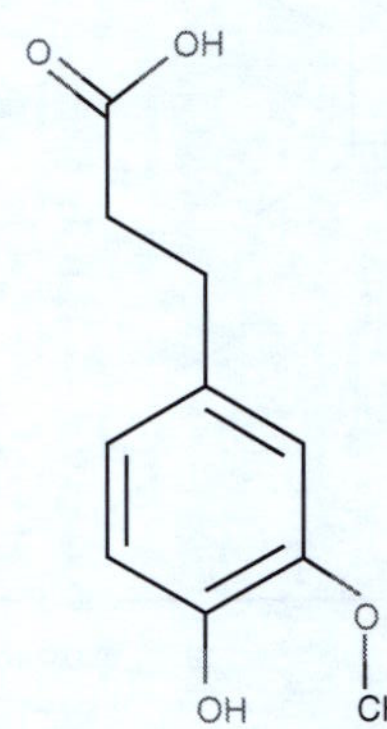	Urine	196.20	van Dorsten et al. (2010)	—	—

(*Continued*)

Table 4.1 (Continued)

Family	Recommended name[a] (common name)	Structure	Detected	MW (g/mol)	References[b]	Organs detected (HMD)[c]	Pathways that could be involved (KEGG)[d]
	3-(3′-hydroxy-4′-methoxyphenyl) propanoic acid (dihydroisoferulic acid)	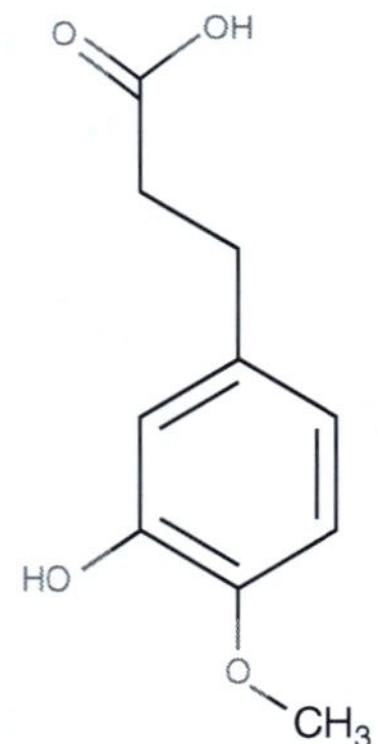	Urine	169.20	Roowi et al. (2009)	—	—
Phenylhydracrylic acid	3-hydroxy-3-(3′-hydroxyphenyl) propanoic acid	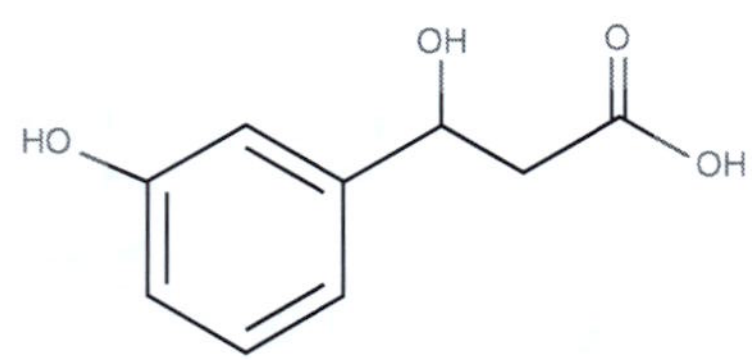	Urine	182.18	Del Rio et al. (2010)	—	—

BW, body weight; COMT, catechol-O-methyltransferase; DOPAC, 3′,4′-dihydroxyphenylacetic acid; MAO, monoamine oxidase.

[a] Recommended names for dietary (poly)phenol metabolites according to Kay et al. (2020). Missing recommended names were filed using analogy.

[b] References of molecular weight, and detected (poly)phenols in the urine or plasma.

[c] The Human Metabolome Database (HMDB) (http://www.hmdb.ca/) was used for searching available information concerning metabolite source (endogenous or by exposure) and the organs were being described to be detected [https://doi.org/10.1093/nar/gkx1089].

[d] Kyoto Encyclopedia of Genes and Genomes (KEGG) database (https://www.kegg.jp/kegg/) was used to localize the metabolite in the KEGG pathways using the organism-specific pathway: *Homo sapiens* (human) [https://doi.org/10.1093/nar/gkv1070]. The presence of the enzymes in humans was confirmed in BRENDA database (https://www.brenda-enzymes.org) (Chang et al. 2021).

that (poly)phenol metabolites do not need to reach the brain to potentially have an effect (Cunha et al. 2019) and multi-organ interactions might also lead to such a conclusion (Schaffer and Halliwell 2012). However, the potential effects of (poly)phenol metabolites are still only marginally addressed. Still, once in circulation, its distribution to target tissues, such as the brain, relies on its ability to cross the blood-brain barrier (BBB), which is still largely unknown.

4.3 (Poly)phenol Metabolites and Their Brain Permeability

The brain is known as the center of information in the human body. Signals from the external environment are integrated into the brain and will trigger a specific response to ensure the homeostasis of the whole organism. Considering the intense activity that occurs in our neuronal cells, it is essential that its chemical environment stays controlled and regulated. One important structure that ensures this regulation is the BBB. BBB is a very important and highly semipermeable barrier, which is found in most of the blood vessels that irrigate the brain (Dando et al. 2014). It regulates the passage of more or less complex molecules that intend to penetrate the brain, counting many protein channels distributed across the barrier. In those vessels, we can find cerebrovascular endothelial cells that coat their walls and maintain attachment with each other by the presence of tight junctions and small proteins that are crucial to restrict the passage by diffusion across the blood vessels (Greene and Campbell 2016). In addition to these, there are the pericytes, which participate in the formation of this barrier covering the endothelial cells, having the ability to contract themselves helping in the regulation of the blood flow, and the astrocytes that also play an important role by projecting their end-feet to the walls of vessels (Figure 4.3) (Sofroniew and Vinters 2010; Greene and Campbell 2016). In normal conditions, all these players ensure the regulation of the passage of ions, molecules, and cells between the circulating blood and the extracellular space of the CNS. This prevents the entrance of toxins, pathogenic agents, and other potentially harmful molecules that could impair brain homeostasis. On the other hand, in a chronic disease context, usually there is an increase in the inflammatory status that could promote the infiltration of viruses, bacteria, and an unbalance of certain ions and molecules in the brain because of the increase of BBB permeability (Li et al. 2015).

After ingestion of food rich in (poly)phenols, phenolic metabolites that result from phase I and II metabolism and reach the systemic circulation, as many other molecules toward the brain will face this semipermeable barrier that acts as a border between them and the neuronal tissue. However, to exert direct effects, they must cross the BBB.

Several evidences have been emerging showing the potential of several (poly)phenol metabolites in reaching the brain. Through different approaches ranging from *in silico*, *in vitro*, and *in vivo* studies, metabolite permeation has been explored. In the literature, we can find some examples where the studies *in silico* were validated in both *in vitro* and *in vivo* approaches. Yet, for most of them, some results have been published pointing in different ways, relying on their experimental designs. In this sense, a direct comparison between studies may not be made. Regarding *in silico* models, we should be aware that they just consider passive permeation as the only way to cross the BBB, but it is described that other possibilities should also be considered *in vivo*. Nevertheless, *in vivo* studies,

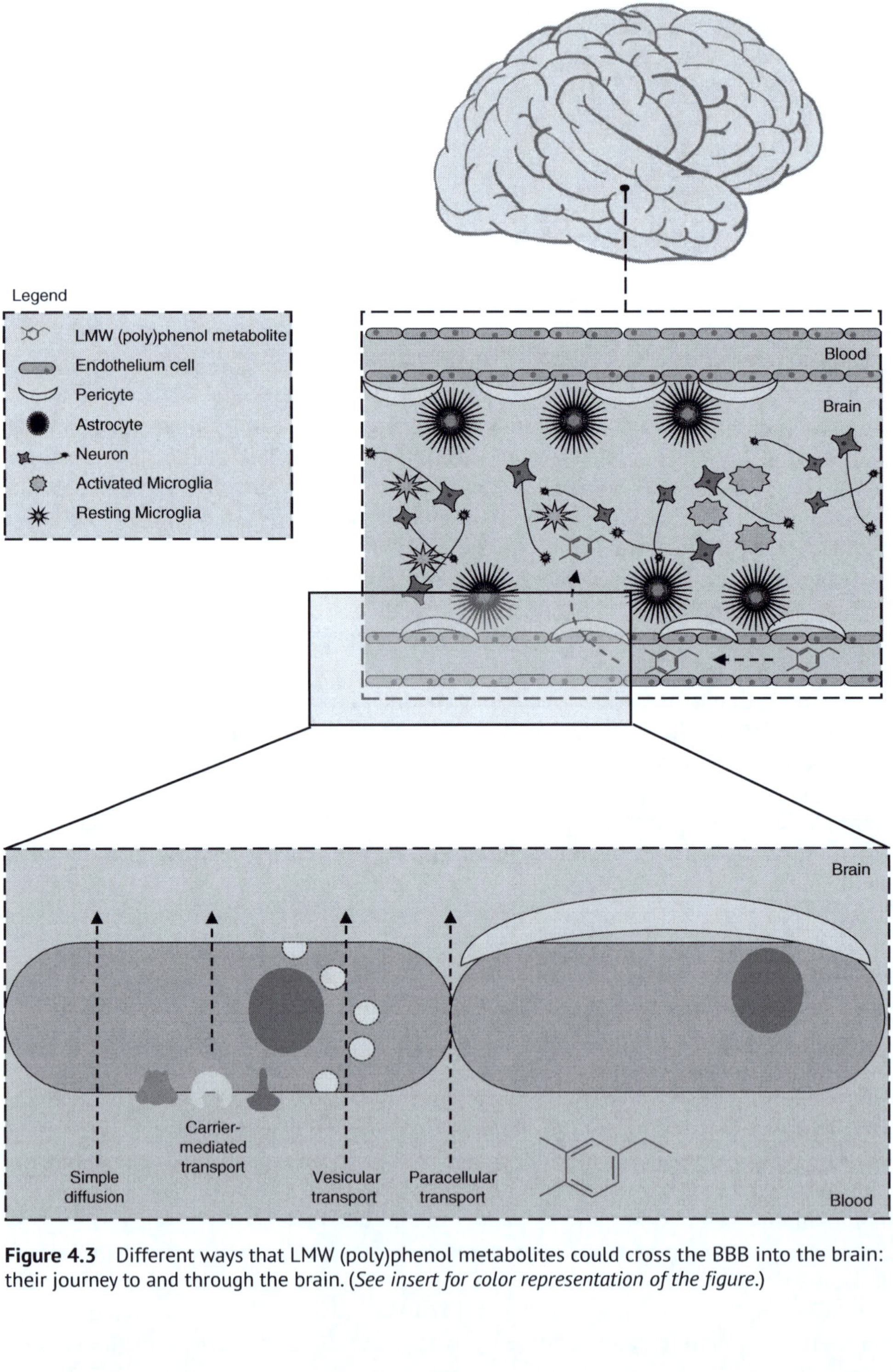

Figure 4.3 Different ways that LMW (poly)phenol metabolites could cross the BBB into the brain: their journey to and through the brain. (*See insert for color representation of the figure.*)

although they represent those that fully recapitulate the complexity of the BBB, they differ in several aspects of the study design: source of (poly)phenols, the type of administration, the time of collecting the tissues for analysis after the last (poly)phenol administration, the concentrations used, the frequency of the procedures and the strain and sex of the animals. In Table 4.2, a summary of available data is illustrated for several (poly)phenol metabolites related to their ability to cross the BBB, considering *in silico*, *in vitro*, and *in vivo* approaches.

The mechanisms by which phenolic metabolites permeate the BBB into the brain are not fully understood. In fact, Youdim and co-workers have shown that the multiple structures provide phenolic metabolites with different ways of crossing the barrier, depending on their level of lipophilicity (Youdim et al. 2003, 2004). Passive permeation and carrier-mediated transport could constitute potential mechanisms by which these metabolites reach the brain (Figure 4.3). Several transporters have been described in the BBB, being part of the ATP-binding cassette (ABC) transporters family, such as P-pg (ABCB1), MRP1 (ABCC1), MRP2 (ABCC2), MRP4 (ABCC4), and BCRP (ABCG2) (Mahringer and Fricker 2016). As these ABC transporters were already identified as one of the mechanisms used by flavonoids to be transported, it is possible that some of them could be one of the mechanisms also used by (poly)phenol metabolites to reach the brain (Lu et al. 2014; Stieger et al. 2017). Beyond these, interactions between phenolic metabolites and several other transporters have also been described like phenolic acids and organic anion transporters (OATs) (Ohtsuki et al. 2002; Wang and Sweet 2012). Such observations could reveal important mechanisms by which these small phenolic metabolites cross the BBB and reach the cerebrospinal fluid (Mori et al. 2003; Kusuhara and Sugiyama 2005). Other types of transport should also be considered, especially in a disease context. Both vesicular and paracellular transport are usually strongly regulated in the BBB. However, increased inflammation could lead to a decrease in cellular junctions leading to an increase in BBB permeability to small molecules (De Bock et al. 2016; Varatharaj and Galea 2016).

Additionally, the presence of several phase I and II enzymes in BBB cells and inside the brain induces further metabolic reactions over the metabolites. In fact, some evidences have shown the ability of endothelial cells to metabolize (poly)phenol metabolites into novel compounds (Figueira et al. 2017) that might be closer of producing biological effects on neuronal cells.

4.4 LMW (Poly)phenol Metabolites as Effectors for Attenuating Neuroinflammation

Among the common ND hallmarks mentioned earlier, neuroinflammation is one of the key processes and is a double-edged sword, being essential for recovery from several conditions, but when it is chronic, it triggers different pathways contributing to disease progression (Carregosa et al. 2019). This process is controlled by microglial cells, the innate immune cells of the CNS that can act as a nervous system protector or contribute to progressive mechanisms that could lead to neurotoxic events (Qin et al. 2007; London et al. 2013).

Table 4.2 LMW (poly)phenol metabolites: *In silico* studies for passive permeation across the BBB, *in vitro* and *in vivo* evidences of transport in cellular models of the blood-brain barrier based on brain endothelial cells.

Class	Recommended name[a] (common name)	*In silico* permeability predictions	*In vitro* permeability studies (% of transport)	*In vivo* brain uptake		
				Dosage	Units	Brain uptake[b] (mmol/g of tissue)
Benzene diols and triols	Benzene-1,2-diol (catechol)	Probable (Li et al. 2017)	Not described	5.5×10^{5c} (intraperitoneal injection)	mmol/Kg BW	290[d] (Rogers et al. 1968) 304[e] (Angel et al. 1969)
	Phenol-2-sulfate (catechol-sulfate)	Probable (Figueira et al. 2017)	7.7 ± 0.6 (Figueira et al. 2017)	Not described		
	Benzene-1,2-diol-3-sulfate (pyrogallol-1-sulfate)	Probable (Figueira et al. 2017)	4.5 ± 0.1 (Figueira et al. 2017)	Not described		
	2-Hydroxy-1-methoxybenzene-3-sulfate (2-methylpyrogallol-3-sulfate)	Probable (Figueira et al. 2017)	1.3 ± 0.4 (Figueira et al. 2017)	Not described		
	2-Methoxyphenol-3-sulfate (2-metylpyrogallol-3-sulfate)	Probable (Figueira et al. 2017)	4.2 ± 0.4 (Figueira et al. 2017)	Not described		
	4-Methyl-phenyl-2-sulfate (4-methylcatechol-2-sulfate)	Probable (Figueira et al. 2017)	12.6 ± 1.9 (Figueira et al. 2017)	Not described		
	Benzene-1,2,3-triol (pyrogallol)	Not described	2.84 ± 0.27 (Unno et al. 2017)	1.1×10^{6f} (intraperitoneal injection) 4.8×10^{5g} (intraperitoneal injection)	mmol/Kg BW	350[h] (Rogers et al. 1968) 220[i] (Angel et al. 1969)
	Benzene-1,3-diol-2-sulfate (pyrogallol-2-sulfate)	Not described	1.91 ± 0.29 (Unno et al. 2017)	Not described		

Benzaldehydes	3,4-Dihydroxybenzaldehyde (protocatechualdehyde)	Probable (Tang et al. 2015)	Not described	Not described		
Benzoic acids	Benzoic acid	Probable (Li et al. 2017; Oliveira et al. 2019)	Not described	125 250 375 1000 (single intragastric injection) 100 (once daily oral administration for 12 weeks)	mmol/Kg BW	15.63 ± 2.40 2.53 ± 1.40 18.88 ± 2.96 11.00 ± 3.95 (Margalef et al. 2015a) 1.466 ± 0.289 (Margalef et al. 2015b)
	2-Hydroxybenzoic acid (salicylic acid)	Not probable (Carpenter et al. 2014)	Not described	Not described		
	3-Hydroxybenzoic acid	Not described	Not described	125 250 375 1000 (single intragastric injection) 100 (once daily oral administration for 12 weeks)	mmol/Kg BW	0.61 ± 0.13 0.43 ± 0.18 0.44 ± 0.04 1.06 ± 0.06 (Margalef et al. 2015a) 0.635 ± 0.081 (Margalef et al. 2015b)

(Continued)

Table 4.2 (Continued)

Class	Recommended name[a] (common name)	*In silico* permeability predictions	*In vitro* permeability studies (% of transport)	*In vivo* brain uptake		
				Dosage	Units	Brain uptake[b] (mmol/g of tissue)
				25 250 (once daily oral administration for 12 weeks)		1.36 ± 0.14 1.75 ± 0.30 (Wang et al. 2015)
Benzoic acids	4-Hydroxybenzoic acid	Probable (Li et al. 2017; Verma and Pandey 2019)	Not described	100 (intravenous injection)	nmol	0.408 (Gasperotti et al. 2015)
	3,4-Dihydroxybenzoic acid (protocatechuic acid)	Probable (Tang et al. 2015; Li et al. 2017) Not Probable (Verma and Pandey 2019)	Not described	2 g of protocatechuic acid mixed with 98 g of diet (oral supply for 12 weeks)	g	0.26 ± 0.10
				4 g of protocatechuic acid mixed with 96 g of diet (oral supply for 12 weeks)		0.41 ± 0.14 (Lin et al. 2011)
				5 (single oral supply)	mL/Kg BW	~ 0.6 ± 0.5[j] (Zhang et al. 2011)
	4-Hydroxy-3-methoxybenzoic acid (vanillic acid)	Probable (Tang et al. 2015; Li et al. 2017)	Not described	3 (single intragastric injection)	g/Kg BW	1.8 1.7 2.0 (Serra et al. 2012)
		Not probable (Verma and Pandey 2019)		50 (intravenous injection)	nmol	0.385 (Gasperotti et al. 2015)

Class	Recommended name[a] (common name)	*In silico* permeability predictions	*In vitro* permeability studies (% of transport)	*In vivo* brain uptake		
				Dosage	Units	Brain uptake[b] (mmol/g of tissue)
Benzoic acids	3-Methoxybenzoic acid-4-sulfate (vanillic acid-sulfate)	Probable (Figueira et al. 2017)	2.5 ± 1.2 (Figueira et al. 2017)	Not described		
	3,4,5-Trihydroxybenzoic acid (gallic acid)	Not probable (Verma and Pandey 2019)	6.5 ± 0.6 (Pervin et al. 2017)	50, 100 and 150 (intragastric repeated dose-escalation for 10 days)	mg/Kg BW	Detected (trace levels) (Ferruzzi et al. 2009)
				900 (intravenous injection)	nmol	0.611 (Gasperotti et al. 2015)
	3,5-Dihydroxy-4-methoxybenzoic acid (4-methylgallic acid)	Probable (Figueira et al. 2017)	4.4 ± 2.1 (Figueira et al. 2017)	Not described		
	3-Hydroxy-4-methoxybenzoic acid-5-sulfate (4-methylgallic acid-sulfate)	Probable (Figueira et al. 2017)	7.9 ± 0.7 (Figueira et al. 2017)	Not described		
	3,5-Dimethoxy-4-hydroxybenzoic acid (syringic acid)	Not probable (Verma and Pandey 2019)	Not described	Not described		
	3′,4′-Dihydroxycinnamic acid (caffeic acid)	Probable (Tang et al. 2015)	Not described	60 (intravenous injection)	nmol	0.0038 (Gasperotti et al. 2015)
	4′-Hydroxy-3′methoxycinnamic acid (ferulic acid)	Probable (Tang et al. 2015; Verma and Pandey 2019)	Medium capacity[k] (Lardeau and Poquet 2013)	20 pills ground into powder (single intragastric injection)	nmol	~2[l] (Wu et al. 2014)
				60 (intravenous injection)	nmol	0.027 (Gasperotti et al. 2015)

(Continued)

Table 4.2 (Continued)

Class	Recommended name[a] (common name)	*In silico* permeability predictions	*In vitro* permeability studies (% of transport)	*In vivo* brain uptake		
				Dosage	Units	Brain uptake[b] (mmol/g of tissue)
Cinnamic acids	4′-Hydroxycinnamic acid (p-coumaric acid)	Probable (Verma and Pandey 2019)	Not described	Not described		
	3′,5′-Dimethoxy-4′-hydroxycinnamic acid (sinapic acid)	Not probable (Verma and Pandey 2019)	Not described	Not described		
	Cinnamic acid	Probable (Verma and Pandey 2019)	Not described	Not described		
Phenylacetic acids	4′-Hydroxyphenylacetic acid	Probable (Li et al. 2017)	Not described	Not described		
	4′-Hydroxy-3′-methoxyphenylacetic acid (homovanillic acid)	Not described	Not described	125 250 375 1000 (single intragastric injection)	mg/Kg BW	2.36 ± 0.48 2.11 ± 0.23 1.84 ± 0.44 2.39 ± 0.40 (Margalef et al. 2015a)
				100 (once daily oral administration for 12 weeks)		0.528 ± 0.072 (Margalef et al. 2015b)

Phenylacetic acids				1 10 100 (single intragastric injection)		0.39 ± 0.048^{m} 0.32 ± 0.028^{m} 0.40 ± 0.05^{m} (Hazas et al. 2015)
				50 (intravenous injection)	nmol	0.433 (Gasperotti et al. 2015)
	3′,4′-Dihydroxyphenylacetic acid (homoprotocatechuic acid, DOPAC)	Not described	Not described	75 (intravenous injection)	nmol	0.501 (Gasperotti et al. 2015)
Phenylpropionic acids	3-(4′-Hydroxy-3′-methoxyphenyl) propanoic acid (dihydroferulic acid)	Not described	Medium capacity[k] (Lardeau and Poquet 2013)	Not described		
	3-(3′,4′-Dihydroxyphenyl) propanoic acid (dihydrocaffeic acid)	Not described	Medium capacity[k] (Lardeau and Poquet 2013)	Not described		
	3-(3′-Hydroxyphenyl)propanoic acid	Not described	Not described	25 250 (Intragastric repeated once daily for 11 days)	mg/Kg BW	1.49 ± 0.14 2.53 ± 0.68 (Wang et al. 2015)
				450 (intravenous injection)	nmol	0.193 (Gasperotti et al. 2015)
	3-(4′-Hydroxyphenyl)propanoic acid	Not described	Not described	5 (single intragastric injection)	g/Kg BW	15 ± 01.1 (Serra et al. 2011)

(*Continued*)

Table 4.2 (Continued)

Class	Recommended name[a] (common name)	*In silico* permeability predictions	*In vitro* permeability studies (% of transport)	*In vivo* brain uptake		
				Dosage	Units	Brain uptake[b] (mmol/g of tissue)
Hippuric acids	Hippuric acid	Not described	Not described	125	mg/Kg BW	0.32 ± 0.10
				250		0.56 ± 0.18
				375		0.42 ± 0.21
				1000		0.22 ± 0.01
				(single intragastric injection)		(Margalef et al. 2015a)
	4′-Hydroxyhippuric acid	Not described	Not described	250	nmol	0.020
				(intravenous injection)		(Gasperotti et al. 2015)

HBMEC, human brain microvascular endothelial cells; BBB, blood-brain barrier; PCA, 3,4-dihydroxybenzoic acid.

[a] Standard recommended names of dietary (poly)phenol metabolites (Kay et al. 2020).

[b] Original value of brain uptake in nmol/g of tissue after oral administration and either intravenous or intraperitoneal injection of the metabolites.

[c] The original value (60 mg/Kg BW) presented by the authors was converted to nmol/Kg BW considering the molecular weight of benzene-1,2-diol (110.1 g/mol).

[d] The original value (32 μg/g of tissue) presented by the authors was converted to nmol/g of tissue considering the brain density of 1 g/cm^3 and the molecular weight of benzene-1,2-diol (110.1 g/mol).

[e] The original value (33.5 μg/g of tissue) presented by the authors was converted to nmol/g of tissue considering the brain density of 1 g/cm^3 and the molecular weight of benzene-1,2-diol (110.1 g/mol).

[f] The original value (120 mg/Kg BW) presented by the authors was converted to nmol/Kg BW considering the molecular weight of benzene-1,2,3-triol (126.1 g/mol).

[g] The original value (60 mg/Kg BW) presented by the authors was converted to nmol/Kg BW considering the molecular weight of benzene-1,2,3-triol (126.1 g/mol).

[h] The original value (44 μg/g of tissue) presented by the authors was converted to nmol/g of tissue considering the brain density of 1 g/cm^3 and the molecular weight of benzene-1,2,3-triol (126.1 g/mol).

[i] The original value (28.4 μg/g of tissue) presented by the authors was converted to nmol/g of tissue considering the brain density of 1 g/cm^3 and the molecular weight of benzene-1,2,3-triol (126.1 g/mol).

[j] The original value (0.09 ± 0.07 μg of 3,4-dihydroxybenzoic acid acid/mL of microdialysis brain sample) presented by the authors was converted to nmol/g of tissue considering the brain density of 1 g/cm^3 and the molecular weight of PCA (154.12 g/mol).

[k] According to the authors, compounds were classified into 3 categories considering their ability to cross the BBB, based on their permeability coefficient (Pe): low capacity ($Pe < 1 \times 10^{-3}$ cm/min), medium capacity (1×10^{-3}cm/min < Pe < 2×10^{-3} cm/min) or high capacity (Pe > 2×10^{-3} cm/min). Pe values were calculated for 4′-hydroxy-3′-methoxycinnamic acid (1.31 ± 0.20 Pe ($\times 10^{-3}$ cm/min), 3-(4′-hydroxy-3′-methoxyphenyl)propanoic acid 1.20 ± 0.24 Pe ($\times 10^{-3}$cm/min) and 3-(3′,4′-dihydroxyphenyl)propanoic acid 1.01 ± 0.1 Pe ($\times 10^{-3}$cm/min).

[l] The original value (0.4 μg of ferulic acid/g of wet tissue) presented by the authors was converted to nmol/g of tissue considering the molecular weight of 4′-hydroxy-3′methoxycinnamic acid (194.18 g/mol).

[m] 4′-hydroxy-3′-methoxyphenylacetic acid was also detected in the vehicle group and only for the highest dose (HT-100) 4′-hydroxy-3′-methoxyphenylacetic acid presented significant increases in the brain.

The primary role of microglial cells is to maintain the normal function of the CNS and surrounding microenvironment to maintain homeostasis and neuronal integrity ensuring the continuous surveillance of the neuronal system. Microglial cells with a short and moderate signal behave as a protective phenotype and support anti-inflammatory functions and cell renewal. However, under chronic activation signal, microglial cells lose their surveillance phenotype and their ramified morphology and became neurotoxic, feeding instead a vicious cycle of toxicity. It is known that stimulation with either toxic protein aggregates, such as amyloid β 42 (Aβ 42) and α-synuclein (αsyn), glutamate or lipopolysaccharide (LPS) (*in vitro* studies), increases the chronic activation signal. These inflammatory responses secrete an array of pro-inflammatory cytokines and reactive oxygen and nitrogen species (ROS and NOS) that may impair neuronal activity. Nevertheless, low concentrations of cytokines such as interferon-γ (IFN-γ), interleukin (IL)-4 or IL-10, provide neuroprotection by regulating the levels of both insulin-like growth factor 1 (IGF-1) and tumor necrosis factor-α (TNF-α) (London et al. 2013; Rangarajan et al. 2016).

Recently, studies have described that adherence to the Mediterranean diet, rich in (poly) phenols, is associated with a reduction of markers of inflammation and improvement of the neuroinflammation status of the brain (Mayr et al. 2018). These studies support the hypothesis that (poly)phenol-rich foods or supplements have a positive impact toward NDs.

Evidence predicts that molecular and cellular pathways modulated by parent (poly)phenols which, though not always nutritionally relevant, give us possibilities about the model of action of (poly)phenol metabolites in a nutritional context. Some studies have shown that flavonoids inhibit the production of pro-inflammatory cytokines such as TNF-α, IL-6, and IL-1 in microglial cells, suggesting close involvement in pathways such as nuclear factor kappa-light-chain-enhancer of activated B cells (NF-κB) or MAPK (Poulose et al. 2012). In another study, Lau and colleagues discovered strong evidence that blueberry (poly)phenols inhibit the production of nitric oxide (NO), IL-1β, and TNF-α in activated microglial cells (Lau et al. 2007). In addition, some studies highlight the potential of caffeic acid phenethyl ester (CAPE) to modulate NF-κB activation in microglial cells when previously stimulated with TNF-α and significantly attenuate TNF-α-induced toll-like receptor (TLR)2 expression (Syed et al. 2007). Recently, a comparative study between different *in vitro*-digested fractions of raspberries reveals that a fraction richer in ellagitannins and its derivatives are major anti-inflammatory (poly)phenols and promising compounds to alleviate neuroinflammation contrasting to an anthocyanin-rich fraction (Garcia et al. 2020). This fraction enriched in ellagitannins, their degradation products and ellagic acid, attenuated pro-inflammatory markers and mediators CD40, NO, TNF-α, and intracellular superoxide via NF-κB, MAPK, and NFAT pathways. Importantly, they also increased the release of the anti-inflammatory cytokine IL-10 (Garcia et al. 2020).

Several studies have reported the ability of (poly)phenol metabolites to reduce neuroinflammatory markers and, in some cases, the underlying involved pathways have been proposed. Using the BV-2 microglial cell line, Zheng and co-workers reported the beneficial effects of benzene-1,2-diol, and its conjugates in decreasing LPS-induced NO and TNF-α. Furthermore, these compounds also inhibited the expression of inducible nitric oxide synthase (iNOS), as well as nuclear translocation of p65 subunit of NF-κB, kinase (IκB) degradation and phosphorylation of p38 MAPK (Zheng et al. 2008). Another study using benzene-1,2-diol and benzene-1,2,3-triol conjugates (both methyl and sulfates) has

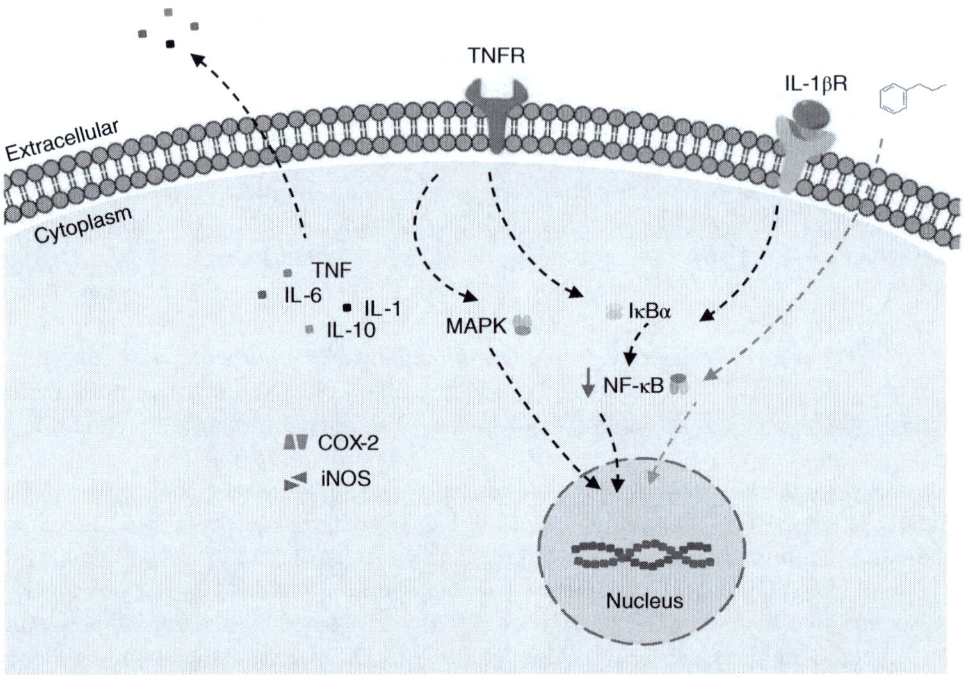

Figure 4.4 Possible molecular mechanisms of LMW (poly)phenols attenuation of neuroinflammation. Multiple receptors are capable of activating NF-kB leading to the activation and translocation of NF-kB to the nucleus and consequent expression of cytokines and inflammatory response enzymes like COX and iNOS. Several LMW (poly)phenol metabolites are known to impact NF-kB, blocking NF-kB translocation into the nucleus, thus blocking the inflammatory response. COX-2, cyclooxygenase 2; IL, interleukin; iNOS, nitric oxide synthase; IκBα, inhibitor of nuclear factor kappa B; MAPK, mitogen-activated protein kinase; NF-κB, nuclear factor kappa-light-chain-enhancer of activated B cells; TNFR, tumor necrosis factor receptors; TNFα, tumor necrosis factor-alpha. (*See insert for color representation of the figure.*)

shown improvement in cellular response to oxidative and inflammatory injuries via modulation of the NF-κB pathway. Moreover, the one hydroxyl of difference between the molecules seemed to make a significant impact on NF-κB pathway modulation (Figueira et al. 2017).

Furthermore, 3,4,5-trihydroxybenzoic acid was also studied by Mi-Jeong Kim and colleagues as an attenuator of neuroinflammation. They demonstrated its neuroprotective effect using the BV-2 cell line and primary microglial cells prior to Aβ stimulation (Kim et al. 2011). 3,4,5-trihydroxybenzoic acid prevented Aβ neurotoxicity by inhibiting the expression levels of iNOS, cyclooxygenase 2 (COX-2), IL-1β, and TNFα. Moreover, in this study, 3′,4′-dihydroxyphenylacetic acid had potent anti-inflammatory effects in BV-2 LPS-stimulated microglia (Kim et al. 2011). 4′-hydroxy-3′methoxycinnamic acid and cinnamic acid (Table 4.2) were also effective metabolites that seem to contribute to the modulation of neuroinflammatory processes, by attenuation of inflammatory markers (Chakrabarti et al. 2018; Rehman et al. 2019).

Overall some clues are already on the horizon regarding the molecular mechanisms by which (poly)phenols attenuate neuroinflammation (Figure 4.4). Nevertheless, more comprehensive studies, testing their potential are still lacking. Several metabolites still need to be evaluated and others require more comprehensive studies in animal models to fully understand the true potential of these molecules and their function in the human organism.

4.5 Concluding Remarks

With increased life expectancy, the incidence of NDs has been increasing and treatments to cure NDs have failed until now. (Poly)phenols have been in the spotlight for their potential for years, although no major breakthrough in human trials has led them to the clinic.

So far, attention has been paid to the effects of dietary phenolics, even if they are not the most representative or abundant compounds in circulation. Now the attention shifted to the LMW (poly)phenol metabolites with higher concentrations found in circulation. However, it is still missing a comprehensive understanding of the mechanisms by which LMW (poly)phenol metabolites reach the brain and possible biotransformation once inside the brain. Moreover, it still requires more evidences and relevant animal and human studies to define their real impact on brain inflammation, a key pathological process in NDs. Filling these gaps will be essential to designing nutritional and pharmacological approaches to create a link between dietary factors and the treatment/prevention of NDs in the future.

Acknowledgments

This work has received funding from the European Research Council (ERC) under the European Union's Horizon 2020 research and innovation programme under grant agreement No. 804229. iNOVA4 Health Research Unit (LISBOA-01-0145-FEDER-007344), which is cofunded by Fundação para a Ciência e Tecnologia (FCT)/Ministério da Ciência e do Ensino Superior, through national funds, and by FEDER under the PT2020 Partnership Agreement, is acknowledged. Authors would also like to acknowledge FCT for the financial support of RC (PD/BD/135492/2018), DC (2020.04630.BD), and DM (2021.05505.BD).

References

Angel, A., Lemon, R.N., Rogers, K.J., et al. (1969). The effect of polyhydroxyphenols on brain ATP in the mouse. *Experimental Brain Research*, **7**: 250–257.

Anson, N.M., Aura, A., Selinheimo, E., et al. (2010). Bioprocessing of wheat bran in whole wheat bread increases the bioavailability of phenolic acids in men and exerts antiinflammatory effects ex vivo. *The Journal of Nutrition*, **141**: 137–143.

Baldrick, F.R., McFadden, K., Ibars, M., et al. (2018). Impact of a (poly)phenol-rich extract from the brown algae Ascophyllum nodosum on DNA damage and antioxidant activity in an overweight or obese population: a randomized controlled trial. *The American Journal of Clinical Nutrition*, **108**: 688–700.

Blacklock, C.J., Lawrence, J.R., Wiles, D., et al. (2001). Salicylic acid in the serum of subjects not taking aspirin. Comparison of salicylic acid concentrations in the serum of vegetarians, non-vegetarians, and patients taking low dose aspirin. *Journal of Clinical Pathology*, **54**: 553–555.

Bonifácio, M.J., Palma, P.N., Almeida, L.L., et al. (2007). Catechol-O-methyltransferase and its inhibitors in Parkinson's disease. *CNS Drug Reviews*, **13**: 352–379.

Boto-Ordóñez, M., Urpi-Sarda, M., Queipo-Ortuño, M.I., et al. (2013). Microbial metabolomic fingerprinting in urine after regular dealcoholized red wine consumption in humans. *Journal of Agricultural and Food Chemistry*, **61**: 9166–9175.

Bowtell, J.L., Aboo-Bakkar, Z., Conway, M., et al. (2017). Enhanced task related brain activation and resting perfusion in healthy older adults after chronic blueberry supplementation. *Applied Physiology, Nutrition, and Metabolism*, **42**: 773–779.

Brettschneider, J., Del Tredici, K., Lee, V.M., and Trojanowski, J.Q. (2015). Spreading of pathology in neurodegenerative diseases: a focus on human studies. *Nature Reviews*, **16**: 109–120.

Carpenter, T.S., Kirshner, D.A., Lau, E.Y., et al. (2014). A method to predict blood-brain barrier permeability of drug-like compounds using molecular dynamics simulations. *Biophysical Journal*, **107**: 630–641.

Carregosa, D., Carecho, R., and Santos, C.N. (2019). Low-molecular weight metabolites from polyphenols as effectors for attenuating neuroinflammation. *Journal of Agricultural and Food Chemistry*, **68**: 1790–1807.

Chakrabarti, S., Jana, M., Roy, A., and Pahan, K. (2018). Upregulation of suppressor of cytokine signaling 3 in microglia by cinnamic acid. *Current Alzheimer Research*, **15**: 894–904.

Chang, A., Jeske, L., Ulbrich, S., et al. (2021). BRENDA, the ELIXIR core data resource in 2021: new developments and updates. *Nucleic Acids Research*, **49**: 498–508.

Cialdella-kam, L., Nieman, D.C., Sha, W., et al. (2013). Dose–response to 3 months of quercetin-containing supplements on metabolite and quercetin conjugate profile in adults. *British Journal of Clinical Pharmacology*, **109**: 1923–1933.

Cladis, D.P., Li, S., Reddivari, L., et al. (2020). A 90 day oral toxicity study of blueberry polyphenols in ovariectomized sprague-dawley rats. *Food and Chemical Toxicology*, **139**: 1–10.

Clifford, M.N., Copeland, E.L., Bloxsidge, J.P., and Mitchell, L.A. (2000). Hippuric acid as a major excretion product associated with black tea consumption. *Xenobiotica*, **30**: 317–326.

Cunha, L.R., Muniz-Junqueira, I.M., and dos Borges, T.K.S. (2019). Impact of polyphenols in phagocyte functions. *Journal of Inflammation Research*, **12**: 205–217.

Curzon, G. and Pratt, R.T.C. (1964). Origin of urinary resorcinol sulphate. *Nature*, **204**: 383–384.

Czank, C., Cassidy, A., Zhang, Q., et al. (2013). Human metabolism and elimination of the anthocyanin, cyanidin-3-glucoside: a 13C-tracer study. *The American Journal of Clinical Nutrition*, **97**: 995–1003.

Dando, S.J., Mackay-sim, A., Norton, R., et al. (2014). Pathogens penetrating the central nervous system: infection pathways and the cellular and molecular mechanisms of invasion. *Clinical Microbiology Reviews*, **27**: 691–726.

Day, A.J., Cañada, F.J., Díaz, J.C., et al. (2000). Dietary flavonoid and isoflavone glycosides are hydrolysed by the lactase site of lactase phlorizin hydrolase Andrea. *FEBS Letters*, **468**: 166–170.

De Bock, M., Haver, V.H., Vandenbroucke, R.E., et al. (2016). Into rather unexplored terrain-transcellular transport across the blood–brain barrier. *Glia*, **64**: 1097–1123.

De Ferrars, R.M., Czank, C., Zhang, Q., et al. (2014). The pharmacokinetics of anthocyanins and their metabolites in humans. *British Journal of Pharmacology*, **171**: 3268–3282.

Del Rio, D., Stalmach, A., Calani, L., and Crozier, A. (2010). Bioavailability of coffee chlorogenic acids and green tea flavan-3-ols. *Nutrients*, **2**: 820–833.

Del Rio, D., Rodriguez-Mateos, A., Spencer, J.P.E., et al. (2013). Dietary (poly)phenolics in human health: structures, bioavailability, and evidence of protective effects against chronic diseases. *Antioxidants and Redox Signaling*, **18**: 1818–1892.

Devore, E.E., Kang, J.H., Breteler, M.M.B., and Grodstein, F. (2012). Dietary intakes of berries and flavonoids in relation to cognitive decline. *American Neurological Association*, **72**: 135–143.

Feliciano, R.P., Boeres, A., Massacessi, L., et al. (2016). Identification and quantification of novel cranberry-derived plasma and urinary (poly)phenols. *Archives of Biochemistry and Biophysics*, **599**: 31–41.

Ferruzzi, M.G., Lobo, J.K., Janle, E.M., et al. (2009). Bioavailability of gallic acid and catechins from grape seed polyphenol extract is improved by repeated dosing in rats: implications for treatment in Alzheimer's disease. *Journal of Alzheimer's Disease*, **18**: 113–124.

Figueira, I., Garcia, G., Pimpão, R.C., et al. (2017). Polyphenols journey through blood-brain barrier towards neuronal protection. *Scientific Reports*, **7**: 1–16.

Garcia, G., Pais, T.F., Pinto, P., et al. (2020). Bioaccessible raspberry extracts enriched in ellagitannins and ellagic acid derivatives have anti-neuroinflammatory properties. *Antioxidants*, **9**: 1–20.

Gasperotti, M., Passamonti, S., Tramer, F., et al. (2015). Fate of microbial metabolites of dietary polyphenols in rats: is the brain their target destination? *ACS Chemical Neuroscience*, **6**: 1341–1352.

Greene, C. and Campbell, M. (2016). Tight junction modulation of the blood brain barrier: CNS delivery of small molecules. *Tissue Barriers*, **4**: 1–10.

Guertin, K.A., Loftfield, E., Boca, S.M., et al. (2015). Serum biomarkers of habitual coffee consumption may provide insight into the mechanism underlying the association between coffee consumption and colorectal cancer. *American Journal of Clinical Nutrition*, **101**: 1000–1011.

Guo, K. and Li, L. (2009). Differential ^{12}C-/^{13}C-isotope dansylation labeling and fast liquid chromatography/mass spectrometry for absolute and relative quantification of the metabolome. *Analytical Chemistry*, **81**: 3919–3932.

Guy, P.A., Renouf, M., Barron, D., et al. (2009). Quantitative analysis of plasma caffeic and ferulic acid equivalents by liquid chromatography tandem mass spectrometry. *Journal of Chromatography B*, **877**: 3965–3974.

Hattori, K., Matsuura, M., Fujiwara, M., and Shiamoto, K. (1969). Inhibition of catechol-O-methyltransferase by hydroxybenzenes and related compounds. *The Japanese Journal of Pharmacology*, **19**: 282–286.

Hazas, M.-C.L., Rubió, L., Kotronoulas, A., et al. (2015). Dose effect on the uptake and accumulation of hydroxytyrosol and its metabolites in target. *Molecular Nutrition & Food Research*, **59**: 1395–1399.

Kanehisa, M., Sato, Y., Kawashima, M., et al. (2016). KEGG as a reference resource for gene and protein annotation. *Nucleic Acids Research*, **44**: D457–D462.

Kay, C.D., Clifford, M.N., Mena, P., et al. (2020). Recommendations for standardizing nomenclature for dietary (poly)phenol catabolites. *The American Journal of Clinical Nutrition*, **112**: 1051–1068.

Kennedy, D.O. (2014). Polyphenols and the human brain: plant "secondary metabolite" ecologic roles and endogenous signaling functions drive benefits. *Advances in Nutricion*, **5**: 515–533.

Kern, S.A.M., Bennett, R.I.N., Mellon, F.A., et al. (2003). Absorption of hydroxycinnamates in humans after high-bran cereal consumption. *Journal of Agricultural and Food Chemistry*, **51**: 6050–6055.

Kim, M., Seong, A., Yoo, J., et al. (2011). Gallic acid, a histone acetyltransferase inhibitor, suppresses b-amyloid neurotoxicity by inhibiting microglial-mediated neuroinflammatio. *Molecular Nutrition & Food Research*, **55**: 1798–1808.

Krikorian, R., Shidler, M.D., Nash, T.A., et al. (2010). Blueberry supplementation improves memory in older adults. *Journal of Agricultural and Food Chemistry*, **58**: 3996–4000.

Kusuhara, H. and Sugiyama, Y. (2005). Active efflux across the blood-brain barrier: role of the solute carrier family. *Journal of the American Society for Experimental NeuroTherapeutics*, **2**: 73–85.

Lardeau, A. and Poquet, L. (2013). Phenolic acid metabolites derived from coffee consumption are unlikely to cross the blood-brain barrier. *Journal of Pharmaceutical and Biomedical Analysis*, **76**: 134–138.

Lau, F.C., Bielinski, D.F., and Joseph, J.A. (2007). Inhibitory effects of blueberry extract on the production of inflammatory mediators in lipopolysaccharide-activated BV2 microglia. *Journal of Neuroscience Research*, **85**: 1010–1017.

Levin, J., Maaß, S., Schuberth, M., et al. (2019). Safety and efficacy of epigallocatechin gallate in multiple system atrophy (PROMESA): a randomised, double-blind, placebo-controlled trial. *The Lancet Neurology*, **18**: 724–735.

Li, M., Fan, Y., Zhang, X., et al. (2014). Fruit and vegetable intake and risk of type 2 diabetes mellitus: meta-analysis of prospective cohort studies. *BMJ Open*, **4**: 1–9.

Li, H., Ye, M., Zhang, Y., et al. (2015). Blood-brain barrier permeability of Gualou Guizhi granules and neuroprotective effects in ischemia/reperfusion injury. *Molecular Medicine Reports*, **12**: 1272–1278.

Li, X., Zhang, S., and Yang, X. (2017). Combination of cheminformatics and bioinformatics to explore the chemical basis of the rhizomes and aerial parts of Dioscorea nipponica Makino. *Journal of Pharmacy and Pharmacology*, **69**: 1846–1857.

Lin, C., Tsai, S., Huang, C., and Yin, M. (2011). Antiglycative effects of protocatechuic acid in the kidneys of diabetic mice. *Journal of Agricultural and Food Chemistry*, **59**: 5117–5124.

Loke, W.M., Jenner, A.M., Proudfoot, J.M., et al. (2009). A metabolite profiling approach to identify biomarkers of flavonoid intake in humans. *The Journal of Nutrition*, **139**: 2309–2314.

London, A., Cohen, M., and Schwartz, M. (2013). Microglia and monocyte-derived macrophages: functionally distinct populations that act in concert in CNS plasticity and repair. *Frontiers in Neuroscience*, **7**: 1–10 https://doi.org/10.3389/fncel.2013.00034.

Lu, X., Chan, T., Xu, C., et al. (2014). The interactions of herbal compounds with human Organic anion/cation transporters. *Journal of Pharmacogenomics & Pharmacoproteomics*, **5**: 1–8.

Macready, A.L., Kennedy, O.B., Ellis, J.A., et al. (2009). Flavonoids and cognitive function: a review of human randomized controlled trial studies and recommendations for future studies. *Genes & Nutrition*, **4**: 227–242.

Mahringer, A. and Fricker, G. (2016). ABC transporters at the blood-brain barrier. *Expert Opinion on Drug Metabolism & Toxicology*, **12**: 499–508.

Margalef, M., Pons, Z., Bravo, F.I., et al. (2015a). Tissue distribution of rat flavanol metabolites at different doses. *The Journal of Nutritional Biochemistry*, **26**: 987–995.

Margalef, M., Pons, Z., Iglesias-carres, L., et al. (2015b). Lack of tissue accumulation of grape seed flavanols after daily long-term administration in healthy and cafeteria diet obese rats. *Journal of Agricultural and Food Chemistry*, **63**: 9996–10003.

Mayr, H.L., Thomas, C.J., Tierney, A.C., et al. (2018). Randomization to 6-month Mediterranean diet compared with a low-fat diet leads to improvement in Dietary Inflammatory Index scores in patients with coronary heart disease: the AUSMED Heart Trial. *Nutrition Research*, **55**: 97–107.

Medina-Remón, A., Tresserra-Rimbau, A., Pons, A., et al. (2015). Effects of total dietary polyphenols on plasma nitric oxide and blood pressure in a high cardiovascular risk cohort. The PREDIMED randomized trial. *Nutrition, Metabolism, and Cardiovascular Diseases*, **25**: 60–67.

Mori, S., Takanaga, H., Ohtsuki, S., et al. (2003). Rat organic anion transporter 3 (rOAT3) is responsible for brain-to-blood efflux of homovanillic acid at the abluminal membrane of brain capillary endothelial cells. *Journal of Cerebral Blood Flow & Metabolism*, **23**: 432–440.

Nehlig, A. (2013). The neuroprotective effects of cocoa flavanol and its influence on cognitive performance. *British Journal of Clinical Pharmacology*, **75**: 716–727.

Nooyens, A.C.J., Bueno-De-Mesquita, H.B., Van Boxtel, M.P.J., et al. (2011). Fruit and vegetable intake and cognitive decline in middle-aged men and women: the Doetinchem Cohort Study. *British Journal of Nutrition*, **106**: 752–761.

Ofosu, F.K., Daliri, E.B., Elahi, F., et al. (2020). New insights on the use of polyphenols as natural preservatives and their emerging safety concerns. *Frontiers in Sustainable Food Systems*, **4**: 1–14.

Ohtsuki, S., Asaba, H., Takanaga, H., et al. (2002). Role of blood – brain barrier organic anion transporter 3 (OAT3) in the efflux of indoxyl sulfate, a uremic toxin: its involvement in neurotransmitter metabolite clearance from the brain. *Journal of Neurochemistry*, **83**: 57–66.

Oliveira, C., Bagetta, D., Cagide, F., et al. (2019). Benzoic acid-derived nitrones: a new class of potential acetylcholinesterase inhibitors and neuroprotective agents. *European Journal of Medicinal Chemistry*, **174**: 116–129.

Olthof, M.R., Hollman, P.C.H., Buijsman, M.N.C.P., et al. (2003). Chlorogenic acid, quercetin-3-rutinoside and black tea phenols are extensively metabolized in humans. *Human Nutrition and Metabolism Chlorogenic*, **133**: 1806–1814.

Ottaviani, J.I., Borges, G., Momma, T.Y., et al. (2016). The metabolome of [2-14C] (−)-epicatechin in humans: implications for the assessment of efficacy, safety, and mechanisms of action of polyphenolic bioactives. *Scientific Reports*, **6**: 1–10.

Perez-Ternero, C., Macià, A., Sotomayor, M.A., et al. (2017). Bioavailability of the ferulic acid-derived phenolic compounds of a rice bran enzymatic extract and their activity against superoxide production. *Food & Function*, **8**: 2165–2174.

Pervin, M., Unno, K., Nakagawa, A., et al. (2017). Blood brain barrier permeability of (−)-epigallocatechin gallate, its proliferation-enhancing activity of human neuroblastoma SH-SY5Y cells, and its preventive effect on age-related cognitive dysfunction in mice. *Biochemistry and Biophysics Reports*, **9**: 180–186.

Pimpão, R.C., Ventura, M.R., Ferreira, R.B., et al. (2015). Phenolic sulfates as new and highly abundant metabolites in human plasma after ingestion of a mixed berry fruit purée. *British Journal of Nutrition*, **113**: 454–463.

Poulose, S.M., Fisher, D.R., Larson, J., et al. (2012). Anthocyanin-rich açai (Euterpe oleracea Mart). Fruit pulp fractions attenuate inflammatory stress signaling in mouse brain BV-2 microglial cells. *Journal of Agricultural and Food Chemistry*, **60**: 1084–1093.

Psaltopoulou, T., Sergentanis, T.N., Panagiotakos, D.B., et al. (2019). Mediterranean diet and stroke, cognitive impairment, depression: a meta-analysis. *Annals of Neurology*, **74**: 580–591.

Qin, L., Wu, X., Block, M.L., et al. (2007). Systemic LPS causes chronic neuroinflammation and progressive neurodegeneration. *Glia*, **55**: 453–462.

Quideau, S., Deffieux, D., Douat-Casassus, C., and Pouységu, L. (2011). Plant polyphenols: chemical properties, biological activities, and synthesis. *Angewandte Chemie, International Edition*, **50**: 586–621.

Rangarajan, P., Karthikeyan, A., and Dheen, S.T. (2016). Role of dietary phenols in mitigating microglia-mediated neuroinflammation. *Neuromolecular Medicine*, **18**: 453–464.

Rehman, S.U., Ali, T., Alam, S.I., et al. (2019). Ferulic acid rescues LPS-induced neurotoxicity via modulation of the TLR4 receptor in the mouse hippocampus. *Jounal of Molecular Neurobiology*, **56**: 2774–2790.

Rendeiro, C., Rhodes, J.S., and Spencer, J.P.E. (2015). The mechanisms of action of flavonoids in the brain: direct versus indirect effects. *Neurochemistry International*, **89**: 126–139.

Rodriguez-Mateos, A., Heiss, C., Borges, G., and Crozier, A. (2013). Berry (Poly)phenols and cardiovascular health. *Journal of Agricultural and Food Chemistry*, **7**: 3842–3851.

Rodriguez-Mateos, A., Vauzour, D., Krueger, C.G., et al. (2014). Bioavailability, bioactivity and impact on health of dietary flavonoids and related compounds: an update. *Archives of Toxicology*, **88**: 1803–1853.

Rogers, K.J., Angel, A., and Butterfield, L. (1968). The penetration of catechol and pyrogallol into mouse brain and the effect on cerebral monoamine levels. *In Journal of Pharmacy and Pharmacology*, **20**: 727–729.

Roowi, S., Mullen, W., Edwards, C.A., and Crozier, A. (2009). Yoghurt impacts on the excretion of phenolic acids derived from colonic breakdown of orange juice flavanones in humans. *Molecular Nutrition & Food Research*, **53**: S68–S75.

Schaffer, S. and Halliwell, B. (2012). Do polyphenols enter the brain and does it matter? Some theoretical and practical considerations. *Genes & Nutrition*, **7**: 99–109.

Serra, A., Macià, A., Romero, M.-P., et al. (2011). Distribution of procyanidins and their metabolites in rat plasma and tissues after an acute intake of hazelnut extract. *Food & Function*, **2**: 562–568.

Serra, A., Rubió, L., Borràs, X., et al. (2012). Distribution of olive oil phenolic compounds in rat tissues after administration of a phenolic extract from olive cake. *Molecular Nutrition & Food Research*, **56**: 486–496.

Sofroniew, M.V. and Vinters, H.V. (2010). Astrocytes: biology and pathology. *Acta Neuropathology*, **119**: 7–35.

Soto, C. (2003). Unfolding the role of protein misfolding in neurodegenerative diseases. *Nature Reviews*, **4**: 49–60.

Spencer, J.P.E. (2008). Food for thought: the role of dietary flavonoids in enhancing human memory, learning and neuro-cognitive performance. *Proceedings of the Nutrition Society*, **67**: 238–252.

Stalmach, A., Edwards, C.A., Wightman, J.D., and Crozier, A. (2012). Gastrointestinal stability and bioavailability of (poly)phenolic compounds following ingestion of Concord grape juice by humans. *Molecular Nutrition & Food Research*, **56**: 497–509.

Stieger, B., Mahdi, Z.M., and Jager, W. (2017). Intestinal and hepatocellular transporters: therapeutic effects and drug interactions of herbal supplements. *Annual Review of Pharmacology and Toxicology*, **57**: 399–416.

Syed, M.M., Phulwani, N.K., and Kielian, T. (2007). Tumor necrosis factor-alpha (TNF- a) regulates Toll-like receptor 2 (TLR2) expression in microglia. *Journal of Neurochemistry*, **103**: 1461–1471.

Tang, C., Ye, Y., Feng, Y., and Quinn, R.J. (2015). TCM, brain function and drug space Chunping. *Natural Product Reports*, **3**: 6–25.

Toshimitsu, N., Kenji, M., Toyokazu, O., et al. (1981). A gas chromatographic-mass spectrometric analysis for phenols in uremic serum. *Clinica Chimica Acta*, **110**: 51–57.

Unno, K., Pervin, M., Nakagawa, A., et al. (2017). Blood-brain barrier permeability of green tea catechin metabolites and their neuritogenic activity in human neuroblastoma SH-SY5Y cells. *Molecular Nutrition & Food Research*, **61**: 1–18.

Van der Hooft, J.J.J., De Vos, R.C.H., Mihaleva, V., et al. (2012). Structural elucidation and quantification of phenolic conjugates present in human urine after tea intake. *Analytical Chemistry*, **84**: 7263–7271.

Van Dorsten, F.A., Grun, C.H., Van Velzen, E.J.J., et al. (2010). The metabolic fate of red wine and grape juice polyphenols in humans assessed by metabolomics. *Molecular Nutrition & Food Research*, **54**: 897–908.

Varatharaj, A. and Galea, I. (2016). The blood-brain barrier in systemic inflammation. *Brain, Behavior, and Immunity*, **60**: 1–12.

Verma, S. and Pandey, A.K. (2019). An In-silico exploration of possible nutraceutical properties of rice against multidirectional attack of cerebral ischemia: a molecular docking study of MMP9 and MMP2 inhibition. *EC Pharmacology and Toxicology*: 494–522.

Wang, L. and Sweet, D.H. (2012). Potential for food – drug interactions by dietary phenolic acids on human organic anion transporters 1 (SLC22A6), 3 (SLC22A8), and 4 (SLC22A11). *Biochemical Pharmacology*, **84**: 1088–1095.

Wang, D., Ho, L., Faith, J., et al. (2015). Role of intestinal microbiota in the generation of polyphenol derived phenolic acid mediated attenuation of Alzheimer's disease β-amyloid oligomerization. *Molecular Nutrition & Food Research*, **59**: 1025–1040.

Welsch, F. (2008). Routes and modes of administration of resorcinol and their relationship to potential manifestations of Thyroid gland Toxicity in animals and man. *International Journal of Toxicology*, **27**: 59–63.

Williams, R.J. and Spencer, J.P.E. (2012). Flavonoids, cognition, and dementia: actions, mechanisms, and potential therapeutic utility for Alzheimer disease. *Free Radical Biology and Medicine*, **52**: 35–45.

Williamson, G., Kay, C.D., and Crozier, A. (2018). The bioavailability, transport, and bioactivity of dietary flavonoids: a review from a historical perspective. *Free Radical Biology and Medicine*, **17**: 1054–1112.

Wishart, D.S., Feunang, Y.D., Marcu, A., et al. (2018). HMDB 4.0: the human metabolome database for 2018. *Nucleic Acids Research*, **46**: D608–D617.

Wu, K., Wang, Z., Liu, D., and Qi, X. (2014). Pharmacokinetics, brain distribution, release and blood-brain barrier transport of Shunaoxin pills. *Journal of Ethnopharmacology*, **151**: 1133–1140.

Youdim, K.A., Dobbie, M.S., Kuhnle, G., et al. (2003). Interaction between flavonoids and the blood – brain barrier: in vitro studies. *Journal of Neurochemistry*, **85**: 180–192.

Youdim, K.A., Qaiser, M.Z., Begley, D.J., et al. (2004). Flavonoid permeability across an in situ model of the blood–brain barrier. *Free Radical Biology and Medicine*, **36**: 592–604.

Zamora-Ros, R., Achaintre, D., Rothwell, J.A., et al. (2016). Urinary excretions of 34 dietary polyphenols and their associations with lifestyle factors in the EPIC cohort study. *Scientific Reports*, **6**: 1–9.

Zhang, Y., Wu, L., Zhang, Q., et al. (2011). Pharmacokinetics of phenolic compounds of Danshen extract in rat blood and brain by microdialysis sampling. *Journal of Ethnopharmacology*, **136**: 129–136.

Zhang, H., Hassan, Y.I., Liu, R., et al. (2020). Molecular mechanisms underlying the absorption of aglycone and glycosidic flavonoids in a Caco-2 BBe1 cell model. *ACS Omega*, **5**: 10782–10793.

Zheng, L.T., Ryu, G., Kwon, B., et al. (2008). Anti-inflammatory effects of catechols in lipopolysaccharide-stimulated microglia cells: inhibition of microglial neurotoxicity. *European Journal of Pharmacology*, **588**: 106–113.

5

Deciphering Complex Natural Mixtures through Metabolome Mining of Mass Spectrometry Data

The Plant Specialized Metabolome as a Case Study

Justin J.J. van der Hooft[1], Madeleine Ernst[2], Daniel Papenberg[1,3], Kyo Bin Kang[4], Iris F. Kappers[3], Marnix H. Medema[1], Pieter C. Dorrestein[5], and Simon Rogers[6]

[1] *Bioinformatics Group, Plant Sciences Group, Wageningen University and Research, Wageningen, the Netherlands*
[2] *Section for Clinical Mass Spectrometry, Department of Congenital Disorders, Danish Center for Neonatal Screening, Statens Serum Institut, Copenhagen, Denmark*
[3] *Laboratory of Plant Physiology, Plant Sciences Group, Wageningen University and Research, Wageningen, the Netherlands*
[4] *College of Pharmacy, Sookmyung Women's University, Seoul, Korea*
[5] *Collaborative Mass Spectrometry Innovation Center, Skaggs School of Pharmacy and Pharmaceutical Sciences, University of California San Diego, La Jolla, CA, USA*
[6] *School of Computing Science, University of Glasgow, Glasgow, UK*

5.1 Introduction

Plant biochemistry and the specialized metabolite arsenal play pivotal roles in growth and health, for example, in defence against herbivores. By illuminating the plant chemical diversity, mass spectrometry-based metabolomics workflows are revolutionizing the natural products discovery field (Wolfender et al. 2019). Metabolomics was coined as the scientific field that aims to comprehend the complete ensemble of small molecules or metabolites, i.e. the metabolome, in an organism (Hall et al. 2002; Tugizimana et al. 2018). Fuelled by this rather ambitious goal, technological advances in equipment have resulted in increasingly information-dense data files from which relevant chemical information needs to be extracted. This higher-throughput acquisition of data requires more streamlined analysis methods (Tugizimana et al. 2020). Consequently, the last decade has witnessed the rise of computational metabolomics workflows supporting and enabling all parts of metabolomics analyses from preprocessing to data interpretation and visualization (Misra and van der Hooft 2016). For example, peak picking and alignment tools like XCMS (Tautenhahn et al. 2012), MZmine (Pluskal et al. 2010), and others (Nothias et al. 2020) have boosted large-scale comparative analyses by tracing metabolite features within samples and subsequently align them between samples. In the following steps in the metabolomics analysis pipeline, statistical tools guide the discovery of sample-type or phenotype-specific metabolite patterns (Tugizimana et al. 2018; Ivanisevic and Want 2019).

Recent Advances in Polyphenol Research, Volume 8, First Edition. Edited by Juha-Pekka Salminen, Kristiina Wähälä, Victor de Freitas, and Stéphane Quideau.

During mass spectrometry-based metabolomics experiments, these statistical analyses typically result in an ensemble of metabolite features (mass and retention time combinations) that are of interest to pursue. One or more of these metabolite features in a metabolomics profile may represent a unique plant-specialized metabolite. However, the chemical space of small molecules is vast (over 10^{60} possible carbon-based small molecules may exist (Bohacek et al. 1996)), and, as a consequence, most molecules have never been described (currently, there are ~10^8 molecules in PubChem (Kim et al. 2016), which is one of the most extensive structure databases), and even much fewer chemical standards are available for laboratory-based studies. Furthermore, it is estimated that only ~2.5% of known specialized metabolites across all kingdoms of life have a reference mass spectrum available in public mass spectral libraries (van der Hooft et al. 2020); thus, as a result, the structural annotation of metabolites present in metabolomics profiles remains limited. In this chapter, we will focus on the structural annotation and prioritizing of known and novel chemistry with the help of computational metabolomics tools.

The plant specialized metabolome is very diverse and large and still contains many completely unknown metabolites (Nakabayashi and Saito 2020). Therefore, unsurprisingly, during plant metabolomics analyses, metabolite annotation and identification of the metabolite features remains a difficult challenge. Yet, it is an essential step to link structural information to metabolomics data to enable biochemical interpretation and make a system-wide analysis possible (Tugizimana et al. 2018). The use of tandem mass spectrometry fragmentation approaches coupled to liquid chromatography (LC-MS/MS) has boosted the annotation power of methods as the resulting MS/MS spectra can be seen as "bar codes" or "fingerprints" of metabolites, thus aiding in discrimination between closely related structures and increasing confidence in annotations together with other analytical information such as retention time order. Furthermore, the increased implementation of MS/MS-based metabolomics approaches has made library matching to reference MS/MS spectra the cornerstone of confident metabolite annotation (Johnson and Lange 2015). However, the current annotation rates through library matching remain low; therefore, computational metabolomics workflows are being developed to support annotation pipelines.

Here, we highlight two main types of structural annotation workflows: (i) metabolome mining (Figure 5.1a) and (ii) *in silico* metabolite annotation approaches (Figure 5.1b). Firstly, metabolome mining strategies like molecular networking (Wang et al. 2016) and substructure pattern discovery (van der Hooft et al. 2016) work through comparing many spectra to each other at a large scale. Such approaches are now widely used to form molecular networks consisting of clusters of metabolite features usually termed molecular families. These metabolome approaches aim to create structure in large and complex metabolomics profiles by grouping molecules according to their mass spectral similarities. Such groupings can then be exploited during structural annotation. Furthermore, they also allow to perform annotations at the substructural level by analysing the groupings in molecular networks or by structural assignment of substructure patterns found by MS2LDA unsupervised substructure discovery.

Secondly, metabolome *in silico* annotation strategies like SIRIUS-CSI:FingerID (Dührkop et al. 2015, 2019), CANOPUS (Dührkop et al. 2020), MAGMa (Ridder et al. 2012; Ridder et al. 2013), DEREPLICATOR (Mohimani et al. 2018), MetFrag (Wolf

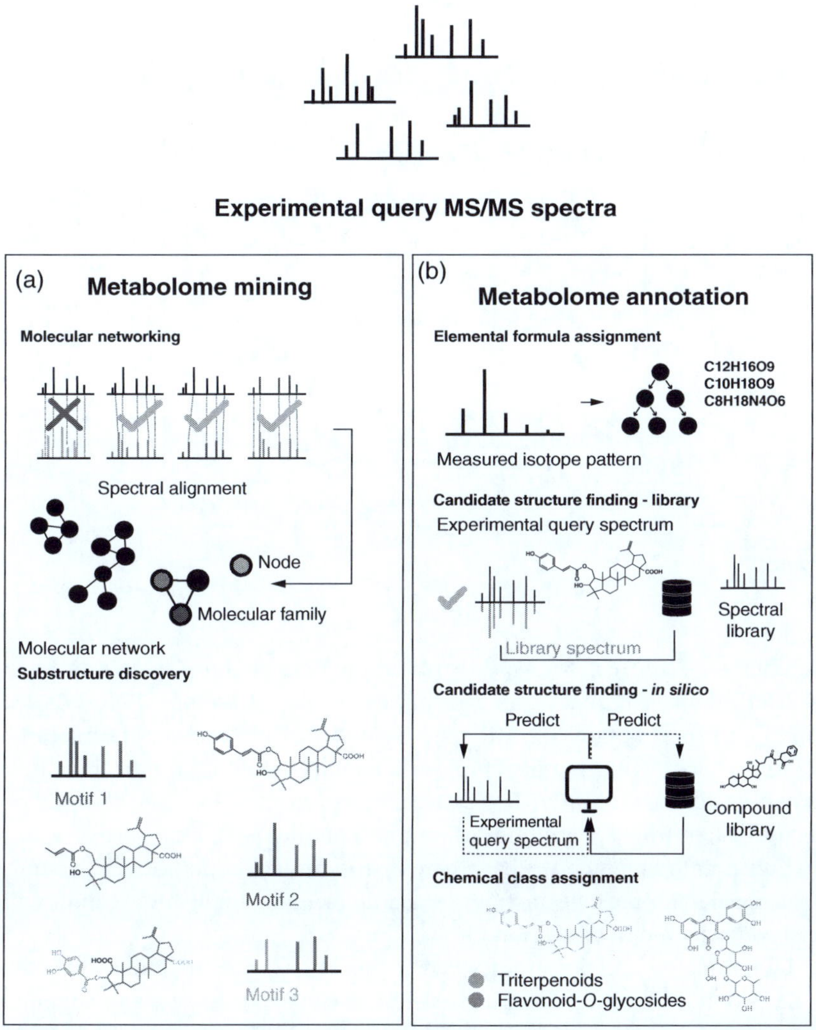

Figure 5.1 The chemical and structural annotation of complex natural mixtures displayed from metabolome mining to annotation tools. (a) Molecular networking and substructure discovery. All mass spectra from the experimental data are compared to each other and metabolites are represented by mass spectral features. These features are the nodes in the molecular network: they can form molecular families with structurally similar molecules (based on their spectral similarity) and an ensemble of molecular families is a molecular network. In unsupervised substructure discovery, topic modelling discovers often co-occurring mass fragments or neutral losses that together could represent metabolite building blocks as schematically depicted here with a hypothetical example where the colored mass fragment peaks correspond to the colored substructures in the displayed metabolites. (b) Elemental formula assignment, library matching, *in silico* candidate structure finding and chemical class assignment. In elemental formula assignment, accurate isotope patterns and the calculation of fragmentation trees based on MS/MS spectra enable accurate elemental formula assignments for many metabolite features. Library matching relies on the presence of reference mass spectra in spectral libraries whereby a mass spectral similarity score (as used for molecular networking in (a) is used for matching these reference data to the experimental data. *In silico* candidate structure finding mostly relies on compound libraries that supply the candidate structures matching to the masses in the metabolomics profiles, after which various approaches are used to rank the often-various possibilities, thereby considering the MS/MS data. Chemical class assignment is possible through direct prediction based on the MS/MS data or through chemical class annotation of candidate structures. (c) Combination of multiple metabolome mining and annotation tools into one output. The schematic figure highlights how for one molecular family in a molecular network chemical compound class information can be combined with substructure annotation and library matches to gain higher-level and in-detail chemical information about the mixture. The figure parts were inspired by those presented in (Dührkop et al. 2020; Ernst et al. 2019b; Aron et al. 2020) and other papers explaining metabolomics mining and annotation tools. (*See insert for color representation of the figure.*)

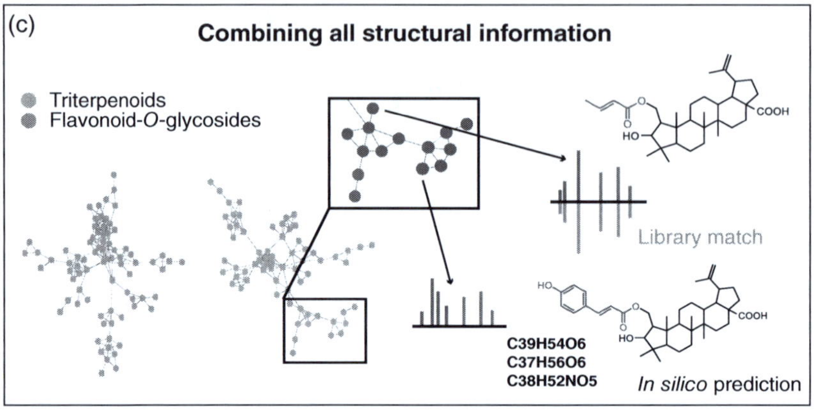

Figure 5.1 (Continued)

et al. 2010; Ruttkies et al. 2016), Network Annotation Propagation (NAP) (da Silva et al. 2018), MolNetEnhancer (Ernst et al. 2019b), and others (Blaženović et al. 2018; Naake et al. 2020) aim to annotate MS/MS spectra with full metabolite structures or chemical class terms. Each of these tools employs a slightly different approach to link structural information to spectral data. Typically, they result in either a *ranked list* of candidate molecules taken from spectral libraries or compound databases (Sorokina and Steinbeck 2020) to annotate a metabolite feature or in structural information (i.e., chemical compound class terms) for most of the fragmented molecules. Some of these tools will be further described in the methods section.

As illustrated by the plethora of tools described earlier, and the many more that exist (Misra and van der Hooft 2016), it becomes a challenge to keep an overview and combine the structural information obtained through these various tools. Therefore, workflows to combine the outputs of metabolite mining and annotation tools in one view have started to emerge (Figure 5.1c), for example, by adding various pieces of structural information to the molecular network through MolNetEnhancer (Ernst et al. 2019b). Taken together, computational metabolomics workflows transform metabolomics spectral data into structural information ranging from chemical compound classes, elemental formulas, and substructures, to entire candidate structures. In this chapter, we will showcase these recent developments with the results of three case studies that investigate the role of plant chemical diversity in ecology, taxonomy, and herbivore resistance.

This chapter highlights how recently developed computational workflows allow the fields of metabolomics and mass spectrometry-based analyses of plant extracts to transition toward the era of large-scale omics analyses (Tugizimana et al. 2020). Based on current developments, we also discuss multi-omics approaches to structurally and functionally annotate the plant specialized metabolome, and finish with a forward-looking perspective on how differential (Nothias et al. 2020), chemically informed (Tripathi et al. 2020), and repository-wide (Jarmusch et al. 2020; Wang et al. 2020) analyses could take place in the future in combination with the mining and annotation tools discussed here.

5.2 Materials and Methods

This chapter focuses on how computational metabolomics workflows are revolutionizing the plant metabolomics field in supporting biochemical interpretation through structural annotation of mass spectra. Therefore, we focus on explaining the computational tools used in the case studies, and their key contributions to metabolomics analyses, and refer to resources where more information on their functions and implementations can be found. An overview of the main tools used during the analyses of the case studies, and their key properties, can be found in Table 5.1.

5.2.1 Case Studies

High-resolution tandem mass spectrometry (LC-MS/MS) data sets were acquired from extracts of the Rhamnaceae family Juss (Kang et al. 2019) and *Euphorbia* genus L. (Ernst et al. 2019a), as well as four accessions of the *Capsicum annuum* L. following the protocols as described previously in (De Vos et al. 2007; van der Hooft et al. 2012a).

Metabolome mining (GNPS Molecular Networking, MS2LDA substructure finding) and a number of annotation tools (i.e. spectral library matching, SIRIUS, NAP through MetFrag) were applied followed by the MolNetEnhancer workflow (Ernst et al. 2019b) to extract and link chemical knowledge to plant taxonomy and herbivore resistance phenotypes. The formed molecular families were annotated with chemical class information providing a high-level chemical overview, whereas substructure information was mapped onto the molecular networks to detect metabolite subfamilies and as an aid in a structural annotation. In the sections later, each tool is described in more detail.

5.2.2 Metabolome Mining Tools

5.2.2.1 Molecular Networking

Molecular networking (Wang et al. 2016) is a powerful tool that aims to overcome one of the fundamental bottlenecks present when working with metabolomics data: the relatively small size of spectral libraries that contain mass spectra of known metabolites. Molecular networking is implemented within the Global Natural Product Social Molecular Networking Analysis Platform (GNPS; https://gnps.ucsd.edu), although other standalone implementations are available (Olivon et al. 2018). To start using molecular networking in your own study based on data obtained from your own platform, this recent protocol serves as a good starting point (Aron et al. 2020).

The fundamental idea behind molecular networking is the grouping of MS/MS spectra based on their similarity (Figure 5.1a). This clustering process groups MS/MS spectra into molecular families within which the connected (fragmented) molecules are assumed to be related through the sharing of structural properties. If one member of a family has been annotated via a spectral match, then it may be possible to infer structural properties of its neighbors through the propagation of structural annotations.

Central to molecular family formation in molecular networking is the use of a modified cosine score to define similarity. Where the standard cosine score aims to give high scores

Table 5.1 Overview of key metabolome mining and annotation tools mentioned in chapter.

Tool//info	Purpose	Main concept	Input	Output	Requirements/notes
SIRIUS/Zodiac (for larger molecules)	Elemental formula (EF) assignment	Computes fragmentation trees to determine most likely EF	LC-MS/MS files in open format	EF-annotated metabolite features	Reliable isotope patterns support EF assignments
CSI:FingerID	Metabolite feature structure annotation	Uses fragmentation trees as input to search best scoring candidates in structure databases	LC-MS/MS files in open format	Structure-annotated metabolite features	CSI:FingerID is nowadays part of the SIRIUS workflow
MZmine	LC-MS metabolite feature finding	Extracts ion traces, detects chromatographic peaks, and aligns them across samples	LC-MS/MS files in open format	Metabolite feature quantification table, and MGF file with MS1 and MS2 spectra of features	Other widely used open source peak picking software tools are XCMS, MetAlign, and MS-DIAL.
GNPS Molecular Networking	Mass spectral networking of LC-MS/MS dataset into molecular network	Creates nodes of very similar spectra and connects nodes based on spectral similarity score threshold	LC-MS/MS files in open format	Structured LC-MS/MS dataset into molecular network with molecular families	A network visualization tool like Cytoscape is key for inspection and analysis
Feature-Based Molecular Networking	Combined quantitative and qualitative LC-MS/MS analysis	Links feature finding to molecular networking	MS2 spectra (i.e. in MGF file format) & feature quantification table – with linked Feature IDs	Molecular network with quantified nodes, also allowing for statistical analysis	With FBMN, isomers are typically resolved in the resulting molecular networks

Network Annotation Propagation	metabolite feature structure annotation	uses network topology to rerank structure candidate lists obtained from compound databases	GNPS molecular networking job ID	structure-annotated metabolite features	choose the appropriate structure databases matching your sample
MS2LDA Substructure Discovery	unsupervised substructure motif finding in metabolite features	groups often co-occurring mass fragments/neutral losses into Mass2Motifs representing substructures	LC-MS/MS files in open format or GNPS molecular networking job	Mass2Motif-annotated metabolite features	MotifDB contains annotated substructure patterns
CANOPUS	systematic chemical compound classification of metabolite features	uses predicted molecular fingerprints from fragmentation trees to predict chemical compound class	LC-MS/MS files in open format	chemical compound-class annotated metabolite features	works with ClassyFire and NPClassifier chemical compound class ontologies
MolNetEnhancer	high-level chemical overview and detailed insights into scaffolds and substructures	combines output of various metabolome mining and annotation tools in one place	GNPS molecular networking job ID as minimum input – adding more annotations will improve chemical class predictions	enhanced molecular network with node and edge annotations	the combined outputs of various tools facilitate their synergetic use

to identical spectra, the modified cosine score aims to give high scores to spectra from pairs of molecules differing by single modifications, such as the addition of a hexose or deoxyhexose. It does this by allowing fragment ions in one of the spectra to be matched in their original position or shifted by the difference between the precursor masses of the fragmented metabolites.

Building on this, nowadays referred to as "Classical Molecular Networking" workflow, is the "Feature-Based Molecular Networking" workflow (Nothias et al. 2020) that allows combining the qualitative mining and annotation analyses with quantitative analyses. The linking of the outcome of peak picking software tools with molecular networking allows for the separation of isomers in the mass spectral networks. Moreover, statistical analyses of metabolite features captured by the network can now rely on the more reliable peak areas instead of spectral counts. Finally, it has also sparked the development of novel workflows such as ion identity molecular networking (IIMN) that decomplexify molecular networks by grouping nodes that originate from the same metabolite, i.e. molecular ions, isotopes, adducts, and/or in-source fragments (Schmid et al. 2021). These developments greatly assist in providing a more accurate view of the true chemical space covered by the chosen metabolomics approach.

5.2.2.2 Substructure Discovery

Many metabolites are modular, being built from a set of discrete building blocks or substructures. One limitation of molecular networking is the implicit assumption that all spectra belong to a single molecular family. Given their modular build, metabolites are not always best represented by discrete clusters and can instead be better represented by decomposition into these building blocks. MS2LDA (van der Hooft et al. 2016) is a technique that performs this unsupervised decomposition from observed spectra. All N spectra in the dataset are assumed to be built from a set of M structural Mass2Motifs (where $M<<N$). Given a set of spectra, MS2LDA performs a decomposition of the N spectra into M Mass2Motifs, learning a set of Mass2Motifs in the process. Mass2Motifs are often easier to structurally annotate than complete molecules as they tend to represent basic molecular building blocks, thus allowing for rapid partial annotation of many more spectra than can typically be annotated via spectral matching to spectral libraries (Figure 5.1a).

In the standard use of MS2LDA, a new set of Mass2Motifs is learnt in each analysis. However, MotifDB (Rogers et al. 2019) has been proposed – an open database of characterized Mass2Motifs that can be added to analyses as well as detecting novel ones. In much the same way as GNPS promotes the sharing of metabolomics data and annotated mass spectra therein, it is hoped that MotifDB will promote the sharing of annotated Mass2Motifs. The best way to get started with MS2LDA is to use it from within the GNPS molecular networking pipeline (https://ccms-ucsd.github.io/GNPSDocumentation/ms2lda/) and then upload your MS2LDA model into ms2lda.org for further exploration (Wandy et al. 2018).

5.2.3 Metabolome Annotation Tools

5.2.3.1 Elemental Formula Assignment – SIRIUS and ZODIAC

A key step in the annotation of molecules is the assignment of their molecular formulas (Figure 5.1b). This is typically done from the observed MS1 m/z. Methods vary in their complexity but normally make use of the MS1 *m/z* as well as the isotope distribution (both to approximate the number of, for example, C atoms) and to estimate the charge state (isotope shifts provide direct evidence of the number of units of charge of an ion). For any given intact ion observed for a given molecule, there are always many possible molecular formulas that can be assigned. MS/MS data holds additional structural information over MS1 to assign an accurate molecular formula of the intact molecule and has been leveraged in recent computational tools.

One widely used tool that performs molecular formula assignments based on MS/MS data is SIRIUS (Dührkop et al. 2019). It relies on the calculation of fragmentation trees from MS/MS spectra (Dührkop et al. 2015) and makes use of an efficient knapsack algorithm to find all combinations of elements that can make formulas within a certain *m/z* range and rank the best matches and has proven to be very accurate in assigning molecular formulas. Recently, Zodiac (Ludwig et al. 2020), which employs a Gibbs sampling-based network strategy was introduced to solve one aspect of SIRIUS' performance: in the higher mass range (>500 Dalton), the elemental formulas were not always correctly assigned. Zodiac re-ranks molecular formula candidates by considering the corresponding fragments and losses between the fragmentation trees. Most importantly, Zodiac predicts the molecular formula *de novo* without predefined ranges of atomic compositions. Its substantially better performance than SIRIUS is very promising for plant extracts that are typically full of relatively large compounds.

Both SIRIUS and Zodiac have limitations in their performances; they perform poorly with metabolites with uncommon atoms and although they still work, they are less effective with negative mode predictions. These limitations are not due to the algorithm but rather the limited amount of data that is available for training. There are a limited number of reference spectra with unique atomic compositions (e.g. sulfur, phosphate, arsenate, and some halogens) and there is limited availability of negative ionization mode reference spectra. Adding more reference spectra to public libraries will allow retraining and eventually solve these bottlenecks. This would further improve the breadth of spectra for which the molecular formulas can be accurately assigned, but the accumulation of fully annotated spectra with metadata in public mass spectral libraries remains a relatively slow process.

5.2.3.2 Candidate Structure Finding – Library Matching

The most common and reliable method for finding candidate structures is matching observed MS2 spectra to databases of standard spectra (Figure 5.1b). Various open databases are available including Massbank EU (Oberacher et al. 2020), MassBank (Horai et al. 2010), MassBank of North America (https://mona.fiehnlab.ucdavis.edu), ReSpect (Sawada et al. 2012), GNPS (Wang et al. 2016), HMDB (Wishart et al. 2018), and natural products specific mass spectral libraries such as MIADB (Fox Ramos et al. 2019), as well as commercial databases such as Metlin (Xue et al. 2020) and NIST 2020. When searching with a particular query spectrum, the database is initially filtered to find all the library spectra with a precursor *m/z* within some small tolerance of the query precursor. Then, the similarity between the query spectrum and the filtered spectra is computed. This is typically done with a cosine-based similarity score in which fragment ions from the two spectra are matched and the product of the matched intensities is calculated. In some cases, the *m/z* value is also used in the calculation of the product (fragment ions with higher *m/z* are often considered more informative).

Cosine scoring functions typically require the definition of various parameters including (i) the precursor *m/z* tolerance, (ii) the fragment ion tolerance (to define a match), (iii) the score above which to define a hit, and (iv) the minimum number of matched fragment peaks that must be found in order to consider a score to be legitimate. Exact settings are dependent on many factors including the samples, sample matrix, and mass spectrometry platforms used – more information and guidelines can be found in (Aron et al. 2020).

Interpreting the scores obtained via database matching is challenging. Unlike in proteomics where false-discovery-rate (FDR) procedures are well established, it is not standard to apply an FDR procedure during metabolomics annotations, although there has been some effort in that direction in the development of algorithms, there is still a need for improvement (Scheubert et al. 2017; Wang et al. 2018; Li et al. 2020).

5.2.3.3 Candidate Structure Finding – CSI:FingerID

The small size of spectral libraries has inspired considerable research into the use of machine learning to generalize beyond existing library spectra and also exploit compound libraries (Figure 5.1b). Such approaches can be broadly classified into two classes: (i) those that extend spectral libraries by predicting spectra for structures that are not present in the library, and (ii) those that attempt to predict features from spectra that can then be matched to features derived from structures in compound databases. The second category has proved more successful, within which the most popular approach is CSI:FingerID (Dührkop et al. 2015) (now packaged as part of the SIRIUS software (Dührkop et al. 2019)). CSI:FingerID makes use of a large number of machine learning classifiers (support vector machines) that each predict the presence or absence of a single molecular fingerprint feature from the query spectrum.

The resulting fingerprint prediction vector can then be compared against fingerprints derived from any chemical structure (an easy process using, e.g., the CDK software library (Willighagen et al. 2017)). The support vector machine classifiers make use of various features derived from the query spectrum, mainly derived from a fragmentation tree that has been inferred from the spectrum. One drawback of CSI:FingerID is that it is relatively slow. The IOKR algorithm (Brouard et al. 2016) adopts a very similar approach to CSI:FingerID (using the same input features) but rather than many support vector machine classifiers, poses the fingerprint prediction problem as a vector regression problem. In doing so, it provided a significant speed-up over CSI:FingerID.

5.2.3.4 Chemical Class Assignment – ClassyFire, NPClassifier, and CANOPUS

Plant scientists traditionally work by classifying the plant specialized metabolites into compound classes that can be linked to function and bioactivity. Thus, when presented with metabolomics data of extracts of one or more plants, it is very useful to focus on a specific subset of the data for in-detailed measurements based on chemical compound class information (Figure 5.1c). One route to make this possible is to look at the (top) candidate structures for each metabolite feature in the dataset using the annotation tools described earlier and then assign chemical compound classes to them. Manually, this would be impracticable for large datasets, and an automated solution is needed to make this route functional.

ClassyFire was the first automated tool coined to do exactly that: based on the textual representation of the candidate structures (using SMILES (O'Boyle, 2012; Southan, 2013) or InchiKeys(O'Boyle, 2012)), ClassyFire (Djoumbou Feunang et al. 2016) returns a hierarchical compound class annotation based on the presence and absence of particular substructures. Assigned chemical taxa terms include many terms relevant to plant chemistry including triterpenoid, flavonoid, or hydrolyzable tannin. Using ClassyFire annotations, researchers are now able to focus on a subset of metabolite features that belong to the chemical class that they study, for example, or to recognize metabolites that are likely to be completely novel as no known chemical class could be assigned to them.

To increase the usability of such an approach for natural products scientists, recently, NPClassifier (Kim et al. 2021) was proposed where machine learning was used to link historical natural product classifications to structures. When provided with a structure in the form of a SMILES, NPClassifier returns this natural product vocabulary and the tool is thus very helpful to prioritize (groups of) structural candidates in plant extracts and to link metabolomics-based measured spectra to genomic information or other sources of information as this can happen at the chemical compound class level. To further assist the selection of *plausible* candidate structures, various approaches have been proposed that link sample information (metadata) to compound information. For example, taxonomically informed metabolite annotation or taxonomic pondering in short was coined by Rutz et al. who demonstrated that considering the taxonomical distance between the analyzed source and the biological source of candidate structures allows an efficient re-ranking of the candidate structures leading to a systematic enhancement of the metabolite annotation (Rutz et al. 2019).

Finally, CANOPUS (Dührkop et al. 2020) was introduced that uses a deep neural network to predict 2,497 ClassyFire compound classes from fragmentation spectra. This workflow does not depend on annotated candidate structures but can assign ClassyFire and NPClassifier chemical compound classes directly to MS/MS spectra if a fragmentation tree can be calculated.

5.2.3.5 Combining All Structural Information – MolNetEnhancer

The multitude of mass spectral metabolome mining tools and manifoldness of different output formats and analysis platforms (Table 5.1), impeding the easy visualization of output from complementary tools within one data file or platform, have fostered the development of MolNetEnhancer (Ernst et al. 2019b) (Figure 5.1c). MolNetEnhancer is a workflow that maps output from spectral library matching and *in silico* structural annotation onto a GNPS mass spectral molecular network.

Broad chemical structural information is summarized by calculating the most predominant chemical classes retrieved for all matches per molecular family at each hierarchical level of the ClassyFire chemical ontology onto the network. Finally, shared substructural motifs retrieved from MS2LDA in between mass features of the same molecular family are visualized through additional edges connecting the nodes. This allows the natural product researcher to investigate and strengthen structural hypotheses by collecting information from several complementary metabolome mining tools. Instead of comparing tables and output from multiple platforms, information from mass spectral molecular networking, *in silico* structure annotation, and substructure discovery can be easily visualized within one data file in Cytoscape (Shannon et al. 2003).

MolNetEnhancer can conveniently be run through the GNPS platform, currently supporting *in silico* structural input from DEREPLICATOR (Mohimani et al. 2018), a tool to annotate (mainly) peptidic molecules as well as some other types like polyketides and flavonoids, and NAP through MetFrag (Ruttkies et al. 2016; da Silva et al. 2018) that exploits the molecular network topology to rerank candidate structures found by MetFrag based on the expected consistency of annotations within molecular families of connected components (https://ccms-ucsd.github.io/GNPSDocumentation/molnetenhancer/). In addition, customizable Jupyter notebooks, which allow tool-independent input of (*in silico*) structural information, are available in python and R (https://github.com/madeleineernst/pyMolNetEnhancer, https://github.com/madeleineernst/RMolNetEnhancer).

5.3 Results and Discussion

As highlighted in the Material and Methods section, a suite of computational metabolomics workflows has been developed to address the information-dense metabolomics data generated by modern mass spectrometers. Here, for three case studies, we show how computational workflows that combine molecular networking with *in silico* annotation, substructure identification, and metabolite classification, support structural annotation of metabolite features in plant extracts to enable biochemical interpretation to gain an insight into the structurally diverse plant chemical space (Figure 5.1).

5.3.1 Rhamnaceae Case Study

Firstly, the chemical diversity of two major clades of the Rhamnaceae plant family (the Ziziphoid and Rhamnoid) was compared by analyzing an enhanced molecular network of 70 plant extracts (Kang et al. 2019). The family Rhamnaceae is a cosmopolitan plant family including about 900 species. Phytochemical investigations revealed the presence of various specialized metabolites such as anthraquinones, flavonoids, triterpenoids, and cyclopeptides in this family, but little has been known about the chemotaxonomical aspect.

LC-MS/MS data from 70 species were processed into a molecular network consisting of 2,268 mass spectral nodes organized into 141 independent molecular families (two or more connected nodes of a graph) via the GNPS platform using the feature-based molecular networking workflow in conjunction with MZmine 2 for LC-MS peak quantification (Pluskal et al. 2010; Nothias et al. 2020). GNPS spectral library matching provided annotations to 51 spectral nodes, which was about 2.5% of the total number of spectral nodes in the molecular network. *In silico* annotation was retrieved by NAP including public structural databases such as ChEBI as well as the Dictionary of Natural Products (DNP) (Sorokina and Steinbeck 2020), and candidate structures were assigned to 2,187 spectral nodes.

Based on all of these *in silico* annotations, class-level annotations for each molecular family were provided. Chemical compound classes for every candidate structure were given by ClassyFire, and the most predominant classes per each molecular family were retrieved. This allowed the authors to give putative chemical classifications to most molecular families in the dataset. Based on this, the normalized distribution pattern of different classes was retrieved, and it was revealed that Rhamnoid species exhibit various flavonoids, carbohydrates, and anthraquinones while Ziziphoid species show more diversified triterpenoids and triterpenoid glycosides.

The two major Rhamnaceae clades also show differentiation in a principal coordinate analysis (PCoA) plot enabled by the GNPS environment in conjunction with QIIME2 (Bolyen et al. 2019) where the samples are presented as a lower-dimension representation of their dissimilarities based on Bray–Curtis distance between their MS feature contents and colored according to their clade membership (Figure 5.2a). In contrast, coloring the samples according to their initial extraction solvent (methanol or 95% ethanol) does not show an obvious pattern (Figure 5.2b), thus indicating that this did not affect the overall analysis. Furthermore, molecular networking combination with statistical analyses can reveal interesting trends and information on both molecular family and individual metabolite levels.

Figure 5.2c shows how a triterpenoid family contains quite some matches to GNPS library spectra, in particular, for those specialized metabolites that (i) are more widely present across the Rhamnaceae plants, and (ii) are more abundant. This makes sense as there will be a higher chance that such molecules were previously isolated and purified from crude plant extracts. Figure 5.2d shows how the quantitative information obtained from MZmine can be used to discover the presence of specialized metabolites across the two major Rhamnaceae clades or across the studied genera.

Finally, this study also highlighted that the integration of MS2LDA with molecular networking can provide information on substructural differences. Mapping the distribution of Mass2Motif on the molecular network revealed subfamily-level differences within a molecular family; for example, spectral nodes annotated as flavonoid glycosides could be further estimated to have aglycone substructures of kaempferol, quercetin, and myricetin, enabled by the MS2LDA substructure discovery. The annotated motifs from the study are also available in MotifDB for reuse in negative ionization mode-based plant metabolomics studies (Rogers et al. 2019).

5.3.2 Euphorbia Case Study

Ernst and collaborators (Ernst et al. 2019a) have used a computational metabolomics workflow that integrates molecular networking, *in silico* chemical structural annotation, substructure recognition, and automated metabolite classification to unravel that *Euphorbia* diterpenoids may be possible evolutionary drivers of specialized metabolite diversity at a global scale involved in herbivore resistance.

With over 2,000 species, the plant genus *Euphorbia* is among the most diverse and species-rich plant genera on Earth with a near-cosmopolitan distribution and extraordinary chemical diversity (Horn et al. 2012, 2014; Vasas and Hohmann 2014). Although chemotaxonomic relevance and, in particular, chemical diversity of highly toxic and likely

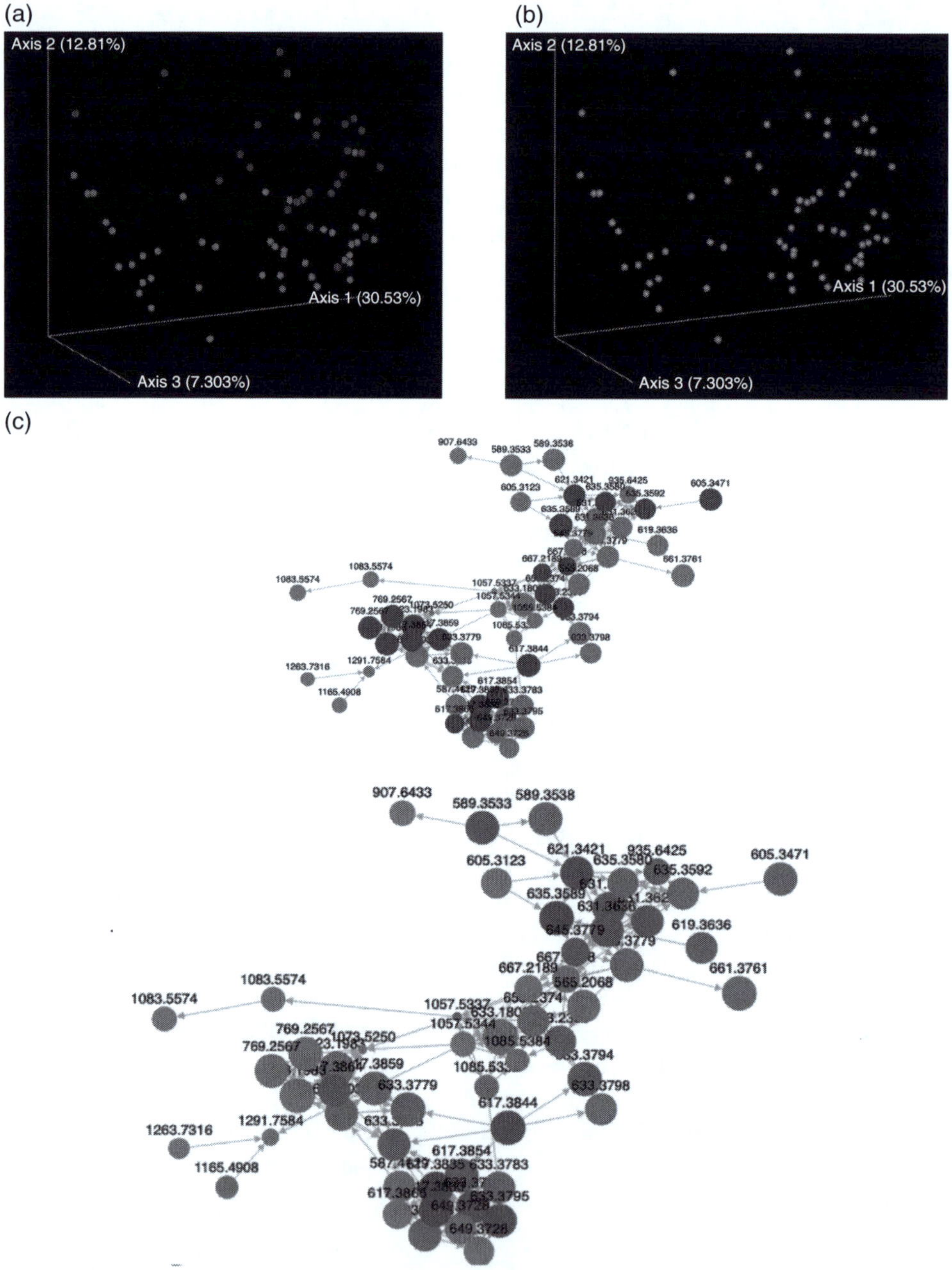

Figure 5.2 (a) Principal coordinate analysis (PCoA) plot based on Bray–Curtis distance between 71 Rhamnaceae plants colored according to their clade-level phylogeny: Rhamnoid (Red) and Ziziphoid (Blue) with 30.5% variation captured in PC1 [Axis 1], 12.8% in PC2, and 7.3% in PC3. (b) Same PCoA plot as in (a) but now colored according to initial extraction solvent: methanol (orange) and 95% ethanol (green). (c) Triterpenoid molecular family with spectral nodes sized proportionally to the number of files in which they occurred and colored (top) blue if a library match was found and (bottom) in a spectrum from light red to dark brown proportional to the summed precursor intensities of the spectra clustered in one node. It can be seen that many smaller and lighter nodes do not have a direct library match, but display sufficient spectral similarity to nodes that do have library matches to allow for the propagation of structural information. (d) Feature area box plots for the node with a precursor *m/z* of 617.3830 ([M−H]-) that matched the pentacyclic triterpenoid 2-*O*-*E*-*p*-Coumaroyl alphitolic acid in the GNPS library, distributed (top) over the two major clades and (bottom) the 15 occurring genera in 71 Rhamnaceae plants. It can be observed how this triterpenoid is mainly present across several genera that belong to the Ziziphoid clade, with only a few noticeable exceptions. (*See insert for color representation of the figure.*)

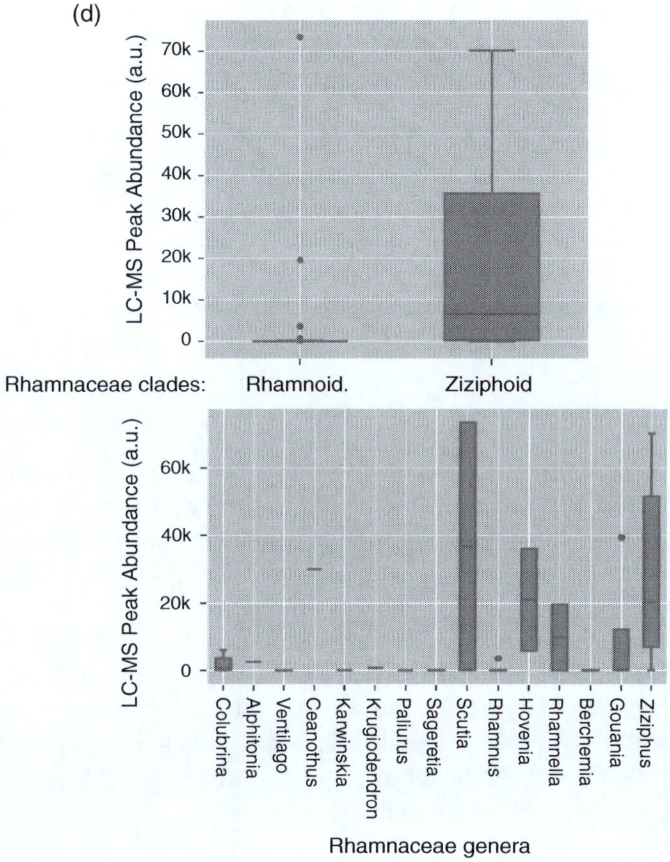

Figure 5.2 (Continued)

feeding deterrent macro- and polycyclic diterpenoids biosynthetically derived from a head-to-tail cyclization of the tetraprenyl pyrophosphate precursor (*Euphorbia* diterpenes) within the genus is known (Batra 1983; Hundsdoerfer et al. 2005; Shi et al. 2008; Huang et al. 2014; Vasas and Hohmann 2014; Appendino 2016; Hua et al. 2017; Ernst et al. 2019a), little is known on how geographic separation over evolutionary time may have impacted chemical differentiation. The integrated metabolomics workflow developed by Ernst and collaborators facilitated the identification of different diterpene spectral fingerprints clustered within several molecular families of 43 plant extracts representing the genus' global genetic diversity and biogeographic history.

A molecular network was created through the GNPS platform, and spectral library hits and *in silico* chemical structural annotations retrieved from Sirius+CSI:FingerID and NAP through a custom *Euphorbia* diterpenoid database were mapped. Furthermore, to identify subtle substructural differences of diterpenoid skeletons or side chains, mass spectral features sharing the same substructures as retrieved from MS2LDA were mapped as independent edges within the network. Finally, to obtain a broad overview of retrieved library matches and *in silico* annotations, chemical classes were recovered for all matches using ClassyFire and the most predominant chemical classes at each level of the chemical taxonomy were calculated at the molecular family level.

Using this approach, Ernst and co-workers showed that specialized metabolite diversity and the number of feeding-deterrent Euphorbia diterpenoids is significantly reduced in phylogenetic clades native to the Americas, compared to species native to the Afro-Eurasian continents. This finding suggested a coevolutionary arms race in species native to the Afro-Eurasian continents and chemo-evolutionary adaptation to reduced herbivory pressure in the Americas. The approach deployed by Ernst and collaborators in this study, alongside the Rhamnaceae study described here, later led to the development of the MolNetEnhancer workflow.

5.3.3 Pepper Case Study

Capsicum is a genus of the nightshade (*Solanaceae*) family of which several species are grown as hot and sweet pepper, which belong to the most grown vegetable crops worldwide (FAOSTAT; Data Productions Crops 2020). Domestication has unintentionally led to the loss of natural resistance toward insects, thereby hampering global pepper cultivation. For instance, western flower thrips'

(*Frankliniella occidentalis* Pergande) infestation in pepper results in leaf deformation, fruit deterioration, and virus transmission (van Haperen et al. 2019; Visschers et al. 2019). As the use of insecticides resulted in the emergence of resistant thrips' populations and as these are a threat to beneficial insects such as pollinating bees and biocontrol agents, identifying the host-plant resistance mechanisms in pepper has become an urgent need (Steenbergen et al. 2018). Resistance to insects may be based on anti-xenosis (non-preference) or anti-biosis (increased mortality or reduced growth, development, and reproduction of the insect) mechanisms and partial thrips resistance has been reported in *Capsicum* (Maharijaya et al. 2011).

In both described mechanisms, the encountering of insects to the specialized metabolites present in the leaves or fruits is of relevance for their performance. For example, abundantly present acyclic diterpene glycosides were linked to thrips resistance in pepper (Macel et al. 2019; Maharijaya et al. 2019). Recent studies suggest that leaf-based thrips resistance fluctuates with plant developmental stages (Macel et al. 2019; Visschers et al. 2019) and with leaf age (van Haperen et al. 2019). These and other studies included *C. annuum* and *C. chinense* genotypes, while less domesticated and wild *Capsicum* species are mostly overlooked.

By using the comprehensive MolNetEnhancer workflow, leaf metabolic profiles across different species in the *Capsicum* genus and among accessions with contrasting thrips susceptibility in different plant- and leaf-developmental stages were investigated for metabolite markers related to thrips resistance. The chemical diversity of the *Capsicum* genus was compared by analyzing LC-MS/MS data from leaf extracts of 35 species to construct a molecular network. As a primer for a broader study, Figure 5.3 shows how positive ionization mode LC-MS/MS data was transformed into informative molecular families that could be annotated with chemical compound classes to assist in pinpointing relevant and

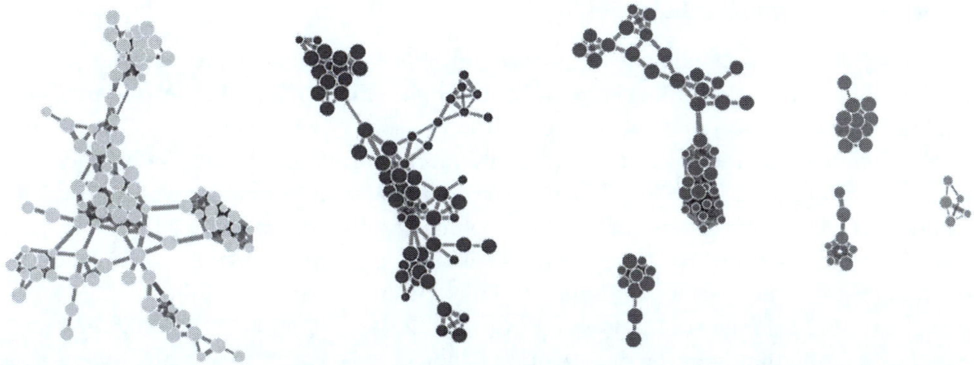

Figure 5.3 MolNetEnhancer provides direct insight into the non-volatile chemical space of sweet pepper (*Capsicum annuum* L.). The most abundant annotated ClassyFire chemical compound class subclass terms were as follows for the positive ionization mode Feature-Based Molecular Network: amino acids, peptides, and analogues (435), diterpenoids (172), linoleic acids and derivates (164), sesquiterpenoids (89), and terpene glycosides (67). Here, example molecular families are shown colored by their annotated chemical subclass: linoleic acids and derivatives (green), terpene glycosides (black), diterpenoids (purple), sesquiterpenoids (red), and depsipeptides (pink). The size of the nodes is proportional to the number of MS/MS spectra obtained for that particular metabolite feature, whereas the thickness of the edge is proportional to the modified cosine score between the connected MS/MS spectra. (*See insert for color representation of the figure.*)

interesting chemistry, as well as to gain an overview of commonly occurring (i.e., diterpenoids) and rarer (i.e., depsipeptides) chemistry.

The integrated metabolomics workflow based on the GNPS platform (Wang et al. 2016; Nothias et al. 2020) combined with MZmine 2 (Pluskal et al. 2010) for compound abundance quantification as developed by Ernst and co-workers (Ernst et al. 2019b) allowed for the identification of different spectral fingerprints among *Capsicum* species and multiple compounds of various biochemical origin were found to have chemotaxonomical relations. For instance, the leaf metabolite signature of species in the more genetically distinct *pubescence* clade substantially differs from those in the *annuum* and the *baccatum* clades. Furthermore, comparative metabolomics resulted in the annotation of various specialized metabolites, including those belonging to the chemical compound classes of flavonoids, diterpenoids, and triterpenoids, which were correlated to leaf susceptibility toward thrips and to developmental changes between vegetative and generative plants. Multiple diterpenoids and triterpene glycosides strongly correlated to herbivore fending, whereas isoflavones and sesquiterpenoids correlated with plant development.

As herbivore feeding often results in changes in metabolite abundances, further insight into how these metabolites are related at the biochemical level will aid to obtain mechanistic insights for biochemical processes that are relevant for metabolome adaptation upon herbivory. Taken together, the use of the computational metabolomics tool MolNetEnhancer remarkably facilitated the characterization and annotation of metabolites in the leaf phytochemistry of pepper, which could assist in future breeding programs toward thrips feeding resistance in pepper.

5.3.4 Other Plant Metabolomics Studies

The computational metabolomics workflows introduced in this chapter are increasingly used in plant chemical profiling studies. For example, *Alnus* spp. (Betulaceae) were investigated for the chemical diversity between different species and plant parts. The metabolomics data was integrated with tyrosinase inhibitory assay data to suggest gallic acid as an active ingredient of *A. firma* based on the correlation between its presence and bioactivity (Kang et al. 2020). Other application areas include plant-pathogen interactions (Zanzarin et al. 2019; DesRochers et al. 2020; Silva et al. 2020), the role of the plant hormone methyl-jasmonate and its analogs in plant-microbe consortium formation (Adaikpoh et al. 2020), plant-based food storage and processing (Xu et al. 2019; Gauglitz et al. 2020), and plant-based food fermentation by the microbiome (Raghuvanshi et al. 2019). Finally, an example of large-scale molecular networking and phylogenetic analyses to investigate chemo-evolutionary diversification that involves specialized diterpenoid chemistry in 291 specimens of *Eremophila* and allied genera was also proposed (Gericke et al. 2020), where the authors show that novel specialized metabolites could be discovered and that chemical compound classes can be linked to functional annotations such as leaf morphology. Indeed, the number of plant metabolomics applications is growing, and this is showing that – driven by computational advances – metabolomics is starting to become a generic screening profile to assess the impact of changing conditions on the plant metabolome.

5.4 Current Limitations

Driven by computational metabolomics workflows, large-scale metabolome profiling is now possible as illustrated by the case studies in this chapter. Furthermore, annotation rates have significantly improved over the last few years. However, especially in natural products' discovery and plant metabolism analysis, there are still limitations of current approaches. For example, spectral libraries with relevant MS/MS spectra are still relatively small. The estimated low amount of about 2.5% of natural products of which reference spectra are currently available in the public libraries of GNPS (van der Hooft et al. 2020) hamper the quick confident structural annotation of many metabolite features in plant metabolomics profiles.

Moreover, candidate structure databases are incomplete. Indeed, various collections of natural products exist – some more dedicated than others – and it was recently shown they together sum up to about 400,000 unique structures (Sorokina and Steinbeck 2020), which resulted in the launch of the novel agglomerated database COCONUT (Sorokina et al. 2021). Furthermore, faced with the complex metabolite mixtures that plant extracts typically produce, it is very unlikely that all constituents are part of the known natural product space. In other words, we do not know yet the existence of many of the molecules that we measure. This makes the elucidation of their structures even more challenging as often a combination of multiple analytical tools, and thus expertise, will be needed to solve the puzzle.

Another current limitation is that the integration of tools and platforms is often hampered by various file types and formats that the tools require. Currently, we are witnessing how various "global metabolomics ecosystems" are emerging: platforms and tools that work together to analyze metabolomics data. Within such an ecosystem, the tools do work together well and most basic functionalities such as peak picking, peak alignment, and library matching are covered; however, there are also complementary tools, which now often prove difficult to use outside their own ecosystem. For plant metabolomics to truly enter the "big data" era, it is inevitable that these "silos" will have to "communicate" with each other, thereby truly democratizing metabolomics approaches by making them universal and globally interoperable. Instrument manufacturers can become major drivers in this effort through building communication channels directly from their software and/or allowing direct export of the data in compatible formats as possible solutions.

Finally, when presented with "big data" or massive amounts of files, many applications get stuck during data processing or visualization. For example, GNPS has till today processed 100,000s of files altogether and the visualization of large-scale repository-wide analysis remains a challenge. Whilst improvements are being made, it proves to be hard for scientific groups to acquire funding to create "proper solutions" for large-scale analyses leaving many tools out there that work for 10–100s of files but when faced with more data, they tend to collapse. We expect that the publication of impactful large-scale studies will stimulate the (available funding for) development of more scalable and sustainable solutions.

5.5 Conclusions

In this chapter, we demonstrate with three plant-based case studies how computational metabolomics workflows are changing the face of plant mass spectrometry-based natural product discovery. The large-scale comparisons between sample types and conditions that are now possible will assist in further charting the plant natural product space. Furthermore, as highlighted by each case study, they allow for new biochemical questions that we can start to ask by relating plant chemical diversity to ecology, taxonomy, or herbivore resistance. We will finish the chapter with an outlook toward promising ways to tackle some of the current limitations that we identified and other exciting directions that will impact mass spectrometry-based natural product discovery.

5.6 Outlook

5.6.1 Extended Natural Product Candidate Structure Space

The vast and especially yet uncharted part of the plant specialized metabolome remains difficult to connect to mass spectral features in metabolomics data since both structures and their spectra are lacking from current databases. To overcome this challenge, one could identify common building blocks and build hypothetical metabolites from them. Alternatively, strategies like MINEs (Jeffryes et al. 2015) and BioTransformer (Djoumbou-Feunang et al. 2019), and EMMF (Hassanpour et al. 2020), use known reactions and biotransformations to create an extended candidate search space of biochemically feasible metabolites, but then adapted for plant metabolism. For example, PlantMAT is one of the first proof-of-principle strategies employing the building block approach (Qiu et al. 2016). PlantMAT uses informed phytochemical knowledge for the prediction of plant natural products such as saponins and glycosylated flavonoids through combinatorial enumeration of aglycone, glycosyl, and acyl building blocks. A sophisticated analytical approach based on LC-MS/MS and NMR was then employed to provide MS/MS spectra to recognize the building blocks and NMR spectra to confirm the annotations and confirm the linkage of the building blocks.

Other examples that show how metabolism rules aid in structural annotation of the plant-based food metabolome in human samples include DAC-Met and MAGMa (Ridder et al. 2014; Ohbuchi et al. 2020). Finally, one can imagine that the chemical space could be built based on known specialized metabolites as a kind of "anchors"; and machine learning techniques have been started to use in this area: for example, DeepSMILES (O'Boyle and Dalke 2018) and SELFIES (Krenn et al. 2020) were proposed to produce novel molecular structures. However, it remains a challenge at this stage to produce molecules with relevant chemical properties for a specific experiment following a generic approach.

Nevertheless, witnessing the current progress in machine learning and deep learning, we can expect improved models in the near future that assist in expanding the candidate space in a smart manner with molecules having relevant physicochemical properties, including those structures that we cannot imagine yet. Altogether, no systematic approach to map the complete plant specialized metabolome is currently available; however, several computational approaches are starting to make inroads into its vast and chemical diverse space.

5.6.2 Improved Mass Spectral Similarity Scoring

Many metabolomics analyses including library matching, analog searching, and mass spectral network analysis such as molecular networking rely on the comparison of two spectra and the assignment of a mass spectral similarity score. These scores are, for example, used to define a threshold to indicate whether a match to a spectral library is relevant or not, or whether or not to connect to fragmented metabolites in a network. Current scores rely on the overlap between mass fragments across two spectra, which works well for nearly identical relatively small structures that yield very similar spectra (Huber et al. 2021a). However, plant specialized metabolites are generally large molecules that do share common scaffolds but typically harbor two or more subtle modifications; and, as a result, the number of overlapping peaks between the mass spectra of such plant metabolites decreases – sometimes to almost zero, whereas the majority of their structures completely overlap (Huber et al. 2021a). Consequently, to assess their validity, spectral matches between two mass spectra should be carefully considered for example by studying the number of overlapping mass fragments and their intensity patterns, as well as by considering the mass difference between the parent ions that could relate to possible biochemical reactions that connect the two metabolites.

In this light, the development of alternative mass spectral similarity scores that do not fully rely on overlapping mass fragments is highly encouraging for the plant natural product discovery field. For example, Spec2Vec, inspired by natural language processing (Huber et al. 2021a), shows better performance for library-matching tasks and good performance especially for larger molecules, likely due to the increased number of mass fragments that they typically produce, but more research should be done to pinpoint this phenomenon. In the light of the growing spectral libraries and especially the available experimental data for repository-wide analysis, another advantage of Spec2Vec is its scalability: once a model is learnt that captures the relationships between mass fragments and neutral losses across many mass fragmentation spectra, determining the spectral similarity score between two spectra using the learnt mass spectral embedding is very fast. Ultimately, combinations of classical (cosine score-based) and novel (machine learning-based) mass spectral similarity scores, such as those facilitated by Spec2Vec and by supervised scores (Huber et al. 2021b), may prove to work best for the entire range of specialized metabolites.

5.6.3 Combined Genome and Metabolome Analyses

Metabolomics analyses have greatly increased our insights into the plant specialized metabolome. Other omics analyses such as genomics can provide complementary structural as well as functional information about plant specialized metabolites. In some cases, groups of chromosomally clustered biosynthetic genes in the plant genome together encode for the production of a metabolite: PlantiSmash (Kautsar et al. 2017, 2018), PhytoClust (Töpfer et al. 2017), PlantClusterFinder (Schläpfer et al. 2017), and CycloNovo (Behsaz et al. 2020) recognize such gene clusters based on specific rules that detect core biosynthetic enzymes that are responsible for the creation of basic scaffolds such as terpenes and peptides. "Unclustered" pathways can be predicted through the identification of coexpression modules from transcriptomics data, which has been shown to constitute a powerful approach to identifying biosynthetic genes involved in the same pathway (Wisecaver et al. 2017). After assigning these predicted pathways from coexpression and/or colocalization information, these predicted biosynthetic gene sets can be used to link genomic information to the metabolomics data, for example, on the stereochemistry of particular substructures that are encoded in the genome.

On a higher level, one can try to link metabolic products in complex extracts to the plants that biosynthesize them: for example, at the level of chemical compound classes that can be recognized in the genome and metabolome such as terpenes and peptides (van der Hooft et al. 2020). Additional functional information on plant metabolites may then come from associated gene ontology (GO) terms (Hill et al. 2008) or gene annotations. For example, the Antibiotic Resistant Target Seeker, ARTS (Mungan et al. 2020), has shown for bacterial specialized metabolites to be able to recognize putative antibacterial metabolites and assign their mode of action based on the presence of specific resistance genes in the close proximity of the biosynthetic gene cluster. Finally, the other omics tools, transcriptomics and proteomics may further assist in prioritizing "active metabolites" and linking them to their biosynthetic genes by measuring RNA and/or protein production under various experimental conditions; for example, the biosynthetic pathway for falcarindiol in tomato was elucidated by identifying multiple specific stress conditions and time points in which the metabolite was produced and found co-abundant with the expression of candidate enzyme-coding genes that were prioritized based on key structural features of this metabolite (Jeon et al. 2020).

It must be added here that to have all four omics data types available in one study is currently still a rarity, mostly due to the required expertise and costs associated with such an effort. Still, multi-omics studies are becoming more and more feasible now that the costs to run omics profiles are decreasing and the data analysis workflows that process the raw omics data into annotated and quantified units are increasingly in place. For example, Belcour et al. used a combination of metabolite profiling and genome-scale metabolic networks to infer biochemical reactions allowing them to annotate novel metabolites in promiscuous biosynthetic pathways (Belcour et al. 2020). Such multi-omics approaches seem promising to fulfil missing and unknown parts of specialized metabolomic pathways.

5.6.4 Linking Complementary Analytical Tools

Mass spectrometry-based metabolomics can provide a wealth of structural information. The use of complementary analytical tools can assist in reducing the number of possible elemental formulas or structures for metabolite features by providing constraints on the elemental composition or atomic connections. For example, it was shown how the use of stable isotope labelling greatly reduced the number of elemental compositions to consider and also yielded elemental formulas not yet present in any database (Nakabayashi and Saito 2020) (Tsugawa et al. 2019). Through comparison of mass spectra from ^{13}C-labelled and non-labelled extracts, the exact number of carbon atoms could be easily established for most metabolite features. That puts massive restrictions on the rest of the elements as well making elemental formula assignment much easier.

The use of hyphenated platforms facilitating linked analysis of MS/MS and NMR has been advocated (van der Hooft et al. 2011, 2012b) to put restrictions on the atomic bonds, for example, by establishing how plant metabolite building blocks are connected; indeed, these hyphenated approaches result in the full structural elucidation of a number of abundant metabolite features in natural extracts. The MS and MS/MS signals then cover the main components of the metabolite whereas specific NMR signals help (i) establish on which atoms decorations like hydroxylations or methylations took place, and (ii) identify how the building blocks are glued together, for example, where glycosylation took place. However, in practice, these approaches tend to be relatively expensive to run. Therefore, computational workflows that prioritize the metabolite features to study with such sophisticated platforms are indeed needed to make such approaches worth pursuing.

5.6.5 Future Perspective: Chemically Informed Repository-Scale Analyses

The demonstrations in this chapter were based on metabolomics datasets that were obtained to answer specific biochemical questions. However, their analyses heavily relied on previously obtained data being available to the community: reference MS/MS spectra to structurally annotate spectra. Moreover, to assess if the chemistry found in your sample is novel, the option to compare your metabolomics profiles to other profiles obtained previously is key. The reason that rutin has been "rediscovered" so many times is mostly that it has been impossible to effectively compare metabolomics profiles from various sources. It is likely that this will change in the nearby future driven by incentives like FAIR (findable, accessible, interoperable, and reusable) data sharing that is increasingly a prerequisite to publishing in scientific literature. Moreover, reporting standards such as those described by Sumner et al. are also increasingly used to make metabolomics studies more reproducible (Sumner et al. 2007). Indeed, this should be a key and integral part of metabolomics and science in general; however, it is clear that as a field, we still observe large differences in adherence to reporting standards making reproducing results and analyses too often too difficult (Considine et al. 2017). A good example of making metabolomics analyses reproducible is the GNPS job system where each job gets a unique URL that allows any user to inspect the results and study the parameters used to do the analysis. If run on public data, each user can also rerun the experiment with the same settings or even "clone" the job and modify parameters before rerunning it.

With metabolomics now truly entering the field of "big data," repository-wide analyses become possible that may uncover patterns in the data that could not be seen before. Examples of such workflows are MASST (Wang et al. 2020) and ReDU (Jarmusch et al. 2020). MASST, analogous to the BLAST tools in genomics, allows researchers to search if and where their measured molecule was found before by sieving through all public LC-MS/MS data available. This greatly assists in the assessment of whether a molecule is novel or contains novel chemistry and also helps start explaining its source. For example, if a molecule measured in plant extracts is found in fungal isolates as well, then the molecule may be of fungal origin. Moreover, ReDU is a community-minded approach to finding and reusing public data containing MS/MS spectra at a repository scale. ReDU is the bridge between GNPS and MassIVE, a public data repository for mass spectrometry data. The key to making this work is a systematic way of capturing basic sample information about each file contributed to ReDU. By adhering wherever possible to ontologies (a controlled vocabulary), comparison across experiments and samples becomes possible. For example, currently >5,800 samples with curated metadata are of plant origin and could be used to create a large network to map plant chemical diversity.

Another recent development that will assist in improved chemical comparisons across large numbers of plant samples is chemically informed sample comparisons using QemisTree (Tripathi et al. 2020). Indeed, many chemicals are unique for a plant species; however, if we look from a slightly broader perspective, many plant species do contain *similar chemistry* as different flavonoids with kaempferol as aglycone still have the kaempferol substructure in common. We could consider a plant with a unique flavonoid core as chemically more distantly related to its kaempferol siblings; however, a plant with special peptides not found in the other plants is even less related. It is exactly this kind of *chemical awareness* that QemisTree allows researchers to use during sample comparisons. Consequently, using QemisTree, the kaempferol siblings from above will end up much closer to each other than the other two plants despite the fact that they may only share a few metabolites between themselves.

Finally, the consistent recording of metadata will fuel approaches like the earlier introduced taxonomic pondering (Rutz et al. 2019) that make use of the taxonomic distribution of metabolites across the kingdom of life to assist in candidate metabolite ranking. Altogether, we expect that current incentives to make metabolomics data publicly available together with the development of computational metabolomics workflows that allow for large-scale analyses will pave the way for plant-based natural product discovery to enter the "big data" era.

Acknowledgments

JJJvdH and MHM were funded by an ASDI eScience grant, ASDI.2017.030, from the Netherlands eScience Center. KBK was supported by the National Research Foundation of Korea (NRF) grant funded by the Korean government (MIST) (No. 2020R1C1C1004046).

PCD was supported by USA National Science Foundation grant IOS-1656481. IFK was supported by the Netherlands Organisation for Scientific Research (NWO-TTW) grant no. 13551. SR was supported by EPSRC EP/R018634/1 and BBSRC BB/R022054/1.

References

Adaikpoh, B.I., Akbar, S., Albataineh, H., et al. (2020). Myxobacterial response to methyljasmonate exposure indicates contribution to plant recruitment of micropredators. *Frontiers in Microbiology*, **11**. doi: 10.3389/fmicb.2020.00034.

Appendino, G. (2016). Ingenane diterpenoids. In: *Progress in the Chemistry of Organic Natural Products. Vol.* **102**. doi: 10.1007/978-3-319-33172-0_1.

Aron, A.T., Gentry, E.C., McPhail, K.L., et al. (2020). Reproducible molecular networking of untargeted mass spectrometry data using GNPS. *Nature Protocols*, **15** (6): 1954–1991.

Batra, S.W.T. (1983). Establishment of Hyles euphorbiae (L.)(Lepidoptera: Sphingidae) in the United States for Control of the Weedy Spurges Euphorbia esula L. and E. cyparissias L. *Journal of the New York Entomological Society*, **91** (4): 304–311.

Behsaz, B., Mohimani, H., Gurevich, A., et al. (2020). De novo peptide sequencing reveals many cyclopeptides in the human gut and other environments. *Cell Systems*, **10** (1): 99–108.

Belcour, A., Girard, J., Aite, M., et al. (2020). Inferring biochemical reactions and metabolite structures to understand metabolic pathway drift. *iScience*, **23** (2): 100849.

Blaženović, I., Kind, T., Ji, J., and Fiehn, O. (2018). Software tools and approaches for compound identification of LC-MS/MS data in metabolomics. *Metabolites*, **8** (2). doi: 10.3390/metabo8020031.

Bohacek, R.S., McMartin, C., and Guida, W.C. (1996). The art and practice of structure-based drug design: a molecular modeling perspective. *Medicinal Research Reviews*, **16** (1): 3–50.

Bolyen, E., Rideout, J.R., Dillon, M.R., et al. (2019). Reproducible, interactive, scalable and extensible microbiome data science using QIIME 2. *Nature Biotechnology*, **37** (8): 852–857.

Brouard, C., Shen, H., Dührkop, K., et al. (2016). Fast metabolite identification with input output kernel regression. *Bioinformatics*, **32** (12): i28–i36.

Considine, E.C., Thomas, G., Boulesteix, A.L., et al. (2017). Critical review of reporting of the data analysis step in metabolomics. *Metabolomics*, **14** (1): 7.

De Vos, R.C.H., Moco, S., Lommen, A., et al. (2007). Untargeted large-scale plant metabolomics using liquid chromatography coupled to mass spectrometry. *Nature Protocols*, **2** (4): 778–791.

DesRochers, N., Walsh, J.P., Renaud, J.B., et al. (2020). Metabolomic profiling of fungal pathogens responsible for root rot in American ginseng. *Metabolites*, **10** (1). doi: 10.3390/metabo10010035.

Djoumbou Feunang, Y., Eisner, R., Knox, C., et al. (2016). ClassyFire: automated chemical classification with a comprehensive, computable taxonomy. *Journal of Cheminformatics*, **8** (61). doi: 10.1186/s13321-016-0174-y.

Djoumbou-Feunang, Y., Fiamoncini, J., Gil-de-la-Fuente, A., et al. (2019). BioTransformer: a comprehensive computational tool for small molecule metabolism prediction and metabolite identification. *Journal of Cheminformatics*, **11** (1) no. 2. doi: 10.1186/s13321-018-0324-5.

Dührkop, K., Fleischauer, M., Ludwig, M., et al. (2015). Searching molecular structure databases with tandem mass spectra using CSI:FingerID. *Proceedings of the National academy of Sciences, USA*, **112** (41): 12580–12585.

Dührkop, K., Nothias, L.-F., Fleischauer, M., et al. (2019). SIRIUS 4: a rapid tool for turning tandem mass spectra into metabolite structure information. *Nature Methods*, **16** (4): 299–302.

Dührkop, K., Shen, H., Meusel, M., et al. (2020). Systematic classification of unknown metabolites using high-resolution fragmentation mass spectra. *Nature Biotechnology*, **39**: 462–471.

Ernst, M., Nothias, L.-F., van der Hooft, J.J.J., et al. (2019a). Assessing specialized metabolite diversity in the cosmopolitan plant genus L. *Frontiers in Plant Science*, **10**. doi: 10.3389/fpls.2019.00846.

Ernst, M., Kang, K.B., Caraballo-Rodríguez, A.M., et al. (2019b). MolNetEnhancer: enhanced molecular networks by integrating metabolome mining and annotation tools. *Metabolites*, **9** (7). doi: 10.3390/metabo9070144.

FAOSTAT (2020). Crops and livestock products. http://www.fao.org/faostat/en/#data/QC (accessed 2 March 2020).

Fox Ramos, A.E., Le Pogam, P., Fox Alcover, C., et al. (2019). Collected mass spectrometry data on monoterpene indole alkaloids from natural product chemistry research. *Scientific Data*, **6** (1) no. 15. doi: doi.org/10.1038/s41597-019-0028-3.

Gauglitz, J.M., Aceves, C.M., Aksenov, A.A., et al. (2020). Untargeted mass spectrometry-based metabolomics approach unveils molecular changes in raw and processed foods and beverages. *Food Chemistry*, **302**: 125290. doi: 10.1016/j.foodchem.2019.125290.

Gericke, O., Fowler, R.M., Heskes, A.M., et al. (2020). Navigating through chemical space and evolutionary time across the Australian continent in plant genus *Eremophila*. *The Plant Journal*. doi: 10.1111/tpj.15448.

Hall, R., Beale, M., Fiehn, O., et al. (2002). Plant Metabolomics. *The Plant Cell*: 1437–1440. doi: 10.1105/tpc.140720.

van Haperen, P., Voorrips, R.E., van Loon, J.J.A., and Vosman, B. (2019). The effect of plant development on thrips resistance in Capsicum. *Arthropod-Plant Interactions*: 11–18. doi: 10.1007/s11829-018-9645-6.

Hassanpour, N., Alden, N., Menon, R., et al. (2020). Biological filtering and substrate promiscuity prediction for annotating untargeted metabolomics. *Metabolites*, **10** (4) no. 160. doi: doi.org/10.3390/metabo10040160.

Hill, D.P., Smith, B., McAndrews-Hill, M.S., and Blake, J.A. (2008). Gene Ontology annotations: what they mean and where they come from. *BMC Bioinformatics*, **9** (5): 1–9.

van der Hooft, J.J.J., Mihaleva, V., de Vos, R.C.H., et al. (2011). A strategy for fast structural elucidation of metabolites in small volume plant extracts using automated MS-guided LC-MS-SPE-NMR. *Magnetic Resonance in Chemistry*, **49** (Suppl 1): S55–S60.

van der Hooft, J.J.J., Vervoort, J., Bino, R.J., and de Vos, R.C.H. (2012a). Spectral trees as a robust annotation tool in LC–MS based metabolomics. *Metabolomics*, 8: 691–703. doi: 10.1007/s11306-011-0363-7.

van der Hooft, J.J.J., Akermi, M., Ünlü, F.Y., et al. (2012b). Structural annotation and elucidation of conjugated phenolic compounds in black, green, and white tea extracts. *Journal of Agricultural and Food Chemistry*, **60** (36): 8841–8850.

van der Hooft, J.J.J., Wandy, J., Barrett, M.P., et al. (2016). Topic modeling for untargeted substructure exploration in metabolomics. *Proceedings of the National academy of Sciences, USA*, **113** (48): 13738–13743.

van der Hooft, J.J.J., Mohimani, H., Bauermeister, A., et al. (2020). Linking genomics and metabolomics to chart specialized metabolic diversity. *Chemical Society Reviews*, **49** (11): 3297–3314.

Horai, H., Arita, M., Kanaya, S., et al. (2010). MassBank: a public repository for sharing mass spectral data for life sciences. *Journal of Mass Spectrometry*, **45** (7). doi: 10.1002/jms.1777.

Horn, J.W., van Ee, B.W., Morawetz, J.J., et al. (2012). Phylogenetics and the evolution of major structural characters in the giant genus Euphorbia L. (Euphorbiaceae). *Molecular Phylogenetics and Evolution*, **63** (2). doi: 10.1016/j.ympev.2011.12.022.

Horn, J.W., Xi, Z., Riina, R., et al. (2014). Evolutionary bursts in Euphorbia (Euphorbiaceae) are linked with photosynthetic pathway. *Evolution; International Journal of Organic Evolution*, **68** (12). doi: 10.1111/evo.12534.

Hua, J., Liu, Y., Xiao, C.J., et al. (2017). Chemical profile and defensive function of the latex of Euphorbia peplus. *Phytochemistry*, **136**. doi: 10.1016/j.phytochem.2016.12.021.

Huang, C.S., Luo, S.H., Li, Y.L., et al. (2014). Antifeedant and Antiviral diterpenoids from the Fresh Roots of *Euphorbia jolkinii. Natural Products and Bioprospecting*, **4** (2). doi: 10.1007/s13659-014-0009-3.

Huber, F., Ridder, L., Verhoeven, S., et al. (2021a). Spec2Vec: improved mass spectral similarity scoring through learning of structural relationships. *PLoS Computational Biology*, **17** (2): e1008724. doi: 10.1371/journal.pcbi.1008724.

Huber, F., van der Burg, S., van der Hooft, J.J.J., and Ridder, L. (2021b). MS2DeepScore: a novel deep learning similarity measure to compare tandem mass spectra. *Journal of Cheminformatics*, **13** (1): 84. doi: 10.1186/s13321-021-00558-4.

Hundsdoerfer, A.K., Kitching, I.J., and Wink, M. (2005). The phylogeny of the Hyles euphorbiae complex (Lepidoptera: Sphingidae): molecular evidence from sequence data and ISSR-PCR fingerprints. *Organisms, Diversity and Evolution*, **5** (3): 173–198.

Ivanisevic, J. and Want, E.J. (2019). From samples to insights into metabolism: uncovering biologically relevant information in LC-HRMS metabolomics data. *Metabolites*, **9** (12): 308. doi: 10.3390/metabo9120308.

Jarmusch, A.K., Wang, M., Aceves, C.M., et al. (2020). ReDU: a framework to find and reanalyze public mass spectrometry data. *Nature Methods*, **17** (9): 901–904.

Jeffryes, J.G., Colastani, R.L., Elbadawi-Sidhu, M., et al. (2015). MINEs: open access databases of computationally predicted enzyme promiscuity products for untargeted metabolomics. *Journal of Cheminformatics*, **7**: 44. doi: 10.1186/s13321-015-0087-1.

Jeon, J.E., Kim, J.-G., Fischer, C.R., et al. (2020). A pathogen-responsive gene cluster for highly modified fatty acids in tomato. *Cell*, **180** (1): 176–187. doi: 10.1016/j.cell.2019.11.037.

Johnson, S.R. and Lange, B.M. (2015). Open-access metabolomics databases for natural product research: present capabilities and future potential. *Frontiers in Bioengineering and Biotechnology*, **3**. doi: 10.3389/fbioe.2015.00022.

Kang, K.B., Ernst, M., van der Hooft, J.J.J., et al. (2019). Comprehensive mass spectrometry-guided phenotyping of plant specialized metabolites reveals metabolic diversity in the cosmopolitan plant family Rhamnaceae. *The Plant Journal*, **98** (6): 1134–1144.

Kang, K.B., Woo, S., Ernst, M., et al. (2020). Assessing specialized metabolite diversity of *Alnus* species by a digitized LC-MS/MS data analysis workflow. *Phytochemistry*, **173**: 112292.

Kautsar, S.A., Suarez Duran, H.G., Blin, K., et al. (2017). plantiSMASH: automated identification, annotation and expression analysis of plant biosynthetic gene clusters. *Nucleic Acids Research*, **45** (W1): W55–W63.

Kautsar, S.A., Suarez Duran, H.G., and Medema, M.H. (2018). Genomic identification and analysis of specialized metabolite biosynthetic gene clusters in plants using plantiSMASH. *Methods in Molecular Biology*, **1795**: 173–188.

Kim, S., Thiessen, P.A., Bolton, E.E., et al. (2016). PubChem Substance and Compound databases. *Nucleic Acids Research*, **44** (Database issue): D1202.

Kim, H., Wang, M., Leber, C., et al. (2021). NPClassifier: a deep neural network-based structural classification tool for natural products. *Journal of Natural Products*, **84** (11): 2795–2807.

Krenn, M., Häse, F., Nigam, A., et al. (2020). Self-referencing embedded strings (SELFIES): a 100% robust molecular string representation. *Machine Learning: Science and Technology*: 045024. doi: 10.1088/2632-2153/aba947.

Li, D., Liu, B., Zheng, H., et al. (2020). XY-Meta: a high-efficiency search engine for large-scale metabolome annotation with accurate FDR Estimation. *Analytical Chemistry*, **92** (8): 5701–5707.

Ludwig, M., Nothias, L.-F., Dührkop, K., et al. (2020). Database-independent molecular formula annotation using Gibbs sampling through ZODIAC. *Nature Machine Intelligence*, **2** (10): 629–641.

Macel, M., Visschers, I.G.S., Peters, J.L., et al. (2019). Metabolomics of thrips resistance in pepper (Capsicum spp.) reveals monomer and dimer acyclic diterpene glycosides as potential chemical defenses. *Journal of Chemical Ecology*, **45** (5-6): 490–501.

Maharijaya, A., Vosman, B., Steenhuis-Broers, G., et al. (2011). Screening of pepper accessions for resistance against two thrips species (*Frankliniella occidentalis* and *Thrips parvispinus*). *Euphytica*, **177**: 401–410. doi: 10.1007/s10681-010-0277-x.

Maharijaya, A., Vosman, B., Pelgrom, K., et al. (2019). Genetic variation in phytochemicals in leaves of pepper (Capsicum) in relation to thrips resistance. *Arthropod-Plant Interactions*: 1–9. doi: 10.1007/s11829-018-9628-7.

Misra, B.B. and van der Hooft, J.J.J. (2016). Updates in metabolomics tools and resources: 2014-2015. *Electrophoresis*, **37** (1): 86–110.

Mohimani, H., Gurevich, A., Shlemov, A., et al. (2018). Dereplication of microbial metabolites through database search of mass spectra. *Nature Communications*, **9** (1): 4035.

Mungan, M.D., Alanjary, M., Blin, K., et al. (2020). ARTS 2.0: feature updates and expansion of the Antibiotic Resistant Target Seeker for comparative genome mining. *Nucleic Acids Research*, **48** (W1): W546–W552.

Naake, T., Gaquerel, E., and Fernie, A.R. (2020). Annotation of specialized metabolites from high-throughput and high-resolution mass spectrometry metabolomics. *Methods in Molecular Biology*, **2104**: 209–225.

Nakabayashi, R. and Saito, K. (2020). Higher dimensional metabolomics using stable isotope labeling for identifying the missing specialized metabolism in plants. *Current Opinion in Plant Biology*, **55**: 84–92.

Nothias, L.F., Petras, D., Schmid, R., et al. (2020). Feature-based molecular networking in the GNPS analysis environment. *Nature Methods*, **17**: 905–908. doi: 10.1038/s41592-020-0933-6.

O'Boyle, N.M. (2012). Towards a Universal SMILES representation – A standard method to generate canonical SMILES based on the InChI. *Journal of Cheminformatics*, **4**: 22. doi: doi.org/10.1186/1758-2946-4-22.

O'Boyle, N. and Dalke, A. (2018). DeepSMILES: an adaptation of SMILES for use in machine-learning of chemical structures. *ChemRxiv*. doi: 10.26434/chemrxiv.7097960.v1.

Oberacher, H., Sasse, M., Antignac, J.-P., et al. (2020). A European proposal for quality control and quality assurance of tandem mass spectral libraries. *Environmental Sciences Europe*, **32** (1): 1–19.

Ohbuchi, K., Sakurai, N., Kitagawa, H., et al. (2020). Differential annotation of converted metabolites (DAC-Met): exploration of Maoto (Ma-huang-tang)-derived metabolites in plasma using high-resolution mass spectrometry. *Metabolomics: Official Journal of the Metabolomic Society*, **16** (5): 63.

Olivon, F., Elie, N., Grelier, G., et al. (2018). MetGem software for the generation of molecular networks based on the t-SNE algorithm. *Analytical Chemistry*, **90** (23): 13900–13908.

Pluskal, T., Castillo, S., Villar-Briones, A., and Orešič, M. (2010). MZmine 2: modular framework for processing, visualizing, and analyzing mass spectrometry-based molecular profile data. *BMC Bioinformatics*. doi: 10.1186/1471-2105-11-395.

Qiu, F., Fine, D.D., Wherritt, D.J., et al. (2016). PlantMAT: a metabolomics tool for predicting the specialized metabolic potential of a system and for large-scale metabolite identifications. *Analytical Chemistry*, **88** (23): 11373–11383.

Raghuvanshi, R., Grayson, A.G., Schena, I., et al. (2019). Microbial transformations of organically fermented foods. *Metabolites*, **9** (8): 165.

Ridder, L., van der Hooft, J.J.J., Verhoeven, S., et al. (2012). Substructure-based annotation of high-resolution multistage MS(n) spectral trees. *Rapid Communications in Mass Spectrometry*, **26** (20): 2461–2471.

Ridder, L., van der Hooft, J.J.J., Verhoeven, S., et al. (2013). Automatic chemical structure annotation of an LC-MS(n) based metabolic profile from green tea. *Analytical Chemistry*, **85** (12): 6033–6040.

Ridder, L., van der Hooft, J.J.J., Verhoeven, S., et al. (2014). In silico prediction and automatic LC-MS(n) annotation of green tea metabolites in urine. *Analytical Chemistry*, **86** (10): 4767–4774.

Rogers, S., Ong, C.W., Wandy, J., et al. (2019). Deciphering complex metabolite mixtures by unsupervised and supervised substructure discovery and semi-automated annotation from MS/MS spectra. *Faraday Discussions*, **218** (0): 284–302.

Ruttkies, C., Schymanski, E.L., Wolf, S., et al. (2016). MetFrag relaunched: incorporating strategies beyond in silico fragmentation. *Journal of Cheminformatics*, **8**: 3.

Rutz, A., Dounoue-Kubo, M., Ollivier, S., et al. (2019). Taxonomically informed scoring enhances confidence in natural products annotation. *Frontiers in Plant Science*, **10**: 1329.

Sawada, Y., Nakabayashi, R., Yamada, Y., et al. (2012). RIKEN tandem mass spectral database (ReSpect) for phytochemicals: a plant-specific MS/MS-based data resource and database. *Phytochemistry*, **82**. doi: 10.1016/j.phytochem.2012.07.007.

Scheubert, K., Hufsky, F., Petras, D., et al. (2017). Significance estimation for large scale metabolomics annotations by spectral matching. *Nature Communications*, **8** (1): 1494.

Schläpfer, P., Zhang, P., Wang, C., et al. (2017). Genome-Wide prediction of metabolic enzymes, pathways, and gene clusters in plants. *Plant Physiology*, **173** (4): 2041–2059.

Schmid, R., Petras, D., Nothias, L.-F., et al. (2021). Ion identity molecular networking for mass spectrometry-based metabolomics in the GNPS environment. *Nature Communications*, **12** (1): 3832.

Shannon, P., Markiel, A., Ozier, O., et al. (2003). Cytoscape: a software environment for integrated models of biomolecular interaction networks. *Genome Research*, **13** (11). doi: 10.1101/gr.1239303.

Shi, Q.W., Su, X.H., and Kiyota, H. (2008). Chemical and pharmacological research of the plants in genus Euphorbia. *Chemical Reviews*, **108** (10). doi: 10.1021/cr078350s.

da Silva, R.R., Wang, M., Nothias, L.-F., et al. (2018). Propagating annotations of molecular networks using in silico fragmentation. *PLoS Computational Biology*, **14** (4): e1006089.

Silva, E., da Graça, J.P., Porto, C., et al. (2020). Unraveling Asian Soybean Rust metabolomics using mass spectrometry and Molecular Networking approach. *Scientific Reports*, **10** (1): 138. doi: 10.1038/s41598-019-56782-4.

Sorokina, M. and Steinbeck, C. (2020). Review on natural products databases: where to find data in 2020. *Journal of Cheminformatics*, **12** (1): 1–51.

Sorokina, M., Merseburger, P., Rajan, K., et al. (2021). COCONUT online: collection of open natural products database. *Journal of Cheminformatics*, **13** (1): 1–13.

Southan, C. (2013). InChI in the wild: an assessment of InChIKey searching in Google. *Journal of Cheminformatics*, **5**: 10. doi: 10.1186/1758-2946-5-10.

Steenbergen, M., Abd-El-Haliem, A., Bleeker, P., et al. (2018). Thrips advisor: exploiting thrips-induced defences to combat pests on crops. *Journal of Experimental Botany*, **69** (8): 1837–1848. doi: 10.1093/jxb/ery060.

Sumner, L.W., Amberg, A., Barrett, D., et al. (2007). Proposed minimum reporting standards for chemical analysis chemical analysis working group (CAWG) metabolomics standards initiative (MSI). *Metabolomics*, **3** (3): 211–221.

Tautenhahn, R., Patti, G.J., Rinehart, D., and Siuzdak, G. (2012). XCMS online: a web-based platform to process untargeted metabolomic data. *Analytical Chemistry*: 5035–5039. doi: 10.1021/ac300698c.

Töpfer, N., Fuchs, L.-M., and Aharoni, A. (2017). The PhytoClust tool for metabolic gene clusters discovery in plant genomes. *Nucleic Acids Research*, **45** (12): 7049–7063.

Tripathi, A., Vázquez-Baeza, Y., Gauglitz, J.M., et al. (2020). Chemically informed analyses of metabolomics mass spectrometry data with Qemistree. *Nature Chemical Biology*. doi: 10.1038/s41589-020-00677-3.

Tsugawa, H., Nakabayashi, R., Mori, T., et al. (2019). A cheminformatics approach to characterize metabolomes in stable-isotope-labeled organisms. *Nature Methods*, **16** (4): 295–298.

Tugizimana, F., Mhlongo, M.I., Piater, L.A., and Dubery, I.A. (2018). Metabolomics in plant priming research: the way forward? *International Journal of Molecular Sciences*, **19** (6). doi: 10.3390/ijms19061759.

Tugizimana, F., Engel, J., Salek, R., et al. (2020). The disruptive 4IR in the life sciences: metabolomics. In: *Lecture Notes in Electrical Engineering*, 227–256. doi: 10.1007/978-3-030-48230-5_10.

Vasas, A. and Hohmann, J. (2014). Euphorbia diterpenes: isolation, structure, biological activity, and synthesis (2008–2012). *Chemical Reviews*, **114** (17). doi: 10.1021/cr400541j.

Visschers, I.G.S., Peters, J.L., van de Vondervoort, J.A.H., et al. (2019). Thrips resistance screening is coming of age: leaf position and ontogeny are important determinants of leaf-based resistance in pepper. *Frontiers in Plant Science*, **10**: 510. doi: 10.3389/fpls.2019.00510.

Wandy, J., Zhu, Y., van der Hooft, J.J.J., et al. (2018). Ms2lda.org: web-based topic modelling for substructure discovery in mass spectrometry. *Bioinformatics*, **34** (2): 317–318.

Wang, M., Carver, J.J., Phelan, V.V., et al. (2016). Sharing and community curation of mass spectrometry data with Global Natural Products Social Molecular Networking. *Nature Biotechnology*, **34** (8): 828–837.

Wang, X., Jones, D.R., Shaw, T.I., et al. (2018). Target-decoy-based false discovery rate estimation for large-scale metabolite identification. *Journal of Proteome Research*, **17** (7): 2328–2334.

Wang, M., Jarmusch, A.K., Vargas, F., et al. (2020). Mass spectrometry searches using MASST. *Nature Biotechnology*, **38** (1): 23–26.

Willighagen, E.L., Mayfield, J.W., Alvarsson, J., et al. (2017). The chemistry development kit (CDK) v2.0: atom typing, depiction, molecular formulas, and substructure searching. *Journal of Cheminformatics*, **9** (1): 1–19.

Wisecaver, J.H., Borowsky, A.T., Tzin, V., et al. (2017). A Global coexpression network approach for connecting genes to specialized metabolic pathways in plants. *The Plant Cell*, **29** (5): 944–959.

Wishart, D.S., Feunang, Y.D., Marcu, A., et al. (2018). HMDB 4.0: the human metabolome database for 2018. *Nucleic Acids Research*, **46** (D1): D608–D617.

Wolf, S., Schmidt, S., Müller-Hannemann, M., and Neumann, S. (2010). In silico fragmentation for computer assisted identification of metabolite mass spectra. *BMC Bioinformatics*, **11**: 148.

Wolfender, J.-L., Nuzillard, J.-M., van der Hooft, J.J.J., et al. (2019). Accelerating metabolite identification in natural product research: toward an ideal combination of liquid chromatography–high-resolution tandem mass spectrometry and NMR profiling, in silico databases, and chemometrics. *Analytical Chemistry*, **91** (1): 704–742. doi: 10.1021/acs.analchem.8b05112.

Xu, S., Wang, J.-J., Wei, Y., et al. (2019). Metabolomics based on UHPLC-Orbitrap-MS and global natural product social molecular networking reveals effects of time scale and environment of storage on the metabolites and taste quality of raw Pu-erh tea. *Journal of Agricultural and Food Chemistry*, **67** (43): 12084–12093.

Xue, J., Guijas, C., Benton, H.P., et al. (2020). METLIN MS molecular standards database: a broad chemical and biological resource. *Nature Methods*, **17** (10): 953–954.

Zanzarin, D.M., Hernandes, C.P., Leme, L.M., et al. (2019). Metabolomics of soybean green stem and foliar retention (GSFR) disease using mass spectrometry and molecular networking. *Rapid Communications in Mass Spectrometry*: e8655.

6

Application of MS-Based Metabolomics to Investigate Biomarkers of Apple Consumption Resulting from Microbiota and Host Metabolism Interactions

Fulvio Mattivi[1,2] *and Maria M. Ulaszewska*[2,3]

[1] *Department of Cellular, Computational and Integrative Biology (CIBIO), University of Trento, Trento, Italy*
[2] *Department of Food Quality and Nutrition, Fondazione Edmund Mach, Research and Innovation Centre, San Michele all'Adige, Italy*
[3] *PROMEFA Facility, San Raffaele Scientific Institute, Center for Omics Sciences, Milan, Italy*

6.1 Introduction

Polyphenols are widely distributed in higher plants and enter the human diet through fruit, vegetables, nuts, and plant-derived beverages. After ingestion, they appear in the circulatory system mainly as phase I and II metabolites. Indeed, substantial quantities of the parent compounds and their metabolites pass into the colon, where they are further metabolized by local microbiota, giving rise principally to low-molecular-weight phenolic acids and other congeners. The latter can be absorbed into the circulation to a varying extent, distributed to organs, and ultimately cleared via renal elimination. This chapter aims to discuss how the investigation of the nutritional fate of polyphenols via mass spectrometry-based untargeted metabolomics can improve our understanding of their bioavailability, and if properly done, provides rich datasets, suitable for inclusion in the multiomic experiments.

6.2 Materials and Methods

6.2.1 Acute Intake Study

An initial nutritional clinical trial was designed to study apple polyphenol metabolism in humans, with a focus on AGER. In this crossover blind trial, healthy volunteers were given 250 mL of cloudy apple juice (CAJ) or the same apple juice enriched with an apple extract powder containing apple polyphenols. The aim of the acute AGER study was to identify the metabolic products of apple polyphenols and to follow their dose-dependent appearance in plasma and excretion in urine after different intakes, over a 5 hours- and 24 hours-period, respectively. A secondary aim was to explore the relations between microbiota and metabolites resulting statistically significantly after enriched juice intake.

Recent Advances in Polyphenol Research, Volume 8, First Edition. Edited by Juha-Pekka Salminen, Kristiina Wähälä, Victor de Freitas, and Stéphane Quideau.

In this experiment, urine was collected at time 0 and between 0 and 2, 2 and 5, 5 and 8, and 8 and 24 hours after juice consumption. Urine samples from 0 to 8 hours were stored at 4 °C immediately after voiding, while 8–24-hours urine samples were collected in 2.5 L plastic bottles containing 9 mL of hydrochloric acid 20% as a preservative. Urine samples were aliquoted and stored at −80 °C until analysis.

A venous blood sample was taken at time 0. Immediately after the first blood collection, participants were provided with the experimental juices. Further blood samples were taken at one, two, three, and five hours after meal consumption. Blood-derived serum was considered since, similarly to serum, it is the most common matrix used in human metabolomic studies, it is relatively easy to collect because the serum metabolome reflects individual changes in metabolism (Ulaszewska et al. 2019).

6.2.2 Long-Term Intake Study

Another nutritional clinical trial was designed to study apple polyphenol metabolism in humans, with a focus on long-term intake (AVAG). In this randomized controlled crossover intervention, forty mildly hypercholesterolemic volunteers consumed two whole apples or a sugar and energy-matched control beverage, daily for eight weeks. This study gave us the opportunity to investigate the biomarkers of long-term apple intake in a free-living population and to explore how apples impacted the human plasma and urine metabolite profiles. Details of the AVAG apple intervention study have been published previously (Koutsos et al. 2020). Briefly, this dietary intervention involved 40 mildly hypercholesterolemic (total cholesterol levels > 5.2 mmol/L) and otherwise healthy volunteers (23 women and 17 men,), aged 29–69 years, with a BMI of 19–33 kg/m^2. Volunteers were recruited from the general population living in Reading, Berkshire, the United Kingdom, and the study was conducted at the Hugh Sinclair Unit of Human Nutrition, University of Reading.

Subjects were randomly allocated to one of two groups to receive daily either two whole apples (Renetta Canada, 340 g containing 990 mg of polyphenols and 8.5 g of total fiber) or a sugar and energy-matched control beverage low in apple polyphenols (100 mL of concentrated apple beverage containing 2.5 mg of polyphenols and no fiber). Subjects were requested to incorporate the intervention products, either apples or the control beverage, into their normal diet without further changing their habitual intake. Biological samples, including fasted plasma, fecal samples and 24-hours urine samples were collected before and after each dietary intervention at weeks 1, 8, 12, and 20, respectively.

6.2.3 Metabolomic Analysis

6.2.3.1 Sample Extraction

In both studies, the metabolome in plasma and urine samples was analyzed via untargeted metabolomics. Urine and plasma samples were thawed at 4 °C overnight. Extraction methods for serum and urine have been described elsewhere (Ulaszewska et al. 2016). Briefly, 50 μL of serum was loaded onto 96-well plates Ostro (Waters), and 50 μL of internal standard in methanol was added (creatinine-labelled 13C), followed by 150 μL of cold acetonitrile with 1% formic acid. The mixture was shaken for five minutes and samples were filtered using a positive pressure-96 manifold (Waters, USA). Elution with acetonitrile and filtration were repeated. Collected eluates were evaporated with a gentle stream of nitrogen to dryness using a Techne Dr-block DB 3D heater at room temperature and redissolved with 50 μL of water and 50 μL of external standard (trans-cinnamic acid-d5).

An aliquot of 100 μL of urine was placed into 96-well plate Millipore PVDF and 100 μL of internal standards in methanol was added. After five minutes of shaking, the mixture was filtered using a positive pressure-96 manifold. Then, 300 μL of external standard (trans-cinnamic acid-d5) was added to the collection plate. The prepared urine and serum samples were submitted for LC–Orbitrap–MS analysis.

Quality control (QC) samples were prepared for each biological fluid by mixing equal volumes of each sample. Samples were fully randomized for extraction and injections, and experiments were blinded to the operator. A double QC injection was performed every eight study samples for monitoring injection performance. A series of 10 QC samples was injected at the beginning of the injection queue to equilibrate the LC column.

6.2.3.2 Chromatography

Samples were analyzed with a hybrid linear ion trap Fourier Transform (LTQ FT) Orbitrap mass spectrometer (Thermo Fisher, Bremen, Germany), interfaced with a Thermo Scientific Dionex UltiMate HPLC system, consisting of an auto-sampler and quaternary gradient HPLC pump. Chromatographic separation of compounds was performed using a Kinetex C18 column (150×2.1 mm ID, particle size 3.5 μm) with pre-column 4.0×2.0 mm ID (Phenomenex Torrance, CA, USA). Details regarding chromatographic conditions for plasma and urine samples have been published elsewhere (Trost et al. 2018a). Briefly, the flow rate was 300 μl/min and the column temperature was maintained at 30 °C. The mobile phases used were: Milli-Q water (Solvent A), and acetonitrile (Solvent B), all with 0.1% formic acid. For urine, 95% of solvent A and 5% of solvent B were maintained for one minute, followed by an increase of solvent B to 45% in 12 minutes and to 80% in two minutes and maintained for two minutes. The initial composition was restored in three minutes. Plasma and blood methanol extracts were analyzed with the following gradient: 95% of solvent A and 5% of solvent B maintained for one minute followed by an increase of solvent B to 100% in 19 minutes. In two minutes, the initial conditions were restored and held for three minutes.

6.2.3.3 Mass Spectrometry

The Orbitrap LTQ was equipped with an electrospray ionization (ESI) probe and operated in both positive and negative ionization modes. The conditions have been described elsewhere (Ulaszewska et al. 2016; Ancillotti et al. 2019). Briefly, samples were injected twice. The first injection was dedicated to the acquisition of full scan spectra at a resolution of 30,000. The second injection was dedicated to the acquisition of high-resolution MS/MS data in data-dependent acquisition (DDA) mode. In DDA mode, the resolving power for MS scans was 7,500 and for MS2 scans 7,500. The static exclusion list consisted of the 300 most abundant ions created by the injection of solvents, which followed the same preparation procedures as the samples. Dynamic exclusion allowed three repeated counts of the same ion in 15 seconds, while the exclusion duration was 45 seconds. The sequences were randomized with regard to participants and treatments.

For every 20 (AGER) or 10 (AVAG) samples, a QC block was analyzed, consisting of solvents and double injection of the QC sample. The protocols for data processing and biomarker identification are described in detail in a previous paper (Ulaszewska et al. 2020a).

6.2.4 Data Processing and Statistical Analysis

All the experiments were done with the relevant controls, such as randomization, injection of pooled QCs, standard mix, and blanks, as recommended for good practice in metabolomics (Ulaszewska et al. 2019) and each dataset passed the QC before being sent to data mining. In specific, the AGER study data were processed with software Sieve 2.0 (Thermo Scientific, USA), which included steps of alignment and peak picking. Following parameters were used for urine and plasma: *m/z* framing window 10 ppm and time window 1 minute ranging. The maximum number of frames was set to 5,000. Positive and negative ionization modes were processed separately. Finally, from data processing, four datasets were obtained, introduced in a further step to statistical analysis performed in R. Urine samples were normalized to urine volume, while plasma samples were adjusted to the baseline (time point zero).

The data from the AVAG study were fully analyzed in R with the XCMS package using the "matchedFilter" peak picking method. A signal to noise ratio cut-off was set to 4, full width at half maximum was set to 10 seconds and a step size of 0.005 Da was used for extracting chromatograms. The resulting feature matrix was annotated using CAMERA to group features corresponding to the same parent ion species.

Slightly different statistical approaches were used for the discovery of *m/z* features indicating short- and long-term apple intake. In the AGER kinetics study, two linear mixed models were fitted for each feature. The first model included treatment (two doses of polyphenols) and time main effect and interactions, whereas the second model only contained a time effect. In both models, a subject-specific random effect was also included. Subsequently, Benjamini and Hochberg test was applied to correct for false discovery rate. In AVAG long-term study, two independent linear mixed models were fitted to each *m/z* feature. In the first model, the concentration of each feature was modelled as a function of meal, sex, and an individual. In the second model, only the time effect was considered. The comparison of the two models allowed to identify the features significantly affected by the meal factor. Also in the AVAG study, the p values were corrected for multiple testing to control the false discovery rate with Benjamini and Hochberg method.

6.2.5 Metabolomic Data Sharing

Untargeted data in mzXML format and metadata are available for download from the MetaboLights repository http://www.ebi.ac.uk/metabolights/ for the AGER study with the persistent unique public identifier MTBLS473 (Trost et al. 2018b).

The AVAG data are deposited in MetaboLights with the persistent unique public identifier MTBLS469, https://www.ebi.ac.uk/metabolights/MTBLS 469 (Ulaszewska et al. 2020b). Metadata for both studies are also available on the DaShiN web-based federated analysis website (http://www.enpadasi.eu/deliverables_final_pdf/D3.2.1.pdf). Metadata in DaShiN are compliant with the need to be extensively annotated in human- and machine-readable form (https://dashin.eu/interventionstudies/study/show/38790886). The persistent unique public identifiers to find the two datasets are https://dashin.eu/interventionstudies/AGER-MELO and https://dashin.eu/interventionstudies/AVAG.

6.3 Results and Discussion

The AGER study was a crossover, blind human trial, where 12 healthy volunteers consumed 250 mL of CAJ, Crispy Pink apple variety, or 250 mL of the same juice enriched with 750 mg of an apple polyphenol extract (PAJ). Plasma and blood were collected at zero, one, two, three, and five hours. Urine was collected at time 0 and 0–2, 2–5, 5–8, and 8–24 hours after juice consumption. Fecal samples were collected once from everyone during the study for 16S rRNA gene profiling. Additional information can be found in the literature (Trost et al. 2018a). The design of the study was largely compliant with the recommended criteria, defined later, for the conduction of acute studies (Ulaszewska et al. 2019), lacking only measurement at 48 hours.

The AVAG trial was a randomized, controlled, crossover, dietary intervention in 40, free-living, mildly hypercholesterolemic, otherwise healthy volunteers, who were randomly allocated to one of two groups to incorporate daily either two whole apples (Renetta Canada) or a sugar- and energy-matched control beverage into their normal diet. Biological samples, including fasted plasma, fecal samples and 24 hours-urine collection were collected before and after each dietary intervention at weeks 1, 8, 12, and 20. Additional information can be found in the literature (Ulaszewska et al. 2020a).

Special effort was made to comply with the principles of Open Science. In the era of molecular system biology and integrative studies, the availability of raw data supported with a rich set of metadata is now compulsory in all omics research (Savoi et al. 2021). The study participants gave written consent to pseudonymized data sharing (Ulaszewska et al. 2020a). The metabolomic dataset and the metadata of the above-described experiments were filed under FAIR principles and are publicly and directly accessible in the MetaboLights repository (Haug et al. 2013, 2020). The link to each dataset is cited here separately from the paper disseminating the results because a persistent, resolvable identifier should be inserted in a citation when the data are published. This has the scope of obtaining indexing of the dataset in the Data Citation Index (DCI) of Clarivate Web of Science, connecting the paper to the associated FAIR data present in the public repository (Savoi et al. 2021).

6.3.1 Lessons from the Acute Study

6.3.1.1 Two Nutrikinetics Patterns

The AGER study allowed us to reliably identify the metabolic transformation products of apple polyphenols using an untargeted metabolomic approach. Additionally, it made it possible to follow their nutrikinetics in plasma and urine after different intakes, considering the dose–response relation. Nutrikinetic curves for each biomarker at the population level were plotted using Statistica 9.0 (StatSoft, USA), based on extracted integrated intensities. Examples of the plasma-integrated intensity-time curves and urine cumulative excretion curves of selected metabolites from the main polyphenol families are shown in Figure 6.1. This series of graphs visualizes the behavior of biomarkers reaching a different intensity of signals between the two doses, emphasizing the common qualitative nutrikinetics in the whole group of volunteers participating in the study. It also allowed the authors to estimate the time required for each metabolite to reach the maximal concentration observed in this acute study (T_{max}) in plasma and/or urine (Trost et al. 2018a).

This multi-compartment approach, and observation of coherent trends for both plasma concentration and urine excretion profiles (at the population level) for the same metabolic pathways, provide experimental evidence of the presence of two main nutrikinetics patterns (Figure 6.1).

A fraction of the native polyphenols is quickly absorbed and metabolized in the upper gut with little or no contribution from human colonic microbiota and is rapidly excreted in urine. This group of catabolites is characterized by an early absorption peak in plasma, which is followed by a rapid decrease in their concentration (within five hours) and a fast appearance in urine. This was the case for the metabolites of dihydrochalcones, chlorogenic acid and cinnamic acids, vanillic acid, epicatechin, feruloyl-quinic and coumaroyl-quinic acids, and quercetin metabolites, all having kinetic patterns with C_{max} between one and two hours postprandial, similarly to N-acetyl tryptophan, one of the main endogenous metabolites of TRP.

A second group of catabolites do not reach their maximum concentrations in plasma within five hours from juice ingestion and are characterized by a delayed appearance in urine, suggesting a prolonged metabolism with likely involvement of the gut microbiota (Figure 6.1). This was the case for catabolites belonging to the classes of derivatives of catechols, valerolactones and valeric acids, propionic and acetic acids derived from hydroxycinnamates (C6-C3), homovanillic acid, and derivatives of hippuric acid. The kinetic profiles of these metabolites were characterized by a slow increase, visible from around the third-hour postprandial in plasma, and two to five hours in urine. C_{max} was not observed in the postprandial samples collected. The kinetic pattern of these polyphenol metabolites was like that of indoxyl sulfate, a pro-inflammatory uremic toxin metabolite of TRP, and of other TRP metabolites observed in the same study (Trost et al. 2018a, supplementary material three).

In total, as many as 110 metabolites were significantly elevated following intake of PAJ, with large inter-individual variations, see examples in Table 6.1. The vast majority (88) were derivatives of polyphenols, while the others were related to the metabolism of tryptophan, fatty acids and other compounds (Trost et al. 2018a).

6.3.1.2 Dose–Response Relationship After Apple Juice Intake

An important aim of this study was to compare the bioavailability of polyphenols when an amount of 250 mL of Crispy Pink CAJ was consumed in natural form, versus the same matrix enriched with 0.75 g of apple extract containing all classes of apple polyphenols. The cloudy juice was a high-quality preparation obtained by cold processing of fresh fruit, excluding pasteurization and stabilized with a high-pressure cold stabilization system (HPP), containing an amount of polyphenols equivalent to the consumption

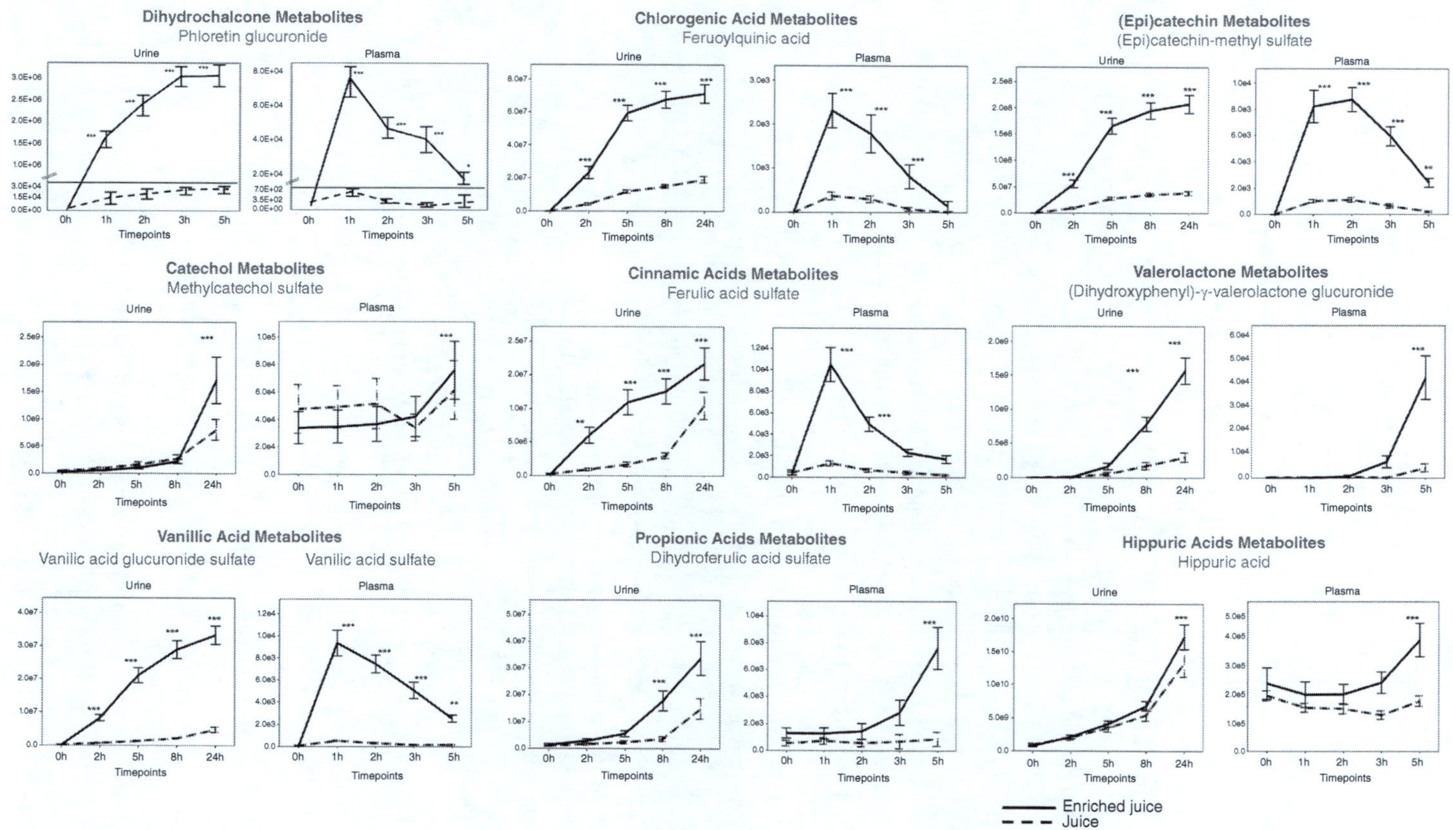

Figure 6.1 Kinetic curves of selected metabolites in plasma and urine, with the composition of the urine and plasma of volunteers who consumed natural cloudy apple juice, (CAJ treatment, dotted line) or the same volunteers who consumed CAJ enriched with 750 mg of mixed apple polyphenol extract (PAJ treatment, continuous line). X axis: time points (hours after intake); Plasma Y axis: peak intensity at the time point (MS response). Urine Y axis: cumulative intensities of peak (MS response). *** p value < .001; ** p value 0.01–0.001; * p value 0.05–0.01. *Source:* Reprinted from Trost et al. (2018a) with permission from Elsevier.

Table 6.1 List of metabolites found in biological fluids after both acute apple juice intake and long-term apple consumption.

Urine		
Acute	**Common metabolites**	**Long-term**
	Phloretin glucuronide – $C_{21}H_{22}O_{11}$ NEG MS/MS^2 449: 273.0746; 315.0849 167.0339; 345.0957; 175.0235 POS MS/MS^2 451: 275.0906; 433.1114; 317.1010; 415.1014;	
	Phloretin glucuronide sulfate – $C_{21}H_{22}O_{14}S$ NEG MS/MS^2 529: 353.0311; 273.0744; 449.1049	
	(Epi)catechin sulfate (two isomers) – $C_{15}H_{13}O_9S$ NEG MS/MS^2 369: 289.0702; 230.9958; 216.9801; 137.0239	
	(Epi)catechin methyl sulfate – $C_{16}H_{16}O_9S$ NEG MS/MS^2 383: 303.0859; 216.9801; 137.0238; 285.0749; 259.0965	
	Hydroxyphenyl-γ-valerolactone sulfate – $C_{11}H_{12}O_6S$ NEG MS/MS^2 271: 271.0268; 191.0704; 147.0810; 207.0648	
	Hydroxyphenyl-γ-valerolactone glucuronide – $C_{17}H_{20}O_9$ NEG MS/MS^2 367: 113.0239; 175.0238; 191.0702; 151.0132; 95.0134; POS MS/MS^2 369: 193.0855; 351.1066; 175.0749; 133.0644;	
	Dihydroxyphenyl-γ-valerolactone methyl glucuronide – $C_{18}H_{22}O_{10}$ NEG MS/MS^2 397: 397.1150; 175.0235; 113.0240; 221.0807	
	Dihydroxyphenyl-γ-valerolactone glucuronide sulfate – $C_{11}H_{12}O_7S$ NEG MS/MS^2 287: 207.0622; 163.0762	
	Dihydroxyphenyl valeric acid – $C_{11}H_{14}O_4$ NEG MS/MS^2 209: 101.0242; 147.081; 191.0705; 107.0500	
	Dihydroxyphenyl valeric acid glucuronide – $C_{17}H_{22}O_{10}$ NEG MS/MS^2 385: 209.0807; 175.0238; 113.0240; 385.1116; 367.1013	
110 annotated metabolites	Hydroxy(dihydroxyphenyl) valeric acid glucuronide – $C_{17}H_{22}O_{11}$ NEG MS/MS^2 401: 225.0757; 175.0240; 401.1068; 113.0241; 167.0368; 207.0655	54 annotated metabolites
	Hydroxy(dihydroxy)phenyl valeric acid sulfate – $C_{11}H_{14}O_8S$ NEG MS/MS^2 305: 225.0752; 207.0469; 163.0752; 101.0242; 123.0448; 96.9599	
	Hydroxy(dihydroxyphenyl) valeric acid methyl glucuronide – $C_{18}H_{24}O_{11}$ NEG MS/MS^2 415: 239.0911; 175.0240; 397.1133; 207.0656; 225.0763; 123.0444	
	Hydroxy(dihydroxy)phenyl valeric acid methyl sulfate – $C_{12}H_{16}O_8S$ NEG MS/MS^2 319: 239.0905; 123.0443; 96.9599	
	Feruloylquinic acid – $C_{17}H_{20}O_9$ NEG MS/MS^2 367: 191.0548; 193.0493; 173.0446; 134.0370	

Table 6.1 (Continued)

	Hydroxycinnamic acid – $C_9H_8O_3$ POS MS/MS^2 165: 147.0436; 123.0436 Hydroxyphenylpropionic acid sulfate – $C_9H_{10}O_6S$ NEG MS/MS^2 245: 165.0556; 121.0660; 119.0497; 96.9598; 79.9571 Dihydroxyphenyl propionic acid two isomers – $C_9H_{10}O_4$ NEG MS/MS^2 181: 121.029; 181.0496; 137.0602 Hydroxyhippuric acid two isomers – $C_9H_9NO_4$ NEG MS/MS^2 194: 100.0037; 93.0344; 155.0554 POS MS/MS^2 196: 121.0279 Hydroxyhippuric acid sulfate – $C_9H_9NO_7S$ NEG MS/MS^2 274: 194.0445;150.0552; 230.0111; 166.0499 Common triggered metabolic pathway: Tryptophan and indole metabolism	
	Plasma	
Acute	**Common metabolites**	**Long-term**
53 annotated metabolites	Hydroxyhippuric acid – $C_9H_9NO_4$ NEG MS/MS^2 194: 100.0037; 93.0344; 155.0554 POS MS/MS^2 196: 121.0279 Hydroxyphenylpropionic acid sulfate – $C_9H_{10}O_6S$ NEG MS/MS^2 245: 165.0556; 121.0660; 119.0497; 96.9598; 79.9571 Common metabolite family: Valeric acids metabolites Common triggered metabolic pathway: Tryptophan and indole metabolism	9 annotated metabolites

of two Crispy Pink apples in 250 mL. The enriched juice was the same matrix, fortified with apple polyphenols so as to contain the quantity of polyphenols present in c. 6–8 apples chosen from elite market cultivars (Farneti et al. 2015). Such values could be considered to represent the maximal natural level observed in *M. domestica* cultivars since they could be reached by consuming just one or two polyphenol-rich apples (Farneti et al. 2015).

Comparison of the average area under the curve (AUC) of circulating metabolites in the plasma and urine of volunteers consuming either the natural (CAJ) or fortified (PAJ) juice demonstrated a stable metabotype, suggesting that an increase in polyphenol concentration in fruit does not limit their bioavailability upon ingestion. In this trial, fortification of CAJ with additional apple polyphenols in the PAJ led to an increase in the total amount of metabolites excreted in urine without any major change in the pattern of metabolites within each chemical class of polyphenol derivatives. This experiment provided evidence that an increase in polyphenols in apples, exploring a concentration range close to the maximal levels observed for table apples (Farneti et al. 2015), would lead to an increase in their concentration in biofluids.

6.3.1.3 Inter-Individual Variability of Metabolic Response and Impact on Bioavailability

In the first part of this chapter, we discussed some factors affecting the time needed to reach the maximal concentration (T_{max}) and maximum MS-intensity (C_{max}) of polyphenol metabolites at the population level. We have not yet mentioned the importance of variations in metabolic potential in individuals, a factor which combines differences in both human genome-encoded metabolic pathways and co-metabolic pathways involving the gut microbiota. Overall bioavailability can be evaluated on the basis of the concentration of the polyphenol intake biomarkers found in urine. The fingerprint of apple polyphenols in urine is richer than in plasma, both in terms of the number of peaks and the intensity of MS-signals, and the AGER experiment allowed us to measure excretion over 24 hours after ingestion. We selected four metabolites for each class of polyphenols in order to visualize their variability. The structures of the metabolites are summarized in Figure 6.2, and the boxplots reporting the intensity of the MS-signals measured in biofluids for each volunteer in the study are displayed in Figures 6.3–6.8. We also computed the cumulative variability for each metabolite, given as the ratio between the C_{max} and C_{min} observed at T_{max}. Please note that the observed T_{max} was not always at the same time point for all volunteers, and in this computation, the experimental C_{max} for each volunteer was used for the computation.

The kinetics of the most abundant epicatechin metabolites are shown in Figure 6.3. Please note that as a true standard is not available, annotation cannot discriminate between diastereoisomers, but it is expected that the most abundant metabolites should derive from the cis isomer, by far the most abundant in the apple, i.e. (−)-epicatechin. The variability in concentrations of epicatechin sulfate, epicatechin methyl sulfate, epicatechin glucuronide, and epicatechin methyl glucuronide was 4.3, 3.8, 5.1, and 5.8, respectively for the enriched juice, and 3.4, 2.6, 2.9, and 9.7 for the cloudy juice.

The most abundant microbial catabolites of the flavonols observed (Figure 6.4) were represented by dihydroxy-γ-valerolactone derivatives (glucuronide and sulfate) and hydroxy-(hydroxyphenyl)-γ-valeric acid derivatives (glucuronide and methyl sulfate). The variability in their MS-signals at T_{max} was respectively 4.2, 6.6, 8.3, and 8.3 following intake of the enriched juice, and 5.0, 3.8,19.3, and 6.1 after intake of cloudy juice. In general, the variability of the γ-phenyl valeric acids was much higher than for γ-valerolactones.

The observed variability in the MS-signals of the most abundant catabolites of the apple hydroxycinnamates (Figure 6.5) was slightly lower, being 4.5, 4.7, 3.0, and 4.5 after intake of enriched juice and 4.1, 4.2, 1.8, and 2.9 following intake of cloudy juice, respectively for ferulic acid sulfate, ferulic acid glucuronide, and the two most intense isomers (I) and (II) of hydroxycinnamates.

One of the volunteers had unusually extreme intensities in the MS-intensity values for C6-C3 hydroxyphenyl propionic acid derivatives (Figure 6.6). Consequently, the overall variability of these compounds was relatively high, respectively 8.9, 11.7, 11.1, and 6.7 for dihydroferulic acid glucuronide, dihydroferulic acid sulfate, hydroxyphenyl propionic

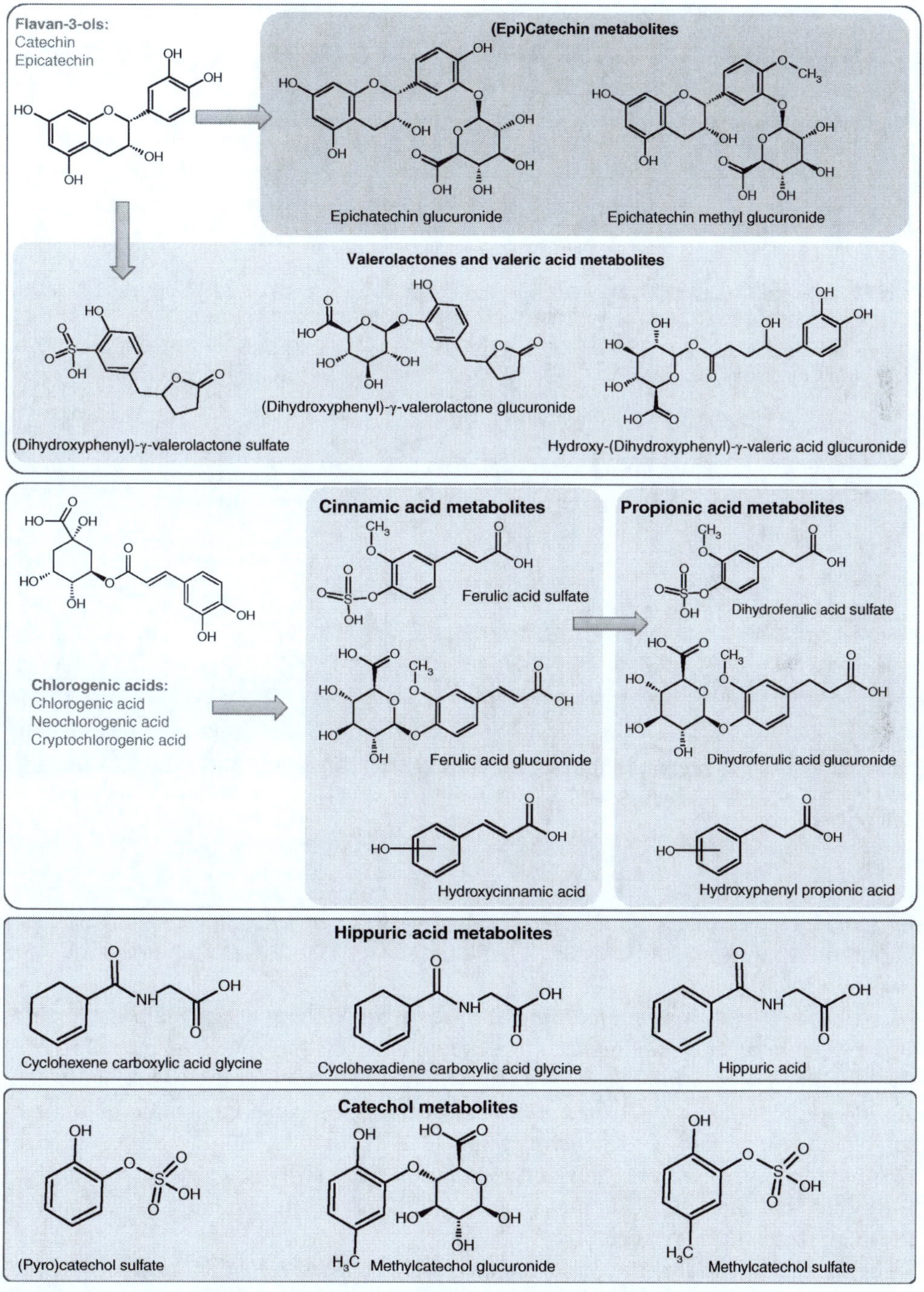

Figure 6.2 Chemical structures of key metabolites found in biological fluids after acute apple juice intake or long-term apple consumption. The yellow background indicates metabolites resulting from co-metabolism phases 1 and 2, while the light blue background indicates metabolism phase 2 only. (*See insert for color representation of the figure.*)

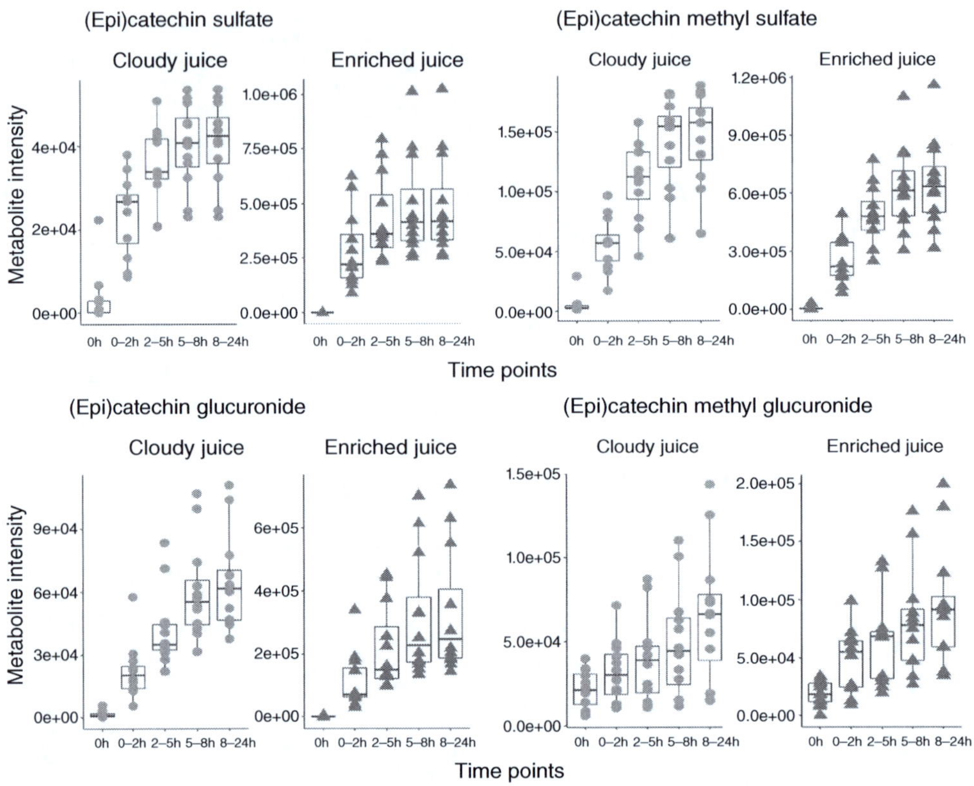

Figure 6.3 Boxplot representing the cumulative excretion in urine (intensity of the MS-signal in each fraction) of the four main epicatechin metabolites after intake of cloudy juice or enriched cloudy juice.

acid, and dihydroxyphenyl propionic acid after intake of enriched juice. Huge variability among the volunteers (respectively, 7.7, 14.4, 4.8, and 4.3) was also observed after the intake of cloudy juice.

Last but not least, it was interesting to observe the kinetics of the main metabolites in the hippurate family, an important catabolite of several polyphenols, including both flavonoids and non-flavonoids (Ulaszewska et al. 2016). Again in this case (Figure 6.7), the observed variability of concentrations following intake of the same amount of polyphenols was relatively extensive, being 11.2, 11.7, 10.4, and 4.5, respectively for cyclohexadiene carboxylic acid glycine, cyclohexene carboxylic acid glycine, hydroxyhippuric acid, and hippuric acid, after intake of the enriched juice, and 3.9, 6.1, 8.8, and 5.6 after intake of cloudy juice.

The last class of metabolites considered was catechols. These important microbial catabolites showed extreme variability, often over one order of magnitude. In this case, the four main substances observed in urine were methylcatechol glucuronide, methylcatechol sulfate, methylcatechol glucuronide sulfate, and pyrocatechol sulfate. The observed cumulative

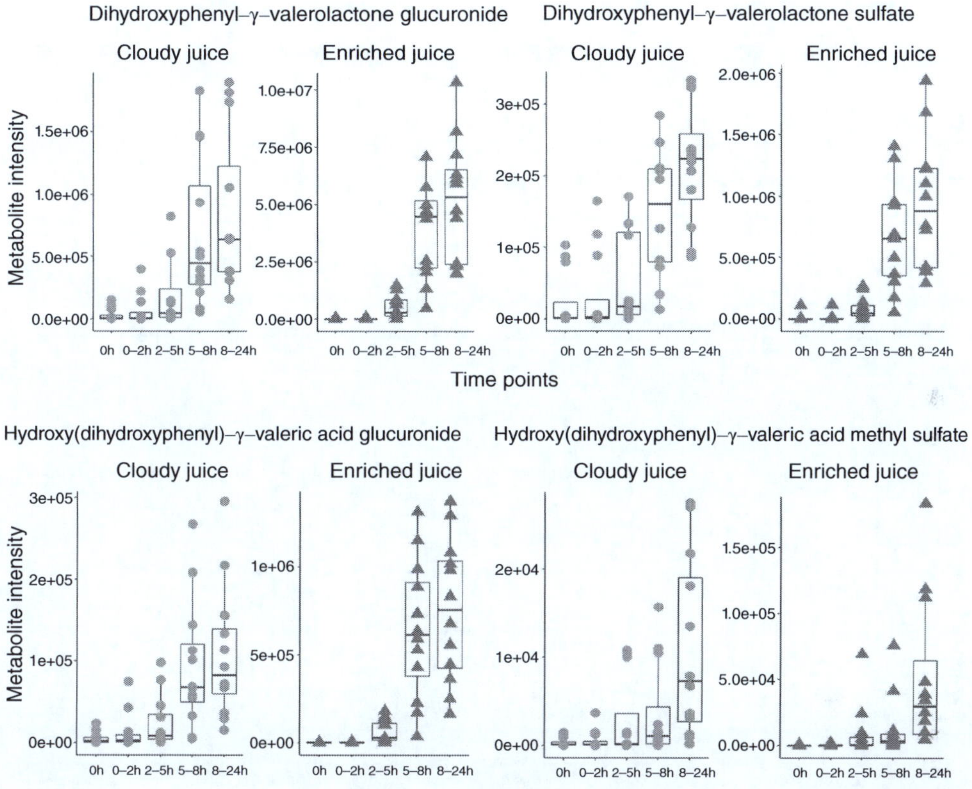

Figure 6.4 Boxplot representing the cumulative excretion in urine (intensity of the MS-signal in each fraction) of the two main γ-valerolactones and γ-valeric acids after intake of cloudy juice or enriched cloudy juice.

variability in their concentrations following intake of enriched juice was 15.9, 11.7,14.4 and 12.8, respectively, while after intake of cloudy juice, it was 7.5, 24.9, 25.8 and 4.5, respectively.

A final consideration results from the general and significant variability in bioavailability observed among the volunteers. To explain or predict diversity among volunteers is still beyond our capability today. However, we can now measure it directly, thanks to the development of nutritional metabolomics. This has obvious implications for any intervention study aimed at evaluating the bioactivity of these metabolites. As already suggested in other studies on γ-valerolactones (Anesi et al. 2019), and extended here to all the classes of apple polyphenols investigated, we suggest that a simple preliminary test should be inserted in any clinical trial, for prescreening of participants to be included in the study. A single intake of a controlled amount of the polyphenols of interest in the chosen matrix, followed by an analysis of 24-hours urine could lead to the selection of groups of participants with a similar metabolic capacity, reducing what is now a critical source of variation, expected to lower the power of the study.

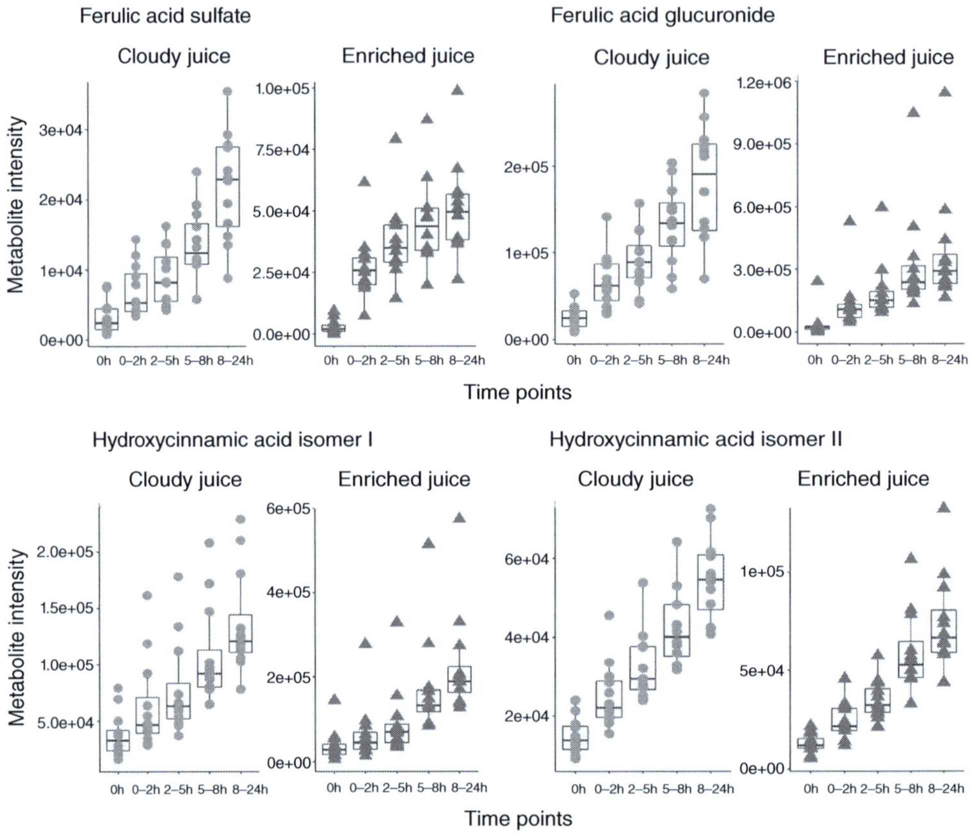

Figure 6.5 Boxplot representing the cumulative excretion in urine (intensity of the MS-signal in each fraction) of the four main metabolites of hydroxycinnamates after intake of cloudy juice or enriched cloudy juice.

6.3.1.4 Is Our Microbiota a Multiplier of Complexity?

On the basis of these results and the literature, we can summarize the main reactions giving rise to the large quantity of diverse polyphenol catabolites (Figure 6.9). Our main focus at present is not on human phase I and phase II metabolism, both producing polar metabolites via oxidation/reduction and hydrolysis, or via methylation, glucuronidation, acetylation and sulfation, which have been the subject of hundreds of studies in the past, but rather on the catabolites requiring host:microbiome co-metabolic processing.

The presence of a highly concentrated source of apple polyphenols in one arm of our acute study (PAJ) allowed us to observe a large number of apple metabolites. Among those processed via the human metabolism, we observed five derivatives of phloridzin, the main apple chalcone, together with 8 derivatives of (epi)catechin, two derivatives of naringenin, one derivative of quercetin, four chlorogenic acid metabolites and six hydroxycinnamic acid metabolites.

Among those requiring the intervention of gut microbiota, in the plasma and urine of the volunteers assuming 250 mL of PAJ, we observed as many as 28 derivatives of the family of valerolactones and valeric acids, 7 methylcatechol derivatives, 14 derivatives of

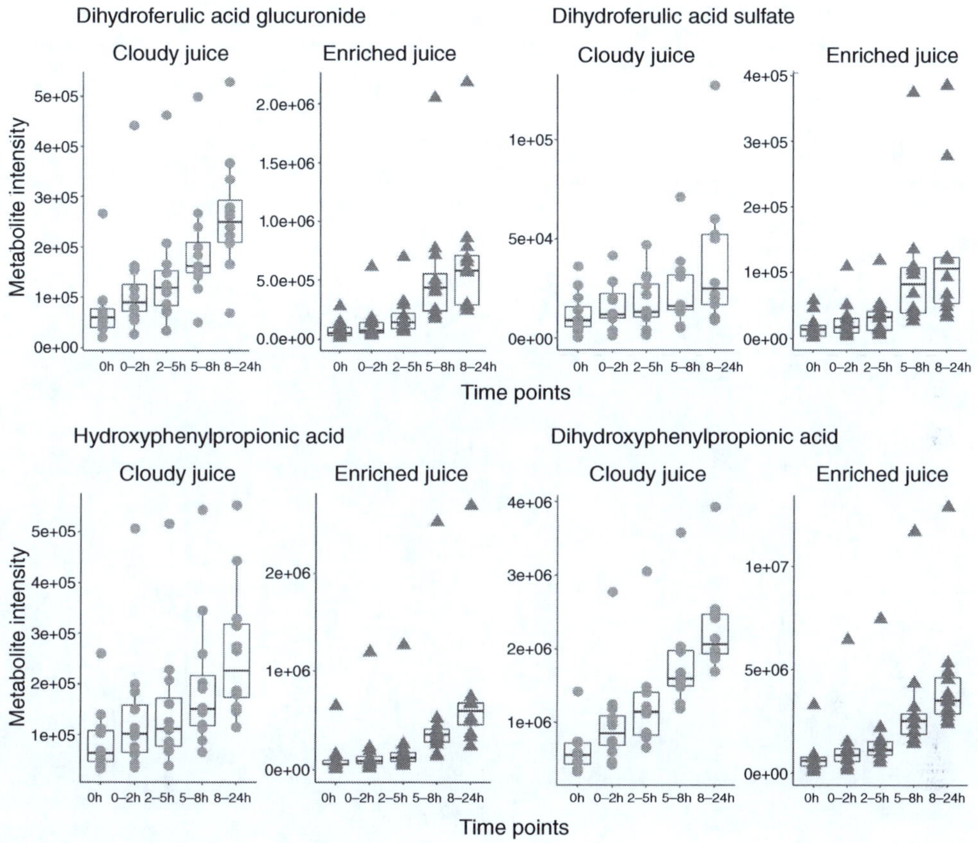

Figure 6.6 Boxplot representing the cumulative excretion in urine (intensity of the MS-signal in each fraction) of the four main hydroxyphenyl propionic acids after intake of cloudy juice or enriched cloudy juice.

phenylpropionic and phenylacetic acid with different hydroxylation patterns, 9 intermediate and final catabolites of hippuric acid, and 6 metabolites associated with vanillic acid.

A question often arising as regards the role of gut microbiota in the metabolism of xenobiotics is whether this can be considered as a multiplier of the number of derivatives in natural polyphenols, adding further complexity to the products of human metabolism. When considering each class of derivatives in isolation, or even a single food source, this is certainly true. In our study of apple polyphenols for example, after intake of a juice containing 12 low molecular weight polyphenols plus oligomeric procyanidins (made up of four building blocks – epicatechin, catechin, epigallocatechin, and epicatechin gallate), we did not find any native compounds in plasma or urine, while we observed as many as 88 derivatives in total. This means complexity increased by seven times, in terms of the number of circulating compounds.

However, if we carefully observe the reaction diagram in Figure 6.9, it can be seen that several metabolites may have derived from a number of different polyphenol families present in the test juices, and these are essentially very similar or the same as those already observed in several other food matrices (see the literature cited in Figure 6.9). Indeed, we

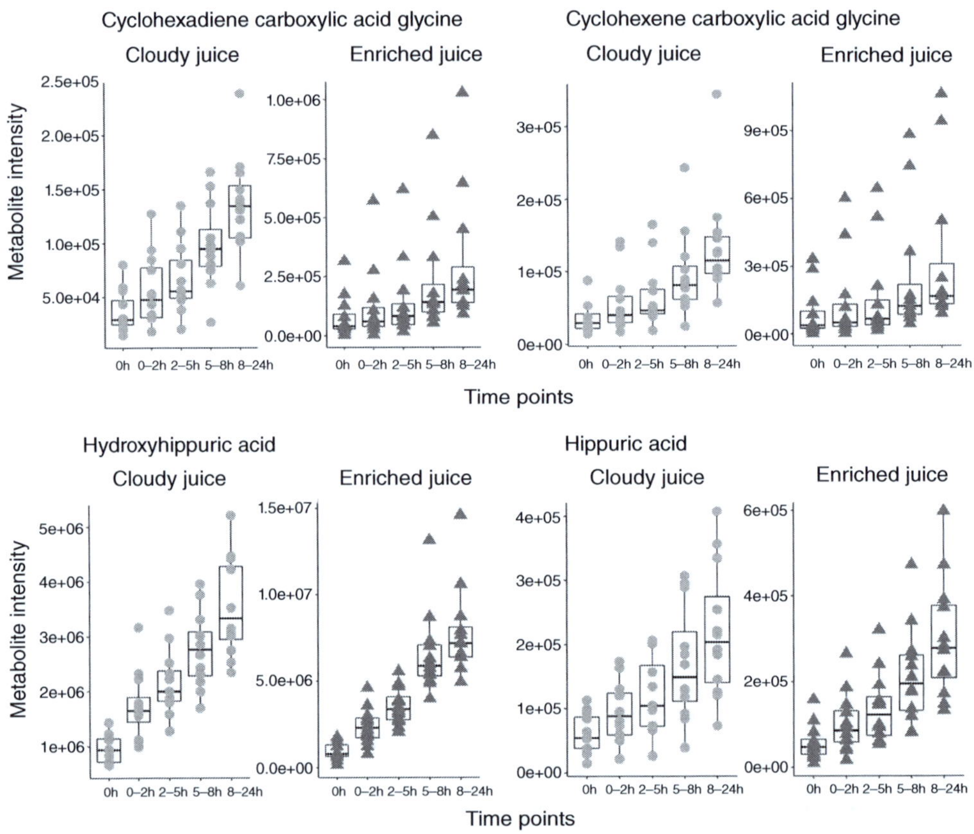

Figure 6.7 Boxplot representing the cumulative excretion in urine (intensity of the MS-signal in each fraction) of the four main hippuric acid metabolites after intake of cloudy juice or enriched cloudy juice.

hypothesize that the global action of the gut microbiota is to cleave several native polyphenols into a relatively limited number of small, common polar catabolites, thus simplifying the final number of circulating metabolites resulting from the complex diet of human omnivores. The human superorganism works by combining the action of our body and the microbiota to serve as a buffer, protecting us from transient, overwhelming amounts of compounds ingested with our plant food.

This hypothesis warrants further study since it has several implications, as has been suggested, for instance, for alkyl catechols and catechol. These catabolites are widely available from the microbial digestion of plant polyphenols and they differ from their (inactive) precursors as potent natural co-factors for activation of the Nrf2 cellular defence pathway both *in vitro* and *in vivo* (Senger et al. 2016).

6.3.1.5 Microbial Catabolites of Flavanols are Persistent in the Body and can Reach High Concentrations

Another consequence of the similarity of the end products of microbial catabolism is that most of the products of the gut metabolism of polyphenols are semi-ubiquitous and lack the specificity to be useful biomarkers of food intake. This is certainly the case for flavan-3-ols.

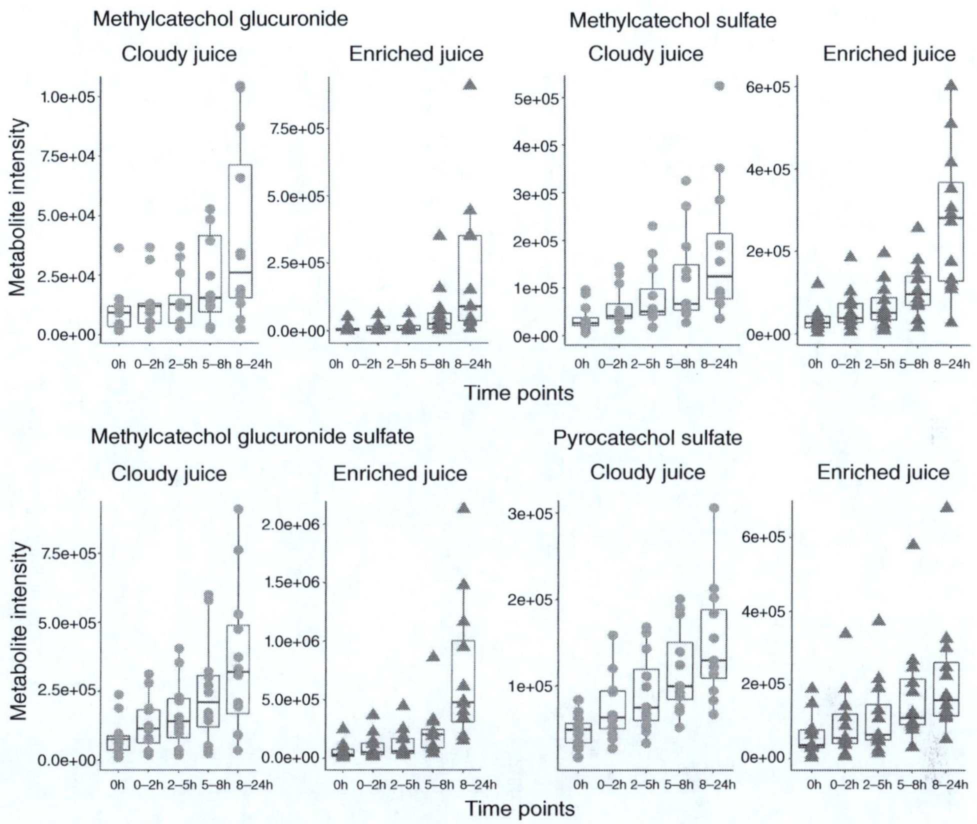

Figure 6.8 Boxplot representing the cumulative excretion in urine (intensity of the MS-signal in each fraction) for the four main catechols after intake of cloudy juice or enriched cloudy juice.

However, this also means that humans are frequently exposed to these semi-ubiquitous metabolites, which, as mentioned before, persist in our bodies for several hours after ingestion. The concentrations are often much higher than those of most derivatives of polyphenols processed via the human metabolism. As an example, very high urinary concentrations (totalling 261 μM) of the microbial catabolites of flavanols have been reported after repeated green tea intake (Brindani et al. 2017). Cumulative urinary excretion of a mixture of eight valerolactones and three phenyl-valeric acids, with a cumulative concentration of up to 524.35 μmol (93% valerolactones and 7% phenyl-valeric acids), has been reported after acute intake of 400 g of Elstar apples in a clinical study performed at the Division of Human Studies at the Max Rubner-Institut in Karlsruhe, Germany (Anesi et al. 2019).

6.3.1.6 Considerations of Bioequivalence Studies

Lastly, if the final compound of the microbial catabolism could be demonstrated to be mechanistically involved in bioactivity, this would give rise to the question of whether the properties are unique to the polyphenols of a given food matrix, or rather more generally associated with the family of catabolites. As an example, the catabolites of flavanols are possibly involved in the improvement of endothelial function following the consumption

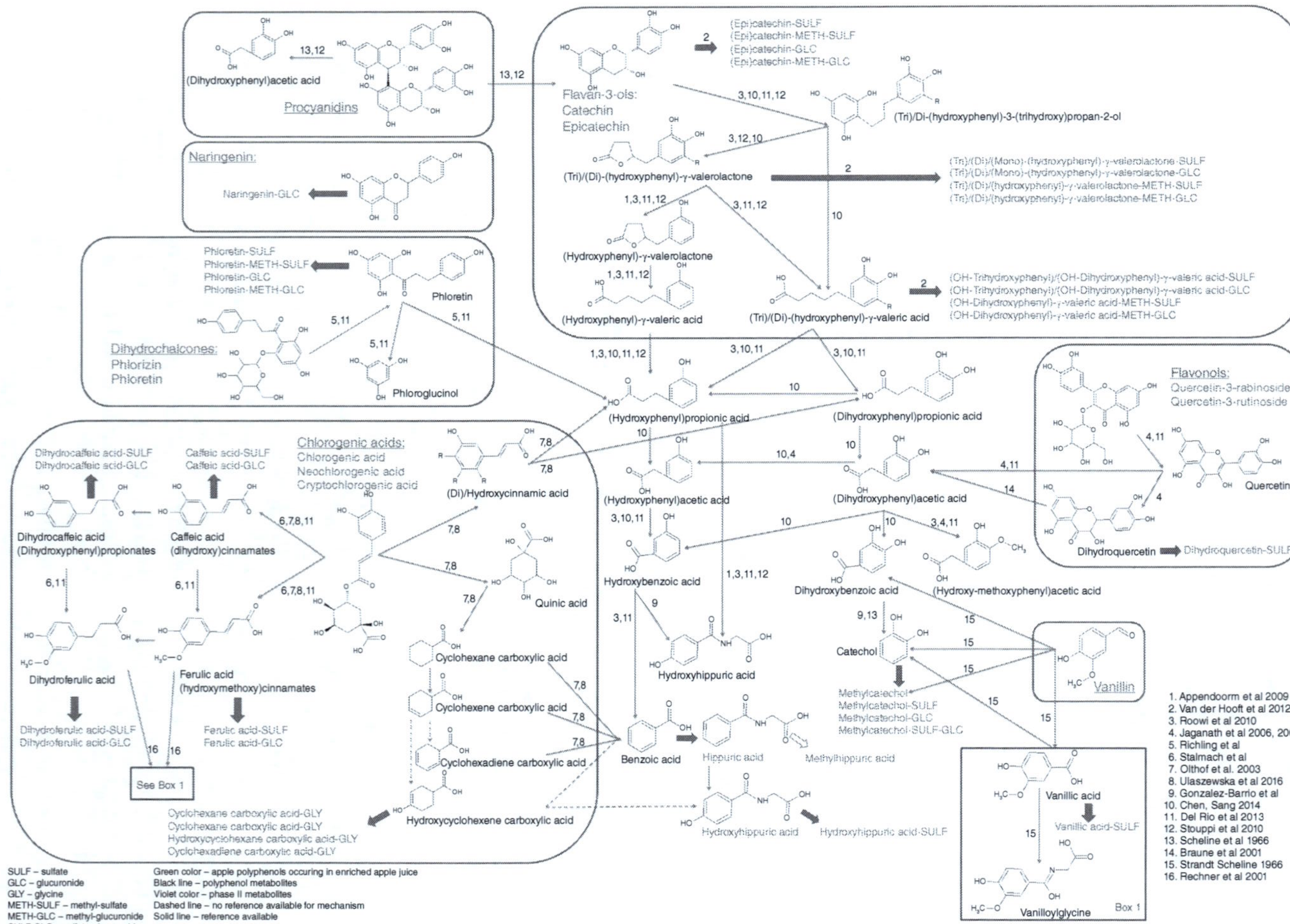

Figure 6.9 A simplified diagram of the metabolic pathways for the main classes of apple polyphenols and their crossing trajectories. *Source:* Reprinted from Trost et al. (2018a), with permission from Elsevier. (*See insert for color representation of the figure.*)

of apples with their peel (Bondonno et al. 2018), and this *in vivo* result seems biologically plausible and coherent with health claims approved by the European Food Safety Authority scientific panel on Dietetic Products, Nutrition and Allergies, due to the role of cocoa procyanidins in maintaining normal endothelium-dependent vasodilation. (EFSA Panel 2012, 2014). At the moment, this claim is accepted only for procyanidins from cocoa. However, in the future, better understanding and modelling of the role of circulating catabolites, which are structurally largely the same for cocoa as for apple procyanidins, for example – both being based on the prevailing presence of epicatechin as the main building block – could raise the question as to whether, in the future, we will be able to precisely define the experiments needed to demonstrate nutritional bioequivalence for the same class of polyphenols obtained from multiple food sources. This would greatly simplify the work of nutritionists and translate into more effective customized advice for the population. For the oral drug products, the bioavailability should be compared between the original and generic products as a part of bioequivalence studies of drug products, where the bioavailability is defined as the rate and extent of absorption of active pharmaceutical ingredients or active metabolites from a product into the systemic circulation (Kuribayashi et al. 2016).

Unfortunately, bioequivalence is not currently defined for natural bioactive compounds from our food, but we envisage that it will be considered in the future, as after all, it requires the straightforward application of standardized protocols already widely applied to assessing the bioequivalence of two or more drugs, essentially based on comparison of relative bioavailability. These protocols for bioequivalence are universally accepted and represent the basis of the economic sustainability of the public health system, and they could and should also be translated into the field of nutrition.

As an example, related to the apple, it is estimated that consumption of a single standard serving of apples in Italy (150 g) would provide, on average, about 102 mg of flavan-3-ols (Vrhovsek et al. 2004). In terms of bioequivalence, consumption of two apples a day could provide the same amount of flavonols as reported by the EFSA NDA panel for cocoa (200 mg) (EFSA Panel 2014), therefore possibly having similar beneficial effects on human health.

6.3.2 Lessons from the Prolonged Exposure Study

6.3.2.1 Renetta Apples and Cardiometabolic Biomarkers

The combination of the data from these complementary studies allowed us to compare the metabolites of apple polyphenols in human biofluids after acute and long-term intake (Table 6.1). Moreover, we observed that long and sustained consumption of apples triggers the gut microbiota to activate tyrosine, tryptophan, and the indole metabolic pathway, and thus the endogenous metabolism.

The design of the long-term study allowed researchers to investigate the changes in the plasma and urine metabolome of 40 healthy, mildly hypercholesterolemic volunteers, consuming either two Renetta apples or a placebo, for eight weeks. At the same time, metagenomic evaluation of the microbiota composition was performed in samples of feces from the volunteers participating in the dietary intervention study. The Renetta apple was chosen because it is very rich in procyanidins, the average daily intake provided being estimated at 854 ± 233 mg. This clinical study provided promising results, finding evidence of the beneficial hypocholesterolemic and vascular effects of daily consumption of PA-rich apples by mildly hypercholesterolemic individuals (Koutsos et al. 2020).

6.3.2.2 Investigating the Importance of Microbial Biodiversity of the Human Gut

The availability of a multiomic and multicompartment dataset with both metabolomic data on the metabolites observed in plasma and urine, and metagenomic data on gut bacteria, and with a reasonable number of participants who completed the study, allowed us to investigate the correlation between gut microbiota and circulating metabolites (Ulaszewska et al. 2020a).

Recently, progress in the field of microbial genomics has provided impressive insight. Scientists have compiled > 200,000 genomes and 170 million protein sequences from more than 4,600 bacterial species in the human gut (Almeida et al. 2021). Databases such as the Unified Human Gastrointestinal Genome collection and the Unified Gastrointestinal Protein catalogue, reveal the tremendous diversity in our gut, and pave the way for further microbiome research. Recovering genomes from shotgun metagenomic sequence data allow detailed taxonomic and functional characterization of individual species or strains in a microbial community, producing metagenome-assembled draft genomes, which can provide invaluable clues about uncultured organisms (Saheb Kashaf et al. 2021). The establishment of such a massive sequence catalogue has highlighted the fact that a large part of species and functional diversity within the human gut microbiome remains uncharacterized (Almeida et al. 2021).

In this context of incredible diversity, we suggest that a simple bottom-up approach aimed at correlating the metabolome with the metagenome could provide a shortlist of the bacteria putatively involved in a specific metabolic step. We used such an approach tentatively with the results of the acute experiment (Trost et al. 2018a). In light of the small scale of the study, this could be considered as a kind of proof-of-principle, where the relative abundance of fecal bacteria was correlated with the AUC observed for each of the microbial catabolites derived from apple polyphenols. In this case, and despite the low number of participants involved in the acute study, this simple statistical technique supported the identification of some previously unpredicted putative associations with specific genera of fecal bacteria. Once these putative associations are discovered from such *in vivo* experiments, the knowledge acquired should help in the design of mechanistic studies aimed at evaluating the impact of relative abundance in the microbial community of given microbial genera (or in the future, also a given strain) on polyphenol metabolism.

Even better results were obtained when correlation analysis between the concentration of metabolites observed in urine and plasma after consumption of apples and the relative abundance of fecal bacteria taxa was explored in a long-term study involving a higher number of participants (40). Again, in this case, the multiomic dataset allowed exploration of the correlations between metabolites modulated by dietary intervention and fecal microbiota species (at genus level), showing, in particular, interesting interactions between the *Granulicatella* genus and the pattern of phenyl-acetic acid metabolites (Figure 6.10). The relative abundance of *Granulicatella* was found to be inversely correlated with

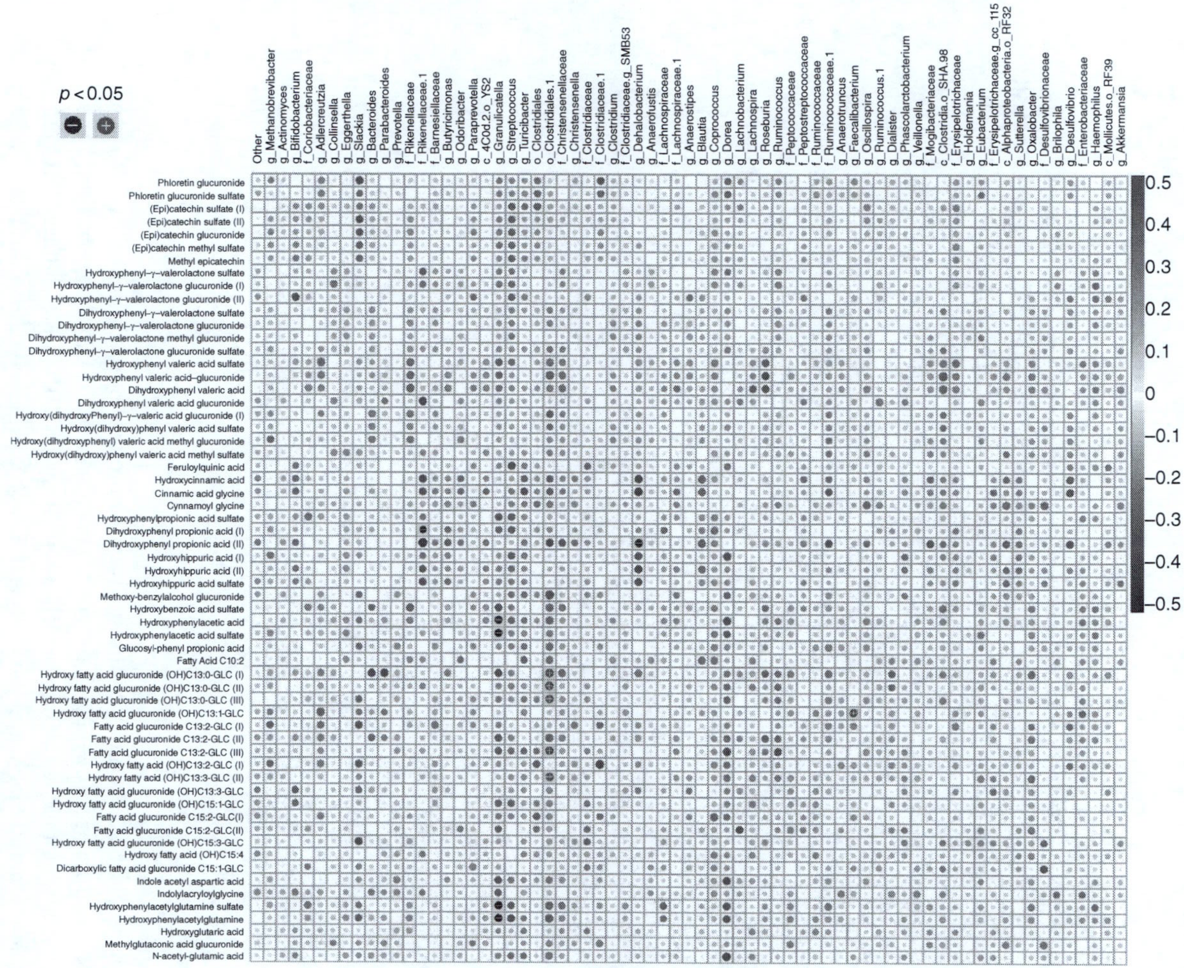

Figure 6.10 Heat map of pairwise correlations between urine metabolites and gut microbiota at the genus level. Correlation coefficients (Spearman's ρ) are represented by color gradient and size of the circles. The significant values (FDR<0.07) are marked with a + and yellow background. For unclassified genera ("g_"), the bacterial family ("f_") or order name ("o_") is reported. *Source:* Reprinted by permission from Ulaszewska et al. (2020a). (*See insert for color representation of the figure.*)

hydroxyphenyl acetyl glutamine, hydroxyphenyl acetylglutamine sulfate, hydroxyphenyl acetic acid, and hydroxyphenyl acetic acid sulfate, a group of structurally similar microbial metabolites containing the hydroxyphenyl acetic moiety, deriving from phenylalanine decarboxylation by the colonic microbiota. Two different isomers of dihydroxyphenyl propionic acid showed a statistically significant, inverse correlation with *Rikenellaceae* and *Dehalobacterium* (Ulaszewska et al. 2020a).

6.4 Conclusion

This work describes the destiny of apple polyphenols in human biofluids and takes a small initial step in linking system-level metabolic processing of dietary polyphenols with microbiome architecture, a necessary move away from limited taxonomic descriptions or measurement of metabolic potential and toward an improved understanding of microbiome metabolic function and nutrikinetics.

In general, recent studies have confirmed the strong involvement of intestinal microbiota in system-level metabolism of complex plant polyphenols ingested with food, and have provided evidence that the late appearance of several small phenolic catabolites of apple polyphenols in blood and urine can be significantly correlated to the relative abundance of different phylogenetically distantly related bacterial genera.

The design of large multi-compartment and multiomic experiments has the power to reveal the putative associations important for the mechanistic understanding of the relationship between gut microbiota and polyphenol metabolism, in a real-life context. This can provide a better prediction of how interindividual differences in the gut microbiota influence the processing of dietary polyphenols, mediating their catabolite-induced toxicity and/or efficacy.

Acknowledgments

Figures 5.1 and 5.9 have been reprinted from Trost et al. (2018a), with permission from Elsevier. Figure 6.10 has been reprinted from Ulaszewska et al. (2020a), with permission from Springer Nature.

Funding

This metabolomic study received funding from the project "Apple fruit quality in the post-genomic era, from breeding new genotypes to post-harvest: nutrition and health," funded by the AGER (Agribusiness and research) with Grant no. 2010–2119, internal funding of Fondazione Edmund Mach, and the European Union's Horizon 2020 research and innovation grant agreement No 696295 ERA-Net Cofund ERA-HDHL "Biomarkers for Nutrition and Health implementing the JPI HDHL objectives" (https://www.healthydietforhealthylife.eu/), project FOODBALL (https://foodmetabolome.org/).

References

Almeida, A., Nayfach, S., Boland, M., et al. (2021). A unified catalog of 204,938 reference genomes from the human gut microbiome. *Nature Biotechnology*, **39**: 105–114.

Ancillotti, C., Ulaszewska, M., Mattivi, F., and Del Bubba, M. (2019). Untargeted metabolomics analytical strategy based on liquid chromatography/electrospray ionization linear ion trap quadrupole/Orbitrap mass spectrometry for discovering new polyphenol metabolites in human biofluids after acute ingestion of *Vaccinium myrtillus* berry supplement. *Journal of the American Society for Mass Spectrometry*, **30** (3): 381–402.

Anesi, A., Mena, P., Bub, A., et al. (2019). Quantification of urinary phenyl-γ-valerolactones and related valeric acids in human urine on consumption of apples. *Metabolites*, **9**: 254.

Bondonno, N.P., Bondonno, C.P., Blekkenhorst, L.C., et al. (2018). Flavonoid-rich apple improves endothelial function in individuals at risk for cardiovascular disease: a randomized controlled clinical trial. *Molecular Nutrition & Food Research*, **62** (3): 1700674.

Brindani, N., Mena, P., Calani, L., et al. (2017). Synthetic and analytical strategies for the quantification of phenyl-γ-valerolactone conjugated metabolites in human urine. *Molecular Nutrition & Food Research*, **61** (9): 6–10.

EFSA Panel on Dietetic Products, Nutrition and Allergies (NDA) (2012). Scientific Opinion on the substantiation of a health claim related to cocoa flavanols and maintenance of normal endothelium-dependent vasodilation pursuant to Article 13(5) of Regulation (EC) No 1924/2006. *EFSA Journal* **10** (7), 10.2903/j.efsa.2012.2809 (accessed 10 November 2021).

EFSA Panel on Dietetic Products, Nutrition and Allergies (NDA) (2014). Scientific Opinion on the modification of the authorisation of a health claim related to cocoa flavanols and maintenance of normal endothelium-dependent vasodilation pursuant to Article 13(5) of Regulation (EC) No 1924/2006. *EFSA Journal*, **12**. https://efsa.onlinelibrary.wiley.com/doi/pdf/10.2903/j.efsa.2014.3654 (accessed 10 November 2021).

Farneti, B., Masuero, D., Costa, F., et al. (2015). Is there room for improving the nutraceutical composition of apple? *Journal of Agricultural and Food Chemistry*, **63** (10): 2750–2759.

Haug, K., Salek, R.M., Conesa, P., et al. (2013). MetaboLights—an open–access general–purpose repository for metabolomics studies and associated meta–data. *Nucleic Acids Research*, **41**: D781–D786.

Haug, K., Cochrane, K., Nainala, V.C., et al. (2020). MetaboLights: a resource evolving in response to the needs of its scientific community. *Nucleic Acids Research*, **48**: D440–D444.

Koutsos, A., Riccadonna, S., Ulaszewska, M.M., et al. (2020). Two apples a day lower serum cholesterol and improve cardiometabolic biomarkers in mildly hypercholesterolemic adults: a randomized, controlled, crossover trial. *American Journal of Clinical Nutrition*, **111** (2): 307–318.

Kuribayashi, R., Takishita, T., and Mikami, K. (2016). Regulatory considerations of bioequivalence studies for oral solid dosage forms in Japan. *Journal of Pharmaceutical Sciences*, **105** (8): 2270–2277.

Saheb Kashaf, S., Almeida, A., Segre, J.A., et al. (2021). Recovering prokaryotic genomes from host-associated, short-read shotgun metagenomic sequencing data. *Nature Protocols*, **16**: 2520–2541.

Senger, D.R., Li, D., Jaminet, S.C., and Cao, S. (2016). Activation of the Nrf2 cell defense pathway by ancient foods: disease prevention by important molecules and microbes lost from the modern western diet. *PLoS One*, **11** (2): e0148042.

Savoi, S., Arapitsas, P., Duchêne, E., et al. (2021). Grapevine and wine metabolomics-based guidelines for FAIR data and metadata management. *Metabolites*, **11** (11): 757.

Trost, K., Ulaszewska, M.M., Stanstrup, J., et al. (2018a). Host: Microbiome co-metabolic processing of dietary polyphenols – An acute, single blinded, cross-over study with different doses of apple polyphenols in healthy subjects. *Food Research International*, **112**: 108–128.

Trost, K., Ulaszewska, M.M., Stanstrup, J. et al. (2018b). Host: Microbiome co-metabolic processing of dietary polyphenols – an acute, single blinded, cross-over study with different doses of apple polyphenols in healthy subjects. *MetaboLights*. https://www.ebi.ac.uk/metabolights/MTBLS473 (accessed 11 November 2021).

Ulaszewska, M.M., Trost, K., Stanstrup, J., et al. (2016). Urinary metabolomic profiling to identify biomarkers of a flavonoid-rich and flavonoid-poor fruits and vegetables diet in adults: the FLAVURS trial. *Metabolomics*, **12**: 1–22.

Ulaszewska, M.M., Koutsos, A., Trošt, K., et al. (2020a). Two apples a day modulate human:microbiome co-metabolic processing of polyphenols, tyrosine and tryptophan. *European Journal of Nutrition*, **59** (8): 3691–3714.

Ulaszewska, M.M., Koutsos, A., Trošt, K. et al. (2020b). Two apples a day modulate human:microbiome co-metabolic processing of polyphenols, tyrosine and tryptophan. *2020. MetaboLights*. https://www.ebi.ac.uk/metabolights/MTBLS469 (accessed 11 November 2021).

Ulaszewska, M.M., Weinert, C.H., Trimigno, A., et al. (2019). Nutrimetabolomics: an integrative action for metabolomic analyses in human nutritional studies. *Molecular Nutrition & Food Research*, **63** (1): 1–38.

Vrhovsek, U., Rigo, A., Tonon, D., and Mattivi, F. (2004). Quantitation of polyphenols in different apple varieties. *Journal of Agricultural and Food Chemistry*, **52**: 6532–6538.

7

Non-Extractable Polyphenols Should be Systematically Included in Polyphenol Analysis

Enrique Báez-García[1,2], Sonia G. Sáyago-Ayerdi[2], and Jara Pérez-Jiménez[1]

[1] *Institute of Food Science, Technology and Nutrition, Spanish Research Council (ICTAN-CSIC), Madrid, Spain*
[2] *Tecnológico Nacional de Mexico/Instituto Tecnológico de Tepic, Tepic, Nayarit, Mexico*

7.1 Introduction: The Concept of Non-Extractable Polyphenols

Research on polyphenols has increased over time, with more than 43,000 scientific papers being published just in the last decade (2009–2019). These studies cover, among others, chemical, biological, and nutritional aspects with the common feature that most of them are based on phenolic compounds extracted from a plant matrix either by traditional solid-liquid extraction or updated techniques such as pressurized hot water extraction (PHWE) or other similar ones. Nevertheless, the residues of these extractions still contain a high proportion of polyphenols, a fact first described more than 20 years ago (Matthews et al. 1997), but which still has not been routinely included in studies on polyphenols. For this reason, this has been defined by some researchers as the "hidden face" of these compounds (Tarascou et al. 2010).

In this way, polyphenols present in foods, besides classifications based on their chemical skeleton (phenolic acids, flavonoids, etc.), may be divided into two important groups: extractable polyphenols (EPP) and non-extractable polyphenols (NEPP) (Saura-Calixto 2012). EPP are mostly low molecular weight compounds (although they also include polymeric structures such as in the case of the different tannin families), belonging to all polyphenol chemical classes, are mostly free in the food matrix and they can be solubilized in mixed aqueous-organic solvents, i.e. they are commonly considered as "total polyphenols." The other groups, which will be explored in this chapter, are NEPP, those polyphenols which are not extracted with the solvents of choice for analyzing these compounds (González-Sarrías et al. 2017). NEPP comprise two kind of structures (Pérez-Jiménez and Torres 2011):

1) high molecular weight polyphenols, mostly non-extractable proanthocyanidins (NEPA), but also other classes, such as non-extractable ellagitannins (García-Villalba et al. 2015) or non-extractable phlorotannins (Sanz-Pintos et al. 2017);

Recent Advances in Polyphenol Research, Volume 8, First Edition. Edited by Juha-Pekka Salminen, Kristiina Wähälä, Victor de Freitas, and Stéphane Quideau.

2) hydrolysable polyphenols (HPP), which are low molecular weight polyphenols linked to cell wall constituents (proteins and polysaccharides) by covalent bonds, hydrogen bonds, and hydrophobic interactions (Pérez-Jiménez et al. 2013). HPP are mostly conformed by phenolic acids such as ferulic acid, sinapic or caffeic acid, and have been studied for long mostly in cereal products, where they were known as bound or insoluble phenolic compounds (John and Shahidi 2010). Nevertheless, they may include other phenolic compounds, such as flavonoid constituents (Arranz et al. 2010). In order to make clear the nomenclature, it should be highlighted that the term HPP is not equivalent to that of hydrolysable tannins, which, indeed, correspond to another category of high molecular weight polyphenols. During last years, several research teams have been using the term HPP for analyzing this kind of small phenolic structure associated with food matrix in very different foods (Faller and Fialho 2009; Durazzo et al. 2017; Gil et al. 2022).

Based on the information currently available, it seems that the main difference between EPP and NEPP relies either on their molecular weight (being higher in NEPP than in EPP) or on their association with other food macromolecules. Regarding this aspect, the linkages that may take place between polyphenols and macromolecules have been reviewed by Le Bourvellec and Renard (2012) and they involve both non-covalent interactions (hydrogen bonds and hydrophobic interactions) and covalent bonds (including ester and ether bonds, as well as the transformation of phenolic structures into carbocations, *o*-quinones and semiquinones), all of them being affected by the structure of both polyphenols and macromolecules. At the same time, although the macromolecules with which NEPP are associated were mostly dietary fiber (DF) and proteins, it should be mentioned that it is known that Maillard reaction products are deeply linked to both DF and polyphenols, constituting the so-called maillardized DFs (Silván et al. 2010); for instance, NEPP release from melanoidin structures during *in vitro* digestion of coffee or bread was recently reported (Alves et al. 2021). These differences between EPP and NEPP mean that, up to this moment, no specific chemical structures have been found to be present in NEPP and absent in EPP. Nevertheless, as it will be discussed later, the analysis of NEPP is based mostly on the characterization of degradation products found after performing extensive hydrolysis of the original materials, in order to break either polymeric phenolic structures or their linkage with the dietary matrix. For this reason, the possibility that NEPP exhibit specific intramolecular linkages or oxidized forms not found yet has not to be disregarded. Another aspect that should be mentioned is that EPP and NEPP may show differential accumulations (both in qualitative and quantitative terms) in the different tissues of the plants, what may be related with their biological actions as protecting agents against several damages (Wang et al. 2020).

Finally, it should be mentioned that NEPP are, as all polyphenols, antioxidant compounds (although they exhibit more mechanisms of action besides their antioxidant capacity). And they are present in foods as macromolecular structures either because they are macromolecules per se (in the case of NEPA) or because they are associated with macromolecules (particularly HPP with DF or with Maillard reaction products, but all of them may be associated with proteins). For these two characteristic facts, in 2015, the term "macromolecular antioxidant" was suggested as équivalent NEPP, considering that it was a term that could be better translated to applications in nutritional context and medical field, since NEPP is sometimes considered as a more chemical concept without biological implications (Pérez-Jiménez and Saura-Calixto 2015; Saura-Calixto 2018). Nevertheless, it should be considered that although most macromolecular antioxidants are polyphenols, the term is not exactly a synonym, since other antioxidants, particularly carotenoids, have been shown to be also closely linked to dietary matrix and, thus, they may also be considered as macromolecular antioxidants (Bohn 2018).

7.2 Analysis of Non-Extractable Polyphenols

During last decades, several efforts have been performed in order to advance in the analytical characterization of NEPP, involving from their release from food matrix to their subsequent chemical analysis and the selection of the most appropriate standards. Nevertheless, as pointed out in this section, NEPP analysis is still far from standardization, what makes difficult its routine determination in a wide number of laboratories. Current situation in this field will be described, while perspectives and suggestions will be explored later in the section on perspectives.

7.2.1 Preparation of Solutions of Non-Extractable Polyphenols

NEPP analysis is divided in three main steps: removal of EPP from the vegetal material, hydrolysis of the remaining material in order to prepare NEPP solutions, and chemical analysis of NEPP solutions; a general scheme of this workflow is depicted in Figure 7.1. Alternatively, it is possible to perform the hydrolysis treatments directly in the whole sample, so EPP and NEPP will be measured together. In both cases, the use of approaches based on NEPP depolymerization or release from food matrix hampers the knowledge of original NEPP structures since they imply breaking of the original bonds in order to obtain either the monomeric units constituting polymeric NEPP or the small phenolic compounds present in HPP. Nevertheless, they constitute the only ones available up to this moment.

For extracting EPP from food matrix, the most common combination is acidic methanol/water (50 : 50 v/v, pH: 2), followed by acetone/water (70 : 30 v/v) (Saura-Calixto et al. 2007; Pérez-Jiménez et al. 2009a). Other authors such as Tow et al. (2011) showed that a solution with 50% ethanol was most effective to extract EPP than using 80% methanol or 80% ethanol in combination with HCl (0.18 N). Generally, aqueous-organic solvents with acid pH are used because polyphenols are more stable at low pH; nevertheless, these are general conditions for all EPP which may have to be optimized when a specific type of phenolic compound is going to be extracted.

Due to the complex nature of NEPP, they are retained in the polysaccharide-lignin network by hydrogen or hydrophobic linkage, thus being not accessible for solvents in normal extraction processes. For this reason, their isolation requires the disruption of macromolecules by enzymatic treatment or chemical hydrolysis (Pérez-Jiménez et al. 2013). Additionally, some emerging extraction techniques have been suggested for NEPP release from food matrix.

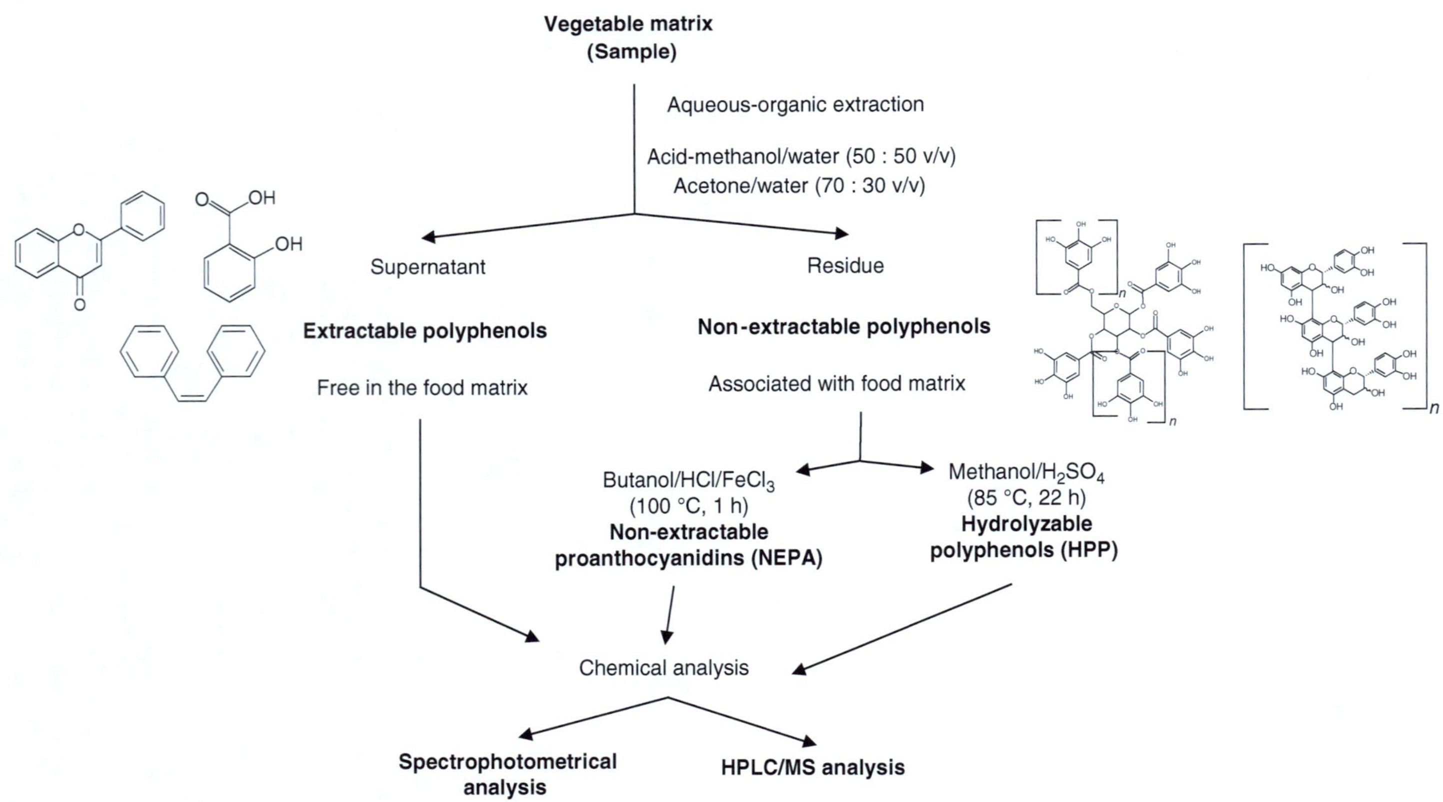

Figure 7.1 Common procedure for polyphenol analysis.

7.2.1.1 Alkaline and Acid Hydrolysis

Alkaline and acid hydrolysis are frequently used for releasing NEPP, either at room temperature or combined with heating. Some examples of conditions for releasing either EPP or NEPA (independent treatments in the residues for EPP removal must be applied) are shown in Figure 7.1. Additionally, a specific procedure for the release of non-extractable ellagitannins was proposed by García-Villalba et al. (2015).

Regarding HPP, in the acid hydrolysis method, inorganic acids such as HCl and H_2SO_4 (Arranz and Saura-Calixto 2010) are used. This treatment can hydrolyze glycosidic bonds of (hemi)cellulose and solubilize sugars (Wu et al. 2018) or cause the depolymerization of hydrolysable tannins producing their degradation into gallic acid and ellagic acid (Pérez-Jiménez and Torres 2011). While alkaline hydrolysis, generally performed with NaOH, is known to break the ester bonds linking NEPP with cell wall. After both treatments, a neutralization step is needed for protecting chromatography columns for extreme pH values. Indeed, in the case of acid hydrolysis, an additional solid phase extraction procedure is recommended when High Performance Liquid Chromatography-Mass Spectrometry (HPLC-MS) analysis is going to be applied in the hydrolysates, in order to remove salts that may damage the ionization source (Pérez-Jiménez and Saura-Calixto 2015).

In some samples, acid hydrolysis has proven to be more effective than alkaline hydrolysis; indeed, it has been suggested that acid hydrolysis increases phenolic compound stability, avoiding quinone formation and deprotonation of hydroxyl groups (Yan and Zhen 2017; Wang et al. 2019a; Wang et al. 2020). Nevertheless, some phenolic compounds can be lost due to acid hydrolysis (92% losses for *trans*-sinapic acid, 87% for caffeic acid and 15% *O*-coumaric acid were observed in oilseeds) (Krygier et al. 1982). Due to these potential losses, there are several studies using antioxidants such as ascorbic acid and a metal chelator like ethylenediaminetetraacetic acid (EDTA) to prevent the degradation of phenolic compounds during alkaline hydrolysis (Ross et al. 2009; Arigò et al. 2018). Overall, for each analysis the pros and cons of chemical acid and alkaline hydrolysis must be considered, in order to choose the most suitable one. Additionally, it has been observed that when consecutive acid and alkaline hydrolyses are used, NEPP release varies depending on the hydrolysis order. Thus, when acid hydrolysis was performed in apple pomace, the main phenolic compounds that were released where 4-hydroxybenzoic acid, protocatechuic acid and isoferulic acid, while in subsequent alkaline hydrolysis, the main phenolic compounds were protocatechuic acid, catechin, and 4-hydroxybenzoic acid. When the order of chemical hydrolysis was inverted, the main phenolic compounds released with alkaline hydrolysis were phloridizin, 4-hydroxybenzoic acid, and quercetin-3-*O*-galactoside, while in subsequent acid hydrolysis, they were 4-hydroxybnezoic acid and isoferulic acid (Li et al. 2020).

Regarding the main parameters to be optimized during acid or alkaline hydrolysis for HPP release, it was reported, in NEPP release from blueberries, that the NEPP yield in alkaline hydrolysis process was significantly affected by the NaOH concentration and liquid/solid ratio, while in the acid hydrolysis process, this yield was significantly affected by temperature, time, and liquid/solid ratio (Cheng et al. 2014). Also, a study in rice found that a 2 M NaOH concentration was more efficient that 4 M NaOH in order to release HPP (Melini and Acquistucci 2017). Irakli et al. (2018) optimized conditions for a maximum recovery (Box-Behnken design) of EPP and NEPP from rice by-products (rice bran and rice husk), analyzing the effect of ethanol concentration, extraction time, temperature, NaOH concentration, hydrolysis time, and temperature, using ultrasound-assisted extraction (UAE). They found that the optimal conditions for NEPP release in rice bran were 3.4 M NaOH at 80 °C for 110 minutes, while for rice husk they were 2.5 M NaOH at 80 °C for 120 minutes.

In the case of NEPA, alkaline hydrolysis has been reported (White et al. 2010), although it is not as efficient as acid hydrolysis. For acid hydrolysis, there are two common procedures. The first one is the Porter method. In this approach, developed some decades ago (Porter et al. 1985), a treatment with butanol/HCl, causes the depolymerization of the structures, releasing monomers as anthocyanins. After first publication, the method has been applied to different samples (Terrill et al. 1992; Matthews et al. 1997) and some updates have been published (Zurita et al. 2012) although the limitations associated with a spectrophotometric method remain (see Section 7.2.2.1). The second approach is performing acid-catalyzed depolymerization in presence of nucleophiles. In this case, the terminal units are released as monomeric flavanols, while the extension units are detected as conjugates with the nucleophile used, e.g. toluene, benzyl mercapthan or phloroglucinol (reviewed in Pérez-Jiménez and Torres 2011). During this process, epimerization of the monomers may take place (Neilson et al. 2016).

Nevertheless, the wide diversity of conditions used in the chemical hydrolysis of NEPP, some of which are shown in Table 7.1, makes rather difficult the direct comparison of results from different studies as well as the suggestion of a standardized methodology for this procedure.

7.2.1.2 Enzymatic Hydrolysis

Enzymatic treatment is another specific and effective way to release NEPP from solid materials. Examples of enzymes used to release NEPP are cellulases, glucuronidases, hemicellulases, pectinases, amylases, glucanases, and tannases. They have been selected for different reasons, e.g. they catalyze specific biochemical reactions, are efficient at low concentrations, active under mild pH and temperature conditions, have high substrate specificity and low toxicity (Wu et al. 2018). Enzymes break down the carbohydrate linkages and hydrolyze the cell wall structure (Matsumura et al. 2016). Some authors have shown that depending on the enzyme used, different NEPP can be released. Thus, Pérez-Jiménez et al. (2009a) reported that, in apple, a combination of cellulose and pectinase exhibited higher yields of NEPA than esterase. Interestingly, electronic microscopy has provided evidence of the structural changes generated by cellulase action, probably due to the degradation of crystalline cellulose components in the primary wall (Mushtaq et al. 2015).

Several factors could have influence on enzymatic hydrolysis yields. For instance, it was reported that a higher cellulose and lignin content in vegetables produced a material highly resistant to enzyme degradation, decreasing the release of NEPP (Domínguez-Rodríguez et al. 2017). Although enzymatic treatment has been considered a novel and effective way to extract bound PC, some drawbacks related to the use of enzymes are their high cost and slow reaction rates (Wu et al. 2018). Nevertheless, in some cases enzymatic hydrolysis was reported to be more effective than alkaline hydrolysis for NEPP release; that was the case

Table 7.1 Examples of chemical hydrolysis for NEPP release in vegetal materials.

Sample	Previous EPP removal	Chemical hydrolysis type	Chemical hydrolysis conditions	References
Apple, grape, and lingonberry	Acetone/methanol/water (2 : 2 : 1, v/v/v). Sonication, 10 min, three times.	Acid	Methanol (1% HCl, 5% benzyl mercaptan) (40°C, 30 min)	Hellström and Mattila (2008)
Barley	Hexane (three times by shaking for 15 min, after centrifugation) 70% acetone/water mixture (v/v) (three times by shaking for 25 min, after centrifugation) Ethyl acetate (three times by shaking for 15 min)	Alkaline and acid	Alkaline hydrolysis 4 M NaOH, EDTA and ascorbic acid (shaking 4 h, 25 °C) followed by neutralization with 35% HCl (pH: 2, shaking three times, 15 min) Acid hydrolysis 1 M HCl (90°C, 45 min) followed by extraction with ethyl acetate (three times, 15 min)	Arigò et al. (2018)
Barley varieties (Falcon, AC Metcalfe, Tyto, Tercel, Phoenix, and Peregrine)	Methanol-acetone-water (7 : 7 : 6, v/v/v) (homogenization at 10,000 rpm for 4 min) Diethyl ether (1 : 1, v/v)	Alkaline	4 M NaOH (4 h, under N_2) followed by neutralization with 6M HCl and diethyl ether extraction	Madhujith and Shahidi (2009)
Cabbage (red) and Brussels sprout waste streams	100% methanol (homogenizer), followed by 80% methanol (homogenizer)	Alkaline	4 N NaOH (80°C, 30–45 min), followed by neutralization with HCl and methanol extraction	Gonzales et al. (2015)
Cocoa beans	1 M NaCl (shaking by orbital shaker at room temperature for 30 min)	Alkaline and acid	Alkaline hydrolysis 2 M NaOH with 2% ascorbic acid and 20 mM EDTA (90 min, 30 °C) followed by an adjustment of pH: 1 with 5 M HCl Acid hydrolysis 10.2 M HCl, 1% ascorbic acid (75 °C, 150 min) followed by neutralization with 10 M NaOH	Oracz et al. (2018)
Couscous	Ethanol/ water (4 : 1, v/v, 15 min in an ultrasonic bath two times)	Alkaline	NaOH 1 M (room temperature overnight, shaking under N_2) followed by neutralization with HCl (pH:2) and extracted with ethyl acetate	Oriente et al. (2020)

(Continued)

Table 7.1 (Continued)

Sample	Previous EPP removal	Chemical hydrolysis type	Chemical hydrolysis conditions	References
Cranberry beans	70% MeOH + 1% HCl (v/v). Sonication for 15 min, rotational mixing at room temperature for 2 h	Alkaline	2 N NaOH. Sonication (15 min), rotational mixing (room temperature, 16 h)	Chen et al. (2015)
Cranberry pomace	Previous hexane defatting, ethyl acetate (room temperature, 10 min, three times)	Alkaline	2N Na OH (60 °C, 15 min), followed by neutralization with HCl	White et al. (2010)
Litchi pulp	We used five types of solvents 80% methanol 80% ethanol 80% acetone 80% ethyl acetate Deionized water All solvents were homogenized and then centrifugated at 5000 rpm for 8 min at 4°C	Alkaline and acid	Alkaline hydrolysis 2 M NaOH (room temperature,18 h, shaking) after neutralized with HCl and ethyl acetate six times Acid hydrolysis H_2SO_4/methanol (90 : 10, v/v, 85 °C, 20 h) after neutralized with NaOH 10 M and ethyl acetate six times	Su et al. (2014)
Oats	Three different extraction (A) 4/1 ethanol/water (v/v) (B) 4/1 methanol/water (v/v) (C) acetone/water (v/v). Followed by centrifugation	Alkaline and acid	Alkaline hydrolysis 2 M NaOH at room temperature for 20 h by shaking under N_2 atmosphere, followed by neutralization with 10 M HCl and extraction with 1/1 diethyl ether acetate (v/v) Acid hydrolysis 96% methanol and 25% HCl (65 °C, 30 min) followed by extraction with 1 : 1 diethyl ether/petroleum (v/v) (40–60 °C).	Verardo et al. (2011)
Pomegranate	Methanol/DMSO/H_2O (40 : 40 : 20, v/v/v, 0.1 HCl) and after centrifugation for 10 min.	Acid	37% HCl (24 h, 90 °C) followed by an adjustment (pH to 2.5) with 10 and 5 M of NaOH and centrifugation for 10 min.	García-Villalba et al. (2015)

Processed fibers derived from the juice industry	Methanol/water (acidified at 0.1% v/v with acetic acid) (50 : 50) in an ultrasonic bath for 30 min Acetone/water (70 : 30) in an ultrasonic bath for 30 min	Alkaline and acid	Alkaline hydrolysis 2 M NaOH (10 mM EDTA, 1% ascorbic acid, shaking, 30 min at 45 °C) after neutralization with HCl 7.2 M Acid hydrolysis *n*-butanol/HCl (95 : 5, v/v) and HCl 2 M (2% ferric ammonium sulphate, 60 min)	Delpino-Rius et al. (2015)
Roselle (*Hibiscus sabdariffa* L.) calyces	Methanol/water (50 : 50, v/v) acidified with HCl 2 N (pH 2) (1 h, room temperature), followed by acetone/water (70 : 30, v/v) (room temperature, 1 h)	Acid	NEPA: Butanol/HCl (97.5 : 2.5) 0.04% $FeCl_3$ (100 °C, 1 h). HPP: Methanol/H_2SO_4 (90 : 10, v/v) (85 °C, 20 h).	Amaya-Cruz et al. (2019a)
Whole grain and bran of white, red, and black rice	80% methanol (by shaking, 30 min at room temperature after a centrifugation 4,000 g for 10 min)	Alkaline	4 M NaOH (room temperature, shaking, 2 h) after adjusted pH: 1.5–2.0 with HCl concentrated and ethyl acetate	Pang et al. (2018)

NEPP: Non-extractable polyphenols, HPP: Hydrolyzable polyphenols, NEPA: Non-extractable proanthocyanidins in *Rubus idaeus* L. leaves and seeds (Wang et al. 2019a). Some studies have used the enzymatic assisted extraction for improving the release of NEPP; for example, de Camargo et al. (2016) applied this kind of extraction in winemaking by-products (cv. Tempranillo) treated with Pronase® and Viscozyme®. Viscozyme® released higher amounts of gallic acid, catechin, and prodelphinidin dimer A as compared with Pronase® treatment. Furthermore, *p*-coumaric acid and caffeic acids, as well as procyanidin dimer B, were extracted with Viscozyme®.

Similar to chemical hydrolysis, enzymatic treatments proposed for NEPP release have used different enzymes and conditions (Table 7.2), which makes it difficult to obtain clear conclusions on the most suitable strategy for this approach.

7.2.1.3 Emerging Extraction Techniques for NEPP

Conventional extraction techniques are being replaced by techniques which offer different advantages. Thus, emerging extraction techniques are faster, automated, and with improved reproducibility and selectivity as compared to conventional extraction techniques. In addition, these techniques use more environmentally sustainable solvents (Herrero et al. 2013). The use of some of these technologies, such as PHWE, UAE, microwave-assisted simultaneous distillation and dual extraction (MDDE), or supercritical fluid extraction (SFE) have been reported in order to recover NEPP from plants (Domínguez-Rodríguez et al. 2017). Some of these technologies are able to break covalent or hydrogen bonds or to alter hydrophobic forces by physical mechanisms, although the effect of temperature on NEPP release is not so clear (Wang et al. 2020).

For instance, UAE was used to release NEPP through hydrolysis from cauliflower, red cabbage, and Brussels sprout waste streams (Gonzales et al. 2015). The recovery was influenced by several factors such as the type of solvent, the ratio of solvents, the pH, or the extraction time. The same technique, UAE, was compared with pressurized liquid extraction (PLE) for NEPP release from barley flour. It was concluded that PLE had higher extraction yields but it was less selective than UAE because of the diffusivity coefficient and the extraction power of liquid increase in PLE (Bonoli et al. 2004).

The use of MDDE conditions for extraction of proanthocyanidins from *Cinnamomi* Cortex was optimized, with the use of different solvents (ethyl ether, dichloromethane, and petroleum ether). The MDDE method developed presented higher extraction yields of proanthocyanidins in shorter times than conventional extraction techniques such as hydrodistillation. Since, in this study, the authors performed the treatment directly in the whole sample, i.e. without a previous removal of EPP, it may be concluded that the obtained results corresponded to the sum of EPP and NEPP, although it is not possible to know the contribution of each fraction (Chen et al. 2016).

Finally, PLE was directly applied to an edible seaweed, *Gracilaria chilensis*, and polyphenol content in the derived extracts was compared with that obtained from aqueous-organic extraction in the seaweed. It was found that, although PLE provided a higher polyphenol content, i.e. it was able to release a fraction of NEPP from the food matrix, it did not provide all NEPP previously measured in the residues of aqueous-organic extraction (Sanz-Pintos et al. 2017). This is relevant since it indicates that although emerging techniques may be directly used in food matrixes in order to increase total polyphenol content (since they will

Table 7.2 Examples of enzymatic hydrolysis for NEPP release in vegetal materials.

Sample	Previous EPP removal	Enzyme	Treatment conditions	References
Apple, nectarine, and peach	Methanol/water (50 : 50, v/v) acidified with HCl 2 N (pH 2) (1 h, room temperature), followed by acetone/water (70 : 30, v/v) (1 h, room temperature)	Independent treatments: • Pectinase • Esterase • Mixture	 37 °C, 24 h 55 °C, 5 h 55 °C, 5 h	Pérez-Jiménez et al. (2009a)
Barley	Water (100 °C, 1 h)	α-Amylase, followed by cellulase	30 °C, 1 h, followed by 30 °C, 10 h	Yu et al. (2001)
Quinoa	10% Methanol (room temperature, 2 h), followed by ultra-sonication, 15 min	Independent treatments: • Xylanase • Pectinase • Feruloyl esterease	 60 °C, 12 h 37 °C, 12 h 60 °C, 12 h	Tang et al. (2016)
Rubus idaeus L. leaves and seeds	70% MeOH + 0.1% HCl (v/v) (room temperature, 2 h)	Cellulase: xylanase: β-glucosidase (1 : 1 : 1)	50°C, 12 h	Wang et al. (2019a)
Sour cherry	Methanol/water (50 : 50, v/v) acidified with HCl 2 N (pH 2) (1 h, room temperature), followed by acetone/water (70 : 30, v/v) (1 h, room temperature)	Independent treatments: • Protease • Pectinase • α-amylase	 60 °C, 1 h Room temperature, 1 h 37 °C, 10 min	Nemer et al. (2018)
Tea (black) infusion residue	Ethanol, followed by water (room temperature)	Independent treatments: • Kemzyme® • Alcalase • Acid cellulase • Pectinex® • Viscozyme®	 45 °C, 90 min 48 °C, 101 min 49 °C, 64 min 57 °C, 100 min 50 °C, 75 min	Mushtaq et al. (2017)

be able to release a fraction of NEPP), additional treatments (with hydrolysis as described earlier or more drastic emerging techniques) will have to be performed in the remaining residues, in order to achieve a comprehensive release of all polyphenols present in the sample. Indeed, some studies have successfully employed this combination of emerging techniques with hydrolysis; for instance, SFE extraction was combined with different enzymes in order to increase NEPP release from the residue of black tea infusion (Mushtaq et al. 2017). In the same way, alkaline hydrolysis combined with sonication was more effective than without sonication (Gonzales et al. 2014); this approach significantly increased

NEPP release from the matrix, especially at longer incubation times. Similar combined approaches should still be tested for other vegetal materials and other emerging extraction techniques.

Overall, these techniques involve the generation of a new liquid medium, where some NEPP have been transferred, but still a solid residue remains. This means that in the context of circular economy and whole use of by-products, the problem that these materials originate is not completely solved by the use of these techniques. In this sense, instant controlled pressure drop (DIC) may emerge as a technique able to improve this situation, since the application of high pressure for a short period of time leads to the recovery of the whole matrix, but with modifications in the internal structures of tissues. Thus, it was shown to generate a transformation of NEPA into extractable proanthocyanidins in grape pomace (Martínez-Meza et al. 2021). A later study with different grape pomace materials showed that DIC effects depend upon experimental conditions and the composition of the starting material (Martínez-Meza et al. 2021, 2022). Further studies may help provide more information about the potential of this technique in the field of polyphenol extraction.

7.2.2 Analysis of the Profile of NEPP

Once NEPP solutions have been obtained by some of the procedures described earlier, the next step is the chemical analysis of these solutions, in order to determine NEPP content, profile or both. This may be performed by spectrophotometry or chromatography techniques.

7.2.2.1 Spectrophotometric Methods for Analysis of NEPP

Spectrophotometric assays have several benefits such as being easy to perform and the low cost of the experiments; however, these methods only provide an estimation of the total phenolic compounds and they do not include quantitative measurement of individual compounds (Ignat et al. 2011). Moreover, in many cases, these methods are not selective and other compounds can react with the reagent used, giving false positives (Domínguez-Rodríguez et al. 2017). The most common spectrophotometric methods, with specific considerations regarding their application for NEPP analysis (since they can also be used for EPP analysis), will be now discussed.

Folin-Ciocalteu Assay In this assay, phenolic compounds reduce the phosphomolybdic-phosphotungstic reagent to produce a blue-colored complex (Folin and Denis 1912; Singleton et al. 1999). The main disadvantage of this assay is the lack of selectivity, as the reagent reacts with all reducing substances, such as sugars, aromatic amines, sulfur dioxide, ascorbic acid, or organic acids (Prior et al. 2005). This may lead to an overestimation of phenolic compounds content (Georgé et al. 2005). For instance, this overestimation was specifically observed for NEPP content in oat samples (Verardo et al. 2011).

Vanillin Assay The vanillin assay consists of the condensation of the nucleophilic site of the phloroglucinol structure (A ring) with the aldehyde reagent in strong acid conditions, producing a colored compound detectable at 500 nm. This assay may be applied to determine NEPA when they have been released from the food matrix either by alkaline or enzymatic hydrolysis. For instance, it was used in chickpea and soybean hulls, finding that NEPA content was similar to that of extractable proanthocyanidins, although they exhibited less antioxidant capacity (Niño-Medina et al. 2017). Several limitations have been described for this method: anthocyanins previously present in the sample contribute to the final value, making necessary a previous subtraction of their content; chlorophyll may interfere in the determination; there are differences in the molar absorption coefficients, for instance, between (+)-catechin and (−)-epicatechin (Hümmer and Schreier 2008).

Butanol-HCl Assay (Porter Method) Once the butanol-HCl procedure previously reported has been applied in order to release NEPA monomers as anthocyanins, their absorbance at 555 nm is measured. This technique has been applied in a wide diversity of foods (Pérez-Jiménez et al. 2009a; Pérez-Jiménez and Saura-Calixto 2018). Nevertheless, the presence of anthocyanins in samples prior to degradation must be accounted for and subtracted from the final results (Neilson et al. 2016). Also, side reactions resulting in other products are known to occur (Scalbert et al. 1989) and other compounds may interfere in the quantitation (Powell et al. 1995). But the main limitation of the method is the fact that the higher the proanthocyanidins are, the more anthocyanins produce, due to the presence of more extension units (Hümmer and Schreier 2008). Some modifications to the original method have been suggested, such as the addition of acetone to the butanol-HCl medium, adding this mixture directly to the sample; in this way, extractable proanthocyanidins and NEPA are measured together (Grabber et al. 2013; Grabber and Zeller 2020).

Dimethylaminocinnamaldehyde (DMACA) Assay Although the reaction mechanism is not clearly defined, some authors mention that the reagent acts in the C8 position of the A ring and reacts only with the terminal units of proanthocyanidins (Oszmianski and Bourzeix 1996; Prior et al. 2010). This procedure was used to determine NEPA in the testa of red wheat seeds, finding that their content increases with maturation since extractable proanthocyanidins are transformed into NEPA (Kohyama et al. 2017). This method has been much less used for NEPA analysis than butanol-HCl, despite it having been recommended as the preferred one because it does not exhibit the interference of other anthocyanins present in the sample (Gullickson et al. 2020). Nevertheless, the low stability of the generated compound has also been pointed out as a drawback of the method (Hüemmer and Schreier 2008).

7.2.2.2 Liquid Chromatography-Mass Spectrometry Analysis of NEPP

Once NEPP have been released from the food matrix by some of the procedures described above, the monomeric units of NEPA (or other polymeric structures present in NEPP, such as non-extractable ellagitannins or non-extractable phlorotannins) or the individual constituents of HPP may be characterized by HPLC analysis, with or without combination with MS analysis. Indeed, this identification of individual constituents is widely recommended, since not all compounds belonging to NEPP fraction may exert the same biological activities; for instance, Irakli et al. (2018) reported that, although NEPP content was lower in rice bran than in rice husk, the former showed the highest antioxidant capacity. The most common system is the reverse phase (RP)-HPLC analysis with a C18 column, usually with a linear gradient elution with a binary solvent system. The most common mobile phase is acidified water (solvent A) combined with an increasing percentage of acidified methanol or acetonitrile (solvent B) or other organic solvents (Ignat et al. 2011; Domínguez-Rodríguez et al. 2017). In the case of thiolysis or phloroglucinolysis, chromatography results allow estimating the mean degree of polymerization and galloylation, although specific corrections must be applied (Neilson et al. 2016). Additionally, direct phase (DP)-HPLC analysis is performed when enzymatic or chemical treatments have not been so drastic as to completely depolymerize all NEPA structures, so the remaining oligomers may be measured depending on their degree of polymerization (Pérez-Jiménez and Torres 2011). A summary of the chromatography conditions used in NEPP analysis in different samples is provided in Table 7.3; in this case, there are more similarities between studies than for chemical or enzymatic hydrolysis conditions. An aspect to be mentioned is hydrophilic interaction liquid chromatography (HILIC), which has shown to be a good strategy for analyzing polyphenols due to its combined mechanism of separation, which is between DP-HPLC and RP-HPLC (Jandera and Hájek 2009), has not been explored for the analysis of NEPP. However, this approach may be particularly useful for assessing oligomeric structures as those found in NEPP; for instance, in a study on EPP in grapevine canes where the authors employed a two-dimensional liquid chromatography-based method (LC × LC) based on the combination of a diol (HILIC) and a C18 (RP) column allowed the identification of up to procyanidin tetramers and resveratrol heptamers (Montero et al. 2018).

When HPLC is combined with an ultraviolet-diode array detector (UV-DAD), NEPP from different subclasses can be identified using UV-DAD. For instance, hydroxybenzoic acid presents an absorbance maximum at around 280 nm due to the A ring benzoyl system, while the maximum peak of hydroxycinnamic acids is around 320 nm, involving the B ring cinnamoyl system (Sáyago-Ayerdi et al. 2016). Due to the lack of commercial standards for some phenolic compounds, in some cases it is only possible to carry out semi-quantitative analyses, expressing contents of the NEPP as equivalents to a component from the same class, normally the monomeric unit (Wu et al. 2018).

Due to these limitations when only HPLC analysis is performed, chromatography analysis coupled to a mass detector (MS) allows performing a more accurate NEPP identification, as it provides structural information on individual compounds in their native forms even in the absence of commercial standards. Thus, HPLC-DAD has been coupled to electrospray ionization (ESI)-MS (operating in ion negative mode, but for anthocyanins, where positive mode is used) for the characterization of NEPP from plant materials and cereals. The evolution in the analysis of NEPP by HPLC-MS has agreed with the advances in the existing MS analyzers. Studies with single quadrupole systems or ion trap systems were based on *m/z* absolute values (without decimals), allowing the identification in different food matrixes of individual compounds present in NEPP fractions such as (epi)catechin gallate benzylthioether, *p*-hydroxybenzoic acid, and caffeoylquinic acid (Hellström and Mattila 2008; Arranz et al. 2009; Arranz and Saura-Calixto 2010). Later, time of flight (TOF) systems have also been used for characterizing NEPP, for instance, allowing the identification, based on exact mass with four decimals, of several phenolic acids and flavonols in

Table 7.3 Examples of chromatography conditions for NEPP hydrolyzates in a food sample.

Sample	HPLC type	Column	Mobile phase	Gradient conditions	References
Almond	Direct phase	260 × 4.60 mm id, 5 μm, Hypersil silica column (ThermoFisher, Bellefonte, PA, USA)	(A) Dichloromethane (B) Methanol (C) Acetic acid/water (50 : 50, v/v)	(min, %B, %C): (0,14,4) (30, 28.4, 4) (45, 39.6, 4) (55, 86, 4) (60, 86, 4) (75, 14, 4)	Xie et al. (2012)
Apples	Reverse phase	250 × 4.6 mm, 5 μm, C18 column (Phenomenex, Bologna, Italy)	(A) Acetic acid solution (2%) (B) Acetonitrile:water:acetic acid (50 : 49.5 : 0.5)	(min, %B): (0–10,10–20) (10–15, 20–40) (15–20,40–80) (25–30,40–80) (20–25, 80–40) (25–30, 40–10)	Veronica et al. (2019)
Apple, grape, lingonberry	Direct phase	150 × 4.0 mm id, 3 μm, Inertsil ODS-3 (GL Sciences Inc., Torrance, CA, USA)	(A) 50 mM aqueous phosphoric acid, pH 2.5 (B) Acetonitrile	(min, %B): (0, 5) (5, 5) (30, 27.5) (32, 50) (38, 50)	Hellström and Mattila (2008)
Barley	Reverse phase	100 × 4.6 mm, 2.6 μm C18 Kinetex™ PFP (Phenomenex, St. Torance, USA)	(A) Acidified water (1% acetic acid) (B) Acetonitrile	(min, %B): (0, 5) (12.5, 30) (17.5, 60) (22, 5)	Gómez-Caravaca et al. (2014)
Canola seed cake	Reverse phase	250 × 4.6 mm, 4 μm, C18 Synergi Column (Phenomenex, Torrance, CA)	(A) Water/formic acid (99.5 : 0.5, v/v) (B) Acetonitrile/formic acid (99.5 : 0.5, v/v)	(min, %B): (50, 99 : 1–50 : 50) (5, 50 : 50–1 : 99)	Zardo et al. (2019)
Industrially processed fibres derived from juice industry	Reverse phase	150 × 2.1 mm, 5 μm dC18 Water Atlantis	(A) Acidified water with 0.1% formic acid (B) Methanol 100% with 0.1% acetic acid	(min, %B): (1.89–17.84, 30) (0.5–15, 30) (15–25, 60) (25–35–80) (35–40, 10)	Delpino-Rius et al. (2015)
Lentil hulls	Reverse phase	250 × 4.6 mm id, 5 μm, Superlcosil LC-18 (Merck, Darmstadt, Germany)	(A) Water, 5% formic acid (B) Methanol/acetonitrile/formic acid (90 : 5 : 5, v/v/v)	(min, %B): (0, 20) (10, 30) (15, 40) (18, 45) (20, 50) (30, 70) (45, 80)	Yeo and Shahidi (2020)

(*Continued*)

Table 7.3 (Continued)

Sample	HPLC type	Column	Mobile phase	Gradient conditions	References
Mixture of grape pomace and pomegranate pomace	Reverse phase	250 × 4.6 mm id, 5 μm Gemini C18 (Phenomenex, Torrance, CA, USA)	(A) Water, 0.1% formic acid (B) Acetonitrile, 0.1% formic acid	(min, %B): (0, 5) (20, 30) (30, 55) (38, 90) (40, 90)	Pérez-Ramírez et al. (2018)
Mushrooms	Reverse phase	250 × 4.6 mm, 5 μm, C18 Intertsil ODS-3 Column	(A) 0.5% acetic acid in water (B) 0.5 acetic acid in methanol	(min, %B): (0–0.01, 0–20) (0.01–2, 20–60) (2–15, 60–80) (30.35, 100–10) (35–40, 10–0)	Çayan et al. (2020)
Mustard grains	Reverse phase	50 × 2.1 mm id, 1.7 μm Acquity BEH C18 column (Waters MA Technologies, Manchester, UK)	(A) Water, 1% formic acid (B) Acetonitrile	(min, %B): (0, 3) (15, 40)	Rasera et al. (2020)
Pomegranate	Reverse phase	250 × 4.0 mm id, 5 μm (Agilent Technologies, Waldbronn, Germany)	(A) Water, 1% formic acid (B) Acetonitrile	(min, %B): (0,5) (20, 30) (30, 55) (38, 90) (40, 90)	García-Villalba et al. (2015)
Sorghum hull	Reverse phase	100 × 2.10 mm, 1.90 μm, Hypeisil GOLD (Agilent Technologies, Germany)	(A) Water/formic acid (99.90 : 0.10, v/v) (B) Acetonitrile	(min, %B): (0–0.5, 4) (0.5–4, 4–7) (4–5, 7–20) (5–10, 20–25) (10–11, 25–50) (11–15, 50–100) (15–20, 100)	Luo et al. (2020)
Whole millet grains	Reverse phase	250 × 4.6 mm, 5 μm, C18 SUPERCOSIL™ column (Darmstad, Germany)	(A) Formic acid (1%) (B) Methanol/acetonitrile/ formic acid (94 : 5 : 1; v/v/v)	(min, %B): (0, 20) (10, 30) (15, 40) (18, 45) (20, 50) (30, 70) (40, 85)	Chandrasekara and Shahidi (2011)

banana by-products (Ramírez-Bolaños et al. 2021). When the TOF analyzer is combined with a quadruple (in a Q-TOF system) it is possible, not only to obtain the exact mass but also the MS/MS results, improving the identification as it is based also on the characteristic fragmentation pattern of each compound; this approach was used, for instance, for identifying phenolic acids, ellagitannins and a flavanone in the NEPP fraction of a grape pomace/pomegranate pomace mixture (Pérez-Ramírez et al. 2018). And, recently, a simple and rapid direct analysis in real-time coupled with high-resolution mass spectrometry (DART-HRMS) was suggested for assessing NEPP content in tropical fruit peels (Domínguez-Rodríguez et al. 2021). The use of triple quadrupole systems, based on the multiple reaction monitoring mode, is the best approach for quantifying compounds, as done from NEPP in seven species of the genus *Camellia* (Luo and Fei 2019).

All HPLC-MS systems imply that a NEPP solution has been previously obtained, thus breaking the original linkages in the food matrix. Furthermore, some MS instruments are limited by their upper mass range (typically around 4,000 Da), which also avoids the characterization of original NEPP complexes (Monagas et al. 2010; Campillo et al. 2015). In this sense, ionization by Matrix Assisted Laser Desorption/Ionization (MALDI), coupled to a mass detector such as TOF could be an alternative since it may be applied in suspensions and it has the possibility to be coupled to analyzers with unlimited mass range (Monagas et al. 2010; Neilson et al. 2016). This technique allows detection of high mass with accuracy and high resolution, being more suitable than ESI analysis, providing a better detection in positive mode for ionization (Domínguez-Rodríguez et al. 2017). However, the application of this technique to NEPP analysis is still pending.

7.2.3 Determination of the Content of Non-Extractable Polyphenols. Which Standard?

An important problem that arises in the evaluation of NEPP content in vegetal material is the subject of the most appropriate standard to be used. In the case of HPP, in spectrophotometry determination, the most common standard is gallic acid; this is useful for comparison purposes since it is the most used standard in Folin-Ciocalteau assay, also for EPP analysis; although, in quantitative terms, ferulic acid may be more abundant in HPP, especially regarding cereals. In the case of HPP analysis by HPLC, when no commercial standard for a particular phenolic compound is available, the most common approach is to use as standard another phenolic compound structurally related (for instance, using ferulic acid for the quantification of all hydroxycinnamic acids) and expressing the results as "equivalents of." Although this approach may imply an under- or overestimation of content, since not every compound exhibits the same dose-response relationship (García-Villalba et al. 2016), it is a common procedure, also for EPP analysis.

But the main problem for NEPP quantification, still not completely solved, is related to the best standard to be used for NEPA quantification. If the treatment used to release these structures has originated from oligomeric structures, there is a lack of commercial standards, which are available only for proanthocyanidin dimers and trimers. In case an exhaustive treatment has been performed, releasing monomeric anthocyanins (in butanol/HCl treatment) or monomeric flavanols, free or conjugated with the nucleophile agent (in the acid-catalyzed depolymerization by nucleophiles), each available standard has *pros* and *cons*. In the first case, the formation of colored anthocyanins depends on the specific structure of proanthocyanidins in the sample (Schofield et al. 2001) and not all of them provide the same response; thus, although, for instance, a NEPA carob concentrate has been suggested as standard (Pérez-Jiménez et al. 2009a), it is difficult to decide which one will be appropriate as a general standard for any vegetal material. In the case of thiolysis or phloroglucinolysis, a non-commercial standard exists for the conjugates originating between terminal units and the nucleophile agent, so they must be in-house synthesized and isolated, which makes difficult the extension of this methodology as a routine approach in a wide number of laboratories.

In an effort to solve the standard problem for NEPA quantification, the preparation of a cranberry proanthocyanidin standard, able to quantify either extractable proanthocyanidins, or NEPA, was recently reported (Gullickson et al. 2020). Nevertheless, as the authors pointed out, even with this standard, it may happen that NEPA content is underestimated and still some of them are not released for their strong association with the food matrix.

7.2.4 Analysis of Dietary Fiber: Connection with Non-Extractable Polyphenols

In the context of the methodologies for evaluating NEPP content in vegetal material, it is worthy to discuss the analysis of DF in these matrixes. This is due to the fact that, although traditionally, DF has been considered a food component independent of polyphenols, scientific evidence proves that NEPP are relevant components of DF (Saura-Calixto 2012). In this way, according to the definition of DF from the *Codex Alimentarius*, the compounds associated with polysaccharides of the cell wall can be included in the concept of DF. Indeed, it was recently suggested that the term "feruloylated arabinoxylans," i.e. ferulic acid structures linked to arabinoxylans, for instance, in maize, should be directly considered synonymous with soluble DF (Herrera-Balandrano et al. 2020). And nuclear magnetic resonance (NMR) spectroscopy analysis has provided novel structural confirmation of the association between some polyphenols and DF as a single complex matrix (Bermúdez-Oria et al. 2020).

Some specific and relevant characteristics of DF are due to the presence of significant quantities of polyphenols associated with the carbohydrate fraction (Saura-Calixto 2011; Macagnan *et al.* 2016). Therefore, the interaction between these two constituents (DF and polyphenols) should be considered both on the determination of DF (since it will include NEPP) and on that of polyphenols because if NEPP are not released from DF and analyzed, an important fraction of polyphenols will be ignored (Saura-Calixto 2011).

The methodology for DF determination in foods called "indigestible fraction" is based on this wider concept of DF, including NEPP as constituents of this complex. The main characteristic of this method, besides adaptations in the enzymatic treatments in order to get more physiological conditions than in other methodologies (Saura-Calixto et al. 2000), is that it includes the evaluation of polyphenols linked to DF. Thus, once soluble or insoluble DF have been isolated, hydrolysis procedures similar to the ones described above are performed, in order to release NEPP associated with DF, which may later be determined either by spectrophotometry or chromatography techniques. This approach has been applied to individual foods (Goñi et al. 2009) and also to whole diets (Tabernero et al. 2007). Interestingly, the indigestible fraction methodology allowed to show that NEPP are also present in common beverages (Díaz-Rubio and Saura-Calixto 2011). Moreover, in NEPP solutions obtained by this methodology, besides the content in these compounds, antioxidant capacity may be determined by different methods (Saura-Calixto 2011; Pérez-Jiménez et al. 2013), in order to obtain a first estimation of the potential biological activities associated with these compounds. Alternatively, to the indigestible fraction method, the official AOAC procedure for DF determination (Latimer 2012) was performed in cereal samples and the obtained soluble and insoluble DF fractions were subjected to alkaline hydrolysis. Phenolic compounds were then measured in the hydrolysates, detecting a diversity of phenolic acid in both fractions of DF (Guo and Beta 2013).

7.3 Why Should Non-Extractable Polyphenols be Systematically Included in Polyphenol Analysis?

7.3.1 Intake of NEPP in Different Populations

As will be discussed later, there is already an accumulation of data on NEPP content in several food products, despite methodologies not being standardized yet. Nevertheless, these data have allowed estimating NEPP intake in different populations, with much less data available than for EPP intake. For instance, a study in the Spanish diet as an example of the Mediterranean diet showed that NEPP contributed to total polyphenol intake in a similar way as EPP (949 vs. 898 mg/p/d), being, however, commonly ignored (Arranz et al. 2010). Fruits and cereals resulted to be the main contributors to total EPP intake, providing 75% of total intake.

Other studies have evaluated NEPP intake from specific food groups. Thus, Pérez-Jiménez and Saura-Calixto (2015) determined NEPP intake derived from the consumption of fruits and vegetables in some European countries: Spain, France, Germany, and the Netherlands. In these four countries, different food groups provided at least 350 mg of daily NEPP intake, with a maximum of nearly 500 mg in Spain. Regarding fruits, the NEPP intake order was Spain > Netherlands > Germany > France, while for vegetables, it was Netherlands > France > Spain > Germany. Another study focused on total polyphenol intake in a Mexican rural population, finding that NEPP intake was three-fold higher than EPP intake (Hervert-Hernández et al. 2011).

Recently, total NEPP intake was estimated in a Spanish population of old people (Goñi and Hernández-Galiot 2019). In particular, it was found that in a functionally independent population aged over 80, NEPP contributed to total antioxidant intake in 71% (46% corresponding to HPP and 25% to NEPA). Regarding the different food groups, NEPA intake came mostly from fruits and legumes, while HPP intake was mainly derived from cereals and fruit consumption.

Also, in order to have a clear estimation of NEPP presence in common diets, it is important to determine their content not only in individual foods but also in combined dishes. In this sense, EPP and NEPP contents were determined in some common Italian dishes (Durazzo et al. 2019).

Finally, a recent study (Gutiérrez-Díaz et al. 2021) evaluated NEPP intake in a sample of 147 Spanish adults with non-declared pathologies, assessing at the same time major phylogenetic types of the intestinal microbiota. After performing a combined analysis by stepwise regression model including DF, EPP, and the different NEPP classes, it was concluded that HPP intake was the best predictor of *Bacteroides-Prevotella-Porphyromonas* group, as well as *Bifidobacterium* levels in feces. Also, HPP were positively associated with butyric acid. To the best of our knowledge, this is the first study on NEPP intake that did not only perform a descriptive exploration of this parameter but also searched for significant associations with indirect health outcomes.

Much research is still needed in order to assess NEPP intake in more populations and to use these data in order to establish a potential association with health outcomes. In this sense, a review on polyphenol intake in different populations (Pinto and Santos 2017) identified limitations or problems to obtain a real approximation of polyphenol intake, being one of these limitations the ignored fraction of NEPP. Nevertheless, it should be mentioned that, recently, one of the existing databases on bioactive compound content in foods (eBASIS), included data for NEPP, which may be used to obtain a rough estimation of the intake of these compounds in different populations (Plumb et al. 2020).

7.3.2 Metabolism of NEPP

For a better understanding of the possible effects associated with the intake of NEPP, as for any bioactive compound, it is necessary to know the different processes that occur along the human gut, which may be explained in the following stages (Pérez-Jiménez et al. 2013):

1) Solubilization of the compounds in the intestinal fluids via different mechanisms, resulting in the compounds becoming bioaccessible. This may be estimated by determining the presence of NEPP in intestinal fluids (small and large intestines).
2) Possible transformation by colonic microbiota of the compounds reaching the colon. The metabolites produced can be determined either *in vitro* (supernatant fractions from *in vitro* fermentation) or *in vivo* (caecal contents or faeces).
3) Absorption, either of the original compounds or the derived metabolites, determined in biological fluids such as urine and blood.
4) Potential accumulation of bioactive metabolites in certain tissues contributing to the local effects of compounds of interest.

All these aspects have been studied in depth for EPP over the last decades, but studies on NEPP are scarce. Nevertheless, from some *in vitro* and preclinical studies on the metabolic fate of HPP (Andreasen et al. 2001; Rondini et al. 2004) and NEPA (Saura-Calixto et al. 2010; Mateos-Martín et al. 2012), some information on the biotransformation of these compounds is already available.

Due to their macromolecule structure, most NEPP reach the colon nearly intact, although limited depolymerization of NEPA may take place in the small intestine (Touriño et al. 2011). Once in the colon, NEPP are subjected to different transformations, depending on their category. In the case of HPP, they may become accessible after the molecules with which they are associated (proteins and carbohydrates) are subjected to the action of some bacterial enzymes able to break covalent bonds, such as esterases (Kroon et al. 1997; Ludwig et al. 2018). Then, the small molecular weight polyphenols released from these associations (mostly phenolic acids) can be directly absorbed and undergo some transformations by the colonic microbiota. For example, ferulic acid is transformed into dihydroferulic acid, which is subsequently metabolized to other phenolic acids.

In the case of NEPA, which are mainly constituted by (epi)catechin, a great variety of compounds are formed due to the action of colonic microbiota. Thus, the breakdown of the basic flavanol skeleton first gives rise to the formation of valerolactones and then to several phenolic acids (hydroxyphenylpropionic acid and hydroxyphenylacetic acid, among others) to later give rise to other molecules of low molecular weight such as hippuric acid. Other compounds such as ellagic acid released from ellagitannins in the gastrointestinal tract are transformed by the microbiota to urolithins, while the gallotannins are hydrolyzed to gallic acid, later transformed into catechol and hydroxybenzoic acid (Cardona et al. 2013; Ludwig et al. 2018).

Therefore, while both the small intestine and the short intestine are key organs for the metabolic transformation of EPP, that of NEPP takes place almost completely in the colon. In this sense, the metabolic fate of NEPP is quite similar to that of DF, a constituent with which, as indicated earlier, NEPP integrates a single complex. Indeed, the influence of DF on the metabolic fate of polyphenols (and vice versa) has been recently reviewed (Nagar et al. 2020). And it was recently reported that the joined fermentation of DF and phenolic compounds does not alter the fermentation rate of both constituents, but it leads to a faster formation of certain phenolic-derived metabolites (Phan et al. 2020). Finally, other food constituents, such as iron ions, may also affect the interactions between these DF and polyphenols and, conversely, their metabolic fate (Chirug et al. 2018).

7.3.3 Beneficial Effects Attributed to NEPP

Regarding the health-related properties of NEPP, they may be classified into three categories, connected with the metabolic fate of these compounds: (i) effects of intact compounds throughout the digestive tube; (ii) local effects in the colon environment and mucose due to metabolites released by the action of colonic microbiota; (iii) systemic effects of NEPP-derived metabolites after colonic absorption. As described for the other aspects of NEPP research, data on this topic, although promising, are still much scarcer than for EPP biological activities. Moreover, since NEPP comprise different structures and diverse associations with food matrix, the observed effects for a particular NEPP category or food matrix, may not be extrapolated to all NEPP constituents. For instance, it was recently shown that NEPA linked to the cell wall by non-covalent bonds showed an anti-inflammatory activity nearly equivalent to that of extractable proanthocyanidins, while this activity was reduced in NEPA associated with the cell wall by covalent bonds (Le Bourvellec et al. 2019).

As regards effects throughout the digestive tube, it must be highlighted that polymeric NEPP still expose their hydroxyl groups to the gastric and intestinal media, so they can counteract free radicals during their passage along the digestive tube. This was, for instance, reported after the *in vitro* digestion of fish products enriched with grape pomace, a material particularly rich in NEPA (Solari-Godiño et al. 2017). The other effect that NEPP may have in the digestive tube is related to the inhibition of digestive enzymes, thus delaying carbohydrate absorption. In this way, the inhibitory capacity of α-glucosidase by NEPP has been recently reviewed (Wang et al. 2020). For instance, in guarana powder, NEPP showed an ability to inhibit this enzyme which was 5.8-fold higher than that of EPP (Pinaffi et al. 2020). Moreover, the digestion process itself may alter the distribution of PP in the food matrix, for instance, generating associations between starch and extractable proanthocyanidins, thus becoming NEPA, and contributing to delay starch digestion; this was observed when cooked adzuki beans and rice were subjected to an *in vitro* digestion process (Takahama et al. 2019). Indeed, the same authors suggested, through *in vitro* experiments, that this mixture was able to reduce salivary nitrite to beneficial nitric oxide (Takahama et al. 2020).

Regarding local effects in the colon when NEPP reach this organ, the first one will be (derived either from original structures or from those released by the action of colonic microbiota) an improvement in local oxidative status. For instance, this has been reported in cecum from rats supplemented with a NEPP-rich product (Goñi and Serrano 2005). Besides this direct increase in antioxidant capacity, preclinical studies with NEPP supplementation also found an enhancement of the endogenous antioxidant system, as well as a decrease in lipid oxidation in colonic mucosa (López-Oliva et al. 2013). Another local effect of NEPP is related to a decrease in inflammation; thus, Cheng et al. (2016) evaluated the effects of EPP and NEPP from blueberries (*Vaccinium corymbosum*) in an *in vitro* model of inflammation, concluding that both classes of polyphenols exerted an anti-inflammatory action via the inhibition of NF-κB (κB nuclear factor). Interestingly, in studies in cell cultures, NEPP showed significantly higher inhibition of the production of nitric oxide in macrophages than EPP, which was accompanied by decreased expression of inducible nitric oxide synthase (iNOS) and increased expression of HO-1 (Han et al. 2019).

The last NEPP mechanism for potential local effects is related to the modulation of microbiota. Thus, it may be expected that those bacterial species containing the enzymes able to release HPP from their association with DF are increased after supplementation with NEPP. This would be the case of genera *Lactobacillus* and *Roseburia* (*Firmicutes*), *Bifidobacterium* (*Actinobacteria*), and *Bacteroides* and *Prevotella* (*Bacteroidetes*) (Chassard et al. 2007; Dodd et al. 2011; Flint et al. 2012). For instance, a study in rats supplemented with feruloylated oligosaccharides generated several beneficial modifications: a decrease in *Firmicutes/Bacteroidetes* ratio, which has been related to a decrease in obesity risk; an increase in *Actinobacteria*, *Bacteroides*, and *Lactobacillus*, associated with resistance against diabetes; and a decrease in *Clostridium* and *Firmicutes,* known to be augmented in a diabetic situation (Ou et al. 2016).

The combination of these local mechanisms of action leads to several biological effects after NEPP supplementation, as observed in preclinical studies. Thus, rats supplemented with NEPP showed other local effects for these compounds, such as upregulation of genes involved in the metabolism of xenobiotics compounds, modulation of plasma glucose, and tumoral suppression, together with down-regulation of genes involved in inflammation, lipid biosynthesis, and proto-oncogenes (Lizarraga et al. 2011). Also, NEPP from foxtail millet bran exhibited anti-proliferative activity in HT-29 cells (Shi et al. 2019). Recently, it was reported that NEPP were able to completely protect from colitis induced by 2,4,6-trinitrobenzene sulfonic (TNBS) acid, via a decrease in the upregulation of GRP94 mRNA triggered by colitis (Maurer et al. 2019), a result that agreed with previous *in vitro* (Shi et al. 2017) and *in vivo* (Fang et al. 2012; Xia et al. 2019) studies on the anti-inflammatory properties of NEPP from cereals.

But the most relevant aspect of local effects of NEPP is associated with colorectal cancer prevention. In this way, studies with cell cultures reported NEPP stronger inhibitory effects on the viability and colony formation capacity of human colon cancer HCT116 cells than those associated with EPP (Han et al. 2019). Moreover, a significant decrease (higher than

50%) in polyps in a model of colorectal cancer was reported after NEPP supplementation (Sánchez-Tena et al. 2013), which connected with the additionally reported reduction of cleaved caspase-3 and minor release of cytochrome C from mitochondria to the cytosol (López-Oliva et al. 2013). And a study in rats with TNBS-induced colitis observed that supplementation with the matrix DF-NEPP obtained from grape peel was able to restore the levels of glutathione and related enzymes, which were disturbed by disease, as well as to prevent the increase of caspase-3 mRNA levels observed in non-supplemented animals (Maurer et al. 2020).

Once absorbed, NEPP-derived metabolites may have different systemic effects. In this sense, it is important to state that, although here specific studies with NEPP will be mentioned, as previously indicated NEPP are not chemically different to EPP; therefore, both polyphenol classes originate from the same derived metabolites, contributing to the pool of circulating bioactive metabolites after polyphenol intake. Therefore, the reported biological effects for EPP-derived metabolites, e.g. inhibition of monocyte adhesion in endothelial cells (Giménez-Bastida et al. 2012; Lee et al. 2017), a decrease of oxidative stress in hepatic cells (Wang et al. 2019b), vasorelaxant effects (Pourová et al. 2018) or stimulation of glucose transport (Houghton et al. 2019), may also be expected after NEPP intake. Indeed, metabolites have been reported as the key responsible for the biological actions attributed to polyphenols (Luca et al. 2020). Specific studies with NEPP have reported that their intake leads to an increase in plasma antioxidant capacity due to the absorption of their derived metabolites (Pérez-Jiménez et al. 2009b; Matsumura et al. 2016; González-Sarrías et al. 2017). Also, *in vitro* studies reported the bile acid-binding capacity associated with NEPP present in several fruits (Hamauzu and Mizuno 2011; Hamauzu and Suwannachot 2019) or anti-obesity effects in adipocytes (Lin et al. 2019). These mechanisms of action, among others, may be underlying the observed improvement in cardiometabolic markers, such as lipid profile, glucose metabolism or inflammation after the consumption of NEPP-rich products, observed either in preclinical (Huang et al. 2018; Amaya-Cruz et al. 2019a; Lin et al. 2019) and in clinical (Pérez-Jiménez et al. 2008; Urquiaga et al. 2015; Vitaglione et al. 2015; Martínez-Maqueda et al. 2018) studies.

A final aspect that should be considered when evaluating the health effects of NEPP is that, due to their intrinsic association with DF, there is a mutual enhancement in the biological activities of these compounds. Thus, it has been suggested that an additional physiological effect of DF would be to behave as a carrier of NEPP throughout the digestive tube (Saura-Calixto 2011). Some authors have highlighted that certain physiological effects attributed to insoluble DF may be partially due to associated phenolic compounds (Macho-González et al. 2017; Macho-González et al. 2018). Regarding the transformation of both food constituents in the colon, it has been reported that proanthocyanidins bound to the cell wall either by covalent or non-covalent linkages led to a higher production by colonic microbiota of both short chain fatty acids (derived from DF fermentation) and phenolic acids (derived from proanthocyanidins transformation), than the fermentation of proanthocyanidins not associated with cell wall (Le Bourvellec et al. 2019). Conversely, several *in vivo* studies have found synergistic effects between polyphenols and DF regarding cardiometabolic markers (Pérez-Jiménez et al. 2008; Vitaglione et al. 2015), although other authors have provided divergent results (Nagar et al. 2020), highlighting the need of a depth exploration of these aspects.

7.4 Relevance of the Determination of Non-Extractable Polyphenols in Quality Control

The increasing evidence on the quantitative and qualitative relevance of NEPP makes it more necessary to systematically include their determination in food samples by combining the methodologies described earlier. Moreover, this evaluation may be relevant in the context of quality control either in an academy or in the food industry. In particular, the routine determination of NEPP in vegetal materials may be useful for a comprehensive characterization of these samples, the identification of new botanical sources with potential applications, comparisons between varieties and evaluation of several processing effects on NEPP content and profile. These aspects will be discussed in the present section.

7.4.1 Comprehensive Characterization of Vegetal Materials

Some studies have included the evaluation of the whole polyphenol profile of a food sample, i.e. not only EPP, as commonly performed, but also NEPP fraction. This has provided interesting information.

For instance, Rodríguez-González et al. (2018) performed a study on peach (*Prunus persica* L.) juices by-products where they identified several individual NEPP, such as ellagic acid glucoside, gallic acid 4-*O*-glucoside, gallic acid ethyl ester, and 5-*O*-galloylquinic acid, apparently associated with insoluble DF. Moreover, NEPP content was found to be higher in those peach juice by-products exhibiting the highest DF content. It is relevant that when these by-products were provided to obese rats, fed with high fat and fructose diet, several health effects related to the management of glucose metabolism were observed.

In another paper, Pérez-Ramírez et al. (2018) characterized a mixture of grape and pomegranate pomace as a potential ingredient for the development of functional foods or dietary supplements. The profile of NEPP in these by-products had not been evaluated. So, in this study, the authors characterized this mixture using HPLC-ESI-QTOF and MALDI-TOF techniques. In the case of NEPP fraction, 61 phenolic compounds were identified, belonging to the sub-classes of HPP and non-extractable ellagitannins. This shows that characterizations of vegetal materials based exclusively on the EPP profile does not cover the whole profile of phenolic compounds present in the sample. At the same, in connection with the methodological aspects previously exposed, there is a need to advance further in the methodology for the unequivocal identification of these compounds especially considering the potential degradation of original structures during drastic hydrolysis, since, among the 61 phenolic compounds in NEPP fraction (based on absorbance wavelength or characteristic MS/MS fragments), identity could be assigned only to 11 of them.

But NEPP characterization should be included not only in the analysis of agri-food by-products but also in that of common food items. For instance, Alu'datt et al. (2017) characterized NEPP in citrus (*Rutaceae* family) fruits. LC/MS-MS analysis allowed the identification of caffeic acid as the main NEPP constituent in pummelo and shamouti, while hesperidin was the major one in lemon and blood orange. The authors performed in all samples both acid and alkaline hydrolyses and, as explained earlier, the release for the different compounds varied depending on the selected procedure. This was especially remarkable for clementine, where ferulic acid was the main NEPP constituent after alkaline hydrolysis, while after acid hydrolysis, it was gallic acid.

In a fruit with specific characteristics due to its high fat content, avocado, NEPP profile was evaluated by HPLC after alkaline and acid hydrolyses, showing the presence of some phenolic acids like protocatechuic acid, *p*-hydroxybenzoic acid, vanillic acid, caffeic acid, ferulic acid, and sinapic acid (Poovarodom et al. 2010).

In the same way, it was recently reported that NEPP content in blue highland barely varieties planted in the Qinghai-Tibet Plateau area was at least similar to that of EPP (Yang et al. 2018). HPLC-DAD analysis allowed the identification of individual constituents of NEPP fraction, showing the presence of several phenolic acids (gallic acid, benzoic acid, syringic acid, 4-coumaric acid, and dimethoxybenzoic acid) and flavonoids (naringenin, hesperidin, quercetin, and rutin). Additionally, the antioxidant capacity associated with NEPP fraction was evaluated.

These selected examples, from different samples, show that, as may be expected, NEPP are as ubiquitous as EPP. Thus, the characterization of any vegetal material with different purposes should definitely include the evaluation of the NEPP fraction, preferably with a detailed characterization of individual constituents. Moreover, in some specific cases, the evaluation of NEPP may be completely needed due to their high contribution to total polyphenol content. Such is the case of the analysis of protective tissues (pericarp, seed coat, hull, etc.), where NEPP content is higher than in nutritional tissues (germ, endosperm, etc.). This is due, on the one hand, to the defense roles that they have in plants and, on the other hand, to the higher content of macronutrients in nutritional tissues (Wang et al. 2020). At the same time, a recent study (Gullickson et al. 2020) showed the relevance of a systematic determination of NEPP when comparing, for instance, different cranberry commercial ingredients. Thus, the authors found that the proportion between extractable proanthocyanidins and NEPA differed significantly between samples, due to a combination of factors (variety, harvest date, location, storage conditions, and processing conditions); and concluded that in a context of increasing interest for the authentication, standardization, and efficacy evaluation of products, the determination of NEPP by validated methodologies in such products would help achieve these objectives.

7.4.2 Identification of New Botanical Sources with Potential Applications

The systematic determination of NEPP in the field of botany, pharmacology, and food science may also help identify understudied or neglected materials with a potential for their use in foods, dietary supplements, or drugs. Thus, recently, a tree commonly found in Latin America (Mexico, Cuba, Guatemala, Honduras, and Brazil) popularly known as "Mutamba" (*Guazuma ulmifolia* L.) was evaluated to know its proximate and phytochemical composition. This product showed a remarkable total EPP content (130.5 mg/100 g fw), while NEPP fraction was much lower (6.6–89.6 μg/g fw). However, this fraction showed remarkable antioxidant capacity. The main individual constituents of the NEPP fraction, evaluated by HPLC-MS analysis with a triple quadrupole apparatus, were (+)-catechin, protocatechuic acid, (−)-epicatechin, and gallic acid. The authors suggested that NEPP can be consumed by means of products using mutamba flour, such as whole bread, and EPP can be consumed as tea (Pereira et al. 2020).

Another promising product for its NEPP content is Roselle (*Hibiscus sabdariffa* L.): the calyces of this plant are used to prepare a soft drink (consumed as a hot or cold beverage, common in Mexico and other countries) (Sáyago-Ayerdi et al. 2007; Mercado-Mercado et al. 2015). This beverage is prepared by boiling dry calyces, leading to the generation of a decoction residue (which is not used), where Amaya-Cruz et al. (2019a) identified by spectrophotometry high contents of NEPP, including both HPP and NEPA. Indeed, when this by-product was provided to rats fed with a high caloric diet, significant effects were observed regarding obesity control and associated complications.

Tropical fruits have also emerged as a relevant source of NEPP. Some studies have showed the relevance of these compounds in mango by-products (peel and pulp) from different varieties, such as "Ataulfo" and "Tommy Atkins" (García-Magaña et al. 2013; Amaya-Cruz et al. 2015). Also, high NEPP concentrations have been reported for banana (Pérez-Jiménez and Saura-Calixto 2015), tropical blackberry (Acosta-Montoya et al. 2010), cashew apple, acerola (Rufino et al. 2010), açaí (Rufino et al. 2011), pequi or souari nut (Leão et al. 2016), or papaya (Velderrain et al. 2016). But there is still a wide amount of tropical fruits, which remain to be explored for a whole characterization of their phytochemical profile, including NEPP content and profile. Examples of these underutilized fruits are bacuri (*Platonia insignis*), camu-camu (*Myrciaria dubia*), cherimoya (*Annona cherimoya*), chicozapote (*Manilkara zapote*), Costa Rican Cas (*Psidium friedrichsthalianum*), nanche (*Byrsonima crassiflora*), naranjilla (*Solanum quitoense*), soursop (*Annona muricata*), among others (Reynoso-Camacho et al. 2018).

Fruit and vegetable by-products have already been mentioned as materials where a comprehensive evaluation of polyphenol profile, including NEPP, has been performed. But these discarded products require much more research due to the potential they have shown. In this sense, Pérez-Jiménez and Saura-Calixto (2018) evaluated the potential of the peels of some fruits such as apple, banana, kiwi, mandarin, mango, melon, nectarine, orange, pear, and watermelon (all of them discarded in food industry) as a source of NEPP. The obtained results showed that NEPP are significant bioactive compounds in fruit peel, and indeed are present in many common fruits in even higher proportions than EPP. This has two implications: (i) for industry, where discarding these materials means removing relevant sources of bioactive compounds with multiple applications; (ii) regarding nutritional recommendations, the significant proportion of NEPP in the peels content of fruits that can be consumed with peel would be an additional reason to promote the consumption of whole fruits, along with EPP and DF contents. This would lead to a substantial increase in current NEPP intake from fruits, which has been estimated for different European countries in the range of 300–400 mg/p/d (Pérez-Jiménez and Saura-Calixto 2015).

In the same way, NEPP content has been evaluated in by-products from the juice industry. Thus, Delpino-Rius et al. (2015) evaluated the total PP content in residues derived from apple, peach, pear, orange, tangerine, lemon, and carrot juice production. Several phenolic acids were identified in the NEPP fraction of these samples, with caffeic acid being the predominant one in tangerine, orange, and carrot, while *p*-coumaric acid and ferulic acid were the most common ones in lemon and apple.

Also, some studies have found relevant NEPP contents in vegetable by-products such as red beet, Brussels sprouts (Gonzales et al. 2015), or cauliflower discarded materials (Gonzales et al. 2014).

7.4.3 Comparison Between Varieties

In the context of quality control, the determination of NEPP may be especially relevant for the comparison between varieties, since they may exhibit similar content in macronutrients, but significant differences in the content or profile of bioactive compounds, including NEPP. For instance, when NEPP were evaluated in the peels of two cultivars of Jaboticaba (*Myrtaceae* family), i.e. *Myrciaria jaboticaba* and *Myrciaria trunciflora*, some differences emerged, with NEPP being 15.3% of total PP content in *Myrciaria jaboticaba* vs. 4.3% in *Myrciaria trunciflora* (Quatrin et al. 2019). Moreover, tetragalloylglucose was a relevant NEPP constituent in both species, while trigalloylglucose was more abundant in *Myrciaria trunciflora* and delphinidin 3-glucoside in *Myrciaria jaboticaba*.

Recently, Rocchetti et al. (2019) worked with 13 commercial tomato cultivars for industrial transformation, screened by UPLC-QTOF-MS for both EPP and NEPP profiles. They found that anthocyanins were the most abundant class among NEPP (being highest in the "Leader F1" and "Defender F1" cultivars), followed by tyrosols (mainly in "Heinz" cultivars). Nevertheless, the results showed that flavones and hydroxybenzoic acids were the most represented discriminant phenolics in the NEPP fraction. As outlined by the authors, these results may be relevant for tomato industry in order to select specific varieties, taking into account the potential bioaccessibility of phenolic compounds present in the raw matter.

As regards fruits, when comparing three varieties of prickly pear (Amaya-Cruz et al. 2019b), it was found that even when NEPP content was similar, as it happened with "Cristalina" (green) and "Selección 2-1-62" (yellow-orange) varieties, the specific NEPP profile was quite different. Thus, "Selección" variety showed a higher amount of hydroxycinnamic acids such as caffeoyl tartaric and 4-sinapoylquinic acids as well as hydroxybenzoic acids such as syringic and benzoic acids than "Cristalina" variety. Regarding peach, some differences between varieties were found in the contribution of NEPP to total antioxidant capacity, although they were not very remarkable; thus, when antioxidant capacity was measured by DPPH (2,2′-diphenyl-1-picrylhydrazyl) assay, the contribution was between 31 and 46%, while for FRAP (Ferric reducing antioxidant power) assay, it was between 12 and 25%.

In the case of some legumes such as beans, an important food in Mexico and other countries of Latin America, Cárdenas-Castro et al. (2020) evaluated NEPP content in two varieties of beans: "Negro Jamapa" and "Azufrado." It was found that while HPP content was rather similar in both varieties, NEPA content was two-fold higher in "Negro Jamapa" than in the "Azufrado" variety.

7.4.4 Evaluation of Processing Effects

A complete assessment of NEPP in vegetal materials should also involve the evaluation of how different processes affect these bioactive compounds. This implies, first of all, the study of some physiological processes in foods which affect their composition, such as ripening or germination. Of course, the effect of the transformation of agricultural products, such as fruit and vegetables, by physical or chemical means into other forms, involves activities such as mincing and macerating, liquefaction, emulsification, pickling, pasteurization, and cooking by boiling, broiling, frying or grilling. Overall, it has been reported that according to the treatment used, the content of both EPP and NEPP can either decrease or increase (Herrera-Hernández et al. 2018).

Regarding physiological processes, during germination, hydrated cells activate and release enzymes that act by hydrolyzing macromolecules such as protein and starch to smaller molecules, used for the growth and metabolism of the seeds (Rasera et al. 2020). This would lead to an increase in EPP, derived from a NEPP decrease. However, at the same time, during germination, there is an activation of the reactions of synthesis and transport of phenolic compounds toward the cell wall (Yeo and Shahidi 2015), a process needed for the vegetal material in order to enhance the rigid structure of the cell wall matrix that protects the grain against UV damage (Jansen et al. 2001) and avoids contamination by fungi and other pathogens (Nicholson and Hammerschmidt 1992; McLusky et al. 1999). The net results of these simultaneous processes are an increase in NEPP during germination, as observed in lentils (Yeo and Shahidi 2015) and in white mustard grains (Rasera et al. 2020), but not so clearly in black mustard grains (Rasera et al. 2020), evidencing the need of specific studies for each vegetal material. Indeed, the ratio EPP/NEPP (designated as soluble and insoluble phenolic compounds, respectively) has been suggested as a tool to identify the degree of advance of this physiological process (Yeo and Shahidi 2015).

In the case of ripening, this process seems to affect differentially the HPP and NEPA classes that constitute the NEPP fraction. Thus, a study on peach cultivars observed that HPP content tended to decrease during ripening (Liu et al. 2019). In contrast, NEPA content evaluated in ripe and overripe pears showed an increase in the former ones, due to the formation of hydrogen bonds and the contribution of hydrophobic interactions (Brahem et al. 2019). This affected especially cell walls from whole flesh and stone cells, while ripening did not affect NEPA in cell walls from skin and parenchyma cells. Since overripe pears are used for perry production, the knowledge of such processes may have industrial implications.

Regarding the treatments that may be applied to foods, Table 7.4 shows a summary of the effect on NEPP of some types of processing applied in some foods of vegetal origin, together with the general modifications produced in the food matrix due to each treatment. Thermal treatments are one of the most common food processing techniques, applied to multiple food products. In the case of bran from cereals, it has been stated that thermal processing contributes to the breakdown of the cell wall and cellular constituents, leading to an increase in the bioavailability of micronutrients, as well as to the release of bound phenolic acids (Dewanto et al. 2002). Indeed, thermal processes have a two-way effect on phenolic compounds profile, that must be carefully assessed: if the compounds are in their simple form (as EPP), they may be destroyed during heating, but if present in their esterified form (NEPP), the linkage will be broken during heating, therefore leading to an increase in its content. Thus, there is commonly a loss of EPP, together with a transformation of NEPP to EPP (Pérez-Jiménez and Saura-Calixto 2005). Such processes have been observed in several cereal samples such as wheat bran and oat bran (Câlinoiu and Vodnar 2020), sorghum (Luo et al. 2020), breads, cake, muffins, and cookies (Ou et al. 2019). Nevertheless, it should be reminded that due to the specific characteristics of NEPP, there is an interest in keeping in the food matrix a part of these compounds in their original association with the food matrix.

In the case of fruits, a study evaluating the effect of microwave-bleaching on mango, apple, orange, and banana peelings observed an increase in EPP content, as expected, due

Table 7.4 Effect of processing on NEPP content of some vegetable samples.

Type of processing	Sample	Effect in NEPP content	Effect in food matrix	References
High pressure boiling (10, 20, and 30 min) Microwave boiling (5, 10, and 15 seconds)	Sorghum.	Identified syringic acid, veratric acid, *p*-hydroxybenzoic acid and caffeic acid after thermal processing, due to a release from NEPP. Increase in EPP (6.20–18.70%) due to NEPP release. Increase in extractable procyanidins (35.92–58.87%) due to NEPA release.	Heating process tends to promote fragmentation of tissues and cells, breaking covalent bonds.	Luo et al. (2020)
Microwave-blanched (720, for 1, 3, and 5 min)	Fruit peels.	Identified epicatechin, ferulic acid, caffeic acid, rosmarinic acid, *p*-cumaric acid and gallic acid, due to a release from NEPP. Decrease of EPP.	Thermal treatment tends to dissociate conjugated forms of some phenolic compounds in their free forms. If the phenolic compounds are present in their free form (EPP), they may be destroyed during the heating process and their content will decrease after process.	Dibanda et al. (2020)
Pressure cooking (15 psi, 7 min, and 9 min for varieties "Azufrado" and "Negro Jamapa," respectively), mashing and frying (4 min)	Beans ("Azufrado" and "Negro Jamapa" varieties)	NEPP decrease in samples where pressure cooking, mashing and frying were applied.	Changes in polyphenols structure (oxidation) associated to the high temperatures applied in the cooking and frying process.	Cárdenas-Castro et al. (2020)
Pressure cooking (15 psi, 7 min, and 9 min for varieties "Azufrado" and "Negro Jamapa," respectively) and mashing		NEPP increase in samples were pressure cooking and mashing treatment were applied.		

(*Continued*)

Table 7.4 (Continued)

Type of processing	Sample	Effect in NEPP content	Effect in food matrix	References
Thermal processing	Baking products enriched with by-products of vegetables	The baking process increases the content of EPP such as ferulic acid and coumaric acid in some baking foods such as bread, cookies and muffins due to NEPP release.	EPP showed low thermal stability at long times and high temperatures.	Ou et al. (2019)
Thermal processing (80 °C, 10 min)	Wheat bran and oat by-products	Increase in EPP in wheat bran (22.49%) and oat (25.84%), due to NEPP release. Increase in these individual phenolic compounds: in wheat bran, ferulic acid, vanillic acid, apigenin-glucoside and *p*-coumaric acid; in oat, avenanthramide 2c and dihydroxybenzoic acids.	Hydrolysis of polysaccharides present in dietary fiber.	Călinoiu and Vodnar (2020)

to NEPP release from the cell wall (Dibanda et al. 2020). This study also provided a surprising result, which the authors could not explain: microwave-bleaching for one minute significantly reduced NEPP content in all samples, as expected, but when the process was performed for three to five minutes, a significant increase in NEPP was observed.

Another common food processing is frying. Cárdenas-Castro et al. (2020) evaluated the effect of the two common ways of cooking beans in Mexico, i.e. pressure cooking followed by mashing, or pressure cooking and mashing followed by frying, in phenolic compounds content and profile. They observed that NEPP content (both HPP and NEPA) was lower after pressure cooking-mashing-frying than after pressure cooking-mashing in the two evaluated varieties. The authors attributed this fact to changes in the chemical structure (mostly oxidations) associated with the high temperatures applied.

Another type of non-thermal process applied to foodstuffs is hydrostatic high pressure. Szczepańska et al. (2020) evaluated the influence of static and multi-pulsed hydrostatic pressure processing treatments on the polyphenol profile in carrot juice, and other parameters such as oxidoreductase activity, color, and browning index. The most relevant result regarding NEPP is that this treatment led to an increase in the concentration of selected polyphenols such as 4-hydroxybenzoic acid, vanillic acid, didymin, pyrogallol, and glucoside derivates of 4-hydroxybenzoic acid, protocatechuic, and ferulic acids. This can be explained either by a release of NEPP cell walls or the enzymatic hydrolysis of higher polyphenols into simple phenols, in both cases due to the effect of the high pressure applied.

Finally, cocoa is a specific food product, with roasting as a key process in its manufacturing. The evolution of EPP and NEPP in cocoa during roasting has been evaluated, concluding that, while EPP content decreases during this stage, due to oxidation reactions, NEPP remain similar because cell wall association provides additional protection to these compounds (Mudenuti et al. 2018). Regarding individual phenolic compounds, it was observed that while (+)-catechin, (−)-epicatechin, and (+)-gallocatechin were mainly found as EPP, more than 80% of protocatechuic acid (commonly considered as a minor compound in cocoa) was present as NEPP.

7.5 Perspectives

As explained in the previous headings of this chapter, NEPP have been traditionally much less studied than EPP. This did not happen by chance, but there is a reason for it: first, it was considered that, as it happens with other food constituents, finding the proper solvent combination would allow to obtain solutions containing all polyphenols in a sample. However, that was not the case, which leads to the second reason limiting the research on these compounds: the analytical problem to analyze the high molecular weight compounds or strongly linked to macromolecules that NEPP constituents.

Nevertheless, the research performed on NEPP has already shown the relevance of these compounds from different perspectives (content, intake, and health-related properties), leading even to a whole book on the topic (Saura-Calixto and Pérez-Jiménez 2018). This means that there are more researchers interested in NEPP, and this should be translated into a close solution to traditional problems regarding the study of these compounds. In particular, there is a need to establish a standardized methodology for NEPP analysis. Although, as stated before, each methodology has its *pros* and *cons* and this should be further explored for research purposes, reaching some consensus would be very helpful in order to extend NEPP analysis not only in the academic field but also in the food industry. In this sense, maybe some pragmatic approach should be provided for those non-experts in the field, in order to make NEPP determination something practical. For instance, a recent study reported a protocol for chromatography analysis of NEPP only in ten minutes (Cocuron et al. 2019).

At the same time that some established methods could be provided to be used as a routine method, for instance, in the food industry, other aspects could be explored with further detail in the academic field. Such would be the case of combining emerging extraction techniques with hydrolysis strategies in order to release NEPP, as it has been already successfully performed (Mushtaq et al. 2015). And, especially, the identification of the whole NEPP-DF matrix without the need of performing the hydrolysis step. In this sense, it is especially relevant that a recent study was able to confirm the existence of this complex by solid-state NMR spectroscopy (Bermúdez-Oria et al. 2020). Moreover, some analytical aspects have not yet been explored in the field of NEPP, such as HILIC separation, as discussed earlier, or the use of two-dimensional chromatography.

Also, an important issue that needs to be solved in order to extend routine NEPP analysis is the best standard to be used for NEPA characterization, an aspect that, as discussed earlier, has several problems. It is very relevant that such a standard from cranberry was recently reported and, indeed, it allowed the first determination of NEPA in this fruit (Gullickson et al. 2020). Nevertheless, this approach has some limitations since if, for each particular analysis, a specific standard has to be developed from that food, that will not be practical, particularly at the scale of the food industry, dealing with several different foodstuffs. This connects with this another available method for characterizing NEPA, i.e. acid-catalyzed depolymerization in presence of nucleophiles, a method with much less drawbacks than butanolysis, but which has, as the main limitation, the fact that there are no commercial standards for the conjugates originated between terminal units and the nucleophile agent. Thus, the availability of such conjugates as commercial standards would be a feasible way to increase the number of researchers adhering to this technique, in the same way that the availability of more commercial standards of polyphenol metabolites has improved the analysis of these compounds.

Another important aspect in the field of NEPP research is that, ultimately, the interest in these compounds is related to the potential health benefits that they may exhibit. For this reason, it is very relevant to know how they behave after their intake. For this, studies performing *in vitro* digestion and colonic fermentation of different matrixes rich in NEPP are particularly relevant. In this sense, this approach was used to explore the bioaccessibility and generation of colonic metabolites from mango bars (Hernández-Maldonado et al. 2019) and, specifically, indigestible fraction from this product (Gutiérrez-Sarmiento et al. 2020). This kind of study should be extended, as well as *in vivo* studies on the metabolic fate of NEPP, which are still rather scarce, as discussed earlier. At the same time, an integral approach to the study of NEPP should not disregard the potential existence of adverse effects linked to an excessive NEPP release; it is known that high antioxidant doses may become prooxidants and this aspect should be kept in mind when exploring new alternatives for making easier to release NEPP from food matrixes.

Besides, as previously shown, the study of NEPP is related to several agronomical factors, such as the ripening stage or variety. For this reason, agronomical research focused on strategies for increasing NEPP content, is rather pertinent. In this sense, it is relevant that some studies have shown, in the last decades, that the use of elicitors may allow increasing concentrations of these compounds; this has been shown in potato (Keller et al. 1996), tomato (Mandal and Mitra 2008) and, recently, in eggplant (Mandal and Gupta 2016). This field should be further explored in view of these promising results.

With the same objective, i.e. increasing NEPP content in food products, several technological strategies may be explored. In this sense, some recent studies must be highlighted. In one of them, it was observed that the sequential application of wounding stress and extrusion in carrots induced the accumulation of phenolic compounds, which resulted in a 288% increase in EPP content and 408% increase in NEPP content (Viacava et al. 2020). Another study reported that both extrusion and fungal fermentation significantly increased NEPP content in rice bran (Chen et al. 2019). Thus, further similar approaches should be tested in order to increase the nutritional value of foodstuffs.

This connects with the last remaining problem in the field of NEPP, i.e. their successful development as food ingredients. Although many studies have explored their use as ingredients in several matrixes, and even the incorporation of antioxidant DFs (particularly rich in NEPP) in meat-based products was recently reviewed (Das et al. 2020), sensory parameters cannot be considered fully satisfied. Indeed, the sensory acceptability of these products is still limited, mostly due to texture problems (Solari-Godiño et al. 2017). For this reason, the development of new strategies to incorporate NEPP as food ingredients would be very relevant. To the authors' knowledge, the first effort in this sense was recently published (Pravinata and Murray 2019), consisting of the encapsulation of water-insoluble polyphenols in Ca-alginate microgel particles produced by the Leeds Jet Homogenizer. Much more research is needed in this field, but successful encapsulation of NEPP would mean a step forward in the use of isolates of these food constituents as food ingredients. Similarly, DIC technology may be relevant for the whole use of agricultural by-products, allowing for modifying the proportions between EPP and NEPP depending on specific objectives. In the current context of the circular economy, this kind of approach would allow effective use of many agricultural by-products, which present relevant NEPP content, such as fruit peels (Pérez-Jiménez and Saura-Calixto 2018).

In conclusion, NEPP have shown to be major constituents in the polyphenol fraction of vegetal materials. This means that they are important contributors to polyphenol intake and the generation of polyphenol-derived metabolites, leading ultimately to beneficial biological effects. Important advances in NEPP research have been performed during the last 15 years, but there is still a gap to be completed before these compounds are routinely considered when performing the characterization of vegetal material. For this, advances and standardization in analytical methodologies are key points. This will allow reaching a situation where quality control in the food industry includes NEPP in the list of constituents to be determined, as it currently happens with many other food compounds.

References

Acosta-Montoya, O., Vailant, F., Cozzano, S., et al. (2010). Phenolic content and antioxidant capacity of tropical highland blackberry (*Rubus adenotrichus* Schldtl.) during three edible maturity stages. *Food Chemistry*, **119**: 1497–1501.

Alu'datt, M.H., Rababah, T., Alhamad, M.N., et al. (2017). Profiles of free and bound phenolics extracted from *Citrus* fruits and their roles biological system: content and antioxidant, anti-diabetic and anti-hypertensive properties. *Food & Function*, **8**: 3187–3197.

Alves, G., Lobo, L.A., Domingues, R.M.C., et al. (2021). Bioaccessibility and gut metabolism of free and melanoidin-bound phenolic compounds from coffee and bread. *Frontiers in Nutrition*, **8**: 708928.

Amaya-Cruz, D.M., Rodríguez-González, S., Pérez-Ramírez, I.F., et al. (2015). Juice by-products as a source of dietary fibre and antioxidants and their effect on hepatic steatosis. *Journal of Functional Foods*, **17**: 93–102.

Amaya-Cruz, D., Pérez-Ramírez, I.F., Pérez-Jiménez, J., et al. (2019a). Comparison of the bioactive potential of Roselle (*Hibiscus sabdariffa* L.) calyx and its by-product: phenolic characterization by UPLC-QTOF MS^E and their anti-obesity effect in vivo. *Food Research International*, **126**: 108598.

Amaya-Cruz, D.M., Pérez-Ramírez, I.F., Delgado-García, J., et al. (2019b). An integral profile of bioactive compounds properties of prickly pear (*Opuntus ficus indica* L.) peel with different tonalities. *Food Chemistry*, **278**: 568–578.

Andreasen, M.F., Kroon, P.A., Williamson, G., and Garcia-Conesa, M.T. (2001). Intestinal release and uptake of phenolic antioxidant diferulic acids. *Free Radical Biology & Medicine*, **31**: 304–314.

Arigò, A., Ĉesla, P., Šilarová, P., et al. (2018). Development of extraction method for characterization of free and bonded polyphenols in barely (*Hordeum vulgare* L.) grown in

Czech Republic using liquid chromatography-tandem mass spectrometry. *Food Chemistry*, **245**: 829–837.

Arranz, S. and Saura-Calixto, F. (2010). Analysis of polyphenols in cereals may be improved performing acid hydrolysis: a study in wheat flour and wheat bran and cereals of the diet. *Journal of Cereal Science*, **51**: 313–318.

Arranz, S., Saura-Calixto, F., Shaha, S., et al. (2009). High contents of nonextractable polyphenols in fruits suggest that polyphenol contents of plant foods have been underestimated. *Journal of Agricultural and Food Chemistry*, **57**: 7298–7303.

Arranz, S., Silván, J.M., and Saura-Calixto, F. (2010). Non-extractable polyphenols, usually ignored, are the major part of dietary polyphenols: a study on the Spanish diet. *Molecular Nutrition & Food Research*, **54**: 1646–1658.

Bermúdez-Oria, A., Rodríguez-Gutiérrez, G., Fernández-Prior, A., et al. (2020). Confirmation by solid-state NMR spectroscopy of a strong complex phenol-dietary fiber with retention of antioxidant activity *in vitro*. *Food Hydrocolloids*, **102**: 105584.

Bohn, T. (2018). Metabolic fate of bioaccesssible an non bioaccessible carotenoids. In: Non-extractable Polyphenols and Carotenoids: Importance in Human Nutrition and Health, (eds. Saura-Calixto, F. and Pérez-Jiménez, J.), 165–200. London, UK: Royal Society of Chemistry.

Bonoli, M., Verardo, V., Macorni, E., and Caboni, M.F. (2004). Antioxidant phenols in barley (*Hordeum vulgare* L.) flour: comparative spectrophotometric study among extraction methods of free and bound phenolic compounds. *Journal of Agricultural and Food Chemistry*, **52**: 5195–5200.

Brahem, M., Renard, C.M.G.C., Bureau, S., et al. (2019). Pear ripeness and tissue type impact procyanidin-cell wall interactions. *Food Chemistry*, **275**: 754–762.

Câlinoiu, L.F. and Vodnar, D.C. (2020). Thermal processing for the release of phenolic compounds from wheat and oat bran. *Biomolecules*, **10**: 21.

de Camargo, A.C., Regitano-d'Arce, M.A.B., Biastoso, A.C.T., and Shahidi, F. (2016). Enzyme-assisted extraction of phenolics from winemaking by-products: antioxidant potential and inhibition of alpha-glucosidase and lipase activities. *Food Chemistry*, **212**: 395–402.

Campillo, N., Viñas, P., Férez-Melgarejo, G., et al. (2015). Determination of phenolic acids and hydrolysable tannins in pomegranate fruit and beverages by liquid chromatography with diode array detection and time-of-flight mass spectrometry. *Food Analytical Methods*, **8**: 1315–1325.

Cárdenas-Castro, A.P., Pérez-Jiménez, J., Bello-Pérez, L.A., et al. (2020). Bioaccessibility of phenolic compounds in common beans (*Phaseolus vulgaris* L.) after *in vitro* gastrointestinal digestion: a comparison of two cooking procedures. *Cereal Chemistry*, **97**: 670–680.

Cardona, F., Andrés-Lacueva, C., Tulipani, S., et al. (2013). Benefits of polyphenols on gut microbiota and implications in human health. *The Journal of Nutritional Biochemistry*, **24**: 1415–1422.

Çayan, F., Deveci, E., Tel-Çayan, G., and Duru, M.E. (2020). Identification and quantification of phenolic acid compounds of twenty-six mushrooms by HPLC-DAD. *Journal of Food Measurement and Characterization*, **14**: 1690–1698.

Chandrasekara, A. and Shahidi, F. (2011). Determination of antioxidant activity in free and hydrolyzed fractions of millet grains and characterization of their phenolic profiles by HPLC-DAD-ESI-MS^n. *Journal of Functional Foods*, **3**: 144–158.

Chassard, C., Goumy, V., Leclerc, M., et al. (2007). Characterization of the xylan-degrading microbial community from human faeces. *FEMS Microbiology Ecology*, **61**: 121–131.

Chen, P.X., Tang, Y., Marcone, M.F., et al. (2015). Characterization of free, conjugated and bound phenolics and lipophilic antioxidants in regular- and non-darkening cranberry beans (*Phaseolus vulgaris* L.). *Food Chemistry*, **185**: 298–308.

Chen, F., Du, X., Zu, Y., et al. (2016). Microwave-assisted method for distillation and dual extraction in obtaining essential oil, proanthocyanidins and polysaccharides by one-pot process from Cinnamomi Cortex. *Separation and Purification Technology*, **164**: 1–11.

Chen, Y., Ma, Y., Dong, L., et al. (2019). Extrusion and fungal fermentation change the profile and antioxidant activity of free and bound phenolics in rice bran together with the phenolic bioaccessibility. *LWT- Food Science and Technology*, **115**: 108461.

Cheng, A., Yan, H., Han, C., et al. (2014). Acid and alkaline hydrolysis extraction of non-extractable polyphenols in blueberries: optimization by response surface methodology. *Czech Journal of Food Sciences*, **32**: 218–225.

Cheng, A., Han, C., Fang, X., et al. (2016). Extractable and non-extractable polyphenols from blueberries modulate LPS-induced expression of iNOS and COX-2 in RAW264-7 macrophages via NF-kB signalling pathway. *Journal of the Science of Food and Agriculture*, **96**: 3393–3400.

Chirug, L., Okun, Z., Ramon, O., and Shpigelman, A. (2018). Iron ions as mediators in pectin-flavonols interactions. *Food Hydrocolloids*, **84**: 441–449.

Cocuron, J.C., Casas, M.I., Yang, F., et al. (2019). Beyond the wall: high-throughput quantification of plant soluble and cell-wall bound phenolics by liquid chromatography tandem mass spectrometry. *Journal of Chromatography A*, **1589**: 93–104.

Das, A.K., Nanda, P.K., Madane, P., et al. (2020). A comprehensive review on antioxidant dietary fibre enriched meat-based functional foods. *Trends in Food Science & Technology*, **99**: 323–336.

Delpino-Rius, A., Eras, J., Vilaró, F., et al. (2015). Characterisation of phenolic compounds in processed fibres from the juice industry. *Food Chemistry*, **172**: 575–584.

Dewanto, V., Wu, X., and Lui, R.H. (2002). Processed sweet corn has higher antioxidant activity. *Journal of Agricultural and Food Chemistry*, **50**: 4959–4964.

Díaz-Rubio, M.E. and Saura-Calixto, F. (2011). Beverages have an appreciable contribution to the intake of soluble dietary fibre: a study in the Spanish diet. *International Journal of Food Sciences and Nutrition*, **62**: 715–718.

Dibanda, R.F., Akdowa, E.P., Rani, P.A., et al. (2020). Effect of microwave blanching on antioxidant activity, phenolic compounds and browning behavior of some fruit peelings. *Food Chemistry*, **302**: 125308.

Dodd, D., Mackie, R.I., and Cann, I.K.O. (2011). Xylan degradation, a metabolic property shared by rumen and human colonic Bacteroidetes. *Molecular Microbiology*, **79**: 292–304.

Domínguez-Rodríguez, G., Marina, M.L., and Plaza, M. (2017). Strategies for the extraction and analysis of non-extractable polyphenols from plants. *Journal of Chromatography A*, **8**: 1–15.

Domínguez-Rodríguez, G., Plaza, M., and Marina, M.L. (2021). High-performance thin-layer chromatography and direct analysis in real time-high resolution mass spectrometry of non-extractable polyphenols from tropical fruit peels. *Food Research International*, **147**: 110455.

Durazzo, A., Lisciani, S., Camili, E., et al. (2017). Nutritional composition and antioxidant properties of traditional Italian dishes. *Food Chemistry*, **218**: 70–77.

Durazzo, A., Lucarini, M., Santini, A., et al. (2019). Antioxidant properties of four commonly consumed popular Italian dishes. *Molecules*, **24**: 1543.

Faller, A.L.K. and Fialho, E. (2009). The antioxidant capacity and polyphenol content of organic and conventional retail vegetables after domestic cooking. *Food Research International*, **42**: 210–215.

Fang, H.Y., Chen, Y.K., Chen, H.H., et al. (2012). Immunomodulatory effects of feruloylated oligosaccharides from rice bran. *Food Chemistry*, **134**: 836–840.

Flint, H.J., Scott, K.P., Duncan, S.H., et al. (2012). Microbial degradation of complex carbohydrates in the gut. *Gut Microbes*, **3**: 289–306.

Folin, O. and Denis, W. (1912). On phosphotungstic-phosphomolybdic compounds as color reagents. *Journal of Biological Chemistry*, **12**: 239–243.

García-Magaña, M.L., García, H.S., Bello-Pérez, L.A., et al. (2013). Functional properties and dietary fiber characterization of mango processing by-products (*Mangifera indica* L., cv *Ataulfo* and *Tommy Atkins*). *Plant Foods for Human Nutrition*, **68**: 254–258.

García-Villalba, R., Espín, J.C., Aaby, K., et al. (2015). Validated method for the characterization and quantification of extractable and non-extractable ellagitannins after acid hydrolysis in pomegranate fruits, juices, and extracts. *Journal of Agricultural and Food Chemistry*, **63**: 6555–6566.

García-Villalba, R., Espín, J.C., and Tomás-Barberán, F.A. (2016). Chromatographic and spectroscopic characterization of urolithins for their determination in biological samples after the intake of foods containing ellagitannins and ellagic acid. *Journal of Chromatography A*, **1428**: 162–175.

Gil, J.V., Esteban-Muñoz, A., and Fernández-Espinar, M.T. (2022). Changes in the polyphenols profile and antioxidant activity of wheat bread after incorporating quinoa flour. *Antioxidants*, **11**: 33.

Georgé, S., Brat, P., Alter, P., et al. (2005). Rapid determination of polyphenols and vitamin C in plant-derived products. *Journal of Agricultural and Food Chemistry*, **53**: 1370–1403.

Giménez-Bastida, J.A., González-Sarrías, A., Larrosa, M., et al. (2012). Ellagitannin metabolites, urolithin A glucuronide and its aglycone urolithin A, ameliorate TNF-α-induced inflammation and associated molecular markers in human aortic endothelial cells. *Molecular Nutrition & Food Research*, **56**: 784–796.

Gómez-Caravaca, A.M., Verardo, V., Berardinelli, A., et al. (2014). A chemometric approach to determine the phenolic compounds in different barely samples by two different stationary phases: a comparison between C18 and pentafluorophenyl core shell columns. *Journal of Chromatography A*, **1355**: 134–142.

Goñi, I. and Hernández-Galiot, A.B. (2019). Intake of nutrient and non-nutrient dietary antioxidants. Contribution of macromolecular antioxidant polyphenols in an elderly Mediterranean population. *Nutrients*, **11**: 2165.

Goñi, I. and Serrano, J. (2005). The intake of dietary fiber from grape seeds modifies the antioxidant status in rat cecum. *Journal of the Science of Food and Agriculture*, **85**: 1877–1881.

Goñi, I., Díaz-Rubio, M.E., Pérez-Jiménez, J., and Saura-Calixto, F. (2009). Towards an updated methodology for measurement of dietary fiber, including associated polyphenols, in food and beverages. *Food Research International*, **42**: 840–846.

Gonzales, G.B., Smagghe, G., Raes, K., and Van Camp, J. (2014). Combined alkaline hydrolysis and ultrasound-assisted extraction for the release of non-extractable phenolics from cauliflower (*Brasscia oleracea* var. botrytis) waste. *Journal of Agricultural and Food Chemistry*, **62**: 3371–3376.

Gonzales, G.B., Raes, K., Vanhoutte, H., et al. (2015). Liquid chromatography-mass spectrometry coupled with multivariate analysis for the characterization and discrimination of extractable and non-extractable polyphenols and glucosinolates from red cabbage and Brussels sprout waste streams. *Journal of Chromatography A*, **1402**: 60–70.

González-Sarrías, A., Espín, J.C., and Tomás-Barberán, F.A. (2017). Non-extractable polyphenols produce gut microbiota metabolites that persist in circulation and show anti-inflammatory and free radical-scavenging effects. *Trends in Food Science & Technology*, **69**: 281–288.

Grabber, J.H. and Zeller, W.E. (2020). Direct versus sequential analysis of procyanidin- and prodelphinidin-based condensed tannins by the HCl—butanol—acetone—iron assay. *Journal of Agricultural and Food Chemistry*, **68**: 2906–2916.

Grabber, J.H., Zeller, W.E., and Mueller-Harvey, I. (2013). Acetone enhances the direct analysis of procyanidin and prodelphinidin-based condensed tannins in lotus species by the butanol-HCl-iron assay. *Journal of Agricultural and Food Chemistry*, **61**: 2669–2678.

Gullickson, E.R., Krueger, C.G., Birmingham, A., et al. (2020). Development of a cranberry standard for quantification of insoluble cranberry (*Vaccinium macrocarpon* Ait.) proanthocyanidins. *Journal of Agricultural and Food Chemistry*, **68**: 2900–2905.

Guo, W. and Beta, T. (2013). Phenolic acid composition and antioxidant potential of insoluble and soluble dietary fibre extracts derived from select whole-grain cereals. *Food Research International*, **51**: 518–525.

Gutiérrez-Díaz, I., Salazar, N., Pérez-Jiménez, J., et al. (2021). New players in the relationship between diet and microbiota: the role of macromolecular antioxidant polyphenols. *European Journal of Nutrition*, **60**: 1403–1413.

Gutiérrez-Sarmiento, W., Sayago-Ayerdi, S.G., Goni, I., et al. (2020). Changes in intestinal microbiota and predicted metabolic pathways during colonic fermentation of mango (*Mangifera indica* L.)-based bar indigestible fraction. *Nutrients*, **12**: 683.

Hamauzu, Y. and Mizuno, Y. (2011). Non-extractable procyanidins and lignin are important factors in the bile acid binding and radical scavenging properties of cell wall material in some fruits. *Plant Foods for Human Nutrition*, **66**: 70–77.

Hamauzu, Y. and Suwannachot, J. (2019). Non-extractable polyphenols and *in vitro* bile acid-binding capacity of dried persimmon (*Diospyros kaki*) fruit. *Food Chemistry*, **293**: 127–133.

Han, Y., Huang, M., Li, L., et al. (2019). Non-extractable polyphenols from cranberries: potential anti-inflammation and anti-colon-cancer agents. *Food & Function*, **10**: 7714–7723.

Hellström, J.K. and Mattila, P.H. (2008). HPLC determination of extractable and unextractable proanthocyanidins in plant materials. *Journal of Agricultural and Food Chemistry*, **56**: 7617–7624.

Hernández-Maldonado, L.M., Blancas-Benítez, F.J., Zamora-Gasga, V.M., et al. (2019). In vitro gastrointestinal digestion and colonic fermentation of high dietary fiber and antioxidant-rich mango (*Mangifera indica* L.) "Ataulfo"-based fruit bars. *Nutrients*, **11**: 1564.

Herrera-Balandrano, D.D., Báez-González, J.G., Carvajal-Millán, E., et al. (2020). Alkali-extracted feruloylated arabinoxylans from nixtamalized maize bran byproduct: a synonymous with soluble antioxidant dietary fiber. *Waste and Biomass Valorization*, **11**: 403–409.

Herrera-Hernández, M.G., Miranda-López, R., Esparza-Martínez, F.J., and Guzmán-Maldonado, S.H. (2018). Fruit and vegetable processing: effects of extractable and non-extractable polyphenols. In: Non-extractable Polyphenols and Carotenoids: Importance in Human Nutrition and Health, (eds. Saura-Calixto, F. and Pérez-Jiménez, J.), 111–132. London: Royal Society of Chemistry.

Herrero, M., Castro-Puyana, M., Mendiola, J.A., and Ibañez, E. (2013). Compressed fluids for the extraction of bioactive compounds. *TrAC Trends in Analytical Chemistry*, **43**: 67–83.

Hervert-Hernández, D., García, O.P., Rosado, J.L., and Goñi, I. (2011). The contribution of fruits and vegetables to dietary intake of polyphenols and antioxidant capacity in a Mexican rural diet: importance of fruit and vegetable variety. *Food Research International*, **44**: 1182–1189.

Houghton, M.J., Kerimi, A., Mouly, V., et al. (2019). Gut microbiome catabolites as novel modulators of muscle cell glucose metabolism. *The Federation of American Societies for Experimental Biology Journal*, **33**: 1887–1898.

Huang, J., Wang, X., Tao, G., et al. (2018). Feruloylated oligosaccharides from maize bran alleviate the symptoms of diabetes in streptozotocin-induced type 2 diabetic rats. *Food & Function*, **9**: 1779–1789.

Hümmer, W. and Schreier, P. (2008). Analysis of proanthocyanidins. *Molecular Nutrition & Food Research*, **52**: 1381–1398.

Ignat, I., Volf, I., and Popa, V.I. (2011). A critical review of methods for characterization of polyphenolic compounds in fruits and vegetables. *Food Chemistry*, **126**: 1821–1835.

Irakli, M., Kleisiaris, F., Kadoglidou, K., and Katsantonis, D. (2018). Optimizing extraction conditions of free and bound phenolic compounds from rice by-products and their antioxidant effects. *Food*, **7**: 93.

Jandera, P. and Hájek, T. (2009). Utilization of dual retention mechanism on columns with bonded PEG and diol stationary phases for adjusting the separation selectivity of phenolic and flavone natural antioxidants. *Journal of Separation Science*, **32**: 3603–3619.

Jansen, M.A.K., van den Noort, R.E., Tan, M.Y.A., et al. (2001). Phenol-oxidizing peroxidases contribute to the protection of plants from ultraviolet radiation stress. *Plant Physiology*, **126**: 1012–1023.

John, J.A. and Shahidi, F. (2010). Phenolic compounds and antioxidant activity of Brazil nut (*Bertholletia excelsa*). *Journal of Functional Foods*, **2**: 196–209.

Keller, H., Hohlfeld, H., Wray, V., et al. (1996). Changes in the accumulation of soluble and cell wall-bound phenolics in elicitor-treated cell suspension cultures and fungus-infected leaves of *Solanum tuberosum*. *Phytochemistry*, **42**: 389–396.

Kohyama, N., Chono, M., Nakagawa, H., et al. (2017). Flavonoid compounds related to seed coat color of wheat. *Bioscience, Biotechnology, and Biochemistry*, **81**: 2112–2118.

Kroon, P.A., Faulds, C.B., Ryden, P., et al. (1997). Release of covalently bond ferulic acid from fiber in the human colon. *Journal of Food Agricultural and Food Chemistry*, **45**: 661–667.

Krygier, K., Sosulski, F., and Hogge, L. (1982). Free, esterified, and insoluble-bound phenolic acids. 1. Extraction and purification procedure. *Journal of Agricultural and Food Chemistry*, **30**: 330–334.

Latimer, G.W. (2012). Official Methods of Analysis of the Association of Official Analytical Chemists (AOAC) (19th edn). Gaithersburg, MD, USA: *AOAC International*.

Le Bourvellec, C. and Renard, C.M.G.C. (2012). Interactions between polyphenols and macromolecules: quantification methods and mechanisms. *Critical Reviews in Food Science and Nutrition*, **52**: 213–248.

Le Bourvellec, C., Bagano Vilas Boas, P., Lepercq, P., et al. (2019). Procyanidin—cell wall interactions within apple matrices decrease the metabolization of procyanidins by the human gut microbiota and the anti-inflammatory effect of the resulting microbial metabolome *in vitro*. *Nutrients*, **11**: 664.

Leão, D.P., Franca, A.S., Oliveira, L.S., et al. (2017). Physicochemical characterization, antioxidant capacity, total phenolic and proanthocyanidin content of flours prepared from pequi (*Caryocar brasilense* Camb.) fruit by-products. *Food Chemistry*, **225**: 146–153.

Lee, C.C., Kim, J.H., Kim, J.S., et al. (2017). 5-(3′,4′-Dihydroxyphenyl-γ-valerolactone), a major microbial metabolite of proanthocyanidin, attenuates THP-1 monocyte-endothelial adhesion. *International Journal of Molecular Sciences*, **18**: 1363.

Li, W., Yang, R., Ying, D., et al. (2020). Analysis of polyphenols in apple pomace: a comparative study of different extraction and hydrolysis procedures. *Industrial Crops and Products*, **147**: 112250.

Lin, S., Wang, Z., Lin, Y., et al. (2019). Bound phenolics from fresh lotus seeds exert anti-obesity effects in 3T3-L1 adipocytes and high-fat diet-fed mice by activation of AMPK. *Journal of Functional Foods*, **58**: 74–84.

Liu, H., Jiang, W., Cao, J., et al. (2019). Changes in extractable and non-extractable polyphenols and their antioxidant properties during fruit on-tree ripening in five peach cultivars. *Horticultural Plant Journal*, **5**: 137–144.

Lizarraga, D., Vinardell, M.P., Noé, V., et al. (2011). A lyophilized red grape pomace containing proanthocyanidin-rich dietary fiber induces genetic and metabolic alterations in colon mucosa of female C57BL/6J mice. *The Journal of Nutrition*, **141**: 1597–1604.

López-oliva, ME., Pozuelo, M.J., Rotger, R., et al. (2013). Grape antioxidant dietary fibre prevents mitochondrial apoptotic pathways by enhancing Bcl-2 and BclxL expression and minimising oxidative stress in rat distal colonic mucosa. *British Journal of Nutrition*, **109**: 4–16.

Luca, S.V., Macovei, I., Bujor, A., et al. (2020). Bioactivity of dietary polyphenols: the role of metabolites. *Critical Reviews in Food Science and Nutrition*, **60**: 626–659.

Ludwig, L.A., Rubió, L., Mosele, J.I., and Moltilva, M.J. (2018). Metabolic fate of extractable and non-extractable polyphenols. In: Non-extractable Polyphenols and Carotenoids: Importance in Human Nutrition and Health, (eds. Saura-Calixto, F. and Pérez-Jiménez, J.), 220–240. London: Royal Society of Chemistry.

Luo, F. and Fei, X. (2019). Distribution and antioxidant activities of free, conjugated, and insoluble-bound phenolics from seven species of the genus *Camellia*. *Journal of The American Oil Chemists' Society*, **96**: 159–170.

Luo, M., Hou, F., Dong, L., et al. (2020). Comparison of microwave assisted and high-pressure processing on bound phenolic composition and antioxidant activities of sorghum hull. *International Journal of Food Science & Technology*, **55**: 3190–3202.

Macagnan, F.T., da Silva, L.P., and Hecktheuer, L.H. (2016). Dietary fibre: the scientific search for an ideal definition and methodolgy of analysis, and its physiological importance as a carrier of bioactive compounds. *Food Research International*, **85**: 144–154.

Macho-González, A., Garcimartín, A., López-Oliva, M.E., et al. (2017). Fiber purified extracts of carob fruit decrease carbohydrate absorption. *Food & Function*, **8**: 2258–2265.

Macho-González, A., Garcimartín, A., Naes, F., et al. (2018). Effects of fiber purified extract of carob fruit on fat digestion and postprandial lipemia in healthy rats. *Journal of Agricultural and Food Chemistry*, **66**: 6734–6741.

Madhujith, M. and Shahidi, F. (2009). Antioxidant potential of barley as affected by alkaline hydrolysis and release of insoluble-bound phenolics. *Food Chemistry*, **117**: 615–620.

Mandal, S. and Gupta, C.K. (2016). Inducing cell wall-bound phenolic compounds by elicitors in eggplant (*Solanum melongena*). *Applied Biochemistry and Microbiology*, **52**: 650–656.

Mandal, S. and Mitra, A. (2008). Accumulation of cell wall-bound phenolic metabolites and their upliftment in hairy root cultures of tomato (*Lycopersicon esculentum* Mill.). *Biotechnology Letters*, **30**: 1253–1258.

Martínez-Maqueda, D., Zapatera, B., Gallego-Narbón, A., et al. (2018). A 6-week supplementation with grape pomace to subjects at cardiometabolic risk ameliorates insulin sensitive, without affecting other metabolic syndrome markers. *Food & Function*, **9**: 6010–6019.

Martínez-Meza, Y., Pérez-Jiménez, J., Rocha-Guzmán, N.E., et al. (2021). Modification on the polyphenols and dietary fiber content of grape pomace by DIC. *Food Chemistry*, **360**: 130035.

Martínez-Meza, Y., Pérez-Jiménez, J., Castaño-Tostado, E., et al. (2022). Instant controlled pressure drop as a strategy to modify extractable and non-extractable phenolic compounds: a study in different grape pomace materials. *Journal of Agricultural and Food Chemistry*. https://doi.org/10.1021/acs.jafc.1c04583.

Mateos-Martín, M.L., Pérez-Jiménez, J., Fuguet, E., and Torres, J.L. (2012). Non-extractable proanthocyanidins from grape are a source of bioavailable (epi)catechin and derived metabolites in rats. *British Journal of Nutrition*, **108**: 290–297.

Matsumura, Y., Ito, T., Yano, H., et al. (2016). Antioxidant potential in non-extractable fractions of dried persimmon (*Diospyros kaki* Thumb). *Food Chemistry*, **202**: 99–103.

Matthews, S., Mila, I., Scalbert, A., and Donnelly, D.M.X. (1997). Extractable and non-extractable proanthocyanidins in barks. *Phytochemistry*, **45**: 405–410.

Maurer, L.H., Cazarin, C.B.B., Quatrin, A., et al. (2019). Grape peel powder promotes intestinal barrier homeostasis in acute TNBS-colitis: a major role for dietary fiber and fiber-bound polyphenols. *Food Research International*, **123**: 425–439.

Maurer, L.H., Cazarin, C.B.B., Quatrin, A., et al. (2020). Dietary fiber and fiber-bound polyphenols of grape peel powder promote GSH recycling and prevent apoptosis in the colon of rats with TNBS-induced colitis. *Journal of Functional Foods*, **64**: 103644.

McLusky, S.R., Bennett, M.H., Beale, M.H., et al. (1999). Cell wall alterations and localized accumulation of feruloyl-3'-methoxytyramide in onion epidermis at sites of attempted penetration by *Botrytis allii* are associated with action polarisation, peroxidase activity and suppression of flavonoid biosynthesis. *The Plant Journal*, **17**: 523–534.

Melini, V. and Acquistucci, R. (2017). Extraction of free and insoluble-bound phenolic compounds from pigmented rice by commonly used procedures: a comparative study. *Journal of Food Measurement and Characterization*, **11**: 2151–2159.

Mercado-Mercado, G., Blancas-Benitez, F.J., Velderrain-Rodríguez, G.R., et al. (2015). Bioaccessibilty of polyphenols released and associated to dietary fibre in calyces and decoction residues of Roselle (*Hibiscus sabdariffa* L.). *Journal of Functional Foods*, **18**: 171–181.

Monagas, M., Quintanilla-López, J.E., Gómez-Cordoves, C., et al. (2010). MALDI-TOF MS analysis of plant proanthocyanidins. *Journal of Pharmaceutical and Biomedical Analysis*, **20**: 358–372.

Montero, L., Sáez, V., von Baer, D., et al. (2018). Profiling of *Vitis vinifera* L. canes (poly) phenolic compounds using comprehensive two-dimensional liquid chromatography. *Journal of Chromatography A*, **1536**: 205–215.

Mudenuti, N.V.R., de Camargo, A.C., Shahidi, F., et al. (2018). Soluble and insoluble-bound fractions of phenolic and alkaloids and their antioxidant activities in raw and traditional chocolate: a comparative study. *Journal of Functional Foods*, **50**: 164–171.

Mushtaq, M., Sultana, B., Bhatti, H.N., and Asghar, M. (2015). RSM based optimized enzyme-assisted extraction of antioxidant phenolics from underutilized watermelon (*Citrus lanatus* Thunb.) rind. *Journal of Food Science and Technology*, **52**: 5048–5056.

Mushtaq, M., Sultana, B., Akram, S., et al. (2017). Enzyme-assisted supercritical fluid extraction: an alternative and green technology for non-extractable polyphenols. *Analytical and Bioanalytical Chemistry*, **409**: 3465–3455.

Nagar, E.E., Okun, Z., and Shpigelman, A. (2020). Digestive fate of polyphenols: updated view of the influence of chemical structure and the presence of cell wall material. *Current Opinion in Food Science*, **31**: 38–46.

Neilson, A.P., O'Keefe, S.F., and Bolling, B.W. (2016). High-molecular-weight proanthocyanidins in foods: overcoming analytical challenges in pursuit of novel dietary bioactive components. *Annual Reviews in Food Science and Technology*, **7**: 43–64.

Nemer, A., Szóllósi, E., Stündl, L., et al. (2018). Determination of flavonoid and proanthocyanidin profile of Hungarian sour cherry. *Molecules*, **23**: 3278.

Nicholson, R.L. and Hammerschmidt, R. (1992). Phenolic compounds and their role in disease resistance. *Annual Review of Phytopathology*, **30**: 369–389.

Niño-Medina, G., Muy-Rangel, D., and Urías-Orona, V. (2017). Chickpea (*Cicer arietinum*) and soybean (*Glycine max*) hulls: byproducts with potential use as a source of high value-added food products. *Waste and Biomass Valorization*, **8**: 1199–1203.

Oracz, J., Nebesny, E., and Zyzelewicz, D. (2018). Identification and quantification of free and bound phenolic compounds contained in the high-molecular weight melanoidin fractions derived from two different types of cocoa beans by UHPLC-DAD-ESI-HR-MS^n. *Food Research International*, **115**: 135–149.

Oriente, M., Díaz-de-Cerio, E., Verardo, V., et al. (2020). Assessment of phytochemical compounds in functional couscous: determination of free and bound phenols and alkylresorcinols. *Food Research International*, **130**: 108970.

Oszmianski, J. and Bourzeix, M. (1996). Comparison of methods for determining the content and polymerization of proanthocyanidins and catechins. *Polish Journal of Food Nutrition Science*, **5**: 43–50.

Ou, J.Y., Huang, J.Q., Song, Y., et al. (2016). Feruloylated oligosaccharides from maize bran modulated the gut microbiota in rats. *Plant Foods for Human Nutrition*, **71**: 123–128.

Ou, J., Wang, M., Zheng, J., and Ou, S. (2019). Positive and negative effects of polyphenol incorporation in baked foods. *Food Chemistry*, **284**: 90–99.

Pang, Y., Ahmed, S., Xu, Y., et al. (2018). Bound phenolic compounds and antioxidant properties of whole grain and bran of white, red and black rice. *Food Chemistry*, **240**: 212–221.

Pereira, G.A., Arruda, H.S., de Morais, D.R., et al. (2020). Mutamba (*Gazuma ulmifolia* Lam.) fruit as a novel source of dietary fibre and phenolic compounds. *Food Chemistry*, **310**: 125857.

Pérez-Jiménez, J. and Saura-Calixto, F. (2005). Literature data may underestimate the actual antioxidant capacity of cereals. *Journal of Agriculture and Food Chemistry*, **53**: 5036–5040.

Pérez-Jiménez, J. and Saura-Calixto, F. (2015). Macromolecular antioxidants or non-extractable polyphenols in fruit and vegetables: intake in four European Countries. *Food Research International*, **74**: 315–323.

Pérez-Jiménez, J. and Saura-Calixto, F. (2018). Fruit peels as sources of non-extractable polyphenols or macromolecular antioxidants: analysis and nutritional implications. *Food Research International*, **111**: 148–152.

Pérez-Jiménez, J. and Torres, J.L. (2011). Analysis of non-extractable phenolic compounds in foods: the current state of the art. *Journal of Agricultural and Food Chemistry*, **59**: 12713–12724.

Pérez-Jiménez, J., Serrano, J., Tabernero, M., et al. (2008). Effects of grape antioxidant dietary fiber on cardiovascular disease risk factors. *Nutrition*, **24**: 646–653.

Pérez-Jiménez, J., Arranz, S., and Saura-Calixto, F. (2009a). Proanthocyanidin content in foods is largely underestimated in the literature data. An approach to quantification of the missing proanthocyanidins. *Food Research International*, **42**: 1381–1388.

Pérez-Jiménez, J., Serrano, J., Tabernero, M., et al. (2009b). Bioavailability of phenolic antioxidants associated with dietary fiber: plasma antioxidant capacity after acute and long-term intake in humans. *Plant Foods for Human Nutrition*, **64**: 102–107.

Pérez-Jiménez, J., Diaz-Rubio, ME., Saura-Calixto, F. (2013). Non-extractable polyphenols, a major dietary antioxidant: Occurrence, metabolic fate and health effects. *Nutrition Research Reviews*, **26**: 118–129.

Pérez-Ramírez, I.F., Reynoso-Camacho, R., Saura-Calixto, F., and Pérez-Jiménez, J. (2018). Comprehensive characterization of extractable and non-extractable phenolic compounds by high-performance liquid chromatography-electrospray ionization-quadrupole time of-flight of a grape/pomegranate pomace dietary supplement. *Journal of Agricultural and Food Chemistry*, **66**: 661–673.

Phan, A.D.T., Williams, B.A., Netzel, G., et al. (2020). Independent fermentation and metabolism of dietary polyphenols associated with a plant cell wall model. *Food & Function*, **11**: 2218–2230.

Pinaffi, A.C.C., Samapio, G.R., Soares, M.J., et al. (2020). Insoluble-bound polyphenols released from guarana powder: inhibition of alpha-glucosidase and proanthocyanidin profile. *Molecules*, **25**: 679.

Pinto, P. and Santos, C.N. (2017). Worldwide (poly)phenol intake: assessment methods and identified gaps. *European Journal of Nutrition*, **56**: 1393–1408.

Plumb, J., Durazzo, A., Lucarini, M., et al. (2020). Extractable and non-extractable antioxidants compositions in the eBASIS database: a key tool for dietary assessment in human health and disease research. *Nutrients*, **12**: 3405.

Poovarodom, S., Haruenkit, R., Vearasilp, S., et al. (2010). Comparative characterization of durian, mango and avocado. *International Journal of Food Science & Technology*, **45**: 921–929.

Porter, L.J., Hrstich, L.N., and Chan, B.G. (1985). The conversion of procyanidins and prodelphinidin to cyanidin and delphinidin. *Phytochemistry*, **25**: 223–230.

Pourová, J., Najmanová, I., Vopršalová, M., et al. (2018). Two flavonoid metabolites, 3, 4-dihydroxyphenylacetic acid and 4-methylcatechol, relax arteries *ex vivo* and decrease blood pressure *in vitro*. *Vascular Pharmacology*, **111**: 36–43.

Powell, C., Clifford, M.N., Opie, S.C., and Gibson, C.L. (1995). Use of Porter's reagents for the characterization of thearubigins and other non-proanthocyanidins. *Journal of the Science of Food and Agriculture*, **68**: 33–38.

Pravinata, L.C. and Murray, B.S. (2019). Encapsulation of water-insoluble polyphenols and β-carotene in Ca-alginate microgel particles produced by the Leeds Jet Homogenizer. *Colloids and Surfaces A: Physicochemical and Engineering Aspects*, **561**: 147–154.

Prior, R.L., Wu, X., and Schaich, K. (2005). Standardized methods for the determination of antioxidant capacity and phenolics in foods and dietary supplements. *Journal of Agricultural and Food Chemistry*, **53**: 4290–4302.

Prior, R.L., Fan, E., Ji, H., et al. (2010). Multi-laboratory validation of a standard method for quantifying proanthocyanidins in cranberry powders. *Journal of the Science of Food and Agriculture*, **90**: 1473–1478.

Quatrin, A., Pauletto, R., Maurer, L.H., et al. (2019). Characterization and quantification of tannins, flavonols, anthocyanins and matrix-bound polyphenols from jaboticaba fruit peel: a comparison between *Myrciaria trunciflora* and *M. jaboticaba*. *Journal of Food Composition and Analysis*, **78**: 59–74.

Ramírez-Bolaños, S., Pérez-Jiménez, J., Díaz, S., and Robaina, L. (2021). A potential of banana flower and pseudo-stem as novel ingredients rich in phenolic compounds. *International Journal of Food Science & Technology*, **56**: 5601–5608.

Rasera, G.B., Hilkner, M.H., and de Castro, R.J.S. (2020). Free and insoluble-bound phenolics: how does the variation of these compounds affect the antioxidant properties of mustard grain during germination? *Food Research International*, **133**: 109115.

Reynoso-Camacho, R., Rufino, M.S.M., Amaya-Cruz, D.M., and Pérez, A.M. (2018). Non-extractable polyphenols in tropical fruits: occurrence and health-related properties. In: Non-Extractable Polyphenols and Carotenoids: Importance in Human Nutrition and Health, (eds. Saura-Calixto, F. and Pérez-Jiménez, J.), 88–110. London: Royal Society of Chemistry.

Rocchetti, G., Senizza, B., and Putnik, P. (2019). Untargeted screening of the bound/free phenolic composition in tomato cultivars for industrial transformation. *Journal of the Science of Food and Agriculture*, **99**: 6173–6181.

Rodríguez-González, S., Pérez-Ramírez, I.F., Amaya-Cruz, D.M., et al. (2018). Polyphenol-rich peach (*Prunus persica* L.) by-product exerts a greater beneficial than dietary fiber-rich by product on insulin resistance and hepatic steatosis in obese rats. *Journal of Functional Foods*, **45**: 58–66.

Rondini, L., Peyrat-Maillard, M.N., Marsset-Baglieri, A., et al. (2004). Bound ferulic acid from bran is more bioavailable than the free compound in rat. *Journal of Agricultural and Food Chemistry*, **52**: 4338–4343.

Ross, K.A., Beta, T., and Arntfield, S.D. (2009). A comparative study on the phenolic acids identified and quantified in dry beans using HPLC as affected by different extraction and hydrolysis methods. *Food Chemistry*, **133**: 336–344.

Rufino, M.D.S., Peréz-Jiménez, J., Tabernero, M., et al. (2010). Acerola and cashew apple as sources of antioxidants and dietary fibre. *International Journal of Food Science and Technology*, **45**: 2227–2233.

Rufino, M.D.S., Pérez-Jiménez, J., Arranz, S., et al. (2011). Açaí (*Euterpe oleraceae*) 'BRS Pará': a tropical fruit source of antioxidant dietary fiber and high antioxidant capacity oil. *Food Research International*, **44**: 2100–2106.

Sánchez-Tena, S., Lizárraga, D., Miranda, A., et al. (2013). Grape antioxidant dietary fiber inhibits intestinal polyposis in Apc Min/+ mice: relation to cell cycle and immune response. *Carcinogenesis*, **34**: 1881–1888.

Sanz-Pintos, N., Pérez-Jiménez, J., Buschmann, A.H., et al. (2017). Macromolecular antioxidants and dietary fiber in edible seaweeds. *Journal of Food Science*, **85**: 289–295.

Saura-Calixto, F. (2011). Dietary fiber as a carrier of dietary antioxidants: an essential physiological function. *Journal of Agricultural and Food Chemistry*, **59**: 43–49.

Saura-Calixto, F. (2012). Concept and health-related properties of non-extractable polyphenols: the missing dietary polyphenols. *Journal of Agricultural and Food Chemistry*, **60**: 11195–12000.

Saura-Calixto, F. (2018). The story of the introduction of non-extractable polyphenols into polyphenol research: origin, development and perspectives. In: Non-Extractable Polyphenols and Carotenoids: Importance in Human Nutrition and Health, (eds. Saura-Calixto, F. and Pérez-Jiménez, J.), 1–16. London, U.K.: Royal Society of Chemistry.

Saura-Calixto, F. and Pérez-Jiménez, J. (eds.). (2018). Non-Extractable Polyphenols and Carotenoids: Importance in Human Nutrition and Health. London, U.K: Royal Society of Chemistry.

Saura-Calixto, F., García-Alonso, A., Goñi, I., and Bravo, L. (2000). in vitro determination of the indigestible fraction in foods: an alternative to dietary fiber analysis. *Journal of Agricultural and Food Chemistry*, **48**: 42–47.

Saura-Calixto, F., Serrano, J., and Goni, I. (2007). Intake and bioaccessibility of total polyphenols in a whole diet. *Food Chemistry*, **101**: 492–501.

Saura-Calixto, F., Pérez-Jiménez, J., Touriño, S., et al. (2010). Proanthocyanidin metabolites associated with dietary fibre from in vitro colonic fermentation and proanthocyanidin metabolites in human plasma. *Molecular Nutrition & Food Research*, **54**: 939–946.

Sáyago-Ayerdi, S.G., Arranz, S., Serrano, J., and Goñi, I. (2007). Dietary fiber and associated antioxidant compounds in roselle flower (*Hibiscus sabdariffa* L.) beverage. *Journal of Agricultural and Food Chemistry*, **55**: 7886–7890.

Sáyago-Ayerdi, S.G., Mercado-Mercado, G., Ramos-Romero, S., et al. (2016). Analysis and characterization of polyphenols extracts. In: Advances in Technologies for Producing Food-Relevant Polyphenols, (eds. Cuevas-Valenzuela, J., Vergara-Salinas, J.R., and Pérez-Correa, J.R.), 193–220. New York, U.S.A: CRC Press. Taylor & Francis Group.

Scalbert, A., Monties, B., and Janin, G. (1989). Tannins in wood: comparison of different estimation methods. *Journal of Agricultural and Food Chemistry*, **37**: 1324–1329.

Schofield, P., Mbugua, D.M., and Pell, A.N. (2001). Analysis of condensed tannins: a review. *Animal Feed Science and Technology*, **91**: 21–40.

Shi, J., Shan, S., Li, H., et al. (2017). Anti-inflammatory effects of millet bran derived-bound polyphenols in LPS-induced HT29 cell via ROS/miR-149/Akt/NF-κB signaling pathway. *Oncotarget*, **8**: 74582–74594.

Shi, J., Shan, S., Zhou, G., et al. (2019). Bound polyphenol from foxtail millet bran exhibits an antiproliferative activity in HT-29 cells by reprogramming miR-149-mediated aerobic glycolysis. *Journal of Functional Foods*, **56**: 246–254.

Silván, J.M., Morales, F.J., and Saura-Calixto, F. (2010). Conceptual study on maillardized dietary fiber in coffee. *Journal of Agricultural and Food Chemistry*, **58**: 12244–12249.

Singleton, V.L., Orthofer, R., Lamuela-Raventos, R.M. (1999). Analysis of total phenols and other oxidation substrates and antioxidants by means of folin-ciocalteu reagent. *Methods in Enzymology*, **299**: 152–178.

Solari-Godiño, A., Pérez-Jiménez, J., Saura-Calixto, F., et al. (2017). Anchovy mince (*Engraulis ringens*) enriched with polyphenol-rich grape pomace dietary fibre: *in vitro* polyphenols bioaccessibility, antioxidant and physico-chemical properties. *Food Research International*, **102**: 639–646.

Su, D., Zhang, R., Hou, F., et al. (2014). Comparison of the free and bound phenolic profiles and cellular antioxidant activities antioxidant activities of litchi pulp extracts from different solvents. *BMC Complementary and Alternative Medicine*, **14**: 9.

Szczepańska, J., Barba, F.J., Skapska, S., and Marszalek, K. (2020). High pressure processing of carrot juice: effect of static and multi-pulsed pressure on the polyphenolic profile, oxidoreductase activity and color. *Food Chemistry*, **307**: 125549.

Tabernero, M., Serrano, J., and Saura-Calixto, F. (2007). Dietary fiber intake in two European diets with high (Copenhagen, Denmark) and low (Murcia, Spain) colorectal cancer incidence. *Journal of Agricultural and Food Chemistry*, **55**: 9443–9449.

Takahama, U., Hirota, S., and Yanase, E. (2019). Slow starch digestion in the rice cooked with adzuki bean: contribution of procyanidins and the oxidation products. *Food Research International*, **119**: 187–195.

Takahama, U., Hirota, S., and Morina, F. (2020). Procyanidins in rice cooked with adzuki bean and their contribution to the reduction of nitrite to nitric oxide (•NO) in artificial gastric juice. *International Journal of Food Sciences and Nutrition*, **71**: 63–73.

Tang, Y., Zhang, B., Li, X., et al. (2016). Bound phenolics of quinoa seeds released by acid, alkaline, and enzymatic treatments and their antioxidant and α-glucosidase and pancreatic lipase inhibitory effects. *Journal of Agricultural and Food Chemistry*, **64**: 1712–1719.

Tarascou, I., Souquet, J.M., Mazauric, J.P., et al. (2010). The hidden face of food phenolic composition. *Archives of Biochemistry and Biophysics*, **501**: 16–22.

Terrill, T.H., Rowan, A.M., Douglas, G.B., and Barry, T.N. (1992). Determination of extractable and bound condensed tannin concentrations in forage plants, protein concentrate meals and cereal grains. *Journal of the Science of Food and Agriculture*, **58**: 321–329.

Touriño, S., Pérez-Jiménez, J., Mateos-Martín, M.L., et al. (2011). Metabolites in contact with the rat digestive tract after ingestion of a phenolic-rich dietary fiber matrix. *Journal of Agricultural and Food Chemistry*, **59**: 5955–5963.

Tow, W.W., Premier, R., Jing, H., and Ajilouni, S. (2011). Antioxidant and antiproliferation of extractable and non-extractable polyphenols isolated from apple waste using different extraction methods. *Journal of Food Science*, **76**: T163–T172.

Urquiaga, I., D'Acuña, S., Pérez, D., et al. (2015). Wine grape pomace flour improves blood pressure, fasting glucose and protein damage in humans: a randomized controlled trial. *Biological Research*, **48**: 49.

Velderrain, G., Quiros-Sauceda, A., Mercado-Mercado, G., et al. (2016). Effect of dietary fiber on the bioaccessibility of phenolic compounds of mango, papaya and pineapple fruits by an *in vitro* digestion model. *Food Science and Technology*, **36**: 188–194.

Verardo, V., Serea, C., Segal, R., and Caboni, M.F. (2011). Free and bound minor polar compounds in oats: different extraction methods and analytical determinations. *Journal of Cereal Science*, **54**: 211–217.

Veronica, S., Lilia, N., Giampiero, S., et al. (2019). Response of organic and conventional apples to freezing and freezing pre-treatment: focus on polyphenols content and antioxidant activity. *Food Chemistry*, **308**: 125570.

Viacava, F., Santana-Gálvez, J., Heredia-Olea, E., et al. (2020). Sequential application of postharvest wounding stress and extrusion as an innovative tool to increase the concentration of free and bound phenolics in carrots. *Food Chemistry*, **307**: 125551.

Vitaglione, P., Mennella, I., Ferracane, R., et al. (2015). Whole-grain wheat consumption reduces inflammation in a randomized controlled trial on overweight and obese subjects with unhealthy dietary and lifestyle behaviors: role of polyphenols bound to cereal dietary fiber. *The American Journal of Clinical Nutrition*, **101**: 251–261.

Wang, L., Lin, X., Zhang, J., et al. (2019a). Extraction methods for the releasing of bound phenolics from *Rubus idaeus* L. leaves and seeds. *Industrial Crops and Products*, **135**: 1–9.

Wang, S., Sarriá, B., Mateos, R., et al. (2019b). TNF-α-induced oxidative stress and endothelial dysfunction in EA. hy926 cells is prevented by mate and green coffee extracts, 5-caffeoylquinic acid and its microbial metabolite, dihydrocaffeic acid. *International Journal of Food Sciences and Nutrition*, **70**: 267–284.

Wang, Z., Li, S., Ge, S., and Lin, S. (2020). Review of distribution, extraction methods, and health benefits of bound phenolics in food plants. *Journal of Agricultural and Food Chemistry*, **68**: 3330–3343.

White, B.L., Howard, L.R., and Prior, R.L. (2010). Release of bound procyanindins from cranberry pomace by alkaline hydrolysis. *Journal of Agricultural and Food Chemistry*, **58**: 7572–7579.

Wu, T., Phuong, N.N.M., Van Camp, J., et al. (2018). Analysis of Non-extractable polyphenols (NEPP). In: Non-Extractable Polyphenols and Carotenoids: Importance in Human Nutrition and Health, (eds. Saura-Calixto, F. and Pérez-Jiménez, J.), 1–16. London, UK: Royal Society of Chemistry.

Xia, X., Zhu, L., Lei, Z., et al. (2019). Feruloylated oligosaccharides alleviate dextran sulfate sodium-induced colitis in vivo. *Journal of Agricultural and Food Chemistry*, **67**: 9522–9531.

Xie, L., Roto, A.V., and Bolling, B.W. (2012). Characterization of ellagitannins, gallotannins, and bound proanthocyanidins from California almond (*Prunus dulcis*) varieties. *Journal of Agricultural and Food Chemistry*, **60**: 12151–11256.

Yan, L. and Zhen, G. (2017). Comparing profiles and antioxidant properties of soluble and insoluble phenolics in *Perilla frutescens* seed flour extracts obtained by different extraction/hydrolysis methods. *International Journal of Food Science and Technology*, **52**: 2497–2504.

Yang, X.J., Dang, B., and Fan, M.T. (2018). Free and bound phenolic compound content and antioxidant activity of different cultivated blue highland barley varieties from the Qinghai-Tibet Plateau. *Molecules*, **23**: 879.

Yeo, J. and Shahidi, F. (2015). Critical evaluation of changes in the ratio of insoluble bound to soluble phenolics on antioxidant activity of lentils during germination. *Journal of Agricultural and Food Chemistry*, **63**: 379–381.

Yeo, J. and Shahidi, F. (2020). Identification and quantification of soluble and insoluble-bound phenolics in lentil hulls using HPLC-ESI-MS/MS and their antioxidant potential. *Food Chemistry*, **315**: 126202.

Yu, J., Vasanthan, T., and Temelli, F. (2001). Analysis of phenolic acids in barley by high-performance liquid chromatography. *Journal of Agricultural and Food Chemistry*, **49**: 4352–4358.

Zardo, I., Rodriguez, N.P., Sarkis, J.R., et al. (2019). Extraction and identification by mass spectrometry of phenolic compounds from seed cake. *Journal of the Science of Food and Agriculture*, **100**: 578–586.

Zurita, J., Díaz-Rubio, M.E., and Saura-Calixto, F. (2012). Improved procedure to determine non-extractable polymeric proanthocyanidns in plant foods. *International Journal of Food Science and Nutrition*, **63**: 936–939.

8

Template-Mediated Engineering of Functional Metal–Phenolic Complex Coatings

Steve Spoljaric, J.J. Richardson, Yi Ju, and Frank Caruso

Department of Chemical Engineering, The University of Melbourne, Parkville, VIC, Australia

8.1 Introduction

Over the years, there has been interest in the development of surface coatings with diverse functional moieties, such as thiols (Harris et al. 2001; Gevrek et al. 2017; Resetco et al. 2017), silanes (Haensch et al. 2010; Karakoy et al. 2014), citrate (Rani et al. 2016), and phosphonic acid (Freese et al. 2012; Queffélec et al. 2012), that adhere to a broad class of substrates. These surface functionalities can be incorporated via a range of surface modification strategies (Sperling and Parak 2010); however, they require specific interactions with the substrate to achieve successful linkage, limiting their applicability. Therefore, there has been growing interest in coating/surface modification techniques that are independent of substrate surface chemistry. Two main approaches that are independent of substrate composition and afford surface-adherent films are polydopamine films formed through self-polymerization of dopamine (Lee et al. 2007) and metal ion–polyphenol networks (which are the focus of this chapter) assembled through the deposition of polyphenols and metal ions (Ejima et al. 2013).

In 2013, a method for surface coating based on mixing naturally occurring polyphenols (tannic acid, TA) and metal ions (Fe^{III}) in the presence of a substrate was introduced (Figure 8.1) (Ejima et al. 2013). The adsorption of the polyphenols onto the substrate and simultaneous crosslinking of TA by Fe^{III} leads to amorphous film formation, with the adjacent hydroxyl groups of TA providing chelating sites for Fe^{III} and a large number of gallol groups on TA, allowing for coordination-driven crosslinking, resulting in a stable, three-dimensional framework termed metal–phenolic networks (MPNs). The formation of MPNs is achieved on a range of planar, inorganic, organic, and biological templates, with subsequent template dissolution facilitating the preparation of microcapsules. Film deposition occurs in seconds/minutes, with thicker layers also being possible by using a multistep process, which consists of alternating immersion of the substrate into metal and polyphenol solutions (Rahim et al. 2014). Both Fe^{III} and TA are inexpensive and readily available;

Recent Advances in Polyphenol Research, Volume 8, First Edition. Edited by Juha-Pekka Salminen, Kristiina Wähälä, Victor de Freitas, and Stéphane Quideau.

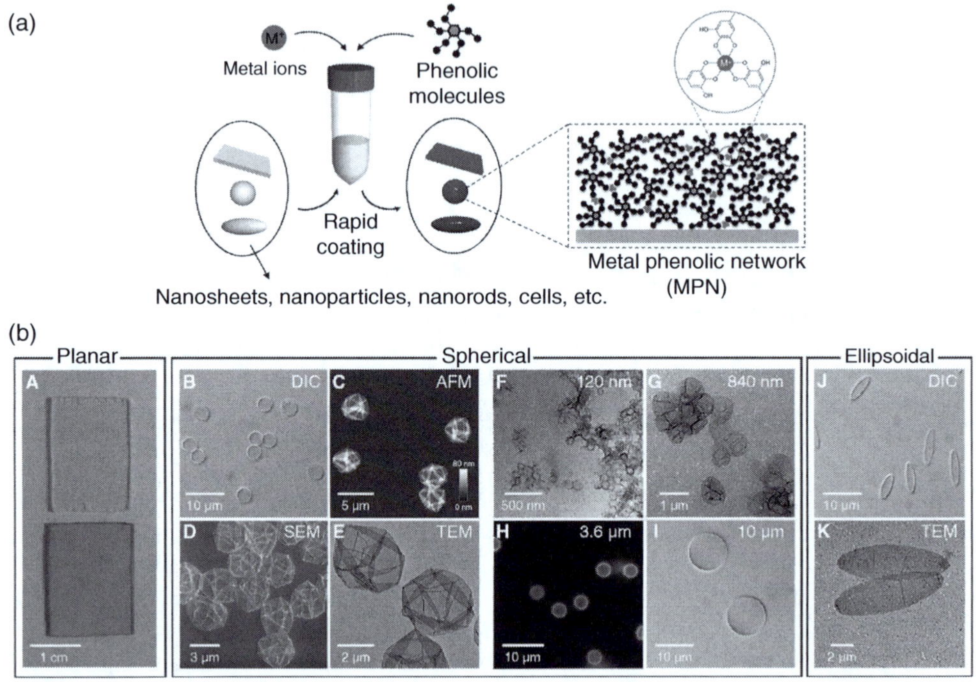

Figure 8.1 (a) Overview of MPN formation and deposition via discrete assembly. *Source:* Ejima et al. (2017), graphical abstract. Reproduced with permission of Elsevier. (b) Fe^{III}-TA films prepared on PS substrates with various shapes (planar, spherical, and ellipsoidal) and sizes (120 nm to 10 mm). Photographs of PS slides before (top) and after (bottom) Fe^{III}-TA coating (A); and differential interference contrast (DIC) images (B, I, J), AFM images (C), TEM images (E, F, G, K), SEM image (D), and fluorescence microscopy image (H) of Fe^{III}-TA capsules. *Source:* Ejima et al. (2013), Reproduced with permission from American Association for the Advancement of Science. (*See insert for color representation of the figure.*)

TA possesses the additional benefit of being renewable and sustainable as it can be extracted from various plant sources.

In the years following the introduction of MPNs, their use has greatly expanded, with a broad range of metal ion–polyphenol ligand combinations being used for coatings, as well as a range of substrates being used for templates (Table 8.1). The breadth of precursor combinations, coupled with the option to adopt different fabrication routes, has expanded the physical characteristics and intrinsic properties that MPNs can possess. This synergy of inexpensive, readily available, and sustainable precursors, with simple, efficient, and rapid fabrication methodologies, and the bespoke nature of MPN characteristics and properties have encouraged the applicability and relevance of MPNs in diverse areas, including but not limited to drug delivery, antibiofouling, biomedical imaging probes, enzyme immobilization, catalysis, functional coatings, and heavy metal ion removal (Ejima et al. 2017).

This chapter describes the techniques employed to fabricate MPNs via the use of templates or substrates (Section 8.2). The various discrete assembly and multistep approaches

Table 8.1 Summary of recent studies of MPN coatings on templates and biointerfaces.

Phenolic ligands	Metal ions	Coated objects	Applications	References
TA	Fe^{III}	PS, PLGA, MF, PDMS, SiO_2, Au, $CaCO_3$, *E. coli*, *S. epidermidis*	Coating	Ejima et al. (2013)
TA	Fe^{III}	PS particle	pH-Responsive capsules	Ejima et al. (2013) and Rahim et al. (2014)
TA	Gd^{III}/Fe^{III}, Cr^{III}/Fe^{III}	PS particle	Hybrid capsules	Ejima et al. (2013)
EGCG	Fe^{III}	PS particle	Coating	Ejima et al. (2013)
TA	Fe^{III}	Graphene oxide	Nano-wrapping	Ozawa and Haga (2015)
TA	Fe^{III}	Graphene oxide	Mechanical enhancement	Liu and Xu (2014)
TA	Fe^{III}	Nanodiamond	Photoluminescence	Bray et al. (2015)
TA, GA, EGCG	Fe^{III}	AuNP	Catalysis	Zeng et al. (2014)
TA	Fe^{III}, Mn^{III}, Gd^{III}	PS particle	MRI imaging	Guo et al. (2014)
TA	$^{64}Cu^{II}$	PS particle	PET imaging	Guo et al. (2014)
TA	Eu^{III}, Tb^{III}	PS particle	Fluorescent capsules	Guo et al. (2014)
TA	Rh^{III}	PS particle	Catalysis	Guo et al. (2014)
TA	V^{III}, Cr^{III}, Co^{II}, Ni^{II}, Cu^{II}, Zn^{II}, Zr^{IV}, Mo^{II}, Ru^{III}, Rh^{III}, Cd^{II}, Ce^{III}	PS particle	Hollow capsules	Guo et al. (2014)
TA	Fe^{III}, Ti^{IV}	$CaCO_3$ particle	Enzyme-immobilized capsules	Yang et al. (2015)
TA	Fe^{III}	$CaCO_3$ particle	Magnetic and enzyme-loaded capsules	Wang et al. (2016)
GA, PyG, PyC	Fe^{III}	PS particle	Hollow capsules	Rahim et al. (2015)
TA	Al^{III}	$CaCO_3$ particle	Drug delivery	Ping et al. (2015)
TA	Fe^{III}	Emulsion	Chlorpyrifos-loaded microcapsules	Li et al. (2016)
Lignin	Fe^{III}	Emulsion	Nanocapsules	Bartzoka et al. (2016)
TA	Fe^{III}, Al^{III}, Eu^{III}	Zein/quaternized chitosan particle	Drug delivery, cell imaging	Liang et al. (2016)

(*Continued*)

Table 8.1 (Continued)

Phenolic ligands	Metal ions	Coated objects	Applications	References
Catechol-functionalized PEG	Fe^{III}	$CaCO_3$ particle	Low-fouling and pH-degradable capsules	Ju et al. (2015)
Catechol-functionalized HA and PEG	Fe^{III}	$CaCO_3$ particle	Cancer cell targeting	Ju et al. (2016a)
Catechol-functionalized HA	Fe^{III}	$CaCO_3$ particle	Protein corona-coated MPN capsules	Ju et al. (2016b)
TA	Fe^{III}	MCM-41	pH- and glutathione-responsive release of curcumin	Kim et al. (2015e)
TA	Fe^{III}	Paclitaxel nanocore	Anticancer therapy	Shen et al. (2016)
TA	Fe^{III}	*S. cerevisiae*	Cytoprotective coating	Park et al. (2014)
TA	Fe^{III}	HeLa, NIH 3T3 fibroblast, Jurkat cells	Cytoprotective coating	Lee et al. (2015)
TA	Fe^{III}	*S. cerevisiae*, *E. coli.*, PC12 cells	Cellular surface engineering	Li et al. (2015)
TA	Fe^{III}	Brome mosaic virus	Enhanced stability, barrier transport	Delalande et al. (2016)
TA	Fe^{III}	Dentinal tubules	Dental hypersensitivity	Oh et al. (2015)
GA	Fe^{III}	Dentinal tubules	Dental hypersensitivity	Prajatelistia et al. (2016)
TA	Ti^{IV}	Chitin microsphere	Enzyme immobilization	Han et al. (2016)
DA	Ti^{IV}	No substrate used	Enzyme immobilization	Yang et al. (2014)
TA	Fe^{III}	$Fe_3O_4@SiO_2$	Protein enrichment for MALDI-TOF MS	Song et al. (2015)
TA	Fe^{III}	Polydopamine-coated glass	Adhesion of marine diatoms	Kim et al. (2015b)
TA	Fe^{III}	Polydopamine-coated surfaces	Marine antifouling coating	Kim et al. (2015a)
TA	Fe^{III}	Zwitterionic poly(MPDSAH) surface	Micropatterned adhesion of proteins and cells	Kim et al. (2015c)
Vegetable tannins	Fe^{III}	No substrate used	Dye-sensitized solar cells	Çakar et al. (2016)
TA	Fe^{III}	MOFs	Etching and surface functionalization	Hu et al. (2016)

Table 8.1 (Continued)

Phenolic ligands	Metal ions	Coated objects	Applications	References
TA	Fe^{III}	Polyamide, RO membrane	Surface modification	Wu et al. (2015)
TA	Fe^{III}	Polyamide, RO membrane	Anti-biofouling	Dong et al. (2017)
TA	Fe^{III}	Poly(ether sulfone) membrane	Heavy metal ion removal	Kim et al. (2015d)
TA	Fe^{III}	Cellulose fibers	Oxygen reduction catalyst	Wei et al. (2016)
TA	Fe^{III}	Melamine sponge	Absorbent	Huang et al. (2015)
TA, polydopamine	Fe^{III}, Al^{III}, Zr^{IV}, Ce^{III}, Zn^{II}	Numerous substrates[a]	Superstructuring of the coated objects	Guo et al. (2016)
Galloyl-functionalized PEG and Pt prodrug	Fe^{III}	Emulsion	Anticancer therapy	Dai et al. (2017)
Quercetin, fisetin, myricetin, luteolin	Fe^{III}	PS particle	Hollow capsules, antioxidants	Bertleff-Zieschang et al. (2017)
TA, EA	Fe^{III}, Zn^{II}, Cu^{II}, Co^{II}	MOFs	Catalysis, drug delivery, carbon precursors	Wang et al. (2017a)
TA	Eu^{III}	$BaGdF_5$ nanoparticles	CT, MRI, and luminescence imaging	Zhu et al. (2017)
TA	Fe^{III}	PS particle	Hollow capsules	Guo et al. (2017)
TA, GA	Fe^{III}	PS particle	Hollow capsules	Rahim et al. (2017)
TA	Cu^{II}	Mesoporous silica particle	Photoresponsive nanocontainers	Park et al. (2017a)
Pectin functionalized with catechin, quercetin, hesperidin, rutin	Fe^{III}, Ca^{II}, Zn^{II}	No substrate used	Antioxidant macromolecules	Ahn et al. (2017)
TA, GA, RA	Fe^{III}	Indium tin oxide-coated QCM electrode	Electrochemical-induced self-assembly	Maerten et al. (2017)
TA	Fe^{III}	AuNP, Fe_3O_4, CdS particle	Core–shell MOFs	Long et al. (2017)
TA, GA, Quercetin	Fe^{III}, Cu^{II}, Al^{III}, Zr^{IV}	Quartz, silicon, PC, PS, PU, PP	UV shielding	Zhong et al. (2018)

(*Continued*)

Table 8.1 (Continued)

Phenolic ligands	Metal ions	Coated objects	Applications	References
GA	Fe^{III}	SnO_2 particle	Lithium storage	Xiao et al. (2018b)
GA	Fe^{III}	PS particle	Redox-induced Fe^{III} release	Cherepanov et al. (2018)
Numerous polyphenols found in green tea	Fe^{III}	PMMA particle	Hollow capsules	Rahim et al. (2018)
TA	Fe^{III}	PS particle, polyester textile	Catalytic and antimicrobial textiles	Yun et al. (2017)
TA	Fe^{III}	Lignin colloidal particle	Environmental remediation	Tardy et al. (2018)
Galloyl-functionalized PEG	Fe^{III}	Emulsion	Low fouling, pH-responsive particles	Besford et al. (2018)
TA	Fe^{III}	PS particle, planar quartz	Origami-like folding behavior, selective degradation	Yun et al. (2018)
Myrica tannins	Ag^{I}/Ti^{IV}	Collagen fiber	Photocatalysis, hierarchical porous nanofibers via thermal transition	Xiao et al. (2018a)
TA, catechin, EGCG, procyanidin	Fe^{III}	No substrate used	Drug delivery, controlled release	Wang et al. (2018a)
TA	Fe^{III}, Ce^{IV}	SEBS film, quartz	Anti-bacterial, biofilm inhibition	Jiang et al. (2018)
TA	Fe^{III}	PCL nanofiber	Tissue engineering scaffolds modified for vascularization	Li et al. (2018b)
TA	Cu^{II}	Au- and Cu-coated QCM resonators, silicon wafer	Underwater oil super repellency	Chen et al. (2018a)
Galloyl-functionalized PEG and Pt prodrug	Fe^{III}	No substrate used	In situ production of hypochlorous acid to enhance platinum drug chemotherapy	Dai et al. (2018a)
Galloyl-functionalized PEG and Pt prodrug	Fe^{III}	No substrate used	Synergistic cancer therapy by encapsulating DOX and Pt prodrugs	Dai et al. (2018b)
Catechol-functionalized PEG	Fe^{III}	$CaCO_3$ particle	Anticancer therapy	Wei et al. (2018)
TA	Fe^{III}, Ca^{II}	Emulsion	Hollow capsules	Li et al. (2018a)

Table 8.1 (Continued)

Phenolic ligands	Metal ions	Coated objects	Applications	References
TA	Fe^{III}	Mesoporous silica particle	pH-Responsive drug delivery, MRI imaging	Chen et al. (2018b)
TA	Zr^{IV}	Poloxamer	Tumor NIFR/PET imaging	Wang et al. (2018c)
GA	Cu^{II}	Numerous substrates[b]	Therapeutic nitric oxide gas generation	Yang et al. (2018)
TA	Fe^{III}	AuNP	Chemiluminescent biosensors	Zou et al. (2018)
TA	Fe^{III}	Poly(ether sulfone) membrane	Molecular separation	Lin et al. (2018)
TA	Fe^{III}, Zn^{II}, Cu^{II}, Co^{II}, Al^{III}	No substrate used	Detection of DNA analog of mIRNA-21	Wang et al. (2018b)
TA, DA, NE	Cu^{II}	Stainless steel disc and foil, PVC film	Anti-inflammatory, antimicrobial and anticoagulant coatings	Li et al. (2018c)
TA	Fe^{III}	Silicon wafer	Salt-induced, continuous MPN deposition	Park et al. (2018)
TA	Fe^{III}	Dendrimer DOX nanocomplex	Anticancer therapy	Guo et al. (2019)
TA	Mg	Mg-2.8% Zn alloy ingots	Corrosion resistance, osteocompatibility	Asgari et al. (2019)
CC, CG	Fe^{III}	$CaCO_3$ particle, PS particle, PMMA particle, glass	Modular assembly via macrocyclic building blocks	Pan et al. (2020)
TA	Fe^{III}	Liposomes	Cell mimicry	Mkam Tsengam et al. (2019)
TA	Fe^{III}	Starch nanoparticles	Antioxidant, antimicrobial, pH sensitive coatings	Qin et al. (2019)
TA	Fe^{III}	Polyether sulfone membrane	High flux nanofiltration	Yang et al. (2019)
TA	Fe^{III}, Al^{III}	Mesoporous silica particle, PS particle, MF	Endosomal escape of nanoparticles	Chen et al. (2019a)
TA	Fe^{III}	Agarose bead suspension	Cell mimicry	Liu et al. (2019b)

(Continued)

Table 8.1 (Continued)

Phenolic ligands	Metal ions	Coated objects	Applications	References
TA, GA, EG, PyG, CH, SA	Fe^{III}	Paper, PVC, latex-painted wall, leather, cotton, nylon cloth, aluminum foil	Visual information storage	Dai et al. (2019)
TA	Fe^{II}, Fe^{III}	PEI-coated quartz and silicon wafers, PC membrane, PU sponge, PS particle	Oxidation-mediated assembly of MPNs	Zhong et al. (2019)
EGCG	Fe^{III}	Green tea catechin NP	Anticancer therapy	Chen et al. (2019b)
p(PEGMA-*co*-DMA)	Fe^{II}, Fe^{III}	Glass, silicon wafer, TiO_2, Ni–Ti alloy, PDMS, PS	Antimicrobial, antifouling	Zheng et al. (2019)
DPGG	Fe^{II}, Fe^{III}, Cu^{II}, Zn^{II}	Self-assembled liposome	Interliposomal aggregation	Park et al. (2019)
TA	Fe^{III}/Co^{II}	Fe_3O_4 NP	Catalysis	Jia et al. (2019)
TA	Fe^{III}	Halloysite nanotubes	Emulsion stabilizers, controlled release	Ojo et al. (2019)
TA, HA, DOX	Fe^{III}	No substrate used	Cancer theranostics	Liu et al. (2019a)
Quercetrin, myricetrin, gallocatechin, catechin, natural eucalyptus leaf extracts	Fe^{III}	Mesoporous silica particle, PMMA particle, quartz	Selective assembly	Lin et al. (2019)
TA	Cu^{II}	UCNP@Si-DOX	pH-Responsive drug release	Hu et al. (2019a)
GA	Cu^{II}	PVC tubing	Antibacterial, anticoagulant blood-contacting devices	Tu et al. (2019)
TA	Co^{II}	AuNP	Inhibition of amyloid fibril formation	Zhang et al. (2019)
EC	Sm^{III}	No substrate used	Anticancer therapy	Li et al. (2019a)
Acacia mearnsii bark extract	U^{IV}	Polyamide microporous membrane	Uranium extraction from seawater	Luo et al. (2019)
TA, GA, PyC, PyG	Fe^{III}, Ga^{III}, In^{III}, Tb^{III}	Planar glass	Tunable mechanical properties	Yun et al. (2019)
TA	Fe^{III}	TiO_2 deposited on a fluorine tin oxide glass	Dye-sensitized solar cells	Çakar and Özacar (2019)

Table 8.1 (Continued)

Phenolic ligands	Metal ions	Coated objects	Applications	References
TA	Zn^{II}	Fe_3O_4	Chlorophenol recovery	Hao et al. (2019)
TA, galloyl-modified (P(EtOx)-Gal)	Fe^{III}	Quartz slides, Au-coated QCM crystals	Protein adsorption	Tardy et al. (2019)
EGCG	Sm^{III}	No substrate use	Anticancer therapy	Li et al. (2019b)
TA	Fe^{III}, Cu^{II}, Al^{III}, Ti^{II}	Titanium, copper, aluminum sheet, PS sphere	Functional coatings	Kang et al. (2019)
TA	Fe^{III}	EVA/APP film	Thermal stability, flame retardancy	Hu et al. (2019b)
TA	Ru^{III}	$BiVO_4$ particle	Photocatalysis	Chen and Hu (2019)
TA	Fe^{III}	Proteins, lipids, nucleic acid, polysaccharide, fingerprints, human blood	Selective deposition	Yun et al. (2020)
TA	Fe^{III}	$CaCO_3$ particle	Controlled pulmonary deposition	Ju et al. (2020)
TA	Fe^{III}	Polyamide membrane	Improved chlorine resistance of nanofiltration membranes	Zhu et al. (2020)

TA, tannic acid; PS, polystyrene; PLGA, poly(lactic-*co*-glycolic acid); MF, melamine formaldehyde; PDMS, polydimethylsiloxane; *S. epidermidis*, *Staphylococcus epidermidis*; EGCG, epigallocatechin gallate; NP, nanoparticle; MRI, magnetic resonance imaging; PET, positron emission tomography; GA, gallic acid; PyG, pyrogallol; PyC, pyrocatechol; PEG, polyethylene glycol; HA, hyaluronic acid; MCM-41, Mobil crystalline material 41; *S. cerevisiae*, *Saccharomyces cerevisiae*; DA, dopamine; MALDI-TOF MS, matrix-assisted laser desorption/ionization time-of-flight mass spectrometry; 3-(methacryloylamino)propyl-dimethyl(3-sulfopropyl)ammonium hydroxide (MPDSAH); MOFs, metal–organic frameworks; RO, reverse osmosis; CT, computed tomography; RA, rosmarinic acid; QCM, quartz crystal microbalance; PC, polycarbonate; PU, polyurethane; PP, polypropylene; EA, ellagic acid; NE, norepinephrine; PVC, polyvinyl chloride; DOX, doxorubicin; CC, cyclodextran catechol; CG, cyclodextrin galloyl; PMMA, poly(methyl methacrylate); EG, ethyl gallate; CH, (+)-catechin hydrate; SA, methylsalicylic acid; p(PEGMA-*co*-DMA), poly(ethylene glycol)methyl methacrylate-*co*-dopamine methacrylamide; DPGG, 1,2-dipalmitoyl-*sn*-glycero-3-galloyl; UCNP, mesoporous silica-coated upconversion nanoparticles; EC, (−)-epicatechin; P(EtOx)-Gal, poly(2-ethyl-2-oxazoline); EVA/APP, ethylene vinyl acetate/ammonium polyphosphate.

[a] Substrates include spherical particles (PS, amino-functionalized SiO_2, magnetic SiO_2, mesoporous SiO_2, MF, $NaYF_4$:Yb/Er, Ag), nanowires (ZnO, AgCN), nanorods (SiO_2, β-FeOOH), $Ni[Ni(CN)_4]$ polygons, Prussian blue cubic, human microvascular endothelial cell line-1.

[b] Substrates include 316 L stainless steel cardiovascular stent and wafer, mesoporous silica nanoparticles, Fe_3O_4 nanoparticles, polyethylene terephthalate wafers, silicon, TiO_2-coated silicon, glass.

Source: Adapted from Ejima et al. (2017), table 1. Elsevier.

are examined (Section 8.2), with the influence of these different approaches on subsequent MPN properties and characteristics detailed (Section 8.3). The interactions between MPNs and various surfaces and biointerfaces are discussed (Section 8.4), along with the applications that arise from these. Finally, the practical considerations and challenges associated with the large-scale synthesis of MPNs (Section 8.5) and process automation (Section 8.6) are highlighted, followed by a summary of the field and a future outlook (Section 8.7).

8.2 Template-Mediated Techniques to Deposit MPNs

MPNs can be prepared using various techniques designed specifically with the unique inherent physicochemical nature of the MPN precursors and/or adopted from other thin film fields (Ejima et al. 2017; Rahim et al. 2019). MPNs deposit on substrates because polyphenols can interact with a range of surfaces via electrostatics, hydrogen bonding, hydrophobic forces, chelation, and other forces (Rahim et al. 2019). Like metal–organic materials made with other ligands, MPNs also form complexes in solution but the affinity of phenolic ligands for surfaces explains their unique ability to form films on different substrates. Owing to the generally rapid formation of MPNs, many techniques adopt both physical and chemical routes to control film formation (Rahim et al. 2019). The formation of MPNs was first reported using a "discrete method" where solutions of Fe^{III} and TA are simply mixed in the presence of substrates (Figures 8.2 and 8.3) (Ejima et al. 2013; Guo et al. 2014). In this discrete assembly method, interfacial assembly rapidly terminates (<60 seconds) but can be repeated to grow films in discrete increments of 10 nm. The thickness of discretely assembled MPNs can be modulated by the ionic strength of the precursor solutions, where salts can shield the chelation complex, yielding rougher and thicker films (Guo et al. 2017). The speed and ease of this method make this a versatile approach to form functional coatings; however, when a finer degree of control over film properties (e.g. thickness and permeability) is desired, other MPN formation techniques can be used. For example, a multistep method (Rahim et al. 2014), using alternating coating/washing cycles of metal and polyphenol, respectively, can be employed to generate ~2 nm-thick layers per cycle (Figure 8.3).

There are numerous solution-based strategies for forming MPNs at interfaces other than discrete assembly, with some requiring specific starting conditions and others requiring external stimulus. For example, in a process using 10–100× higher concentrations of the metal and ligand, termed "diffusion-driven assembly," continuous growth can be achieved on various substrates (Yun et al. 2018). This allows for MPN films to be grown on the order of a micrometer over the course of weeks, with the final film permeability being significantly less than those obtained with discrete assembly. Alternatively, different metal sources can be employed to yield Fe^{III} for continuous MPN assembly to proceed. One of the earlier approaches used to achieve continuous MPN assembly involved rust as the iron source, where the phenolics in the solution chelate the iron in the rust for subsequent deposition onto the templates as MPNs (Figure 8.4) (Rahim

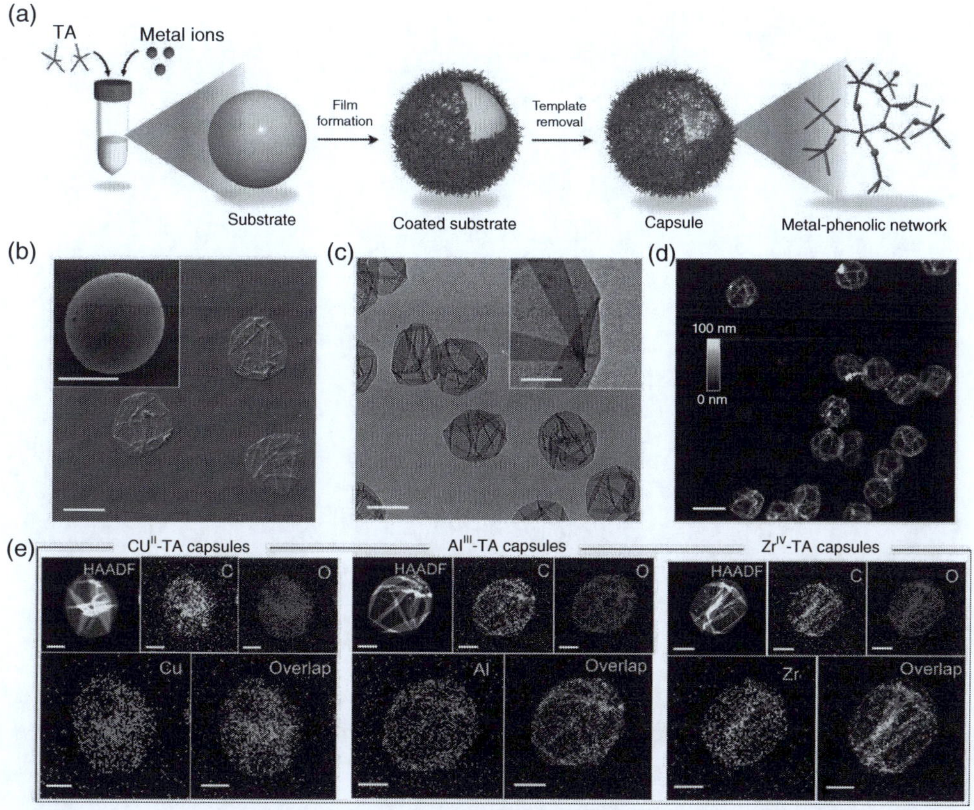

Figure 8.2 (a) Overview of discrete assembly template-mediated fabrication of MPNs and (b–d) associated microscopy images of capsules obtained following template dissolution: SEM (b), TEM (c), and AFM (d) images of Cu^{II}-TA capsules. (e) EDX elemental mapping of Cu^{II}-TA, Al^{III}-TA, and Zr^{IV}-TA capsules. Scale bars are 2 mm in (b–d), 2 mm in the inset of (b), 200 nm in the inset of (c) and 1 mm in (e). *Source:* Guo et al. (2014), Reproduced with permission John Wiley & Sons. (*See insert for color representation of the figure.*)

et al. 2017). This method yields denser films than discrete assembly and is amenable to using crude extracts of phenolics for the formation of MPNs (Cherepanov et al. 2018). Similarly, continuous growth can be achieved by using Fe^{II} and allowing it to oxidize in situ to Fe^{III} (Lee et al. 2018). In earlier studies, dissolved oxygen was used to drive oxidation, which allowed for the continuous formation of thick MPN films at the air–water interface. MPNs could be deposited on substrates either by stamping a substrate onto the MPN at the interface or by adding substrates into the solution during oxidation-mediated coordination (OMC) assembly. To expedite film formation, sonication can be used to form reactive oxygen species (ROS) to rapidly oxidize Fe^{II} to Fe^{III} (Figure 8.5), enabling a ~30× increase in the rate of film growth (Zhong et al. 2019). MPNs can also be continuously grown using an electro-triggered assembly where Fe^{II} is electrically converted into

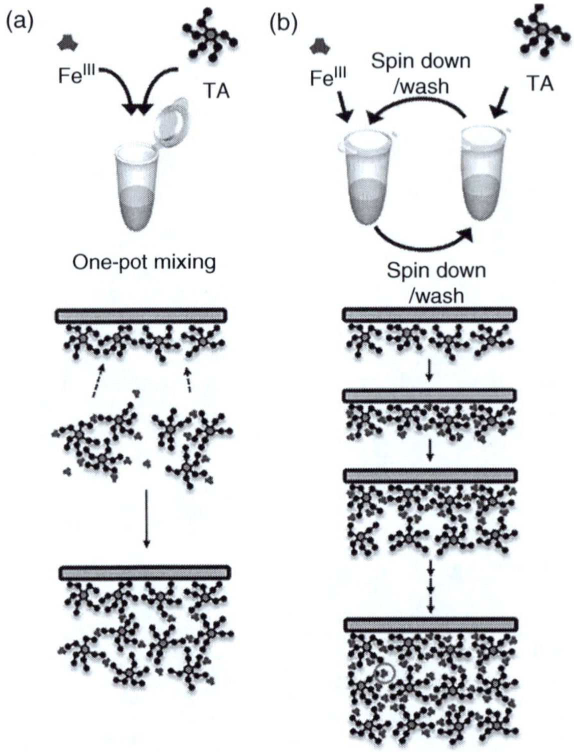

(c)

	One-step	Multistep
Time required	Seconds	Hours
Thickness/cycle	10 nm	2 nm
Modulus	1.0 GPa	0.4 GPa
Molecular Weight Cut-Off	~100 kDa	~10 kDa

Figure 8.3 Comparative overview of (a) discrete assembly and (b) multistep methods used to synthesize MPNs and (c) comparison of time-scale and properties obtained from the two methods. *Source:* Ejima et al. (2017), Reproduced with permission of Elsevier.

Fe^{III} at the electrode in the presence of TA (Maerten et al. 2017). Fe^{II} conversion and film growth can be controlled by the duration and intensity of the current and the metal/ligand ratio.

Common thin film assembly methods from other fields can also be used to form MPNs. For example, MPNs can be formed on various substrates of different length scales using spray assembly (Park et al. 2017b). This approach allows for large, industrially relevant substrates to be coated as the spray nozzle can be manipulated to cover large areas. In addition, spray assembly allows for fine control over MPN film thickness and can even be used to prepare films with gradient thicknesses (Figure 8.6a, b) (Zhong et al. 2018). Similar to spraying, MPNs can be printed with inkjets, allowing for complex patterns and images to

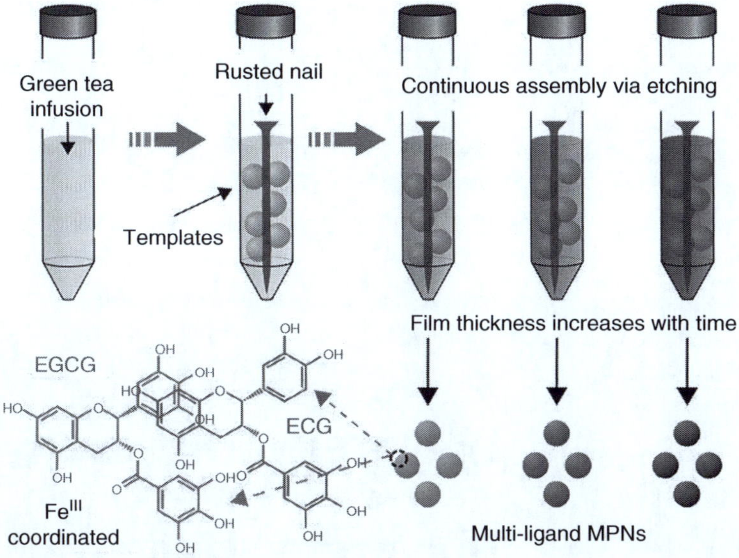

Figure 8.4 Continuous MPN assembly involving rust from a nail as an iron source. *Source:* Rahim et al. (2018), graphical abstract. Reproduced with permission of the American Chemical Society. (*See insert for color representation of the figure.*)

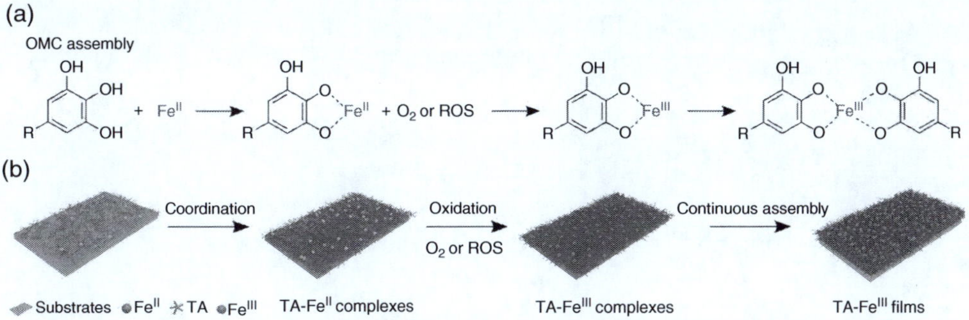

Figure 8.5 (a) Oxidation-mediated coordination (OMC) assembly employing ROS to rapidly oxidize Fe^{II} to Fe^{III} and (b) overview of MPN formation via OMC assembly on a template. *Source:* Zhong et al. (2019)/John Wiley & Sons. (*See insert for color representation of the figure.*)

be printed (Figure 8.6b) (Dai et al. 2019). Moreover, as MPNs have historically been used as inks, they can be used to write, paint, and stamp various patterns on different substrates. To prepare MPN capsules, microfluidic assembly at the interface of emulsions can be used, with the choice of metal salt and solvents being key to achieving controlled film formation in the fluidic system. Various cargos can be loaded into the inner phase (water), such as bacteria and fungi, which allows for cell encapsulation during MPN formation (Kim et al. 2017).

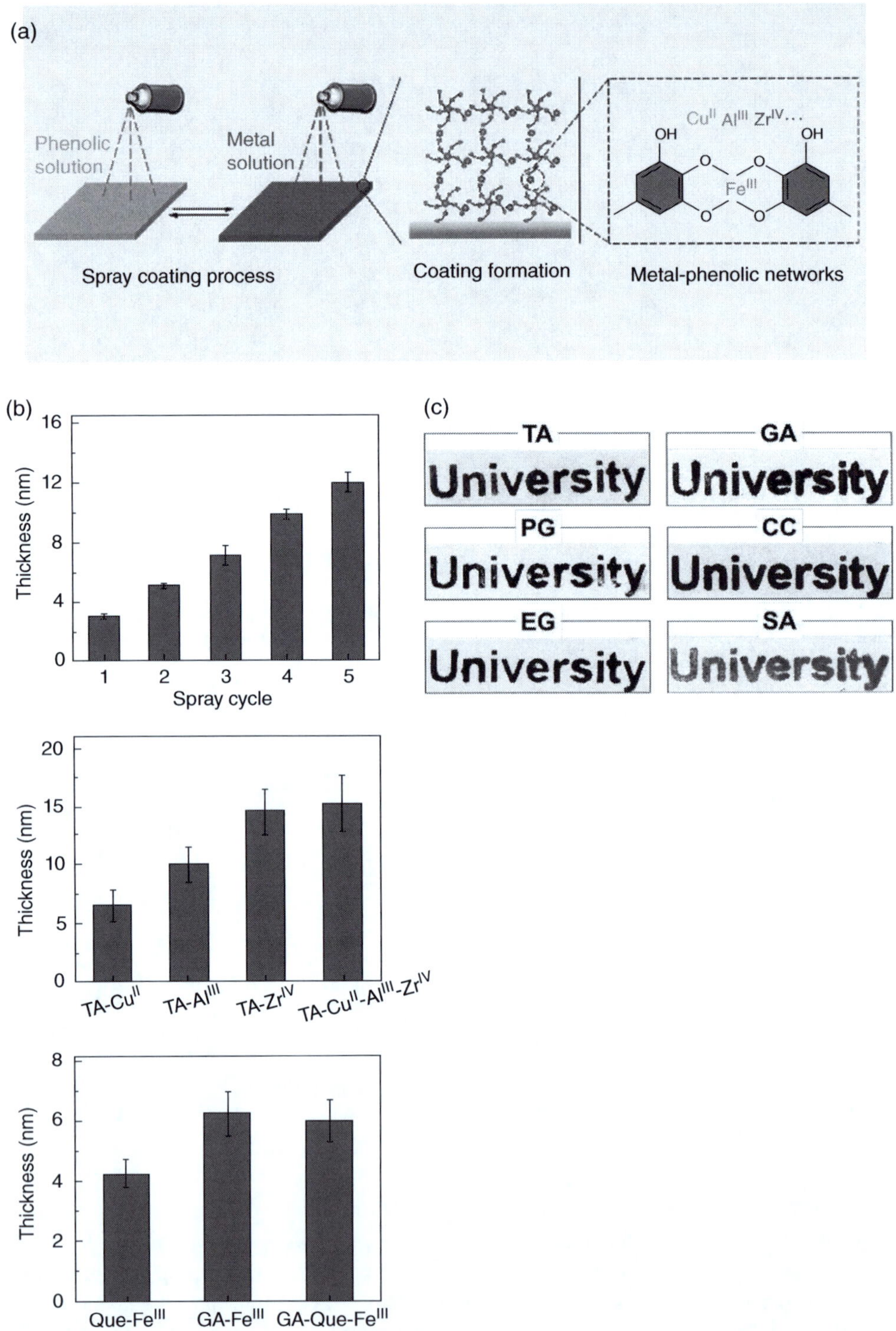

Figure 8.6 (a) Overview of MPN formation via a spray-coating process. *Source:* Zhong et al. (2018), Reproduced with permission of the American Chemical Society. (b) The influence of the number of spray cycles, metal ion and ligand on MPN film thickness. *Source:* Zhong et al. (2018), Reproduced with permission of the American Chemical Society. (c) MPNs prepared via printing, with Fe ink being printed onto paper treated with different polyphenols. *Source:* Adapted from Dai et al. (2019). (*See insert for color representation of the figure.*)

8.3 MPN Film Properties

The properties of MPNs can vary depending on the technique used to assemble the films. The thickness of MPNs can be examined on planar or particle substrates using various characterization methods including atomic force microscopy (AFM) (Ejima et al. 2013). A comparison of oxygen oxidation, diffusion-driven, rust-mediated, and ROS oxidation assembly methods has revealed film growth rates of 5, 7.5, 9.6, and 176 nm/h for these respective methods (Zhong et al. 2019). As the thickness can be easily modulated by various conditions in electro-triggered assembly, it is difficult to directly compare this method to the other continuous assembly methods (Maerten et al. 2017). Similarly, discrete assembly methods, such as spray assembly (2.5 nm/cycle) and one-step assembly (10 nm/cycle), cannot be directly compared to continuous assembly methods (Ejima et al. 2013; Zhong et al. 2018). The permeability of the films varies widely depending on the assembly method used. Discrete assembly generates films with intermediate permeability (~50% of capsules are permeable to 70 kDa fluorescein isothiocyanate (FITC)-dextran (Ejima et al. 2017)), whereas diffusion-driven assembly generates the most permeable capsules (nearly 100% of capsules are permeable to 2000 kDa FITC-dextran) (Yun et al. 2018) and OMC assembly produces the least permeable films (impermeable to 4 kDa FITC-dextran) (Zhong et al. 2019). These findings highlight that even with the same materials, film properties can differ considerably depending on the assembly technique.

To ensure reproducible synthesis of MPNs with a defined property profile, robust synthesis protocols are crucial, particularly at the laboratory/bench scale. The various steps involved during MPN fabrication, including the addition of reagents, centrifugation, vortexing, and washing, can influence the final product if variations in experimental protocols occur. This is particularly common during the onboarding of researchers unfamiliar with MPN synthesis techniques. To overcome these challenges, Björnmalm et al. (2017) published a synthesis protocol for Fe^{III}-TA that outlines each synthesis step in detail, complemented by video footage. This allows researchers to directly view what is required in each step. In particular, centrifugation, washing, supernatant removal, and resuspension were highlighted, as an incorrect technique can result in sample loss, MPN aggregation, and ineffective removal of unbound metal ions and/or phenolic ligand. The influences of single-step versus multistep deposition, pH, ligand:metal ion concentration ratio, and template are also discussed. Spoljaric et al. (2021) later reported replicating the synthesis of Fe^{III}-TA at an increased scale. Reproducible, scalable synthesis of MPNs was achieved, summarized in a detailed protocol with accompanying video footage. The key centrifugation, washing, supernatant removal, and resuspension steps were detailed, with considerations for synthesis at an increased scale provided (the considerations and challenges associated with upscaling MPN synthesis are discussed in detail in Section 8.5). Clear protocols with detailed steps (supplemented by video, commentary, etc.) are key to ensuring reproducible MPN synthesis, irrespective of scale. They facilitate experimental uptake, aid method validation, benchmarking, and protocol development.

In addition, the choice of metal ion/phenolic ligand combinations can determine MPN properties, with numerous internal (bond type, long-range order, and defects) and external (assembly conditions including temperature, pH, and solvent) factors influencing MPN properties (Yun et al. 2019). A broad range of possible metal ion/phenolic ligand combinations is possible, which can be expanded if multiple metal/ligand species are used within a single MPN or if the aforementioned external/experimental factors are modified. Yun et al. (2019) highlighted how these structure–property and experimental factors determine the mechanical properties of MPNs. Most notably, the pH-dependence of MPN self-assembly and complexation state influenced the mechanical behavior of the resulting film owing to the type and extent of bonding throughout the network; these factors varied with metal ion choice. Fe^{III}-TA, In^{III}-TA, Ga^{III}-TA, and Tb^{III}-TA displayed Young's modulus (E_Y) values of 2.9, 5.9, 5.9, and 21.9 GPa, respectively. Furthermore, the complexity and size of phenolic ligands likewise influenced network strength, with the enhanced molecular orientation and tighter packing of smaller and simpler ligands yielding MPNs with higher E_Y values. MPN films of Fe^{III}-gallic acid, Fe^{III}-pyrogallol, and Fe^{III}-pyrocatechol featured E_Y values of 5.1, 11.3, and 7.4 GPa, respectively. Likewise, Guo et al. (2014) demonstrated the influence of metal ion and feed concentration on the disassembly behavior of MPN capsules. Another study demonstrated how the choice of the phenolic ligand could be used to impart binding affinity and selectivity (e.g. hyaluronic acid (HA) exhibits high binding affinity toward CD44 receptors overexpressed on numerous types of cancer cell membranes (Ju et al. 2016b)). In addition, Ju et al. (2015) demonstrated how Fe^{III}-based MPNs displayed superior antifouling behavior and lower nonspecific protein adsorption and cell association when complexed with poly(ethylene glycol) (PEG)-polyphenols than with TA. These findings highlight the importance of the appropriate choice of metal and ligand species, while also exemplifying the diversity in properties/applications and tunable/bespoke nature of MPNs that can be obtained.

8.4 MPN Surface Interactions and Applications

Owing to the distinct properties of metal–phenolic complexes, MPNs can be endowed with multiple functionalities and applied in a variety of applications, including surface functionalization, theranostics, catalysis, and gas and metal separations. TA is recognized as safe (GRAS) by the FDA and is used in the food industry. Given its high biocompatibility and high binding affinities to various substrates, MPNs are widely applied as coating materials for biointerfaces. For example, MPNs have been used as a dental sealant for tooth hypersensitivity treatment (Figure 8.7a) (Oh et al. 2015; Prajatelistia et al. 2016). By simply dipping teeth into an MPN solution for a short period of time (four minutes), the dentin can be coated with MPN films that occlude the exposed dentinal tubes, which are mechanically stable to tooth brushing. In addition, MPNs have been applied as cytoprotective exoskeletons for living cells. Park et al. (2014) demonstrated that the MPN coating

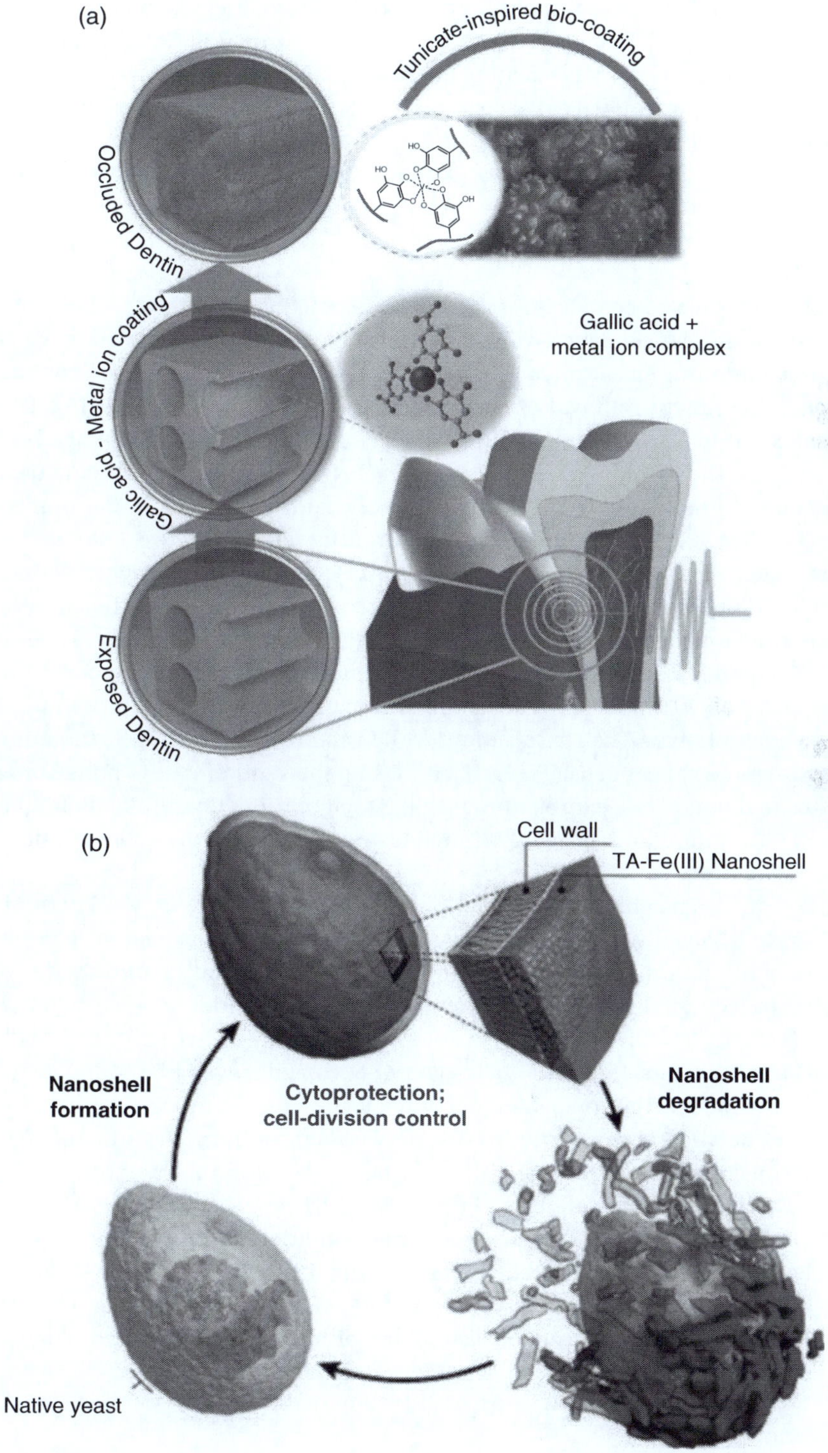

Figure 8.7 Examples of MPNs used as coating materials at biointerfaces: (a) as coating materials for hypersensitive teeth. *Source:* Prajatelistia et al. (2016), Reproduced with permission John Wiley & Sons. (b) as cytoprotective exoskeletons for yeast cells. *Source:* Park et al. (2014), Reproduced with permission of Wiley. (*See insert for color representation of the figure.*)

on individual yeast cells can protect them from multiple external aggressors (including UV-C irradiation, lytic enzymes, and silver nanoparticles) (Figure 8.7b). Cell division was suspended in the presence of the MPN coating but restored after degradation of the MPN shell. Li et al. (2015) further applied MPN coatings to multiple living cell types including yeast cells, *Escherichia coli* (*E. coli*, bacteria), and mammalian cells. Importantly, the MPN shell can be used as a versatile platform for secondary modifications. Small molecule drugs and magnetic resonance imaging (MRI) contrast agents can be incorporated into the MPN shell by exploiting the reactivity of the phenolics to thiols, amines, and boronic acid (Ju et al. 2020).

MPN-based particles have been used for drug delivery, medical imaging, and diagnosis. For example, the anticancer drug doxorubicin (DOX) was encapsulated within MPN capsules by preloading into calcium carbonate ($CaCO_3$) particles, followed by coating with MPNs, and subsequent removal of the particle templates (Ping et al. 2015). The obtained capsules exhibited high drug-loading capacity and acid-triggered disassembly/drug release owing to the pH-dependent stability of MPNs. As a result, the DOX-loaded MPN capsules showed high cytotoxicity to human cervical cancer (HeLa) and human breast cancer (MDA-MB-231) cells due to intracellular drug release (Figure 8.8a). MPNs have also been coated onto hydrophobic drug (paclitaxel) nanoparticles to achieve long-term colloidal stability (Shen et al. 2016). To further enhance drug delivery efficiency, Ju and co-workers functionalized low-fouling materials (PEG) and targeting ligands (HA) with phenolic groups and used them as synthetic polyphenol building blocks to prepare MPN capsules (Ju et al. 2015, 2016a). The obtained capsules showed resistance to non-specific protein adsorption and cell association, while exhibiting high specific targeting to CD44 overexpressing cancer cells (Figure 8.8b). Using the synthetic PEG-polyphenol system, MPN-coated emulsions exhibited low cell association *in vitro*, a blood circulation half-life of 50 minutes *in vivo*, and negligible toxicity in healthy mice (Besford et al. 2018).

MPN capsules can also be potentially used for photothermal therapy, as MPNs exhibit strong near-infrared light absorption and thus generate heat after exposure to near-infrared irradiation (Hu et al. 2016). By varying the chelating metal ligand, the MPN capsules demonstrated metal-specific imaging functionalities, including positron emission tomography (PET, $^{64}Cu^{II}$), MRI (Fe^{III}, Gd^{III}, Mn^{II}), and fluorescence (Eu^{III} and Tb^{III}) (Guo et al. 2014). The radioactivity of $^{64}Cu^{II}$-TA MPN capsules was high enough for tracking their *in vivo* biodistribution, as demonstrated using a mouse model (Figure 8.9a) (Guo et al. 2014). The MRI relaxivity of Mn^{II}-TA MPN capsules was on the order of 60 s/mM and thus considered a promising MRI contrast agent for *in vivo* study (Guo et al. 2014). Furthermore, the chelated Fe^{III} in Fe^{III}-TA MPN capsules could catalyze the conversion of H_2O_2 into O_2 under physiological conditions (Figure 8.9b) (Guo et al. 2015). The generated O_2 microbubbles were rapidly detected by ultrasound imaging systems (Kiessling et al. 2014). Therefore, Fe^{III}-TA MPN capsules can be used to detect elevated levels of H_2O_2 *in vitro* and *in vivo*, an important parameter for pathological diagnosis.

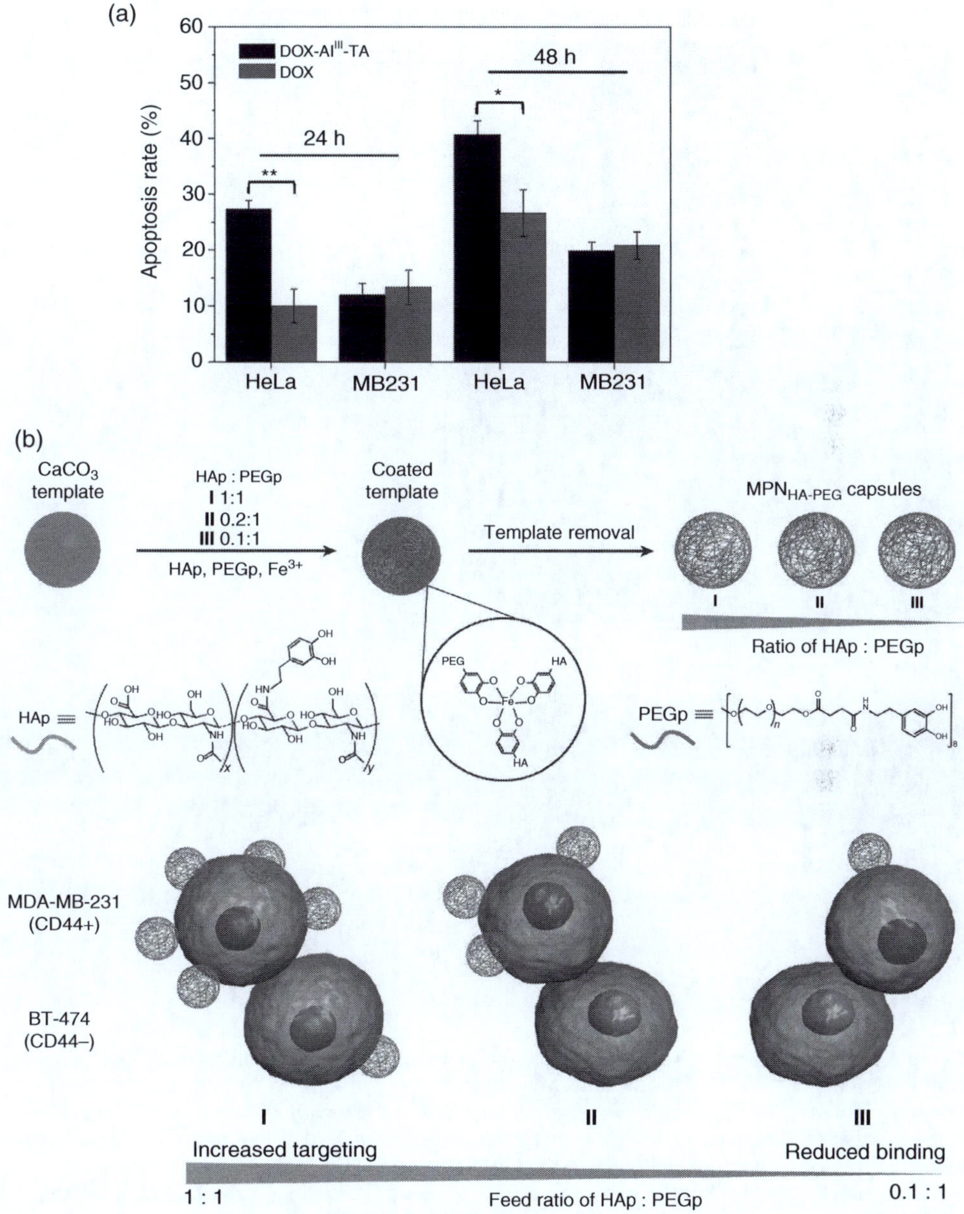

Figure 8.8 (a) Percentage of apoptotic HeLa and MB231 cells after treatment with DOX-Al^{III}-TA MPNs or DOX at different times. *Source:* Adapted from Ping et al. (2015), Reproduced with permission of Wiley. (b) Synthesis of HA-functionalized PEG-Fe^{III} MPNs and the subsequent influence of HA:PEG ratio on specific targeting of CD44 cancer cells and binding efficiency. *Source:* Ju et al. (2016a), Reproduced with permission of the American Chemical Society. (*See insert for color representation of the figure.*)

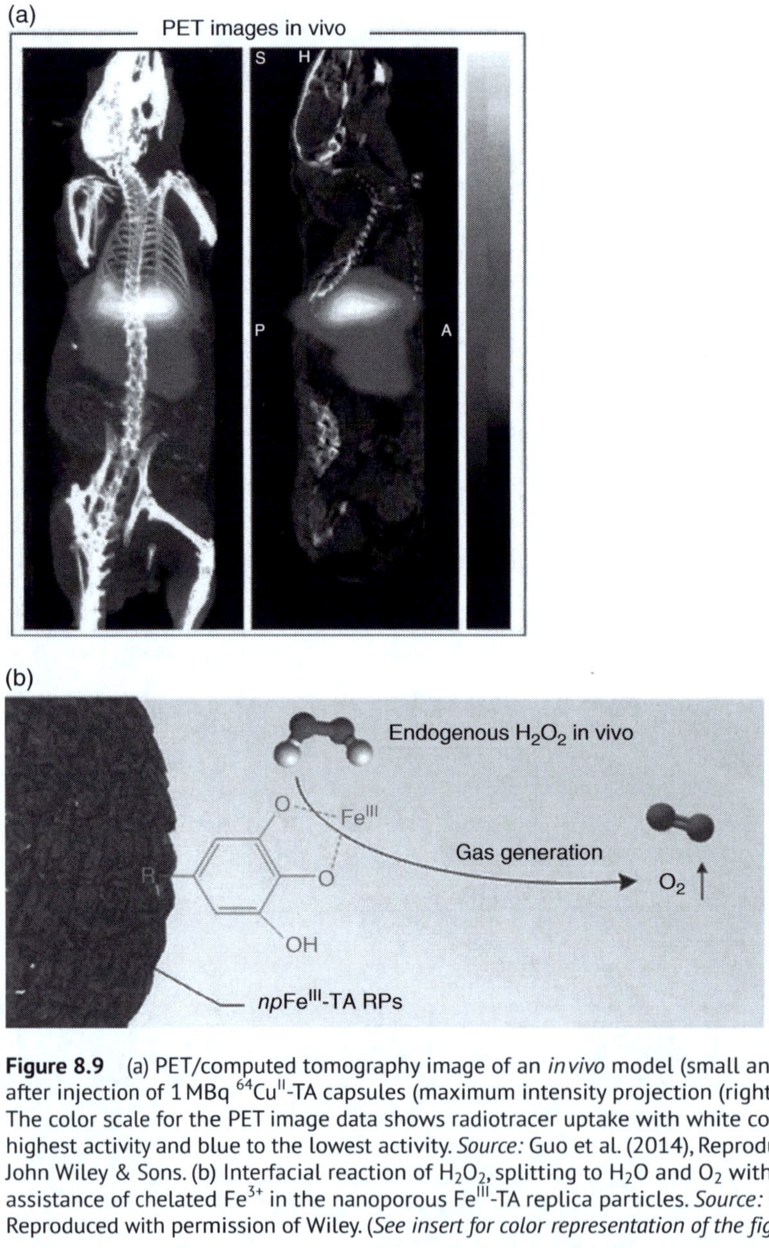

Figure 8.9 (a) PET/computed tomography image of an *in vivo* model (small animal) 30 minutes after injection of 1 MBq $^{64}Cu^{II}$-TA capsules (maximum intensity projection (right), sagittal view (left)). The color scale for the PET image data shows radiotracer uptake with white corresponding to the highest activity and blue to the lowest activity. *Source:* Guo et al. (2014), Reproduced with permission John Wiley & Sons. (b) Interfacial reaction of H_2O_2, splitting to H_2O and O_2 with the catalytic assistance of chelated Fe^{3+} in the nanoporous Fe^{III}-TA replica particles. *Source:* Guo et al. (2015), Reproduced with permission of Wiley. (*See insert for color representation of the figure.*)

MPNs have been recently introduced as promising materials for energy and environment applications. Owing to the intrinsic adherent properties of polyphenols, MPNs can be coated on various surfaces, allowing for surface functionalization. This has encouraged their potential use in applications including dye-sensitized solar cells (Çakar et al. 2016; Çakar and Özacar 2019), lithium storage (Xiao et al. 2018b), and corrosion resistance (Asgari et al. 2019), and as thermal stability and flame retardancy additives (Hu et al. 2019b).

In addition, the polyphenols allow MPNs to chelate diverse metals and interact with different biomolecules (proteins, lipids, nucleic acids, and polysaccharides), which makes MPNs suitable candidates for metal sequestration, catalysis, and water separation. For example, Luo et al. (2019) coated MPNs on microporous polyamide membranes that were applied for uranium extraction from seawater (Figure 8.10). The MPN coating endowed the membrane with excellent hydrophilicity, swelling capacity, and mechanical strength. As a result, the MPN-based membrane demonstrated a significantly high uranium extraction rate of 84% after the stable and long-term processing of 10 L of seawater. Further environmental remediation (Kim et al. 2015d; Tardy et al. 2018) and antifouling (Ju et al. 2015; Kim et al. 2015a;

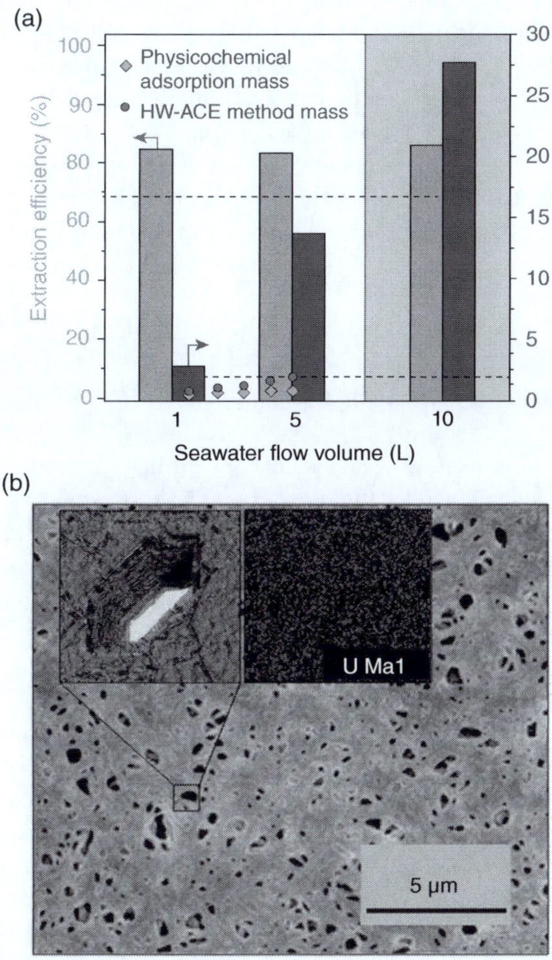

Figure 8.10 (a) Extraction efficiency and mass of uranium from seawater by three layers of MPN-based membranes. *Source:* Luo et al. (2019), figures 3 and 4. Reproduced with permission of the Royal Society of Chemistry. (b) SEM and energy-dispersive X-ray spectroscopy mapping of membranes following uranium deposition. *Source:* Luo et al. (2019), Reproduced with permission Royal Society of Chemistry. (*See insert for color representation of the figure.*)

Dong et al. 2017; Besford et al. 2018; Jiang et al. 2018) applications have been successfully confirmed. In addition to metals, polyphenols exhibit strong interactions with biomolecules. Yun et al. (2020) showed the preferential deposition of MPNs on proteins, lipids, nucleic acids, and polysaccharides than on glass substrates, thus enabling the fabrication of MPN patterns and the visualization of latent fingerprints. Ju et al. (2016b) reported that the protein corona formed from human serum on HA-based MPN capsules reduced non-specific cell interactions while maintaining specific interactions with targeted cells (CD44 overexpressing breast cancer cells).

MPNs have also been applied as catalysts for multiple reactions including radical scavenging (Bertleff-Zieschang et al. 2017), the Fenton reaction (Tardy et al. 2018), the Suzuki–Miyaura coupling reaction (Pd nanoparticle deposited MPNs) (Yun et al. 2017), and hydrogenation of quinoline (Rh-TA capsules) (Guo et al. 2014). Wei et al. (2016) coated cellulose fibers derived from commercial products (i.e. filter paper, tissue paper, and cotton) and polyurethane-foam waste with Fe^{III}-TA MPN ink, followed by carbonization of the composites to yield ecological catalysts for oxygen reduction. Gold nanoparticles were coated with an MPN (Fe-TA) core by Zeng et al. (2014), yielding a novel catalyst system that facilitated the reduction of 4-nitrophenol to 4-aminophenol. The synthesis was accomplished without the need for additional heating, organic solvents, specialized equipment, or additional reagents. Furthermore, the catalysts developed by both Zeng et al. (2014) and Wei et al. (2016) displayed markedly high catalytic activity stemming from the synergistic effects of the substrate and the MPN.

8.5 Upscaling Considerations and Challenges

The ability to reproducibly synthesize large-scale volumes of MPN capsules (and other micro- and nanoparticles in general) is a challenge for the broader applicability and commercialization of these materials. Material characteristics, including but not limited to size (diameter), shape, size distribution, coating thickness, porosity, dispersibility, functionality, and cell targetability, largely determine the desired material performance and properties, especially those destined for biomedical and therapeutic applications, where slight variations in material structure/behavior can have important implications on bioactivity and physiology. Therefore, methodologies employed in the large-scale synthesis of MPNs must ensure that particles with a desired characteristic profile can be obtained reproducibly and consistently. Despite the lack of literature detailing the large-scale synthesis of MPNs, various other nanoparticles have been synthesized at increased volumes using a range of techniques. Although the initial results concerning the large-scale synthesis of MPNs seem encouraging (Yun et al. 2018; Liu et al. 2019a, Spoljaric et al. 2021), there are various factors, considerations, and challenges (such as reagent availability and handling, engineering requirements and possibilities, particle/capsule washing and solvent considerations, human resource and training requirements, and environmental health and safety considerations) that must be addressed before a viable method can be developed, as discussed later.

8.5.1 Reagent Considerations

Generally, the reagents that are reported in the literature for the preparation of MPNs are quite readily available and low- or nontoxic. However, several factors need to be considered as the scale of the reagents used increases, especially when bespoke/in-house functionalization is performed. Availability and storage are two important factors when considering a bulk purchase and the use of reagents. The handling and dispensing of such reagents similarly vary from using gram-scale bottles in a research setting to the use of industrial silos and hoppers to provide kilogram-level yields. The subsequent preparation of solutions, buffers, and various other suspensions requires controls to take into consideration the increased volume of solvents used, as well as ensuring sufficient mixing/shear force is applied to provide adequate dissolution of the MPN precursors (this is discussed further in Section 8.5.2). Quality assurance and subsequent purification of precursors and solutions are also required to increase the longevity of stocks and prevent contamination; various columns, syringe filters/membranes, and dialyzers have been used to purify precursors for MPN synthesis. Although these purification tools are suitable at the laboratory scale, the associated purification methods involved become time-consuming, laborious, and (in the case of dialysis) resource-intensive when larger scales are involved. However, process automation through the use of pumps, vacuum, or fluidic techniques (e.g. microfluidics) can help increase material throughput and purification efficiency. Environmental health and safety practicalities should always be considered, with the risks and controls required changing with the type and amount of materials used. For example, iron(III) chloride is corrosive but can readily be dissolved in water without any serious risk of incident. In contrast, aluminum(III) chloride is not only corrosive, but also a neurotoxin, and when added to water releases hydrochloric acid gas as a by-product. These risks can be readily mitigated through the implementation of appropriate engineering controls (e.g. ventilation and fume cupboards) and safe operating practices. Finally, the disposal of these materials generally requires specialized handling and storage, with limitations on maximum storable volumes in many countries.

8.5.2 Engineering Controls

The preparation of MPNs at the bench scale has been optimized for common laboratory equipment and glassware/consumables (Björnmalm et al. 2017). However, these often have limitations in terms of maximum workable volumes, output, and operational capacity. Alternative, larger-sized devices and laboratory consumables may be manufactured from different materials, have different geometries, or operate under different principles. An example is the difference in the shape of centrifuge vials/tubes, which can impact the effective formation of an MPN capsule or MPN-coated particle pellet. Appropriate levels of mixing/shear force are required across many aspects of MPN synthesis, including solution preparation, MPN fabrication, washing, and template dissolution. Benchtop vortex mixers, magnetic stirrers, and ultrasonic probes are satisfactory at laboratory scale (milligram to gram), while at increased volumes, they may be inappropriate and impractical.

Therefore, sufficient mixing/shear efficiency is required, which can be produced using large-scale ultrasonic probes and mechanical mixers. Similarly, the separation of MPNs from solution is typically achieved via centrifugation: this is severely limited by the maximum capacity of the device. As mentioned previously, the use of bench-top filtration/purification consumables, such as filters/membranes and/or columns, can be streamlined and automated through the use of pumps or vacuum to facilitate the continuous flow of solutions. Appropriate safety controls (i.e. ventilation, fume cupboards) are also required if corrosive, volatile, and/or flammable materials are involved. Aside from implementing specific engineering controls, the use of alternative processing techniques provides an efficient and high-throughput option to potentially prepare large quantities of MPNs. For example, the use of techniques such as microfluidics (Ma et al. 2017), electrospray (Ma et al. 2018), and flash nanoprecipitation (Markwalter and Prud'homme, 2018) provides opportunities for preparing large quantities of MPNs through continuous or semi-continuous processes, while significantly reducing space and labor considerations and costs, as well as potentially eliminating the need for time-consuming/labor-intensive procedural steps (i.e. centrifugation and re-dispersion).

8.5.3 Washing and Solvents

Most processes detailing template-mediated fabrication of MPNs require various volumes of solvents for numerous reasons as exemplified later.

8.5.3.1 Dissolution of Reagents and Preparation of Buffers used in MPN Fabrication

In most cases, reagents and buffers are dissolved in high-purity water (e.g. Milli-Q water) on an as-required basis to reduce wastage and maintain high purity. If required, most metal and polyphenol solutions, as well as buffer systems, can be stored for some periods of time, with filter-based purification performed before commencing synthesis.

8.5.3.2 Synthesis of Mesoporous Templates or Functionalized Polyphenols

A wide variety of templates, metal salts, and polyphenols can be readily purchased from suppliers for MPN synthesis. In addition, these aforementioned precursors can be functionalized or synthesized in-house according to the sought-after characteristics or properties. In many cases (such as the preparation of mesoporous $CaCO_3$ and silica), reactions are conducted in aqueous solutions, and additional solvents such as ethanol and acetone are used for washing, with appropriate engineering and handling controls facilitating their safe use at increased volumes. Other functionalization pathways, such as the preparation of functional PEG-polyphenol, may involve dimethylformamide. Furthermore, purification of functional PEG-polyphenol is achieved via dialysis, with recycling or filtration of the washing effluent providing an avenue to reduce water consumption.

8.5.3.3 Washing of MPN-Coated Templates and MPN Capsules

Similar to the preparation of stock solutions and buffers, washing typically involves the use of Milli-Q water or other high-purity water standards. The primary concern here involves the (potentially) use of large volumes of water during washing processes. Recycling/filtration and particle-removal processes may be viable options to enhance sustainability and reduce overall water requirements.

8.5.3.4 Dissolution of Particle Templates to Obtain MPN Capsules

As discussed earlier, MPNs have been successfully coated onto a broad range of substrates, including particle templates, which upon dissolution (template removal) yield MPN capsules. The three most commonly used particle templates (both solid and mesoporous) are $CaCO_3$, polystyrene (PS), and silica (SiO_2). $CaCO_3$ is readily dissolved with ethylenediaminetetraacetic acid, which provides a safe and cost-effective option for particle template dissolution. Tetrahydrofuran (THF) is the most commonly used reagent to dissolve PS as it is readily available and inexpensive. THF is however volatile and flammable, which could pose serious potential risks if used in larger volumes. However, these risks can be readily mitigated with appropriate engineering controls (i.e. ventilation, safe dispensing controls) and safe operating procedures. To dissolve SiO_2 particle templates, hydrofluoric (HF) acid is required. The highly corrosive nature and toxicity of HF require care and caution during its handling, especially at increased volumes. However, HF is commonly used in the electronics industry to etch silicon surfaces, with industrial-scale etching tanks providing a proven method of safe, large-scale dissolution of SiO_2 (Baldwin and Williams 1998). Alternative dissolution agents such as hot, highly concentrated potassium hydroxide or other alkali metal hydroxides can also be used. However, these solutions are also quite corrosive and may have undesired effects on cargo and/or surface functionalities (for example, alkaline conditions can denature proteins), and their ability to dissolve SiO_2 is generally less effective than HF. A final consideration is that solvents that are used to dissolve inorganic templates can interact with metals and metal ions and therefore could potentially disrupt the MPN structure.

8.5.4 Human Resources and Training

Synthesis techniques designed for bench-top fabrication within academic laboratories do not always necessitate the same time and human resource requirements when scaled toward larger volumes. In many cases, the increased volume of reagents/solvents used and synthesis products generated correspond to an increase in time, concentration, and operation efficiency required for protocol steps (e.g. coating iterations and washing). Existing methodologies optimized for lab-scale synthesis (i.e. multistep MPN fabrication) can be comparatively time-consuming and labor-intensive. The resulting stress and fatigue can lead to reduced output and reproducibility. When designing protocols for large-scale synthesis, the use of automated processes (e.g. centrifugation and pipetting) and/or engineering controls to perform repetitive and more labor-intensive tasks is a practical solution. Alternatively, the use of appropriate ergonomic aids can help reduce potential fatigue, stress, or injury.

Similarly, the training of personnel to perform synthesis techniques is necessary to ensure products can be obtained reproducibly, at desired yields, effectively and safely. Depending on the nature of the protocol, the training of researchers can require a considerable investment of time and experienced personnel. Additionally, the trainee may need to invest considerable time and effort in learning the technique. In the case of successful synthesis possibly requiring multiple attempts, a significant investment in consumables and reagents may also be required. To minimize this learning period and achieve an efficient transition from novice to experienced researcher, clear, unambiguous, and detailed protocols/standard operating procedures are essential. The use of first-person action cameras to record and detail steps of the synthesis process can provide invaluable practical guidance regarding the specifics and intricacies of synthesis steps. This combination of detailed procedures and video instructional footage was highlighted by Björnmalm et al. (2017), applying the strategy to a number of particle and capsule synthesis strategies and detailing key aspects, considerations, challenges, and specific requirements for various synthesis steps (e.g. appropriate centrifugation times and speeds for particle templates of varying diameter). Ideally, the generation of a protocol database containing, for example, detailed procedures, video supplementary information, and reference data to confirm successful synthesis would increase approachability toward these various methods and potentially expand the applicability of the particles and capsules fabricated.

8.5.5 Environmental Health and Safety Considerations

Safe working practices and hazard minimization are at the forefront of protocol design. This is particularly the case with large-scale synthesis, where increased reagent/solvent volumes and time/labor demands can exacerbate existing risks or create new risks. The risks and strategies and their elimination or minimization have been discussed in the preceding subsections. To summarize, automation and engineering controls should be introduced where possible, with ergonomic measures providing an alternative option. Appropriate training and observance of safety guidelines and policies as well as legal regulations can likewise ensure any potential risks are identified and controlled. Concurrent application of these strategies would ensure effective, safe, and reproducible synthesis of MPNs at increased scales.

8.6 Method Automation: Possibilities and Outlook

As mentioned previously, almost all fabrication methods used to prepare MPNs involve several sequential washing and assembly steps. For example, the discrete assembly method for preparing MPNs consists of the following generalized steps: (i) preparation and filtration of stock solutions and reagents; (ii) washing of particle templates; (iii) MPN assembly; (iv) washing of MPN-coated particles (multiple washing/centrifugation/redispersion cycles, as shown in Figure 8.11); (v) dissolution of particle templates (multiple washing/centrifugation/redispersion cycles); and (vi) washing of MPN capsules (multiple washing/centrifugation/redispersion cycles).

Despite being feasible and optimized for bench-scale yields (i.e. micrograms), if larger yields are desired, these aforementioned labor and time considerations will likely be exacerbated owing to the handling of larger volumes of material, waste/washing effluent, and possibly larger-scale equipment. Furthermore, the repetitive washing/centrifugation/redispersion steps can lead to a loss of yield if precise pipetting and transfer of materials are

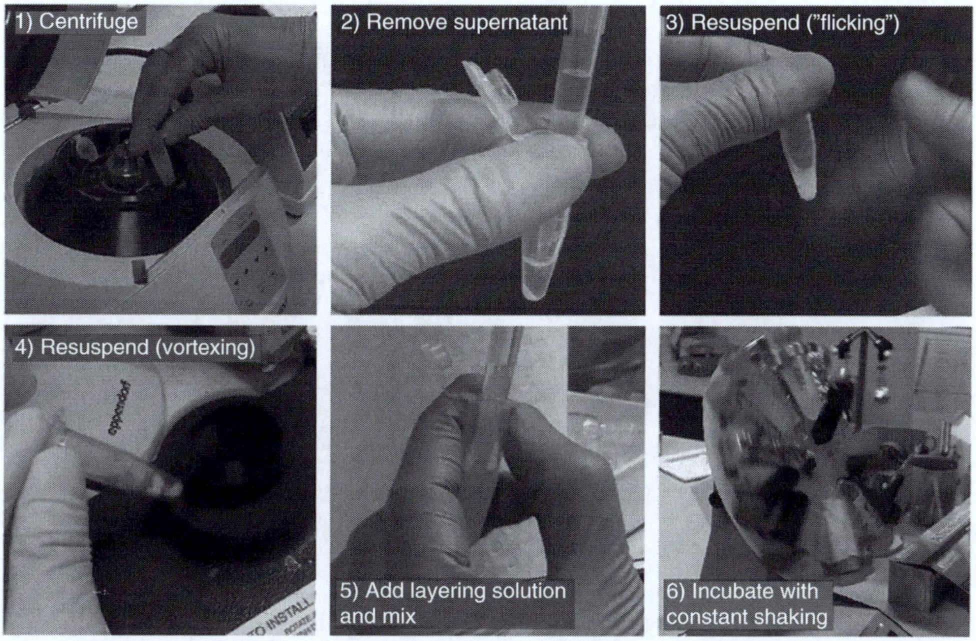

Figure 8.11 Overview of the various steps involved in washing MPN particles and capsules. *Source:* Björnmalm et al. (2017), Reproduced with permission American Chemical Society.

not achieved with each iteration. In addition, occupational health and safety and engineering considerations are important when increased volumes of flammable solvents (e.g. ethanol and/or THF) are involved.

To facilitate these considerations and bridge the transition from bench- to gram-scale production of MPNs, numerous automated and/or mechanized methods have been employed (or could potentially be employed) for MPN assembly and/or particle washing.

8.6.1 Automated Assembly Techniques

8.6.1.1 Robotic/Automated Immersive Assembly

The use of commercially available or custom-built robotic dipping systems to facilitate the immersive assembly of coating layers onto templates (both planar and spherical) has been reported (Jang and Grunlan 2005; Richardson et al. 2013, 2016), with multiple coating iterations allowing for the deposition of thicker coating layers. Variables, including incubation time, number of coating cycles, temperature, and pH, can be readily controlled. Richardson et al. (2013) demonstrated the coating of immobilized silica particles with alternating polymer layers using a robotic dipping device (Figure 8.12). These same principles can be applied to the preparation of MPNs, where alternating immersion cycles in metal, ligand, and washing solutions are involved for depositing MPN layers onto various substrates. Alternatively, the metal and ligand solutions can be added to the same reaction vessel, with the coated template being removed for subsequent washing and/or template dissolution. This principle was demonstrated by Yun et al. (2017) for the deposition of

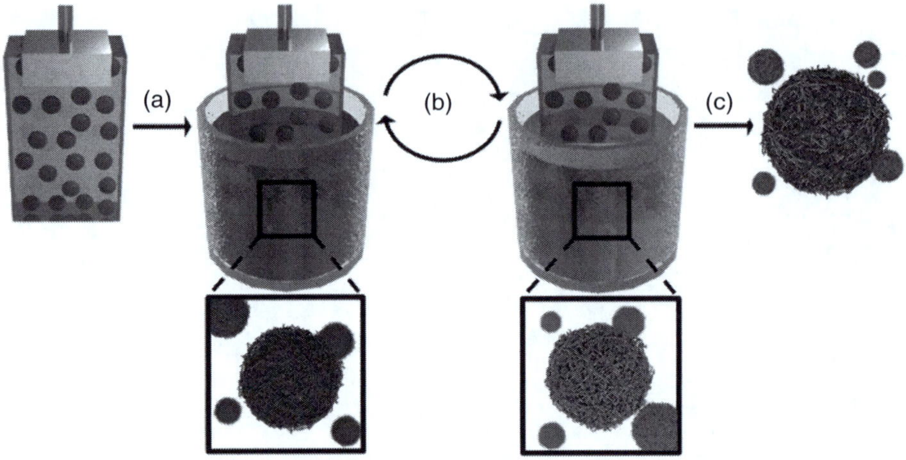

Figure 8.12 Template particle coating via immersive polymer assembly: (a) immobilized particles are immersed in a polymer solution, followed by washing steps (immersion in water); (b) subsequent immersion in a solution of a complementary polymer, followed by water washing steps (the washing steps are not shown for simplicity; this process is repeated until the desired number of layers is attained); and (c) the layered particles are recovered from the polymer agarose, and the template particles can be dissolved to yield capsules. *Source:* Richardson et al. (2013), Reproduced with permission John Wiley & Sons. (*See insert for color representation of the figure.*)

MPN (Fe^{III}-TA) onto polyester textile substrates. Aside from dedicated robotic dipping systems, a more rudimentary approach to MPN formation can be taken by adopting automated pipetting robots for pipetting alternating solutions of metal and ligand solution into multi-well plates. In contrast to relying on fixed immersion times, as is commonly employed for most automated dipping devices, quartz crystal microbalance (QCM) devices can be used to determine the mass deposited per cycle (Shiratori and Yamada 2000). For flexible and more robust/stable templates, spin-coating, roll-to-roll coating (Fujimoto et al. 2004), and drum coating may potentially be adapted to prepare continuous batches of MPN-coated substrates.

8.6.1.2 Robotic/Automated Spray Assembly

As opposed to immersing templates in solution, alternating or simultaneous coating solutions of metal and ligand solution (and washing/rinsing solutions) can be applied directly to the substrate in the form of a spray or aerosol to deposit MPN layers. Morphology, composition, and properties can be modified by controlling, for instance, the solution concentration, spraying time, flow rate, time between coatings, temperature, and humidity. The simplest method of MPN spray assembly involves the manual spraying of alternate metal and ligand solutions using a conventional spray gun, as demonstrated by Zhong et al. (2018) (Figure 8.6a). Automated spray devices have been used for layer-by-layer assembly (Hong and Park 2011), with potential use in MPN fabrication allowing for increased control, accuracy, and reduced manual handling and time constraints. As with immersive assembly techniques, automated spray assembly can be coupled with QCM devices to obtain thin layers with precise thickness and monitor layer evolution in real time (Fukao et al. 2011). The potential for upscaling automated spray coating was investigated by Krogman et al. (2013), who integrated the technique with roll-to-roll processing to deposit multilayers onto bulk flexible substrates. This proof-of-concept study is encouraging for further application of the spray assembly technology, particularly with MPN systems where precise control of both the thickness and number of layers is paramount. Finally, alternative approaches such as surface acoustic waves (Qi et al. 2011) or electrospray techniques (Ma et al. 2018) to atomize droplets for subsequent deposition may provide additional production and purification pathways.

8.6.1.3 Fluidic Assembly

Microfluidics has been exploited to prepare a plethora of micro- and nanoparticles, spheres (hollow and core shell), Janus particles, and vesicles (Capretto et al. 2013; Björnmalm et al. 2014; Ma et al. 2017; Wang et al. 2017b), comprising a range of polymer (Dendukuri et al. 2006; Chu et al. 2007; Xu et al. 2009), organic- (Majedi et al. 2013), inorganic- (Lee et al. 2008), and noble metal-based (Song et al. 2006; Köhler et al. 2007; Duraiswamy and Khan 2009; Zhao et al. 2011) materials. The technique at its most rudimentary consists of a pump/pressure source (a vacuum can also be applied), capillary tubing, and the microfluidic device/chamber generally placed atop a microscope slide (a QCM substrate can also be used to provide real-time information on layer growth). The coating solutions, particle templates, and washing solutions can be fed into the system, with multichannel setups, allowing for numerous coating iterations and the control of particle flow relative to the coating and washing streams (Figure 8.13). Solutions containing cargo molecules can also be introduced into the system, allowing for molecule encapsulation during particle

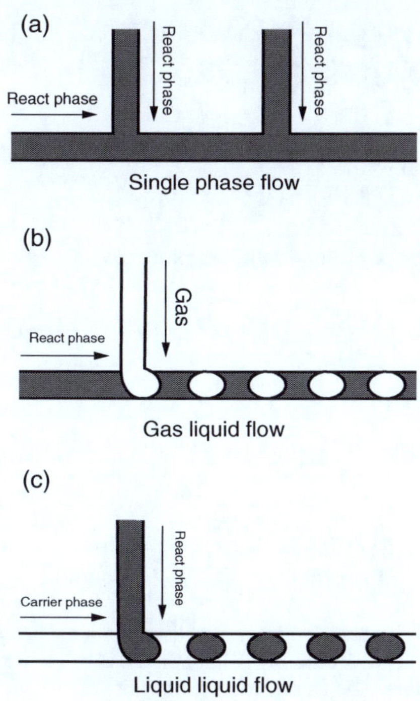

Figure 8.13 Overview of various fabrication techniques involving microfluidics: (a) single-phase flow; (b) gas–liquid flow; and (c) liquid–liquid flow. *Source:* Ma et al. (2017), Reproduced with permission of the Royal Society of Chemistry.

formation (Kolishetti et al. 2010). Aside from "one-pot" strategies in which templates, nanoparticle reagents, and cargo molecules are simultaneously fed through the microfluidics device, pre-synthesized or commercially sourced particles can be mixed with cargo molecules to facilitate the flow-induced incubation of particles. Weiss et al. (2018) demonstrated this strategy where microfluidics was used to adsorb proteins onto mesoporous silica, poly(2-methacryloyloxyethylphosphorylcholine) (PMPC)-coated silica hybrid particles, and PMPC replica particles. The high-throughput capability and versatility of microfluidic techniques offer potential for application toward template-mediated MPN fabrication and molecule loading.

Building on the principles of microfluidics, flash nanoprecipitation (Saad and Prud'homme 2016; Tao et al. 2019) and flash nanocomplexation (Santos et al. 2016; Qiao et al. 2018) are emerging techniques, in which particle formation is triggered by the turbulent mixing of two or more oppositely charged polyelectrolyte solutions within a multi-inlet vortex chamber. The continuous, high-throughput synthesis of reproducible particles, enhanced process controls and increased drug loading, encapsulation efficiency and bioactivity retention of proteins have all been reported using these techniques (Pustulka et al. 2013; Allen et al. 2017; Levit et al. 2019). Recently, Liu et al. (2019a) reported the preparation of hybrid Fe^{III}-TA-HA-DOX MPNs via flash nanocomplexation, as shown in Figure 8.14. The MPNs exhibited a narrow size distribution, high reproducibility, high drug

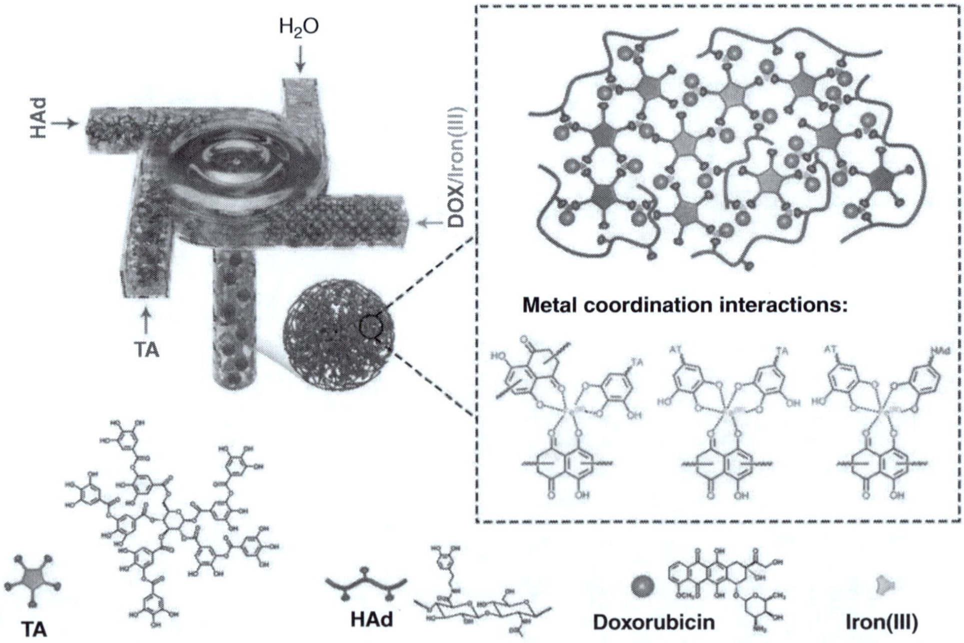

Figure 8.14 MPN formation via flash complexation, with complex formation facilitated via a small, turbulent mixing device into which reagent solutions are fed. *Source:* Liu et al. (2019b), Reproduced with permission Royal Society of Chemistry. Note: The actual composition of TA consists of a complex mixture of different galloylated glucoses (see Sylla et al. 2015). (*See insert for color representation of the figure.*)

loading capacity, and pH-triggered drug release behavior, allowing for MPNs to be synthesized without the need for a particle template. These initial results are encouraging toward the development and implementation of flash nanocomplexation for the high-volume, reproducible production of bespoke MPNs.

8.6.1.4 Automated Washing and Filtration Techniques

A range of existing automated, programable equipment and machinery can be incorporated into MPN preparation value chains to promote yield maximization during particle/capsule washing and resuspension. Automated centrifuges coupled with automated pipetting devices can eliminate manual handling and ergonomic considerations associated with particle washing and re-suspension. However, there may be limitations if pellet agitation/mixing during the addition of fresh washing solution is required. A possible solution to circumvent these limitations is the use of programmable robotic arms to facilitate the transfer of a specimen tube between the centrifuge/pipette and the mixing device. Similarly, columns (e.g. Sephadex®) and filter units (Amicon® and Pegasus®) used for the purification of certain MPN capsules are gravity-driven, yet require careful, manual loading of solutions to minimize sample loss. This can be overcome through continuous pumping of the solution into the column coupled with a sensor to monitor filtration rates. Alternatively, filtration columns may be retrofitted to existing chromatographic devices, allowing for the continuous filtration of MPNs.

8.7 Conclusions

MPNs are a versatile and robust family of materials, facilitated by their broad number of metal–ligand combinations, tunable properties, and diverse range of applications. The use of various templates has facilitated the preparation of MPNs with different and distinct sizes and shapes via an array of different fabrication methods and technologies. In addition to metal ion selection, the choice of ligands and the potential for surface modification/functionalization greatly expand the scope of the resulting materials, particularly regarding biointerface interactions where selective targeting, stealth-like behavior, and low-fouling properties are desirable. Similarly, this versatility in composition has expanded the use of MPNs in non-biomedical applications, including catalysis, marine antifouling coatings, lithium storage, environmental remediation, corrosion resistance, UV-shielding, functional textiles, dye-sensitized solar cells, and emulsion stabilizers. Furthermore, experimental factors, including temperature, pH, and solvent, can influence the self-assembly of MPN materials.

To further expand their applicability and facilitate commercialization, methods for the large-scale preparation of MPNs with reproducible dimensions and properties must be developed. Although most published template-mediated techniques used to fabricate MPNs have been optimized at the laboratory-bench (i.e. milligram) scale, many of the techniques are sufficiently robust and reproducible to warrant upscaling to larger (i.e. gram) scales and beyond. However, the primary bottlenecks and limitations associated with successful large-scale synthesis involve the multistep nature of the different techniques (especially those involving particle washing, pellet formation via centrifugation and subsequent re-dispersion iterations), fabrication times, and manual labor demands on personnel. These challenges can be minimized or overcome through various mechanisms, namely process automation/mechanization, enhancing synthesis ergonomics, or employing automatic, high-throughput techniques (e.g. microfluidics and flash nanoprecipitation). It is anticipated that this potential for reproducible, large-scale synthesis coupled with the promising attributes of MPNs will drive their commercialization and applicability in a range of fields.

References

Ahn, S., Halake, K., and Lee, J. (2017). Antioxidant and ion-induced gelation functions of pectins enabled by polyphenol conjugation. *International Journal of Biological Macromolecules*, **101**: 776–782.

Allen, S., Osorio, O., Liu, Y.-G., and Scott, E. (2017). Facile assembly and loading of theranostic polymersomes via multi-impingement flash nanoprecipitation. *Journal of Controlled Release*, **262**: 91–103.

Asgari, M., Yang, Y., Yang, S., et al. (2019). Mg–phenolic network strategy for enhancing corrosion resistance and osteocompatibility of degradable magnesium alloys. *ACS Omega*, **4**: 21931–21944.

Baldwin, D.G. and Williams, M.E. (1998). Industrial hygiene. In: Semiconductor Safety Handbook, (ed. Bolmen, R.A. Jr.), 204–268: William Andrew Inc.

Bartzoka, E.D., Lange, H., Thiel, K., and Crestini, C. (2016). Coordination complexes and one-step assembly of lignin for versatile nanocapsule engineering. *ACS Sustainable Chemistry & Engineering*, **4**: 5194–5203.

Bertleff-Zieschang, N., Rahim, M.A., Ju, Y., et al. (2017). Biofunctional metal–phenolic films from dietary flavonoids. *Chemical Communications*, **53**: 1068–1071.

Besford, Q.A., Ju, Y., Wang, T.-Y., et al. (2018). Self-assembled metal–phenolic networks on emulsions as low-fouling and pH-responsive particles. *Small*, **14**: 1802342.

Björnmalm, M., Yan, Y., and Caruso, F. (2014). Engineering and evaluating drug delivery particles in microfluidic devices. *Journal of Controlled Release*, **190**: 139–149.

Björnmalm, M., Cui, J., Bertleff-Zieschang, N., et al. (2017). Nanoengineering particles through template assembly. *Chemistry of Materials*, **29**: 289–306.

Bray, K., Previdi, R., Gibson, B.C., et al. (2015). Enhanced photoluminescence from single nitrogen-vacancy defects in nanodiamonds coated with phenol-ionic complexes. *Nanoscale*, **7**: 4869–4874.

Çakar, S. and Özacar, M. (2019). The pH dependent tannic acid and Fe-tannic acid complex dye for dye sensitized solar cell applications. *Journal of Photochemistry and Photobiology, A: Chemistry*, **371**: 282–291.

Çakar, S., Güy, N., Özacar, M., and Fındık, F. (2016). Investigation of vegetable tannins and their iron complex dyes for dye sensitized solar cell applications. *Electrochimica Acta*, **209**: 407–422.

Capretto, L., Carugo, D., Mazzitelli, S., et al. (2013). Microfluidic and lab-on-a-chip preparation routes for organic nanoparticles and vesicular systems for nanomedicine applications. *Advanced Drug Delivery Reviews*, **65**: 1496–1532.

Chen, W. and Hu, Y. (2019). Multiple modifications of $BiVO_4$ through the assembly of metal-phenolic networks for enhanced photocatalytic activity. *Catalysis Communications*, **123**: 124–128.

Chen, Y., Meng, J., Zhu, Z., et al. (2018a). Bio-inspired underwater super oil-repellent coatings for anti-oil pollution. *Langmuir*, **34**: 6063–6069.

Chen, Y., Wang, J., Liu, J., and Lu, L. (2018b). Metal-phenolic encapsulated mesoporous silica nanoparticles for pH-responsive drug delivery and magnetic resonance imaging. *Zeitschrift für Physikalische Chemie*, **232**: 1733.

Chen, J., Li, J., Zhou, J., et al. (2019a). Metal–phenolic coatings as a platform to trigger endosomal escape of nanoparticles. *ACS Nano*, **13**: 11653–11664.

Chen, X., Yi, Z., Chen, G., et al. (2019b). DOX-assisted functionalization of green tea polyphenol nanoparticles for effective chemo-photothermal cancer therapy. *Journal of Materials Chemistry B*, **7**: 4066–4078.

Cherepanov, P., Rahim, M., Bertleff-Zieschang, N., et al. (2018). Electrochemical behavior and redox-dependent disassembly of gallic acid/Fe^{III} metal–phenolic networks. *ACS Applied Materials & Interfaces*, **10**: 5828–5834.

Chu, L.-Y., Kim, J.-W., Shah, R., and Weitz, D. (2007). Monodisperse thermoresponsive microgels with tunable volume-phase transition kinetics. *Advanced Functional Materials*, **17**: 3499–3504.

Dai, Y., Guo, J., Wang, T.-Y., et al. (2017). Self-assembled nanoparticles form phenolic derivatives for cancer therapy. *Advanced Healthcare Materials*, **6**: 1700467.

Dai, Y., Cheng, S., Wang, Z., et al. (2018a). Hypochlorous acid promoted platinum drug chemotherapy by myeloperoxidase-encapsulated therapeutic metal phenolic nanoparticles. *ACS Nano*, **12**: 455–463.

Dai, Y., Yang, Z., Cheng, S., et al. (2018b). Toxic reactive oxygen species enhanced synergistic combination therapy by self-assembled metal-phenolic network nanoparticles. *Advanced Materials*, **30**: 1704877.

Dai, Q., Yu, Q., Tian, Y., et al. (2019). Advancing metal–phenolic networks for visual information storage. *ACS Applied Materials & Interfaces*, **11**: 29305–29311.

Delalande, L., Tsvetkova, I., Zeng, C., et al. (2016). Catching a virus in a molecular net. *Nanoscale*, **8**: 16221–16228.

Dendukuri, D., Pregibon, D.C., Collins, J., et al. (2006). Continuous-flow lithography for high-throughput microparticle synthesis. *Nature Materials*, **5**: 365–369.

Dong, C., Wang, Z., Wu, J., et al. (2017). A green strategy to immobilize silver nanoparticles onto reverse osmosis membrane for enhanced anti-biofouling property. *Desalination*, **401**: 32–41.

Duraiswamy, S. and Khan, S.A. (2009). Droplet-based microfluidic synthesis of anisotropic metal nanocrystals. *Small*, **5**: 2828–2834.

Ejima, H., Richardson, J.J., Liang, K., et al. (2013). One-step assembly of coordination complexes for versatile film and particle engineering. *Science*, **341**: 154–157.

Ejima, H., Richardson, J.J., and Caruso, F. (2017). Metal-phenolic networks as a versatile platform to engineer nanomaterials and biointerfaces. *Nano Today*, **12**: 136–148.

Freese, C., Uboldi, C., Gibson, M.I., et al. (2012). Uptake and cytotoxicity of citrate-coated gold nanospheres: comparative studies on human endothelial and epithelial cells. *Particle and Fibre Toxicology*, **9**: 23.

Fujimoto, K., Fujita, S., Ding, B., and Shiratori, S. (2004). Fabrication of layer-by-layer self-assembly films using roll-to-roll process. *Japanese Journal of Applied Physics*, **44**: L126–L128.

Fukao, N., Kyung, K.-H., Fujimoto, K., and Shiratori, S. (2011). Automatic spray-LBL machine based on in-situ QCM monitoring. *Macromolecules*, **44**: 2964–2969.

Gevrek, T.N., Kosif, R., and Sanyal, A. (2017). Surface-anchored thiol-reactive soft interfaces: Engineering effective platforms for biomolecular immobilization and sensing. *ACS Applied Materials & Interfaces*, **9**: 27946–27954.

Guo, J., Ping, Y., Ejima, H., et al. (2014). Engineering multifunctional capsules through the assembly of metal–phenolic networks. *Angewandte Chemie International Edition*, **53**: 5546–5551.

Guo, J., Wang, X., Henstridge, D.C., et al. (2015). Nanoporous metal–phenolic particles as ultrasound imaging probes for hydrogen peroxide. *Advanced Healthcare Materials*, **4**: 2170–2175.

Guo, J., Tardy, B.L., Christofferson, A.J., et al. (2016). Modular assembly of superstructures from polyphenol-functionalized building blocks. *Nature Nanotechnology*, **11**: 1105–1111.

Guo, J., Richardson, J.J., Besford, Q.A., et al. (2017). Influence of ionic strength on the deposition of metal–phenolic networks. *Langmuir*, **33**: 10616–10622.

Guo, Y., Zhang, X., Sun, W., et al. (2019). Metal–phenolic network-based nanocomplexes that evoke ferroptosis by apoptosis: Promoted nuclear drug influx and reversed drug resistance of cancer. *Chemistry of Materials*, **31**: 10071–10084.

Haensch, C., Hoeppener, S., and Schubert, U.S. (2010). Chemical modification of self-assembled silane based monolayers by surface reactions. *Chemical Society Reviews*, **39**: 2323–2334.

Han, P., Shi, J., Nie, T., et al. (2016). Conferring natural-derived porous microspheres with surface multifunctionality through facile coordination-enabled self-assembly process. *ACS Applied Materials & Interfaces*, **8**: 8076–8085.

Hao, L., Liu, W., Wang, C., et al. (2019). Novel porous Fe_3O_4@C nanocomposite from magnetic metal-phenolic networks for the extraction of chlorophenols from environmental samples. *Talanta*, **194**: 673–679.

Harris, J.M., Martin, N.E., and Modi, M. (2001). Pegylation: a novel process for modifying pharmacokinetics. *Clinical Pharmacokinetics*, **40**: 539–551.

Hong, J. and Park, H. (2011). Fabrication and characterization of block copolymer micelle multilayer films prepared using dip-, spin- and spray-assisted layer-by-layer assembly deposition. *Colloids and Surfaces, A: Physicochemical and Engineering Aspects*, **381**: 7–12.

Hu, M., Ju, Y., Liang, K., et al. (2016). Void engineering in metal–organic frameworks via synergistic etching and surface functionalization. *Advanced Functional Materials*, **26**: 5827–5834.

Hu, F., Liu, B., Chu, H., et al. (2019a). Real-time monitoring of pH-responsive drug release using a metal-phenolic network-functionalized upconversion nanoconstruct. *Nanoscale*, **11**: 9201–9206.

Hu, Z.-Y., Xu, S., Zhao, H.-B., and Wang, Y.-Z. (2019b). Metal-phenolic networks: a biobased synergist for EVA/APP composites toward enhanced thermal stability and flame retardancy. *Journal of Applied Polymer Science*, **136**: 47243.

Huang, S., Li, X., Jiao, Y., and Shi, J. (2015). Fabrication of a superhydrophobic, fire-resistant, and mechanical robust sponge upon polyphenol chemistry for efficiently absorbing oils/organic solvents. *Industrial & Engineering Chemistry Research*, **54**: 1842–1848.

Jang, W.-S. and Grunlan, J.C. (2005). Robotic dipping system for layer-by-layer assembly of multifunctional thin films. *Review of Scientific Instruments*, **76**: 103904.

Jia, X., Wu, J., Lu, K., et al. (2019). Organic–inorganic hybrids of Fe–Co polyphenolic network wrapped Fe_3O_4 nanocatalysts for significantly enhanced oxygen evolution. *Journal of Materials Chemistry A*, **7**: 14302–14308.

Jiang, R.-J., Yan, S.-J., Tian, L.-M., et al. (2018). A biomimetic surface for infection-resistance through assembly of metal-phenolic networks. *Chinese Journal of Polymer Science*, **36**: 576–583.

Ju, Y., Cui, J., Müllner, M., et al. (2015). Engineering low-fouling and pH-degradable capsules through the assembly of metal-phenolic networks. *Biomacromolecules*, **16**: 807–814.

Ju, Y., Cui, J., Sun, H., et al. (2016a). Engineered metal-phenolic capsules show tunable targeted delivery to cancer cells. *Biomacromolecules*, **17**: 2268–2276.

Ju, Y., Dai, Q., Cui, J., et al. (2016b). Improving targeting of metal–phenolic capsules by the presence of protein coronas. *ACS Applied Materials & Interfaces*, **8**: 22914–22922.

Ju, Y., Cortez-Jugo, C., Chen, J., et al. (2020). Engineering of nebulized metal–phenolic capsules for controlled pulmonary deposition. *Advanced Science*, **7**: 1902650.

Kang, J., Bai, G., Ma, S., et al. (2019). On-site surface coordination complexation via mechanochemistry for versatile metal–phenolic networks films. *Advanced Materials Interfaces*, **6**: 1801789.

Karakoy, M., Gultepe, E., Pandey, S., et al. (2014). Silane surface modification for improved bioadhesion of esophageal stents. *Applied Surface Science*, **311**: 684–689.

Kiessling, F., Fokong, S., Bzyl, J., et al. (2014). Recent advances in molecular, multimodal and theranostic ultrasound imaging. *Advanced Drug Delivery Reviews*, **72**: 15–27.

Kim, S., Gim, T., and Kang, S.M. (2015a). Versatile, tannic acid-mediated surface PEGylation for marine antifouling applications. *ACS Applied Materials & Interfaces*, **7**: 6412–6416.

Kim, S., Kim, D.S., and Kang, S.M. (2015b). Enhanced adhesion of marine diatoms on a solid substrate by tannic acid coating. *Bulletin of the Korean Chemical Society*, **36**: 9–10.

Kim, S., Kwak, S., Lee, S., et al. (2015c). One-step functionalization of zwitterionic poly[(3-(methacryloylamino)propyl)dimethyl(3-sulfopropyl)ammonium hydroxide] surfaces by metal–polyphenol coating. *Chemical Communications*, **51**: 5340–5342.

Kim, H.J., Kim, D.-G., Yoon, H., et al. (2015d). Polyphenol/Fe^{III} complex coated membranes having multifunctional properties prepared by a one-step fast assembly. *Advanced Materials Interfaces*, **2**: 1500298.

Kim, S., Philippot, S., Fontanay, S., et al. (2015e). pH- and glutathione-responsive release of curcumin from mesoporous silica nanoparticles coated using tannic acid–Fe(III) complex. *RSC Advances*, **5**: 90550–90558.

Kim, B., Han, S., Lee, K., and Choi, I. (2017). Biphasic supramolecular self-assembly of ferric ions and tannic acid across interfaces for nanofilm formation. *Advanced Materials*, **29**: 1700784.

Köhler, J.M., Romanus, H., Hübner, U., and Wagner, J. (2007). Formation of star-like and core-shell AuAg nanoparticles during two- and three-step preparation in batch and in microfluidic systems. *Journal of Nanomaterials*, **2007**: 98134.

Kolishetti, N., Dhar, S., Valencia, P.M., et al. (2010). Engineering of self-assembled nanoparticle platform for precisely controlled combination drug therapy. *Proceedings of the National Academy of Sciences of the Unites States of America*, **107**: 17939–17944.

Krogman, K., Cohen, R., Hammond, P., et al. (2013). Industrial-scale spray layer-by-layer assembly for production of biomimetic photonic systems. *Bioinspiration & Biomimetics*, **8**: 045005.

Lee, H., Dellatore, S.M., Miller, W.M., and Messersmith, P.B. (2007). Mussel-inspired surface chemistry for multifunctional coatings. *Science*, **318**: 426–430.

Lee, I., Yoo, Y., Cheng, Z., and Jeong, H.-K. (2008). Generation of monodisperse mesoporous silica microspheres with controllable size and surface morphology in a microfluidic device. *Advanced Functional Materials*, **18**: 4014–4021.

Lee, J., Cho, H., Choi, J., et al. (2015). Chemical sporulation and germination: cytoprotective nanocoating of individual mammalian cells with a degradable tannic acid–Fe^{III} complex. *Nanoscale*, **7**: 18918–18922.

Lee, H., Kim, W.I., Youn, W., et al. (2018). Iron gall ink revisited: In situ oxidation of Fe (II)–tannin complex for fluidic-interface engineering. *Advanced Materials*, **30**: 1805091.

Levit, S.L., Walker, R.C., and Tang, C. (2019). Rapid, single-step protein encapsulation via flash nanoprecipitation. *Polymers*, **11**: 1406.

Li, W., Bing, W., Huang, S., et al. (2015). Mussel byssus-like reversible metal-chelated supramolecular complex used for dynamic cellular surface engineering and imaging. *Advanced Functional Materials*, **25**: 3775–3784.

Li, B.X., Guan, L., Wang, K., et al. (2016). Formula and process optimization of controlled-release microcapsules prepared using a coordination assembly and the response surface methodology. *Journal of Applied Polymer Science*, **133**: 42865.

Li, B.X., Li, X.X., Liu, Y., et al. (2018a). Easily tunable membrane thickness of microcapsules by using a coordination assembly on the liquid-liquid interface. *Frontiers in Chemistry*, **6**: 387.

Li, Q., Xiao, W., Zhang, F., et al. (2018b). Tannic acid-derived metal-phenolic networks facilitate PCL nanofiber mesh vascularization by promoting the adhesion and spreading of endothelial cells. *Journal of Materials Chemistry B*, **6**: 2734–2738.

Li, X., Gao, P., Tan, J., et al. (2018c). Assembly of metal–phenolic/catecholamine networks for synergistically anti-inflammatory, antimicrobial, and anticoagulant coatings. *ACS Applied Materials & Interfaces*, **10**: 40844–40853.

Li, K., Dai, Y., Chen, W., et al. (2019a). Self-assembled metal-phenolic nanoparticles for enhanced synergistic combination therapy against colon cancer. *Advanced Biosystems*, **3**: 1800241.

Li, K., Xiao, G., Richardson, J.J., et al. (2019b). Targeted therapy against metastatic melanoma based on self-assembled metal-phenolic nanocomplexes comprised of green tea catechin. *Advanced Science*, **6**: 1801688.

Liang, H., Li, J., He, Y., et al. (2016). Engineering multifunctional films based on metal–phenolic networks for rational pH-responsive delivery and cell imaging. *ACS Biomaterials Science & Engineering*, **2**: 317–325.

Lin, C.-E., Zhou, M.-Y., Hung, W.-S., et al. (2018). Ultrathin nanofilm with tailored pore size fabricated by metal-phenolic network for precise and rapid molecular separation. *Separation and Purification Technology*, **207**: 435–442.

Lin, G., Rahim, M.A., Leeming, M.G., et al. (2019). Selective metal–phenolic assembly from complex multicomponent mixtures. *ACS Applied Materials & Interfaces*, **11**: 17714–17721.

Liu, R.-Y. and Xu, A.-W. (2014). Byssal threads inspired ionic cross-linked narce-like graphene oxide paper with superior mechanical strength. *RSC Advances*, **4**: 40390–40395.

Liu, J., Guo, Z., and Liang, K. (2019a). Biocatalytic metal-organic framework-based artificial cells. *Advanced Functional Materials*, **29**: 1905321.

Liu, Z., Le, Z., Lu, L., et al. (2019b). Scalable fabrication of metal–phenolic nanoparticles by coordination-driven flash nanocomplexation for cancer theranostics. *Nanoscale*, **11**: 9410–9421.

Long, Y., Xiao, L., Cao, Q., et al. (2017). Efficient incorporation of diverse components into metal organic frameworks via metal phenolic networks. *Chemical Communications*, **53**: 10831–10834.

Luo, W., Xiao, G., Tian, F., et al. (2019). Engineering robust metal–phenolic network membranes for uranium extraction from seawater. *Energy & Environmental Science*, **12**: 607–614.

Ma, J., Lee, S.M.-Y., Yi, C., and Li, C.-W. (2017). Controllable synthesis of functional nanoparticles by microfluidic platforms for biomedical applications – a review. *Lab on a Chip*, **17**: 209–226.

Ma, Y., Björnmalm, M., Wise, A.K., et al. (2018). Gel-mediated electrospray assembly of silica supraparticles for sustained drug delivery. *ACS Applied Materials & Interfaces*, **10**: 31019–31031.

Maerten, C., Lopez, L., Lupattelli, P., et al. (2017). Electrotriggered confined self-assembly of metal–polyphenol nanocoatings using a morphogenic approach. *Chemistry of Materials*, **29**: 9668–9679.

Majedi, F.S., Hasani-Sadrabadi, M.M., Hojjati Emami, S., et al. (2013). Microfluidic assisted self-assembly of chitosan based nanoparticles as drug delivery agents. *Lab on a Chip*, **13**: 204–207.

Markwalter, C.E. and Prud'homme, R.K. (2018). Design of a small-scale multi-inlet vortex mixer for scalable nanoparticle production and application to the encapsulation of biologics by inverse flash nanoprecipitation. *Journal of Pharmaceutical Sciences*, **107**: 2465–2471.

Mkam Tsengam, I.K., Omarova, M., Shepherd, L., et al. (2019). Clusters of nanoscale liposomes modulate the release of encapsulated species and mimic the compartmentalization intrinsic in cell structures. *ACS Applied Nano Materials*, **2**: 7134–7143.

Oh, D.X., Prajatelistia, E., Ju, S.-W., et al. (2015). A rapid, efficient, and facile solution for dental hypersensitivity: The tannin–iron complex. *Scientific Reports*, **5**: 10884.

Ojo, O.F., Farinmade, A., Trout, J., et al. (2019). Stoppers and skins on clay nanotubes help stabilize oil-in-water emulsions and modulate the release of encapsulated surfactants. *ACS Applied Nano Materials*, **2**: 3490–3500.

Ozawa, H. and Haga, M.-A. (2015). Soft nano-wrapping on graphene oxide by using metal–organic network films composed of tannic acid and Fe ions. *Physical Chemistry Chemical Physics*, **17**: 8609–8613.

Pan, S., Guo, R., Bertleff-Zieschang, N., et al. (2020). Modular assembly of host–guest metal–phenolic networks using macrocyclic building blocks. *Angewandte Chemie International Edition*, **59**: 275–280.

Park, J.H., Kim, K., Lee, J., et al. (2014). A cytoprotective and degradable metal–polyphenol nanoshell for single-cell encapsulation. *Angewandte Chemie International Edition*, **53**: 12420–12425.

Park, C., Yang, B.J., Jeong, K.B., et al. (2017a). Signal-induced release of guests from a photolatent metal–phenolic supramolecular cage and its hybrid assemblies. *Angewandte Chemie International Edition*, **56**: 5485–5489.

Park, J.H., Choi, S., Moon, H.C., et al. (2017b). Antimicrobial spray nanocoating of supramolecular Fe(III)-tannic acid metal-organic coordination complex: applications to shoe insoles and fruits. *Scientific Reports*, **7**: 6980.

Park, T., Kim, W.I., Kim, B.J., et al. (2018). Salt-induced, continuous deposition of supramolecular iron(III)–tannic acid complex. *Langmuir*, **34**: 12318–12323.

Park, K.H., Yang, S.Y., An, B.-S., et al. (2019). Metal ion-mediated interliposomal aggregation of polydiacetylene liposomes incorporating a phenolic lipid. *Journal of Nanoscience and Nanotechnology*, **19**: 3755–3761.

Ping, Y., Guo, J., Ejima, H., et al. (2015). pH-Responsive capsules engineered from metal–phenolic networks for anticancer drug delivery. *Small*, **11**: 2032–2036.

Prajatelistia, E., Ju, S.W., Sanandiya, N.D., et al. (2016). Tunicate-inspired gallic acid/metal ion complex for instant and efficient treatment of dentin hypersensitivity. *Advanced Healthcare Materials*, **5**: 919–927.

Pustulka, K.M., Wohl, A.R., Lee, H.S., et al. (2013). Flash nanoprecipitation: particle structure and stability. *Molecular Pharmaceutics*, **10**: 4367–4377.

Qi, A., Chan, P., Ho, J., et al. (2011). Template-free synthesis and encapsulation technique for layer-by-layer polymer nanocarrier fabrication. *ACS Nano*, **5**: 9583–9591.

Qiao, D., Liu, L., Chen, Y., et al. (2018). Potency of a scalable nanoparticulate subunit vaccine. *Nano Letters*, **18**: 3007–3016.

Qin, Y., Wang, J., Qiu, C., et al. (2019). Self-assembly of metal–phenolic networks as functional coatings for preparation of antioxidant, antimicrobial, and pH-sensitive-modified starch nanoparticles. *ACS Sustainable Chemistry & Engineering*, **7**: 17379–17389.

Queffélec, C., Petit, M., Janvier, P., et al. (2012). Surface modification using phosphonic acids and esters. *Chemical Reviews*, **112**: 3777–3807.

Rahim, M.A., Ejima, H., Cho, K.L., et al. (2014). Coordination-driven multistep assembly of metal–polyphenol films and capsules. *Chemistry of Materials*, **26**: 1645–1653.

Rahim, M., Kempe, K., Muellner, M., et al. (2015). Surface-confined amorphous films from metal-coordinated simple phenolic ligands. *Chemistry of Materials*, **27**: 5825–5832.

Rahim, M., Bjornmalm, M., Bertleff-Zieschang, N., et al. (2017). Rust-mediated continuous assembly of metal–phenolic networks. *Advanced Materials*, **29**: 1606717.

Rahim, M.A., Björnmalm, M., Bertleff-Zieschang, N., et al. (2018). Multiligand metal–phenolic assembly from green tea infusions. *ACS Applied Materials & Interfaces*, **10**: 7632–7639.

Rahim, M.A., Kristufek, S.L., Pan, S., et al. (2019). Phenolic building blocks for the assembly of functional materials. *Angewandte Chemie International Edition*, **58**: 1904–1927.

Rani, M., Moudgil, L., Singh, B., et al. (2016). Understanding the mechanism of replacement of citrate from the surface of gold nanoparticles by amino acids: a theoretical and experimental investigation and their biological application. *RSC Advances*, **6**: 17373–17383.

Resetco, C., Hendriks, B., Badi, N., and Du Prez, F. (2017). Thiol–ene chemistry for polymer coatings and surface modification – building in sustainability and performance. *Materials Horizons*, **4**: 1041–1053.

Richardson, J.J., Liang, K., Kempe, K., et al. (2013). Immersive polymer assembly on immobilized particles for automated capsule preparation. *Advanced Materials*, **25**: 6874–6878.

Richardson, J.J., Cui, J., Björnmalm, M., et al. (2016). Innovation in layer-by-layer assembly. *Chemical Reviews*, **116**: 14828–14867.

Saad, W.S. and Prud'homme, R.K. (2016). Principles of nanoparticle formation by flash nanoprecipitation. *Nano Today*, **11**: 212–227.

Santos, J.L., Ren, Y., Vandermark, J., et al. (2016). Continuous production of discrete plasmid DNA-polycation nanoparticles using flash nanocomplexation. *Small*, **12**: 6214–6222.

Shen, G., Xing, R., Zhang, N., et al. (2016). Interfacial cohesion and assembly of bioadhesive molecules for design of long-term stable hydrophobic nanodrugs toward effective anticancer therapy. *ACS Nano*, **10**: 5720–5729.

Shiratori, S.S. and Yamada, M. (2000). Nano-scale control of composite polymer films by mass-controlled layer-by-layer sequential adsorption of polyelectrolytes. *Polymers for Advanced Technologies*, **11**: 810–814.

Song, Y., Modrow, H., Henry, L.L., et al. (2006). Microfluidic synthesis of cobalt nanoparticles. *Chemistry of Materials*, **18**: 2817–2827.

Song, Q., Zhao, W.-J., Yin, H.-X., and Lian, H.-Z. (2015). Facile synthesis of Fe^{III}–tannic acid film-functionalized magnetic silica microspheres for the enrichment of low-abundance peptides and proteins for MALDI-TOF MS analysis. *RSC Advances*, **5**: 63896–63902.

Sperling, R.A. and Parak, W.J. (2010). Surface modification, functionalization and bioconjugation of colloidal inorganic nanoparticles. *Philosophical Transactions of the Royal Society, A: Mathematical, Physical and Engineering Sciences*, **368**: 1333–1383.

Spoljaric, S., Ju, Y., and Caruso, F. (2021). Protocols for reproducible, increased-scale synthesis of engineered particles—bridging the "upscaling gap". *Chemistry of Materials*, **33**: 1099–1115.

Sylla, T., Pouységu, L., Da Costa, G., et al. (2015). Gallotannins and tannic acid: first chemical syntheses and in vitro inhibitory activity on Alzheimer's amyloid β-peptide aggregation. *Angewandte Chemie International Edition*, **54**: 8217–8221.

Tao, J., Chow, S.F., and Zheng, Y. (2019). Application of flash nanoprecipitation to fabricate poorly water-soluble drug nanoparticles. *Acta Pharmaceutica Sinica B*, **9**: 4–18.

Tardy, B.L., Richardson, J.J., Guo, J., et al. (2018). Lignin nano- and microparticles as template for nanostructured materials: formation of hollow metal-phenolic capsules. *Green Chemistry*, **20**: 1335–1344.

Tardy, B.L., Richardson, J.J., Nithipipat, V., et al. (2019). Protein adsorption and coordination-based end-tethering of functional polymers on metal–phenolic network films. *Biomacromolecules*, **20**: 1421–1428.

Tu, Q., Shen, X., Liu, Y., et al. (2019). A facile metal–phenolic–amine strategy for dual-functionalization of blood-contacting devices with antibacterial and anticoagulant properties. *Materials Chemistry Frontiers*, **3**: 265–275.

Wang, Y., Zhang, Y., Hou, C., et al. (2016). Facile one-pot assembly of adhesive phenol/Fe^{III}/PEI complexes for preparing magnetic hybrid microcapsules. *New Journal of Chemistry*, **40**: 781–788.

Wang, H., Zhu, W., Ping, Y., et al. (2017a). Controlled fabrication of functional capsules based on the synergistic interaction between polyphenols and MOFs under weak basic condition. *ACS Applied Materials & Interfaces*, **9**: 14258–14264.

Wang, J., Li, Y., Wang, X., et al. (2017b). Droplet microfluidics for the production of microparticles and manoparticles. *Micromachines*, **8**: 22.

Wang, C., Sang, H., Wang, Y., et al. (2018a). Foe to friend: supramolecular nanomedicines consisting of natural polyphenols and bortezomib. *Nano Letters*, **18**: 7045–7051.

Wang, G., Qin, J., Zhou, X., et al. (2018b). Self-template synthesis of mesoporous metal oxide spheres with metal-mediated inner architectures and superior sensing performance. *Advanced Functional Materials*, **28**: 1806144.

Wang, X., Yan, J., Pan, D., et al. (2018c). Polyphenol–poloxamer self-assembled supramolecular nanoparticles for tumor NIRF/PET imaging. *Advanced Healthcare Materials*, **7**: 1701505.

Wei, J., Liang, Y., Hu, Y., et al. (2016). A versatile iron–tannin-framework ink coating strategy to fabricate biomass-derived iron carbide/Fe-N-carbon catalysts for efficient oxygen reduction. *Angewandte Chemie International Edition*, **55**: 1355–1359.

Wei, Y., Wei, Z., Luo, P., et al. (2018). pH-Sensitive metal-phenolic network capsules for targeted photodynamic therapy against cancer cells. *Artificial Cells, Nanomedicine, and Biotechnology*, **46**: 1552–1561.

Weiss, A.C.G., Kempe, K., Förster, S., and Caruso, F. (2018). Microfluidic examination of the "hard" biomolecular corona formed on engineered particles in different biological milieu. *Biomacromolecules*, **19**: 2580–2594.

Wu, J., Wang, Z., Yan, W., et al. (2015). Improving the hydrophilicity and fouling resistance of RO membranes by surface immobilization of PVP based on a metal-polyphenol precursor layer. *Journal of Membrane Science*, **496**: 58–69.

Xiao, G., Chen, W., Tian, F., et al. (2018a). Thermal transition of bimetallic metal–phenolic networks to biomass-derived hierarchically porous nanofibers. *Chemistry – An Asian Journal*, **13**: 972–976.

Xiao, Z., Meng, J., Li, Q., et al. (2018b). General and precise carbon confinement of functional nanostructures derived from assembled metal–phenolic networks for enhanced lithium storage. *Journal of Materials Chemistry A*, **6**: 18605–18614.

Xu, Q., Hashimoto, M., Dang, T.T., et al. (2009). Preparation of monodisperse biodegradable polymer microparticles using a microfluidic flow-focusing device for controlled drug delivery. *Small*, **5**: 1575–1581.

Yang, C., Wu, H., Shi, J., et al. (2014). Preparation of dopamine/titania hybrid nanoparticles through biomimetic mineralization and titanium(IV)–catecholate coordination for enzyme immobilization. *Industrial & Engineering Chemistry Research*, **53**: 12665–12672.

Yang, C., Wu, H., Yang, X., et al. (2015). Coordination-enabled one-step assembly of ultrathin, hybrid microcapsules with weak pH-response. *ACS Applied Materials & Interfaces*, **7**: 9178–9184.

Yang, Z., Yang, Y., Xiong, K., et al. (2018). Metal-phenolic surfaces for generating therapeutic nitric oxide gas. *Chemistry of Materials*, **30**: 5220–5226.

Yang, Y., Li, Y., Li, Q., et al. (2019). Rapid co-deposition of graphene oxide incorporated metal-phenolic network/piperazine followed by crosslinking for high flux nanofiltration membranes. *Journal of Membrane Science*, **588**: 117203.

Yun, G., Pan, S., Wang, T.-Y., et al. (2017). Synthesis of metal nanoparticles in metal-phenolic networks: catalytic and antimicrobial applications of coated textiles. *Advanced Healthcare Materials*, **7**: 1700934.

Yun, G., Besford, Q.A., Johnston, S.T., et al. (2018). Self-assembly of nano- to macroscopic metal–phenolic materials. *Chemistry of Materials*, **30**: 5750–5758.

Yun, G., Richardson, J.J., Biviano, M., and Caruso, F. (2019). Tuning the mechanical behavior of metal–phenolic networks through building block composition. *ACS Applied Materials & Interfaces*, **11**: 6404–6410.

Yun, G., Richardson, J.J., Capelli, M., et al. (2020). The biomolecular corona in 2D and reverse: patterning metal–phenolic networks on proteins, lipids, nucleic acids, polysaccharides, and fingerprints. *Advanced Functional Materials*, **30**: 1905805.

Zeng, T., Zhang, X., Guo, Y., et al. (2014). Enhanced catalytic application of Au@polyphenol-metal nanocomposites synthesized by a facile and green method. *Journal of Materials Chemistry A*, **2**: 14807–14811.

Zhang, W., Christofferson, A.J., Besford, Q.A., et al. (2019). Metal-dependent inhibition of amyloid fibril formation: synergistic effects of cobalt–tannic acid networks. *Nanoscale*, **11**: 1921–1928.

Zhao, C.-X., He, L., Qiao, S.Z., and Middelberg, A.P. (2011). Nanoparticle synthesis in microreactors. *Chemical Engineering Science*, **66**: 1463–1479.

Zheng, H.-T., Bui, H.L., Chakroborty, S., et al. (2019). Pegylated metal-phenolic networks for antimicrobial and antifouling properties. *Langmuir*, **35**: 8829–8839.

Zhong, Q.-Z., Pan, S., Rahim, M.A., et al. (2018). Spray assembly of metal–phenolic networks: formation, growth, and applications. *ACS Applied Materials & Interfaces*, **10**: 33721–33729.

Zhong, Q.-Z., Li, S., Chen, J., et al. (2019). Oxidation-mediated kinetic strategies for engineering metal–phenolic networks. *Angewandte Chemie International Edition*, **58**: 12563–12568.

Zhu, W., Liang, S., Wang, J., et al. (2017). Europium-phenolic network coated $BaGdF_5$ nanocomposites for tri-modal computed tomography/magnetic resonance/luminescence imaging. *Journal of Materials Science: Materials in Medicine*, **28**: 74.

Zhu, X., Cheng, X., Xing, J., et al. (2020). In-situ covalently bonded supramolecular-based protective layer for improving chlorine resistance of thin-film composite nanofiltration membranes. *Desalination*, **474**: 114197.

Zou, Y., Wang, Z., Zhang, H., and Liu, Y. (2018). A novel electrogenerated chemiluminescence biosensor for histone acetyltransferases activity analysis and inhibition based on mimetic superoxide dismutase of tannic acid assembled nanoprobes. *Biosensors and Bioelectronics*, **122**: 205–210.

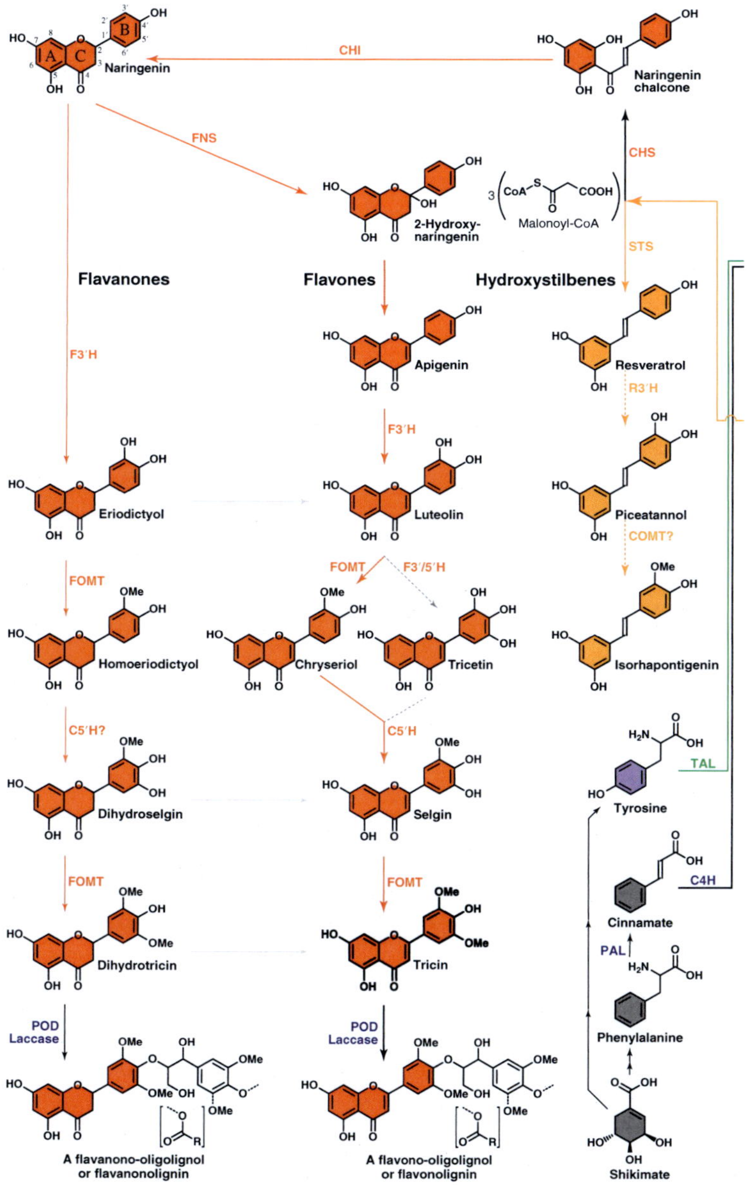

Figure 1.1 Biosynthetic pathways feeding into lignification. The primary biosynthetic pathway leading to the monolignols and the two related hydroxycinnamyl alcohols, bolded and highlighted in yellow (in the colored version of this figure), is to the right and is shown as a metabolic grid as in the old days (Dixon et al. 2001; Higuchi 2006; Matsui et al. 2000); unfortunately, since modifications to the pathway flux have been discovered, it is not possible to order the intermediates in a linear fashion vertically – conversions jump rows in a way that is simply not pleasing (and, were this a circuit-board, would represent bad design but seems unavoidable here). Where possible, the primary pathways are shown with black arrows, and the major pathways are slightly bolder, whereas minor pathways, or those that might not be completely demonstrated are in a light gray; dashed arrows represent pathways (and genes/enzymes) that are not yet known. Note that we do not try to show the pathway from *p*-hydroxybenzoate, via its CoA thioester (not shown) to the monolignol *p*-hydroxybenzoate (and other benzoate) conjugates in the bottom row; the long-anticipated (Ralph 2010; Ralph et al. 2019) transferase enzyme/gene required to produce monolignol *p*-hydroxybenzoates has now been identified by two groups (de Vries et al. 2022;

Recent Advances in Polyphenol Research, Volume 8, First Edition. Edited by Juha-Pekka Salminen, Kristiina Wähälä, Victor de Freitas, and Stéphane Quideau.

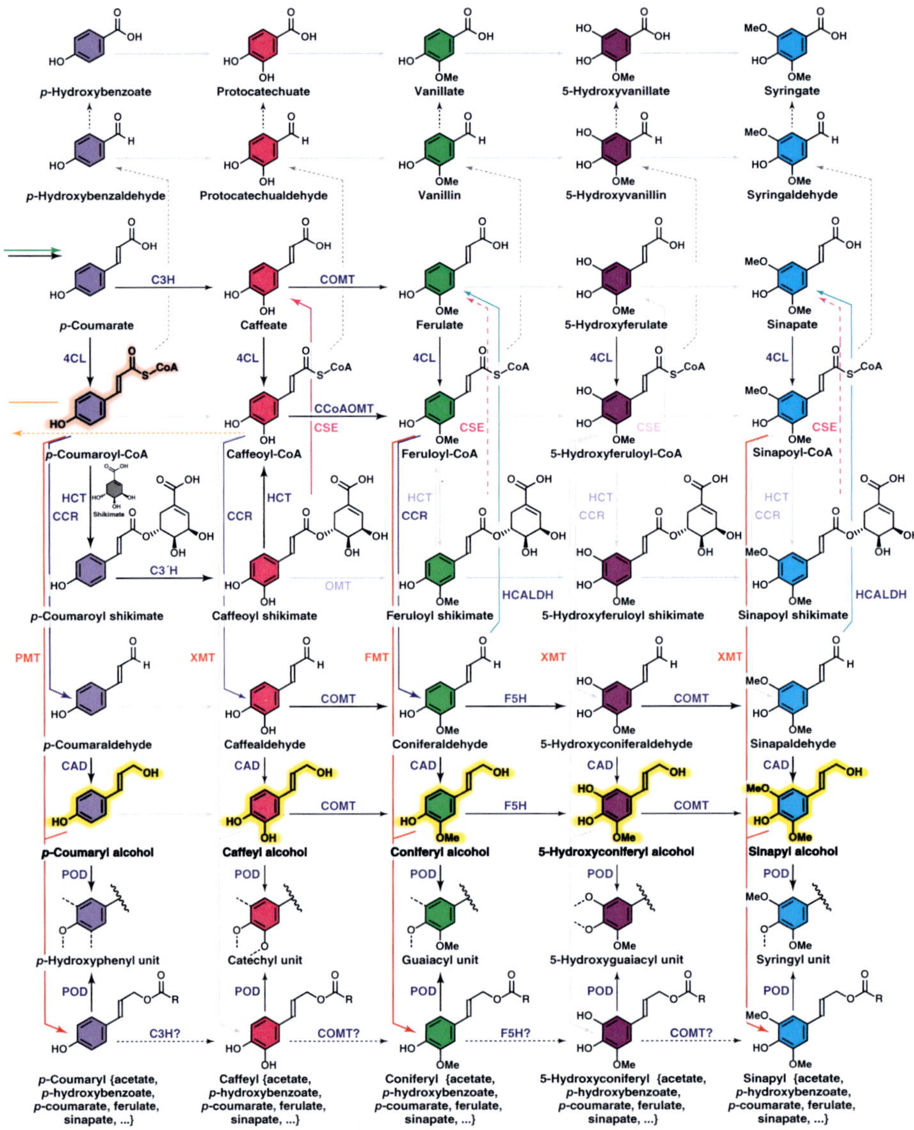

Figure 1.1 (Continued) Zhao et al. 2021). For clarity, the five types of aromatic nuclei, i.e. with the aromatic ring substitution at the *p*-hydroxyphenyl, catechyl, guaiacyl, 5-hydroxyguaiacyl, and syringyl level, in their columns are color-coded in common. To the left are the pathways to the hydroxystilbenes and flavonoids that may also be monomers used in lignification. The gray-ringed structures to the left of center at the bottom are the primary inputs into these pathways with the aromatics deriving from shikimic acid, via the amino-acids phenylalanine and tyrosine (Yoo et al. 2013). A major branchpoint for all three pathways is *p*-coumaroyl-CoA, also bolded, and highlighted in pink. Enzyme abbreviations: Monolignol pathway: PAL, phenylalanine ammonia-lyase; TAL, tyrosine ammonia-lyase (in monocots); C4H, cinnamate 4-hydroxylase; C3H, *p*-coumarate (or *p*-coumaroyl-CoA) 3-hydroxylase; C3′H, *p*-coumaroyl shikimate 3′-hydroxylase; CCoAOMT, caffeoyl-CoA *O*-methyltransferase; 4CL, 4-coumarate: CoA ligase; CCR, cinnamoyl-CoA reductase; HCT, hydroxycinnamoyl-CoA: quinate/shikimate hydroxycinnamoyl transferase; CSE, caffeoyl shikimate esterase; F5H, ferulate 5-hydroxylase; COMT, caffeoyl-CoA *O*-methyltransferase; COMT, caffeic acid *O*-methyltransferase; CAD, cinnamyl alcohol dehydrogenase; HCALDH, hydroxycinnamaldehyde dehydrogenase; PMT, *p*-coumaroyl-CoA: monolignol transferase; FMT, feruloyl-CoA: monolignol transferase; XMT, a general hydroxycinnamoyl-CoA: monolignol transferase; POD, peroxidase (or laccase, not shown except in flavonoid pathway); Hydroxystilbene pathway: STS, stilbene synthase; R3′H, resveratrol 3′-hydroxylase; COMT?, the monolignol pathway COMT or another OMT; Flavonoid pathway: CHS, chalcone synthase; CHI, chalcone isomerase; FNS, flavone synthase; F3′H, flavonoid 3′-hydroxylase; F5′H, flavonoid 5′-hydroxylase or a dual-function 3′/5′-hydroxylase; FOMT, flavonoid-*O*-methyltransferase; C5′H, chrysoeriol 5′-hydroxylase. We heartily thank Ruben Vanholme (VIB) for helping clarify the nuances of the monolignol pathway.

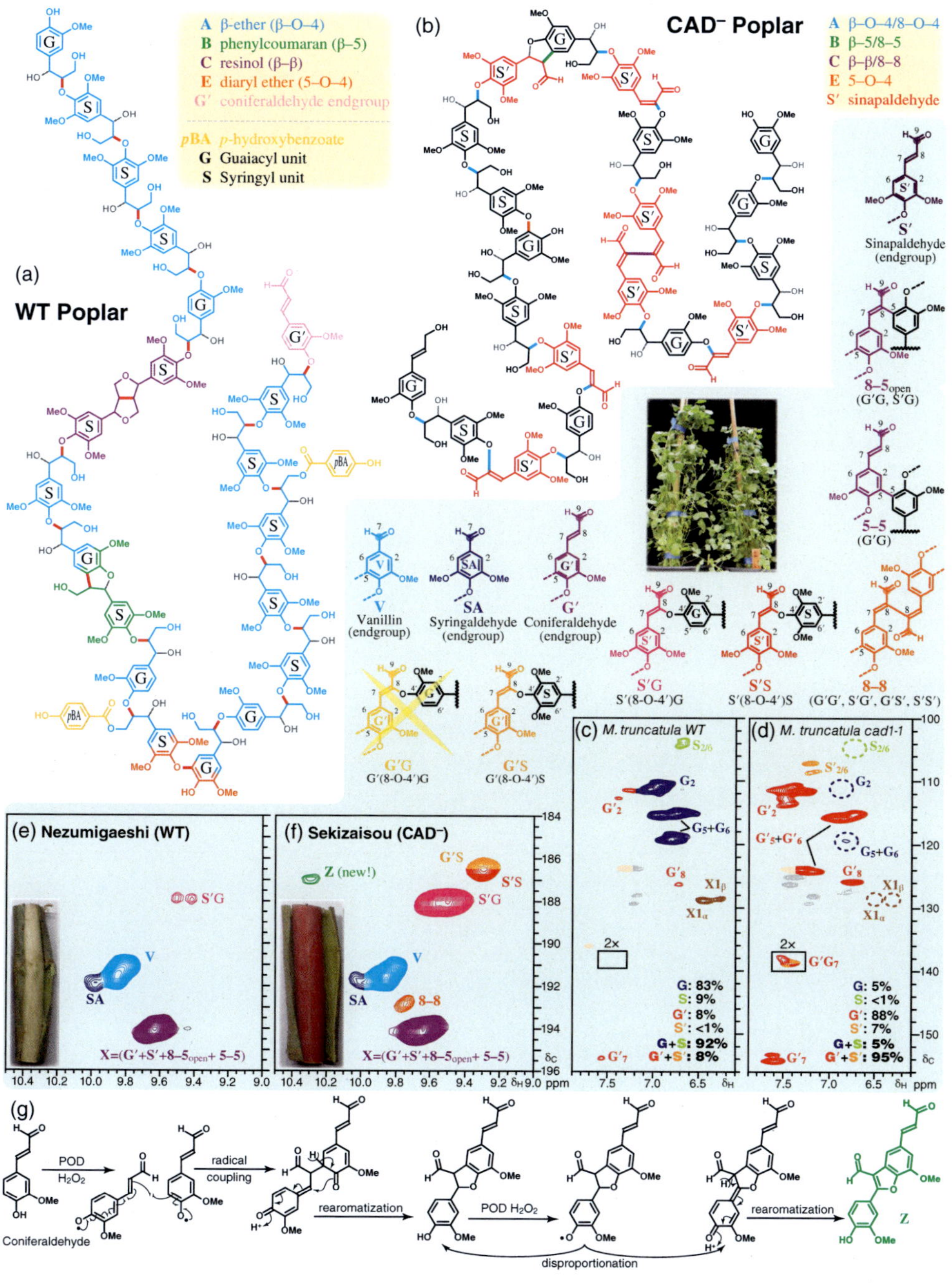

Figure 1.2 Aldehyde-containing lignins. (a) A model of a poplar lignin. *Source:* Ralph et al. (2019)/Elsevier. (b) A model of a poplar CAD-deficient lignin that incorporated mainly sinapaldehyde units. *Source:* Ralph et al. (2019)/Elsevier. (c) The aromatic regions of HSQC NMR spectra from lignins isolated from WT *Medicago truncatula* and (d) a CAD mutant in the same background showing that the lignin is heavily (~95%) hydroxycinnamaldehyde-derived, i.e. that this viable plant has very

Figure 1.2 (Continued) little of its lignin derived from any of the canonical monolignols. *Source:* Zhao et al. (2013)/with permission of Proceedings of the National Academy of Sciences. (e) A mulberry variety (Nezumigaeshi) with normal-colored wood (inset) and in which the NMR aldehyde region shows the typical hydroxycinnamaldehyde **X** and benzaldehyde **SA** and **V**) endgroups along with, as we often detect now (Van Acker et al. 2017; Yan et al. 2019), low levels of the product in which sinapaldehyde has 8–*O*–4-cross-coupled with a G-unit in the growing polymer to become part of the internal backbone of the lignin, structure **S′G**, not just an endgroup. *Source:* Yamamoto et al. (2020)/Oxford University Press. (f) The same aldehyde region from the NMR of the lignin isolated from a wild mutant mulberry (Sekizaisou) discovered to have a red-colored wood and leaves that are particularly suited for silkworm feeding that was recently revealed to be a CAD-deficient mutant, the first reported in a hardwood. *Source:* Yamamoto et al. (2020)/Oxford University Press. Units are derived from the 8–*O*–4-cross-coupling of both sinapaldehyde and coniferaldehyde into the polymer (**S′G** and **G′S**, but not **G′G**); by separate experiments, it is possible to determine that sinapaldehyde cross-couples with either guaiacyl or syringyl units, whereas coniferaldehyde cross-couples with only syringyl units (as is also the case *in vitro*) (Kim et al. 2003). The green peak labeled "**Z** (new)" has been observed previously in spectra from our labs from aldehyde-rich lignins and synthetic lignins incorporating coniferaldehyde but had not previously been structurally assigned. (g) The new product has recently been determined to be an unusual benzofuran **Z** that is clearly not a primary coupling product but arises from 8–5-dimerization of coniferaldehyde followed by further single-electron oxidation and then radical disproportionation to a quinone methide intermediate followed by rearomatization (K. Yoshioka, in preparation). Note: Here and in Figure 1.3, we use the standard convention by which the 3-carbon sidechains on monolignol and lignin units are labeled α, β, and γ (from the aromatic C1 attachment), whereas the benzaldehydes and cinnamaldehydes, as well as other "oxidized units" such as ferulate and *p*-coumarate, use 7, 8, and 9.

Figure 1.3 (Continued) (2012)/with permission of Proceedings of the National Academy of Sciences. (c–e) The stem, pod, and leaf tissues show S/G profiles typical of an angiosperm, with no evidence of C-units. *Source:* Chen et al. (2012)/Proceedings of the National Academy of Sciences. (f) A caffeyl alcohol synthetic lignin (dehydrogenation polymer, DHP) is rather simple, with notable end-units in its HSQC NMR spectrum because of its low molecular weight. The main units are benzodioxanes. *Source:* Adapted from Li et al. (2018). (g) An isolated enzyme lignin (EL) from vanilla seedcoats has a particularly simple HSQC NMR spectrum, showing essentially only benzodioxane units from β–*O*–4-coupling (along with peaks from residual cellulose, C); from the original data used in figure 1a and S1C in Li et al. (2018). The inset is a picture of a sliced-though vanilla seedpod at 10 days post-pollination, showing the black seedcoats containing the C-lignin. *Source:* Chen et al. (2012)/with permission of National Academy of Science. (h) Yield and selectivity of Pd/C-catalyzed hydrogenolysis to produce arylpropanol, arylpropane, and other monomers, for the vanilla seedcoat C-lignin here (in which the arylpropanol monomer is catechylpropanol and the minor arylpropane monomer is catechylpropane) *vs* spruce, birch, and high-S poplar. The C-lignin delivers almost 90% yields of monomers in a product that is 90% one compound, dihydrocaffeyl alcohol (= catechylpropanol); the prior record monomer yield had been from the high-S lignin (Lan et al. 2018a; Shuai et al. 2016). Conditions: 5% Pd/C, 473 K, 15 h, 4 MPa H_2. *Source:* Li et al. (2019)/Royal Society of Chemistry. Picture in the bottom right is part of the picture prepared for a cover issue (Li et al. 2018). Note: Here and in Figure 1.2, we use the standard convention by which the 3-carbon sidechains on monolignol and lignin units are labeled α, β, and γ (from the aromatic C1 attachment), whereas the benzaldehydes and cinnamaldehydes, as well as other "oxidized units" such as ferulate and *p*-coumarate, use 7, 8, and 9.

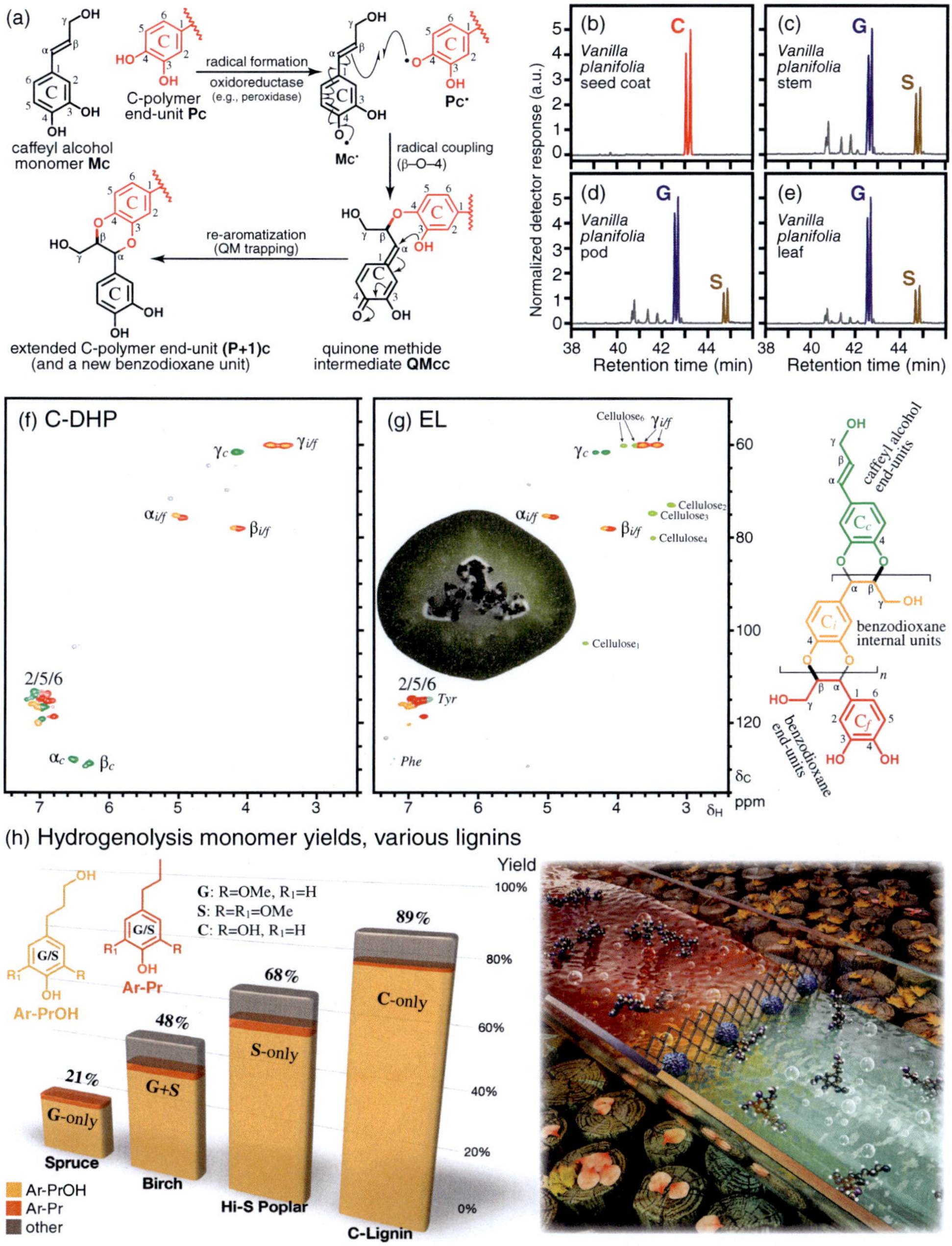

Figure 1.3 C-lignin, derived from caffeyl alcohol. (a) Lignification using caffeyl alcohol as the sole monomer results in an almost pure homopolymer of benzodioxane units arising from β–*O*–4-coupling; although C-units can, in principle, couple with all the options open to G-units, the β–*O*–4-coupling propensity is simply overwhelming such that only low levels of other structures can be found in the polymer. *Source:* Chen et al. (2012)/National Academy of Sciences. (b) Thioacidolysis of vanilla seedcoats releases the typical isomer pair of a sole C-monomer, with no contributions from S- or G-units derived from the canonical monolignols. *Source:* Chen et al.

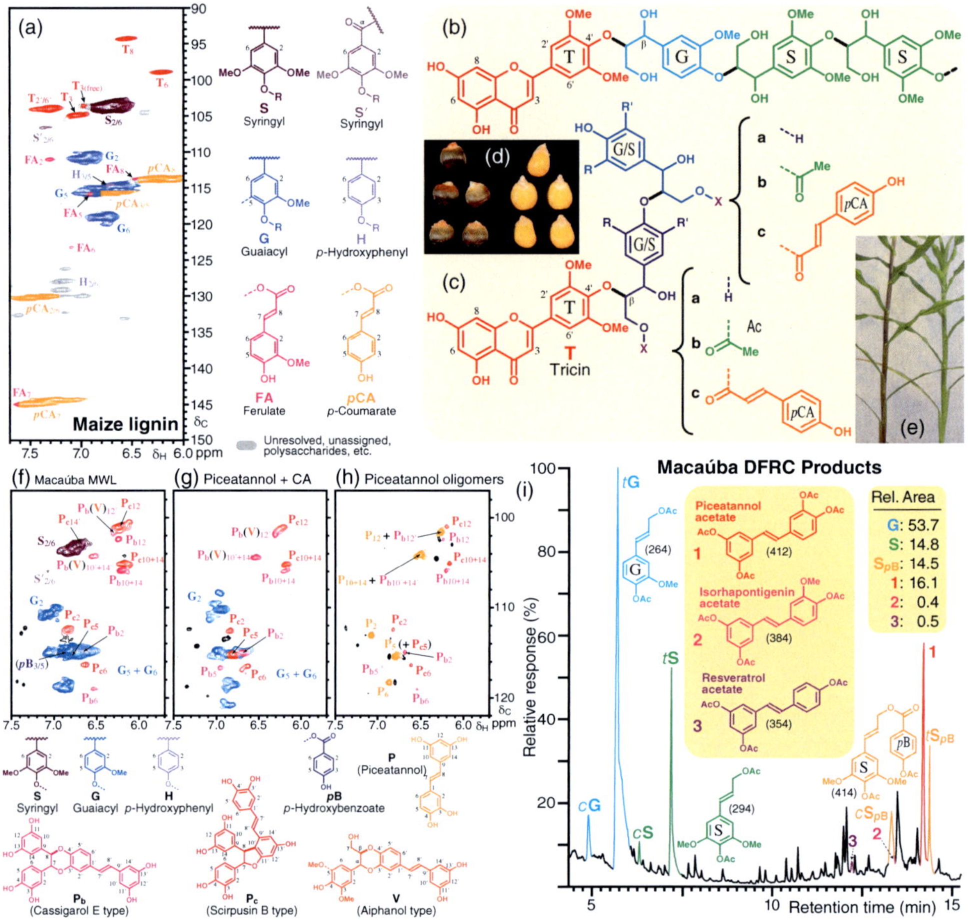

Figure 1.4 Tricin and hydroxystilbenes in lignins. Tricin was first discovered in a wheat straw lignin (del Río et al. 2012). (a) The tricin is clearly evident in "all" grasses, as in the aromatic region of the HSQC-NMR spectrum from maize stem lignin here. *Source:* Lan et al. (2015)/Oxford University Press. (b) Tricin can only start a lignin chain, as shown in this 4-unit model, and is predominantly found coupled to a guaiacyl unit, i.e. from the coupling of coniferyl alcohol with tricin in early lignification, as shown here. (c) Tricin-oligolignol metabolites provided a rich source of information about the combinatorial coupling of tricin with not just coniferyl and sinapyl alcohol, but also with the various acetate and *p*-coumarate conjugates from them, and for the next unit as well. Not all but a great many of the possible combinations indicated have been identified by metabolic profiling. *Source:* Adapted from Lan et al. (2016a). (d, e) Tricin biosynthesis can be completely blocked by knocking out the first committed enzyme (chalcone synthase, CHS, see Figure 1.1) from *p*-coumaroyl-CoA into the flavonoid pathway (Eloy et al. 2017). In a CHS maize mutant, not only is the lignin devoid of tricin (not shown), but the grain and the corn stems lack the dark pigmentation resulting from flavonoids. *Source:* Eloy et al. (2017)/Oxford University Press. (f) Hydroxystilbenes are found in the lignins from various palm fruit endocarp tissues, such as Macaúba, as shown in the aromatic region of HSQC NMR spectra here. *Source:* del Río et al. (2017)/Oxford University Press. (g, h) Synthetic polymers derived from either copolymerizing piceatannol with coniferyl alcohol, or allowing piceatannol alone to polymerize, help identify several of the new peaks in the lignin spectrum in (f). *Source:* del Río et al. (2017)/Oxford University Press. (i) Cleavage of lignin β-ethers using DFRC releases, in addition to the usual G and S monolignols (as their acetates), the acetates of the three hydroxystilbenes piceatannol, isorhapontigenin, and resveratrol, indicating their presence in the lignin as ethers. *Source:* del Río et al. (2017)/Oxford University Press.

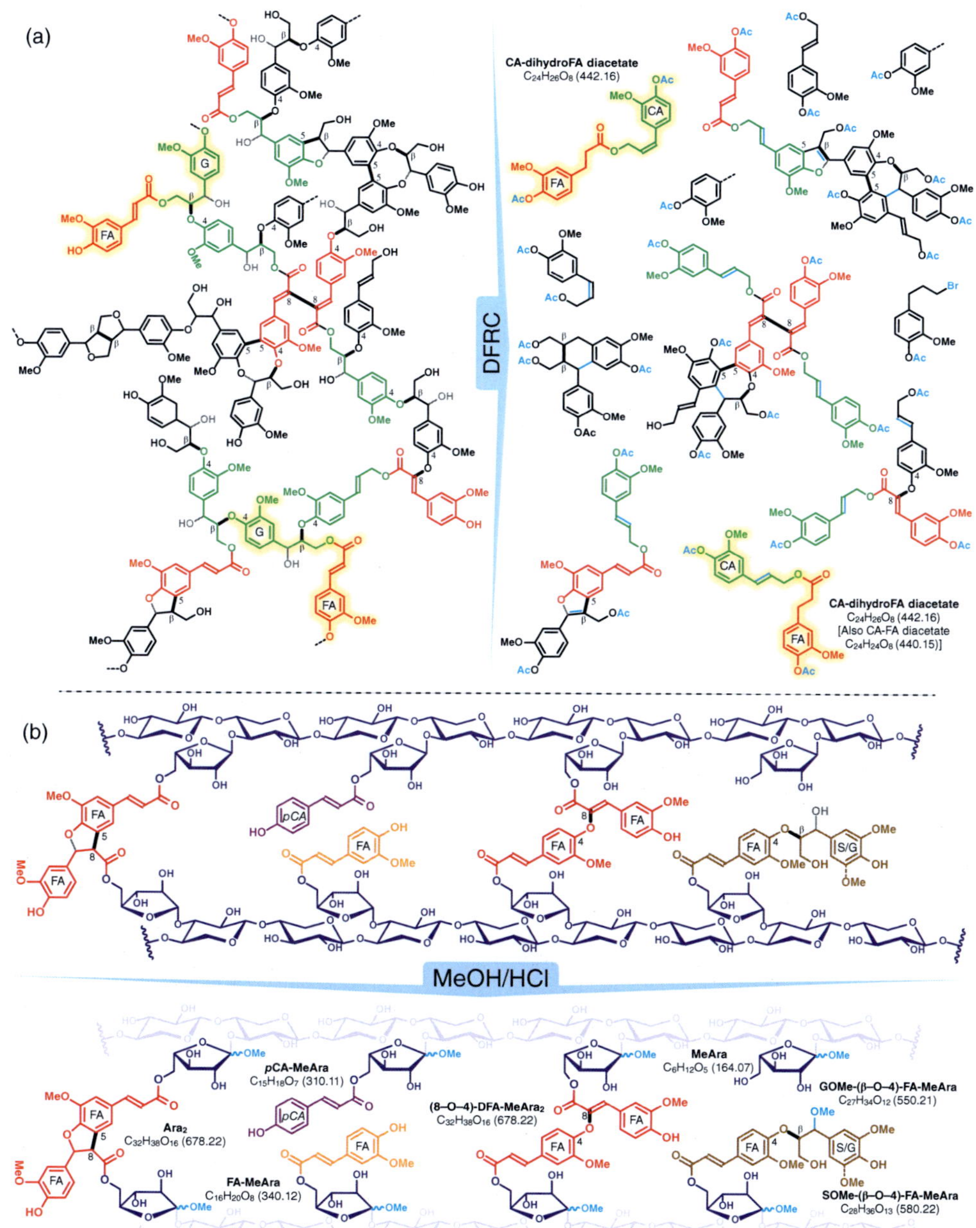

Figure 1.11 DFRC and mild acidolysis, orthogonal methods for determining hydroxycinnamates on lignins *vs* on polysaccharides. (a) A model for a G-lignin derived from coniferyl alcohol (CA) and coniferyl ferulate (CA-FA), i.e. a zip-lignin (*Source:* Adapted from Ralph et al. (2019)) in which a plausible structure derived from the coupling of 7 CA-FA conjugates (green-red in the color figure) and 13 further monolignol moieties (black) is shown. The bonds produced by radical coupling are shown in bold (red in the color figure), and the bonds formed by post-coupling rearomatization reactions (along with the OH from water addition) are in gray in the color figure. Following DFRC, hydroxyls (including those freshly liberated) are acetylated and β-ether bonds are cleaved leaving a cinnamyl

Figure 1.11 (Continued) double-bond signature, but esters remain intact (Wilkerson et al. 2014); some of the products from the depolymerization expected to occur are shown in the figure to the right with new bonds produced during DFRC and new acetate groups shown in cyan (in the colored version). [A great deal of information on DFRC mechanisms and products is attempted to be conveyed here in the products on the right from the various structures found in the lignins. Monomer units linked solely by β-ether bonds will release (acetylated) monolignols (Lu and Ralph 1997b); two are shown here along with two possible units for which the sidechains are not specified. A coniferyl alcohol endgroup unit (halfway down on the right) will release one of a variety of monomers (Lu and Ralph 1999b); the acetylated guaiacylbromopropane is shown. Phenylcoumarans typically dehydrate to benzofurans; two are shown, one derived from CA (lower left), and one from CA-FA (top-middle). Resinols undergo acid rearrangement to produce aryltetralins (Ralph et al. 2004b); one is shown (middle-left). As far as we know, even though they are a kind of β-ether, ferulate-8–*O*–4-ethers may not cleave in DFRC; for similar reasons, they do not cleave even upon high-temperature base treatment either (Grabber et al. 1995). Finally, the 8-membered-ring dibenzodioxocins resulting initially from 5–5-coupling of guaiacyl units appear to undergo a rearrangement to a 7-membered ring structure, at least in model systems (Argyropoulos et al. 2002); two examples are shown here, one from normal lignin guaiacyl units and the other from 5–5-diferulate.] In the model on the left, there are two ferulates deriving from the incorporation of CA-FA conjugates (highlighted in gray or yellow shading on the left) by means that will release quantifiable CA-FA and/or coniferyl dihydroferulate conjugates (similarly highlighted on the right) upon DFRC (Regner et al. 2018; Wilkerson et al. 2014). Other incorporated CA-FA units will not be able to release the simple conjugates; other lower molecular mass fragments will also be released by the β-ether cleavage but as more complex oligomers for which authentic standards and quantitative methods are not yet available. (b) Methanolysis of hydroxycinnamoylated arabinoxylan (Eugene et al. 2020; Lapierre et al. 2018, 2019). Top: Schematic arabinoxylan chains shown with the arabinosyl substitutions on every third unit as revealed recently (Busse-Wicher et al. 2016), along with *p*CA and FA substitutions on arabinosyl units, two representative diferulates (8–*O*–4 and 8–5) formed by radical coupling, and an ML-FA conjugate, all known to be present in grass cell walls; acetate substitutions, also known, are not shown. The bonds produced by radical coupling are shown in bold-black, and the bonds formed by post-coupling rearomatization reactions (along with the OH from water addition in the case of the ML-FA coupling) are in gray. Bottom: The derived products released for structural analysis and quantification upon methanolysis of the arabinosyl unit from the xylan backbone leave the arabinosyl esters largely intact. The original xylan chain is shown with low opacity for easier visualization of the source of the products, and the elemental formulae and exact masses (for MS analysis) are shown. *Source:* Figures and data are adapted from Eugene et al. (2020) and Lapierre et al. (2019).

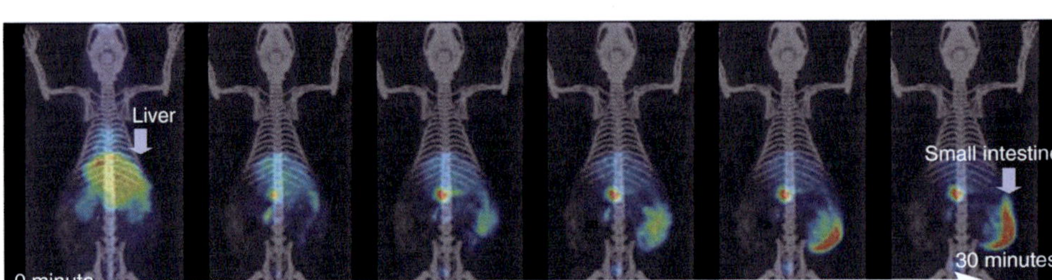

Figure 2.3 Whole-body PET imaging of the distribution of intravenously injected [^{11}C] Me-EGCg. Biodistribution of [^{11}C] Me-EGCg was determined with a small-animal PET system (Clairvivo PET). [^{11}C] Me-EGCg solution was intravenously injected into a Wistar rat via a tail vein, and PET scanning was started immediately. Images acquired from 0 to 30 minutes are shown.

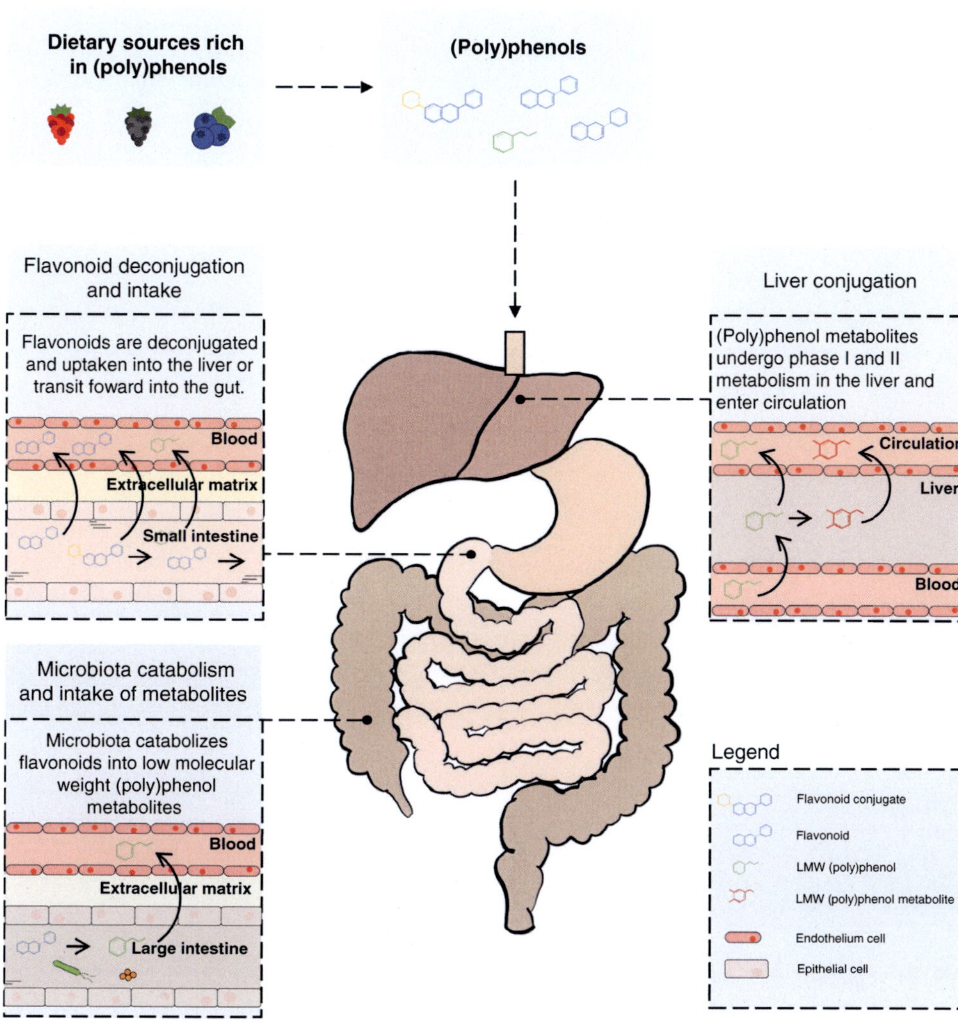

Figure 4.1 (Poly)phenols route for absorption and metabolism in the human gastrointestinal system. (Poly)phenols, represented by flavonoids, enter the gastrointestinal tract from different fruits and vegetables rich in (poly)phenols. (Poly)phenols start by being unconjugated and parts are absorbed in the small intestine. The remaining unconjugated flavonoids are catabolized by the microbiota into low molecular weight (LMW) (poly)phenol metabolites. Both parent compounds and LMW (poly)phenols suffer metabolism in the epithelium and liver and reach circulation. ECM, extracellular matrix; SI, small intestine; LI, large intestine; LMW; low molecular weight.

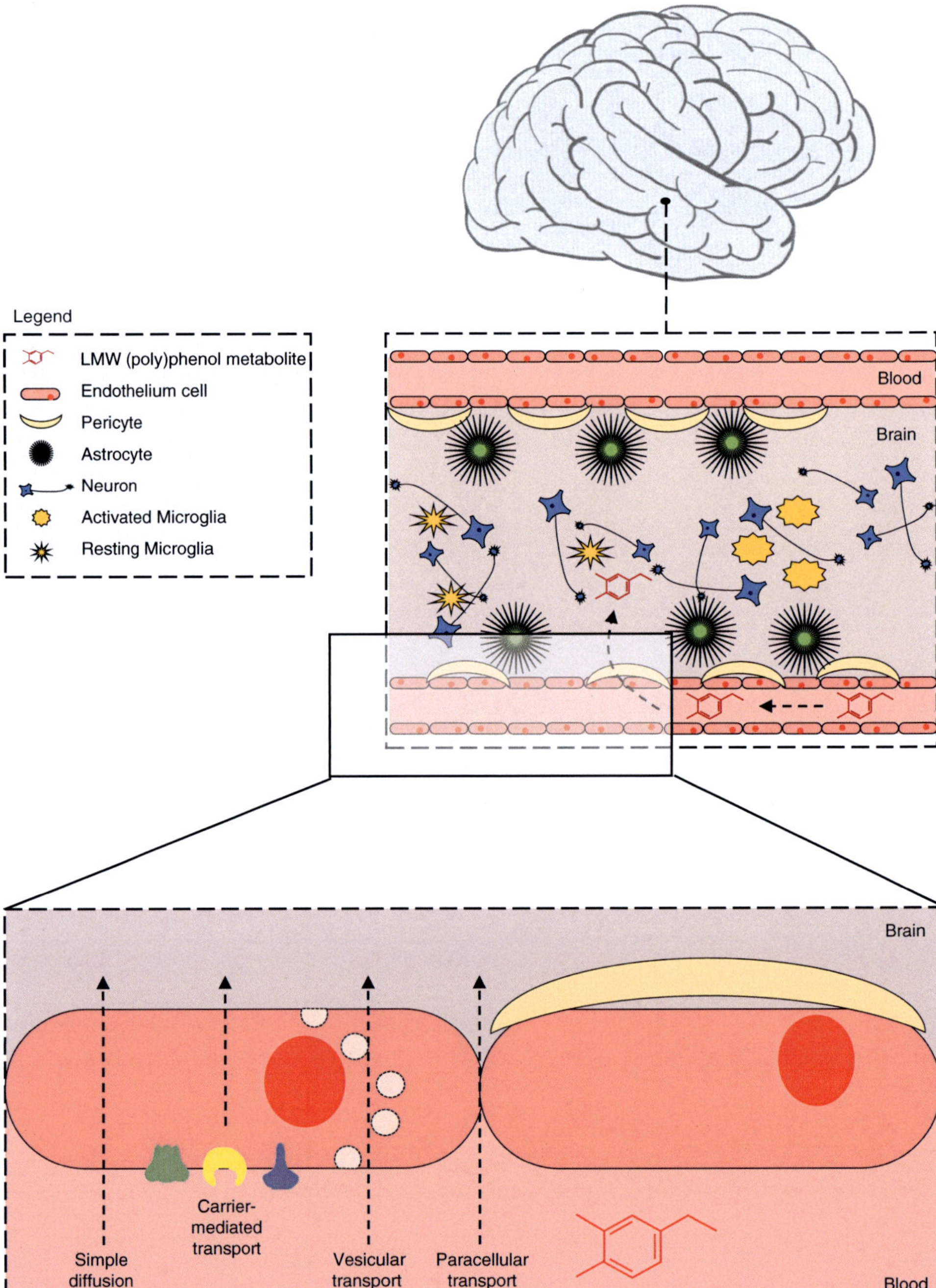

Figure 4.3 Different ways that LMW (poly)phenol metabolites could cross the BBB into the brain: their journey to and through the brain.

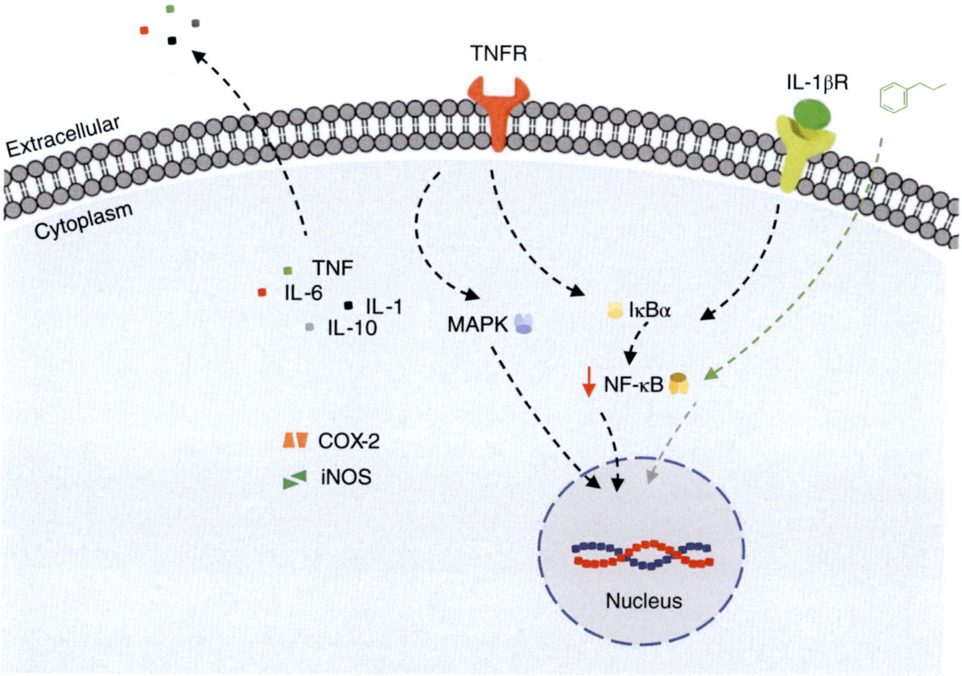

Figure 4.4 Possible molecular mechanisms of LMW (poly)phenols attenuation of neuroinflammation. Multiple receptors are capable of activating NF-kB leading to the activation and translocation of NF-kB to the nucleus and consequent expression of cytokines and inflammatory response enzymes like COX and iNOS. Several LMW (poly)phenol metabolites are known to impact NF-kB, blocking NF-kB translocation into the nucleus, thus blocking the inflammatory response. COX-2, cyclooxygenase 2; IL, interleukin; iNOS, nitric oxide synthase; I**κ**B**α**, inhibitor of nuclear factor kappa B; MAPK, mitogen-activated protein kinase; NF-**κ**B, nuclear factor kappa-light-chain-enhancer of activated B cells; TNFR, tumor necrosis factor receptors; TNF**α**, tumor necrosis factor-alpha.

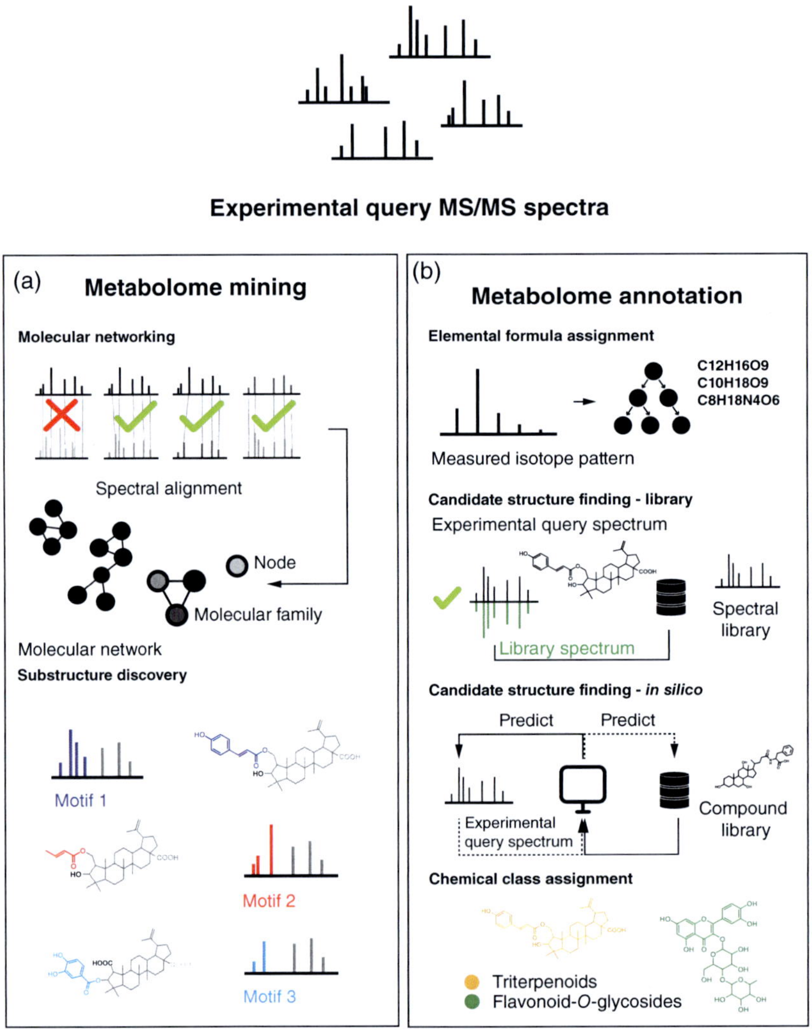

Figure 5.1 The chemical and structural annotation of complex natural mixtures displayed from metabolome mining to annotation tools. (a) Molecular networking and substructure discovery. All mass spectra from the experimental data are compared to each other and metabolites are represented by mass spectral features. These features are the nodes in the molecular network: they can form molecular families with structurally similar molecules (based on their spectral similarity) and an ensemble of molecular families is a molecular network. In substructure discovery, topic modelling discovers often co-occurring mass fragments or neutral losses that together could represent metabolite building blocks as schematically depicted here with a hypothetical example where the colored mass fragment peaks correspond to the colored substructures in the displayed metabolites. (b) Elemental formula assignment, library matching, *in silico* candidate structure finding and chemical class assignment. In elemental formula assignment, accurate isotope patterns and the calculation of fragmentation trees based on MS/MS spectra enable accurate elemental formula assignments for many metabolite features. Library matching relies on the presence of reference mass spectra in spectral libraries whereby a spectral similarity score (as used for molecular networking in (a) is used for matching these reference data to the experimental data. *In silico* candidate structure finding mostly relies on compound libraries that supply the candidate structures matching to the masses in the metabolomics profiles, after which various approaches are used to rank the often-various possibilities, thereby considering the MS/MS data. Chemical class assignment is possible through direct prediction based on the MS/MS data or through chemical

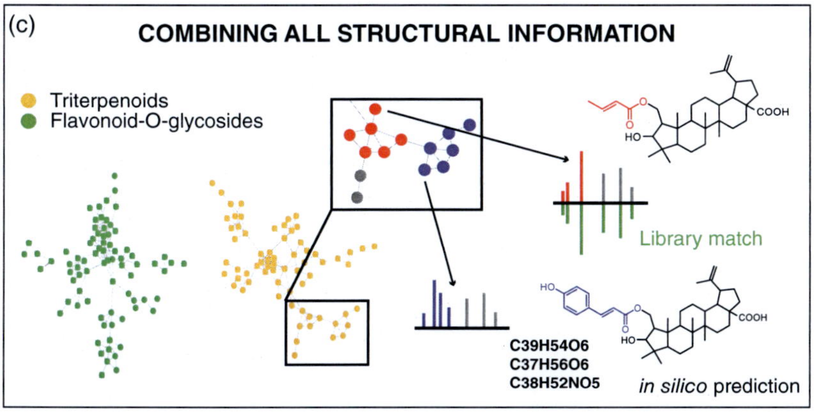

Figure 5.1 (Continued) class annotation of candidate structures. (c) Combination of multiple metabolome mining and annotation tools into one output. The schematic figure highlights how for one molecular family in a molecular network chemical class information can be combined with substructure annotation and library matches to gain higher-level and in-detail chemical information about the mixture. The figure parts were inspired by those presented in (Dührkop et al. 2020; Ernst et al. 2019b; Aron et al. 2020) and other papers explaining metabolomics mining and annotation tools.

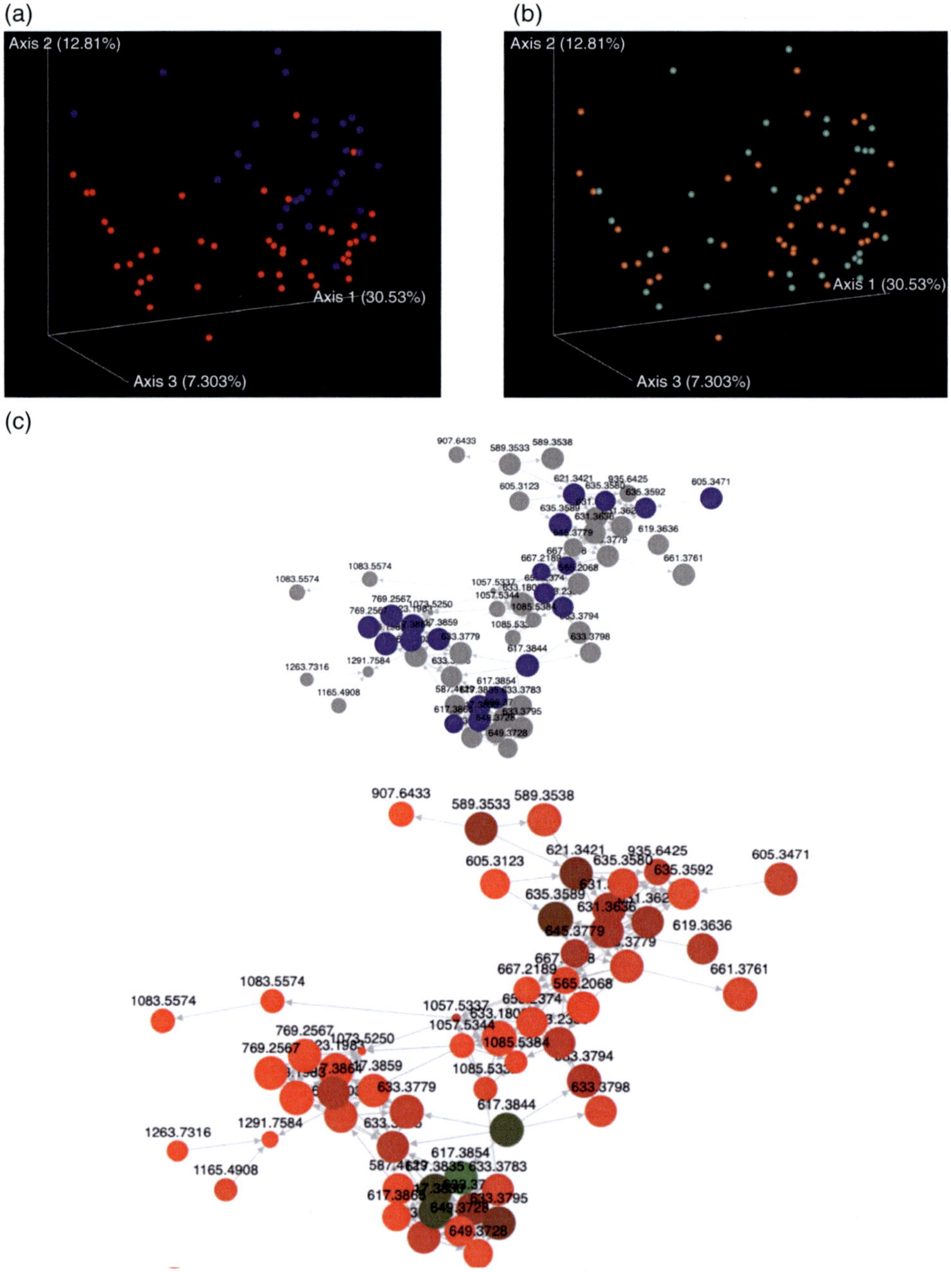

Figure 5.2 (a) Principal coordinate analysis (PCoA) plot based on Bray–Curtis distance between 71 Rhamnaceae plants colored according to their clade-level phylogeny: Rhamnoid (Red) and Ziziphoid (Blue) with 30.5% variation captured in PC1 [Axis 1], 12.8% in PC2, and 7.3% in PC3. (b) Same PCoA plot as in (a) but now colored according to initial extraction solvent: methanol (orange) and 95% ethanol (green). (c) Triterpenoid molecular family with spectral nodes sized proportionally to the number of files in which they occurred and colored (top) blue if a library match was found and (bottom) in a spectrum from light red to dark brown proportional to the summed precursor intensities of the spectra clustered in one node. It can be seen that many smaller and lighter nodes do not have a direct library match, but display sufficient spectral similarity to nodes that do have library matches to allow for the propagation of structural

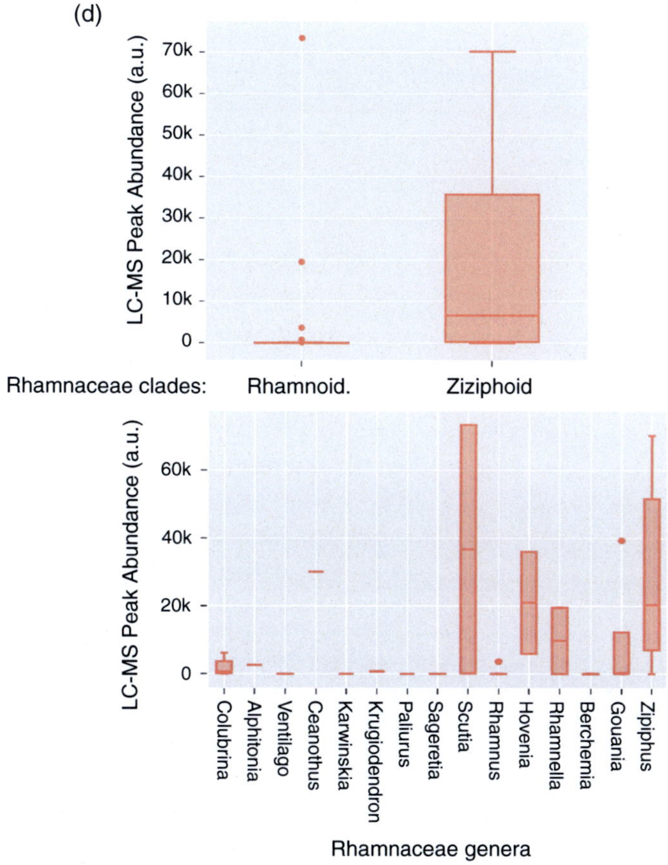

Figure 5.2 (Continued) information. (d) Feature area box plots for the node with a precursor *m/z* of 617.3830 ([M−H]-) that matched the pentacyclic triterpenoid 2-*O*-*E*-*p*-Coumaroyl alphitolic acid in the GNPS library, distributed (top) over the two major clades and (bottom) the 15 occurring genera in 71 Rhamnaceae plants. It can be observed how this triterpenoid is mainly present across several genera that belong to the Ziziphoid clade, with only a few noticeable exceptions.

Figure 5.3 MolNetEnhancer provides direct insight into the non-volatile chemical space of sweet pepper (*Capsicum annuum* L.). The most abundant annotated ClassyFire chemical compound class subclass terms were as follows for the positive ionization mode Feature-Based Molecular Network: amino acids, peptides, and analogues (435), diterpenoids (172), linoleic acids and derivates (164), sesquiterpenoids (89), and terpene glycosides (67). Here, example molecular families are shown colored by their annotated chemical subclass: linoleic acids and derivatives (green), terpene glycosides (black), diterpenoids (purple), sesquiterpenoids (red), and depsipeptides (pink). The size of the nodes is proportional to the number of MS/MS spectra obtained for that particular metabolite feature, whereas the thickness of the edge is proportional to the modified cosine score between the connected MS/MS spectra.

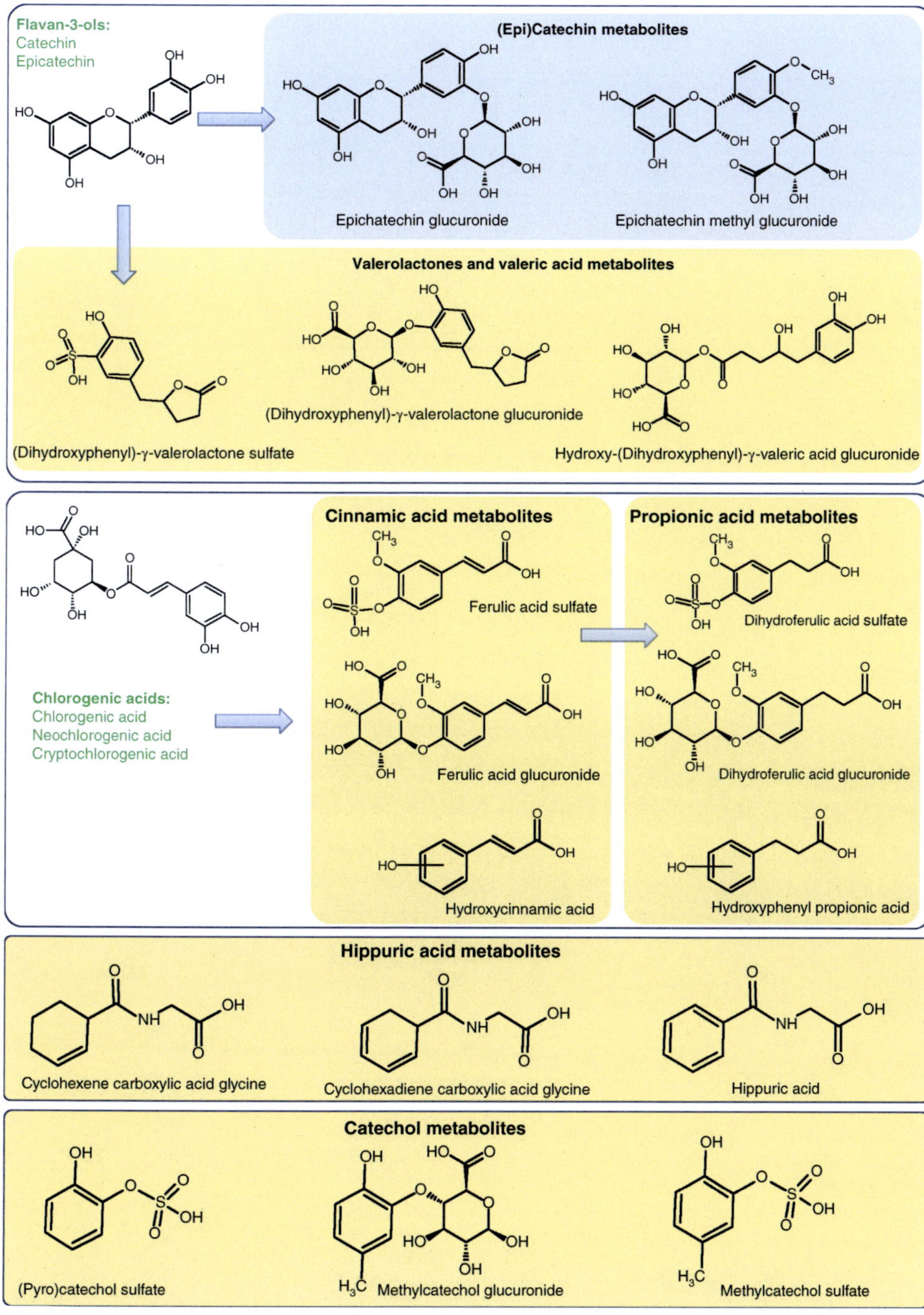

Figure 6.2 Chemical structures of key metabolites found in biological fluids after acute apple juice intake or long-term apple consumption. The yellow background indicates metabolites resulting from co-metabolism phases 1 and 2, while the light blue background indicates metabolism phase 2 only.

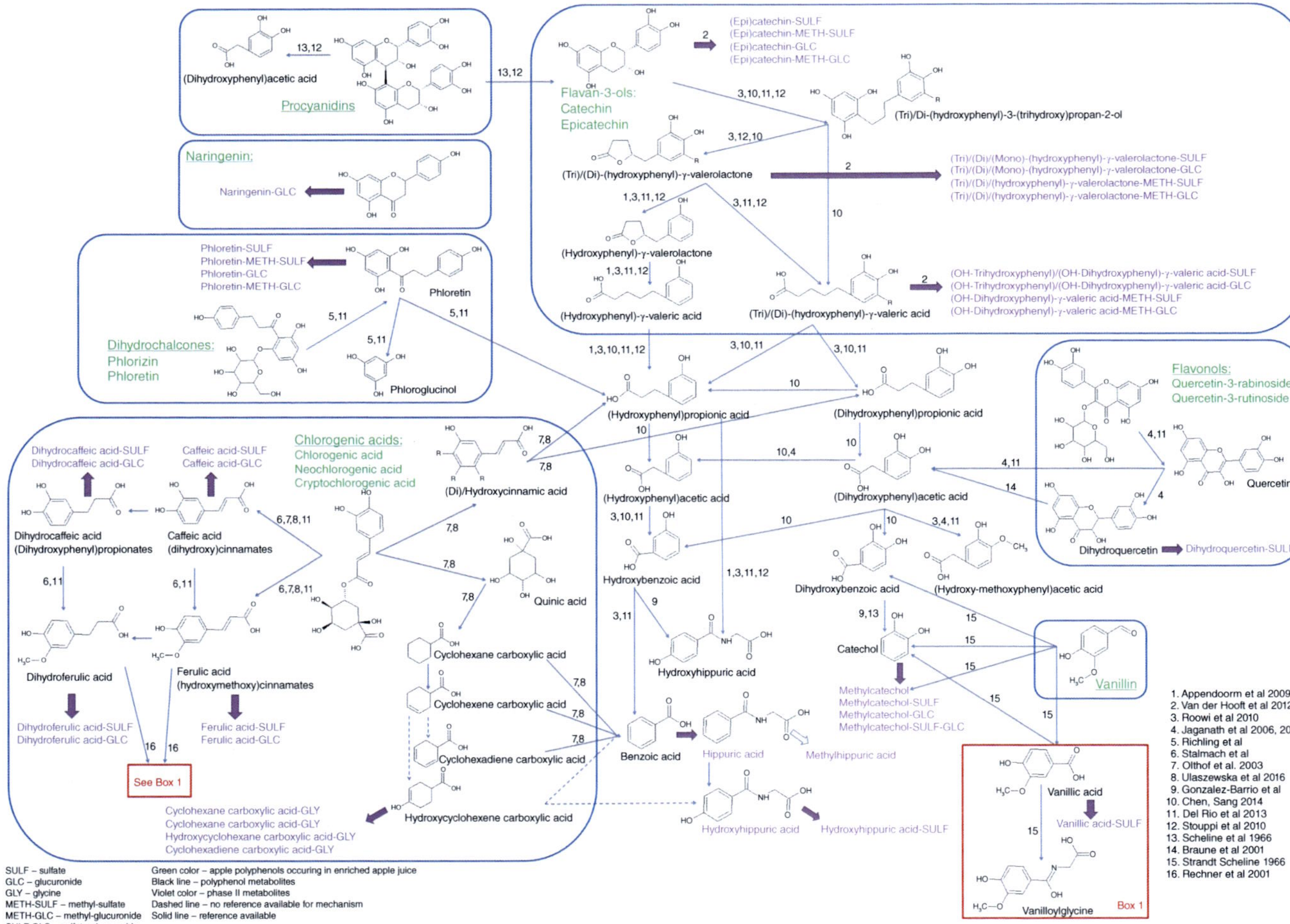

Figure 6.9 A simplified diagram of the metabolic pathways for the main classes of apple polyphenols and their crossing trajectories. *Source:* Reprinted from Trost et al. (2018a), with permission from Elsevier.

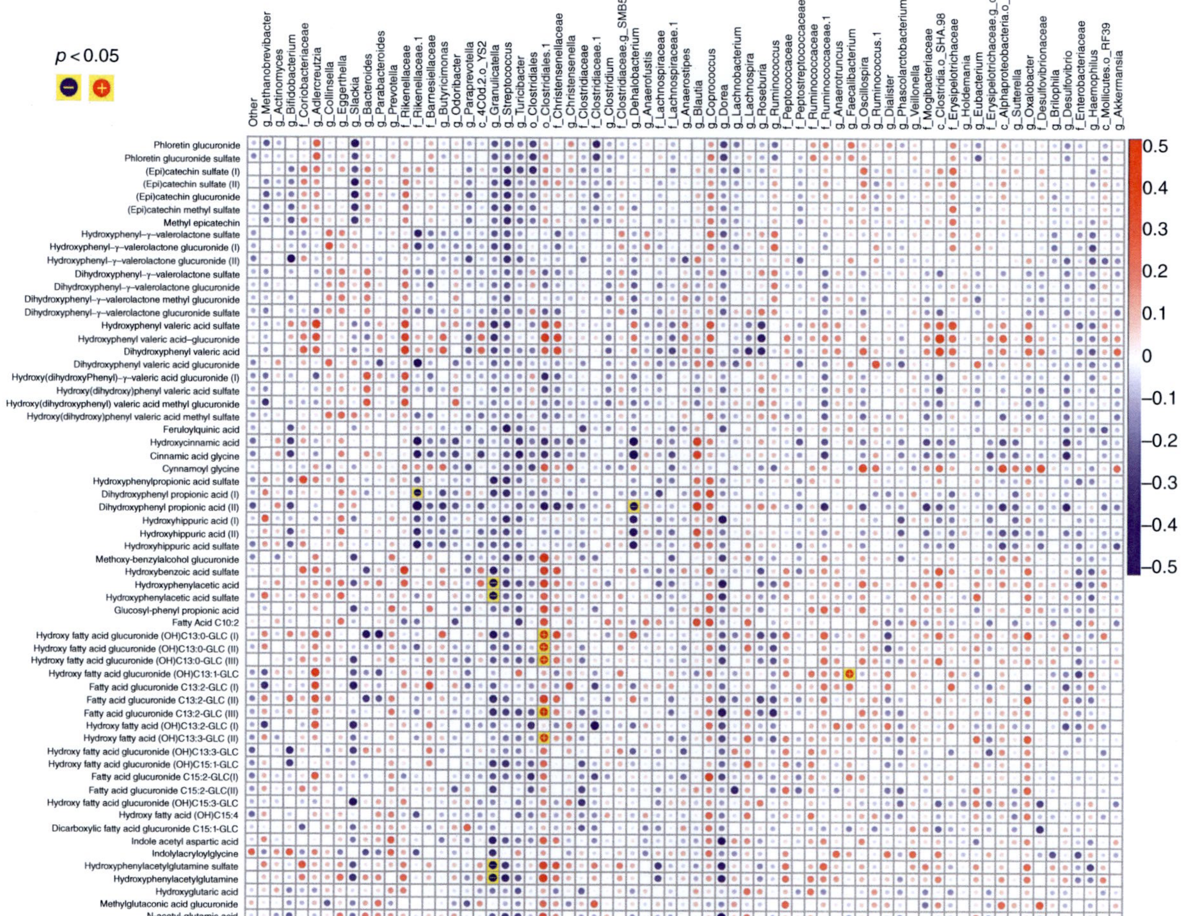

Figure 6.10 Heat map of pairwise correlations between urine metabolites and gut microbiota at the genus level. Correlation coefficients (Spearman's ρ) are represented by color gradient and size of the circles. The significant values (FDR<0.07) are marked with a + and yellow background. For unclassified genera ("g_"), the bacterial family ("f_") or order name ("o_") is reported. *Source:* Reprinted by permission from Ulaszewska et al. (2020a).

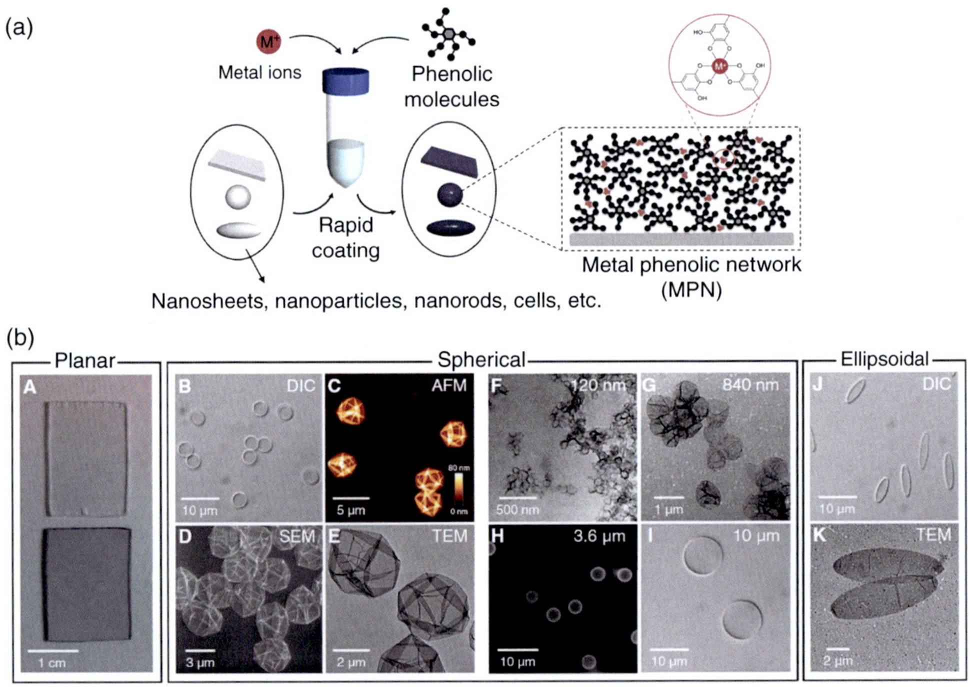

Figure 8.1 (a) Overview of MPN formation and deposition via discrete assembly. *Source:* Ejima et al. (2017), graphical abstract. Reproduced with permission of Elsevier. (b) Fe^{III}-TA films prepared on PS substrates with various shapes (planar, spherical, and ellipsoidal) and sizes (120 nm to 10 mm). Photographs of PS slides before (top) and after (bottom) Fe^{III}-TA coating (A); and differential interference contrast (DIC) images (B, I, J), AFM images (C), TEM images (E, F, G, K), SEM image (D), and fluorescence microscopy image (H) of Fe^{III}-TA capsules. *Source:* Ejima et al. (2013), Reproduced with permission from American Association for the Advancement of Science.

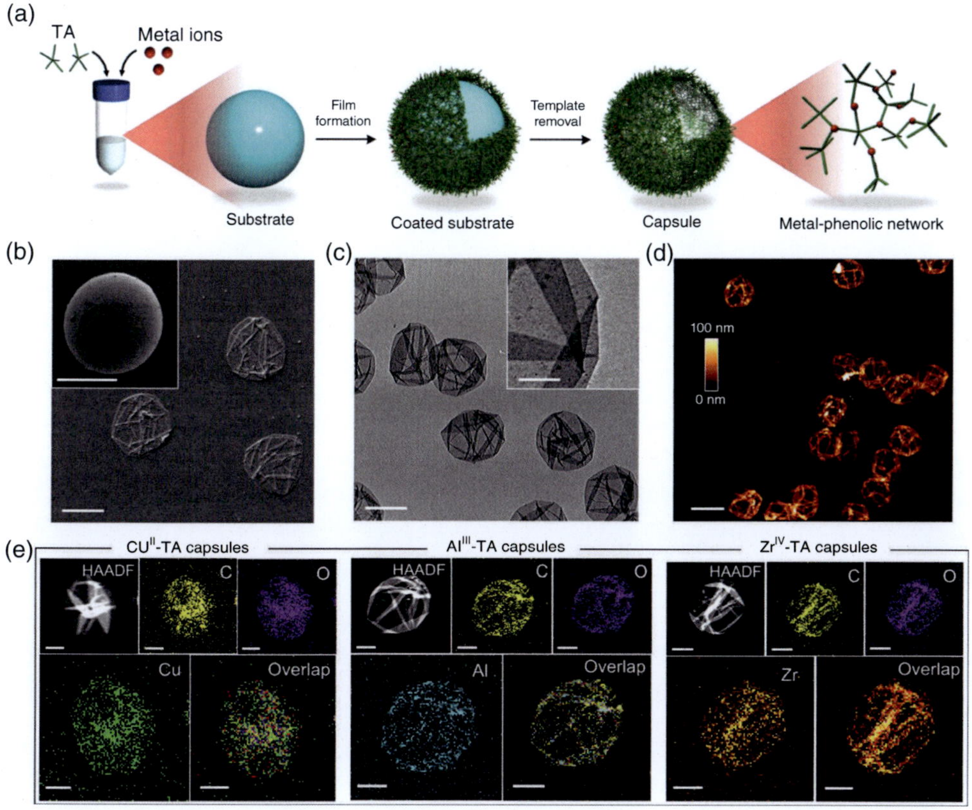

Figure 8.2 (a) Overview of discrete assembly template-mediated fabrication of MPNs and (b–d) associated microscopy images of capsules obtained following template dissolution: SEM (b), TEM (c), and AFM (d) images of Cu^{II}-TA capsules. (e) EDX elemental mapping of Cu^{II}-TA, Al^{III}-TA, and Zr^{IV}-TA capsules. Scale bars are 2 mm in (b–d), 2 mm in the inset of (b), 200 nm in the inset of (c) and 1 mm in (e). *Source:* Guo et al. (2014), Reproduced with permission John Wiley & Sons.

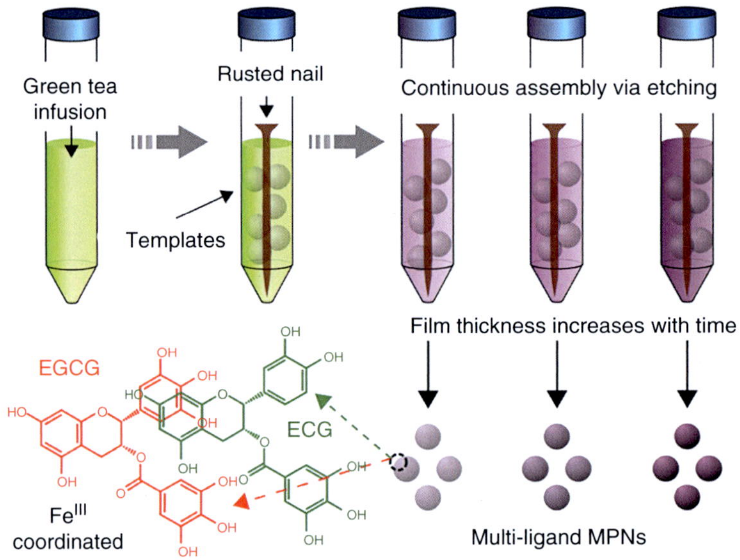

Figure 8.4 Continuous MPN assembly involving rust from a nail as an iron source. *Source:* Rahim et al. (2018), graphical abstract. Reproduced with permission of the American Chemical Society.

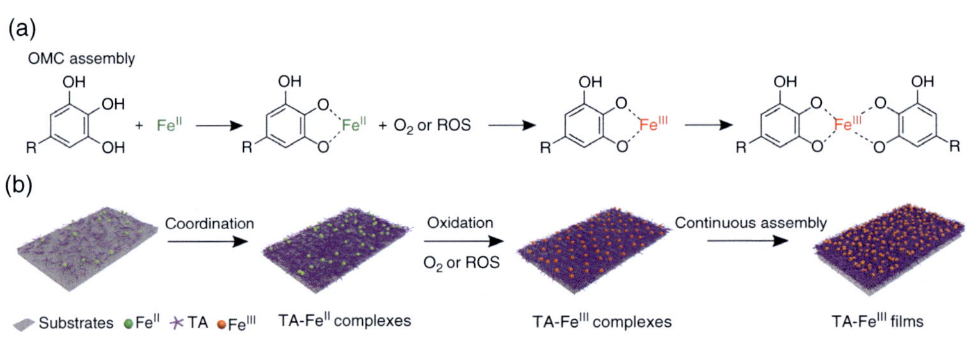

Figure 8.5 (a) Oxidation-mediated coordination (OMC) assembly employing ROS to rapidly oxidize Fe^{II} to Fe^{III} and (b) overview of MPN formation via OMC assembly on a template. *Source:* Zhong et al. (2019)/John Wiley & Sons.

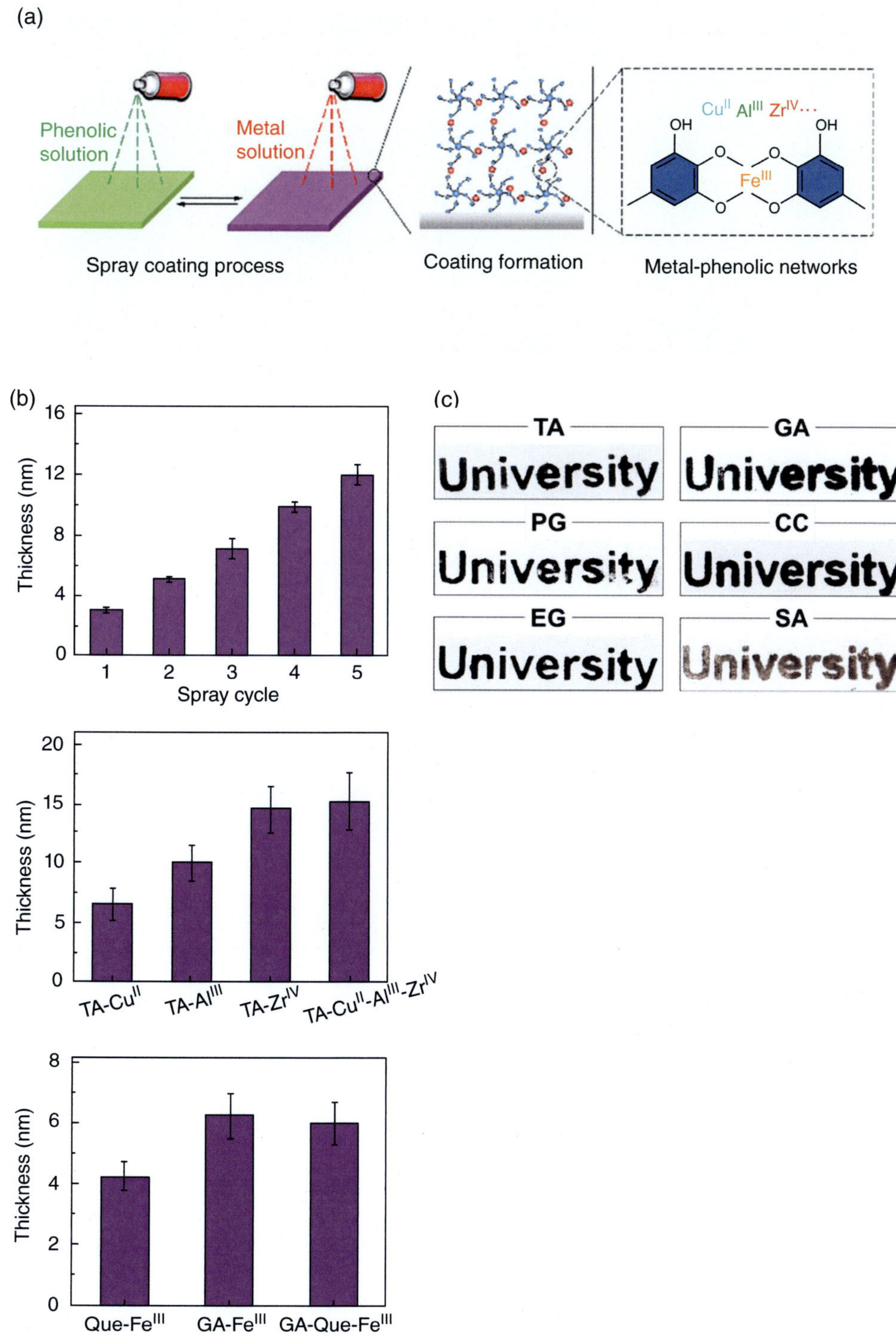

Figure 8.6 (a) Overview of MPN formation via a spray-coating process. *Source:* Zhong et al. (2018), Reproduced with permission of the American Chemical Society. (b) The influence of the number of spray cycles, metal ion and ligand on MPN film thickness. *Source:* Zhong et al. (2018), Reproduced with permission of the American Chemical Society. (c) MPNs prepared via printing, with Fe ink being printed onto paper treated with different polyphenols. *Source:* Adapted from Dai et al. (2019).

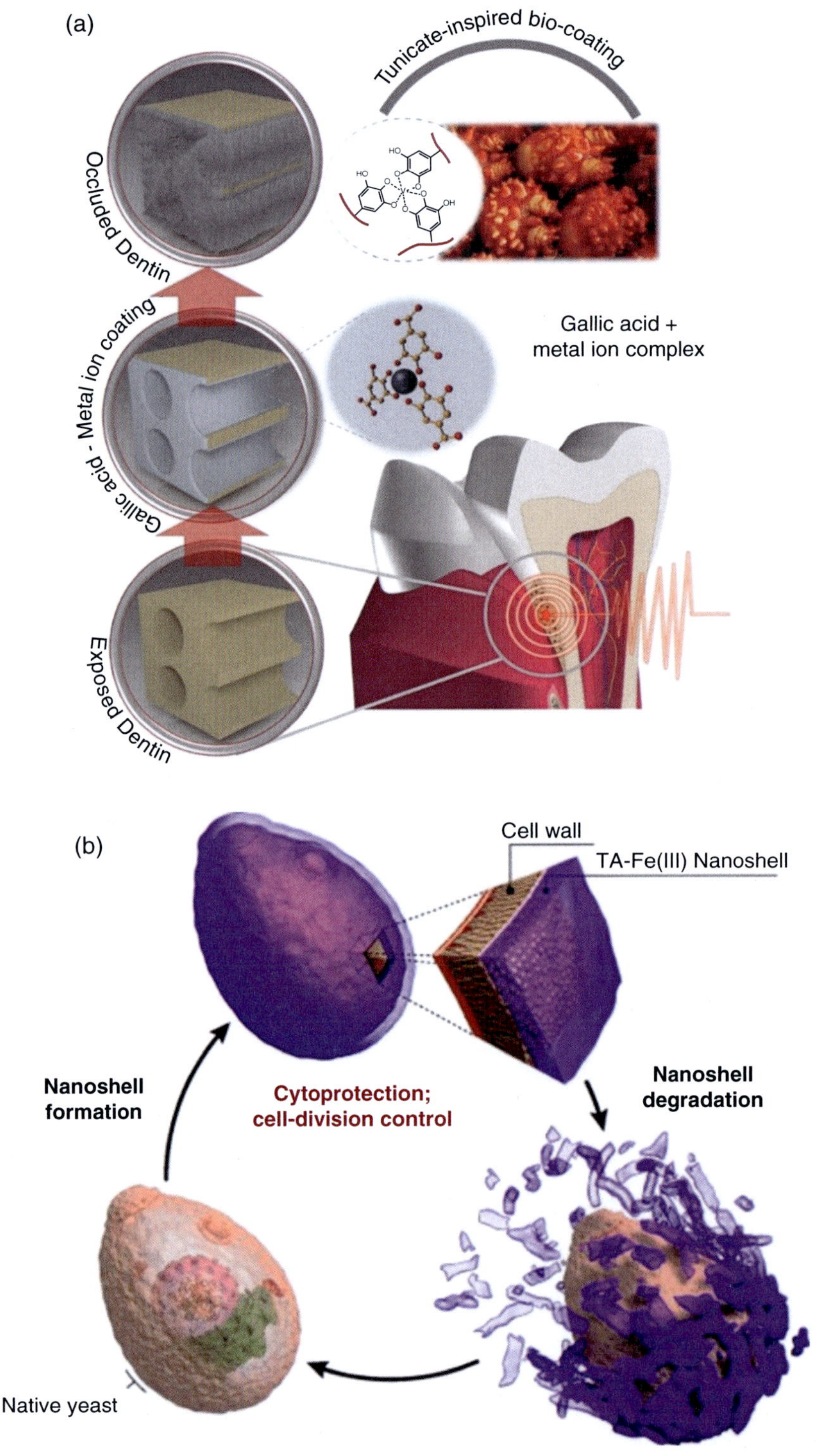

Figure 8.7 Examples of MPNs used as coating materials at biointerfaces: (a) as coating materials for hypersensitive teeth. *Source:* Prajatelistia et al. (2016), Reproduced with permission John Wiley & Sons. (b) as cytoprotective exoskeletons for yeast cells. *Source:* Park et al. (2014), figure 1. Reproduced with permission of Wiley.

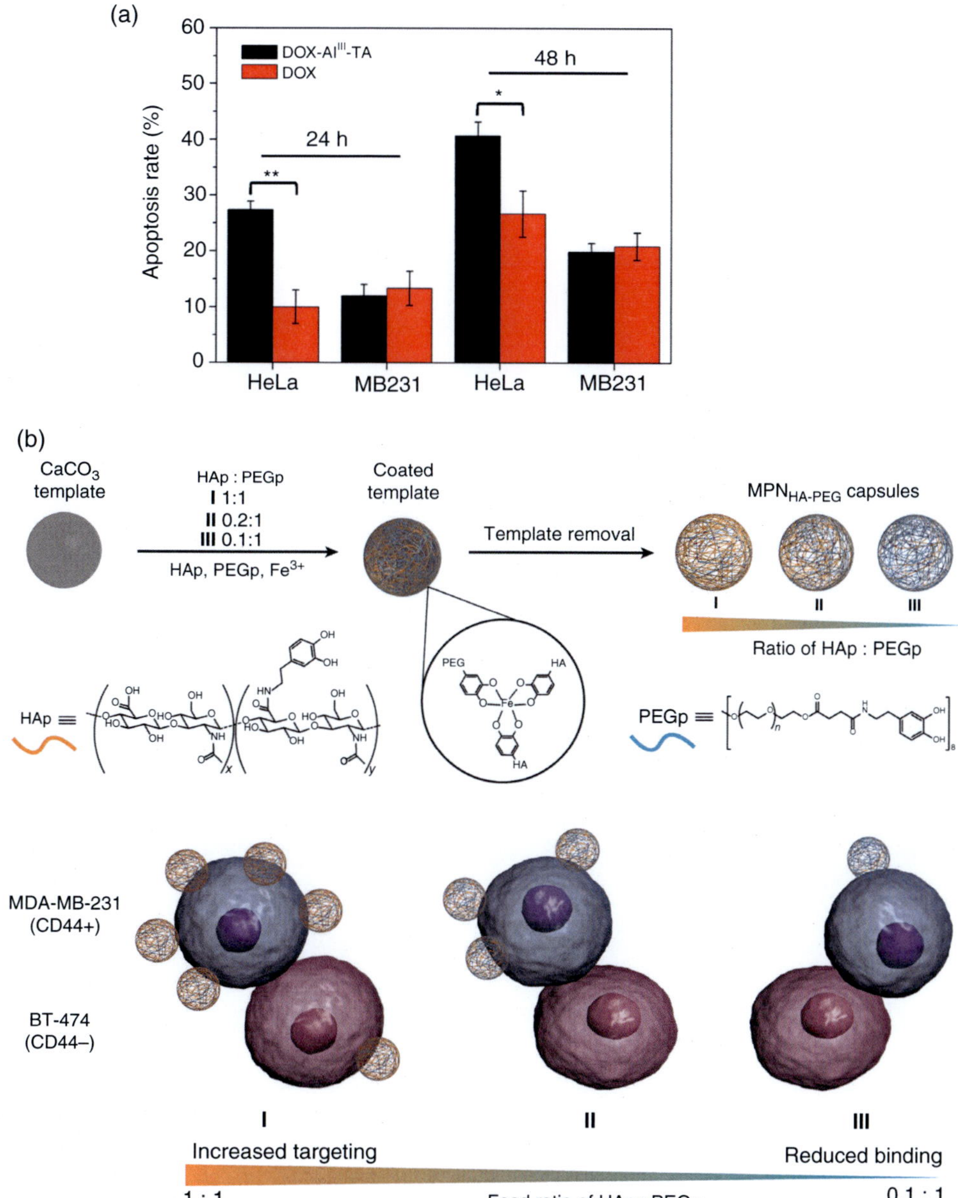

Figure 8.8 (a) Percentage of apoptotic HeLa and MB231 cells after treatment with DOX-Al^{III}-TA MPNs or DOX at different times. *Source:* Adapted from Ping et al. (2015), Reproduced with permission of Wiley. (b) Synthesis of HA-functionalized PEG-Fe^{III} MPNs and the subsequent influence of HA:PEG ratio on specific targeting of CD44 cancer cells and binding efficiency. *Source:* Ju et al. (2016a), Reproduced with permission of the American Chemical Society.

(a)

(b)

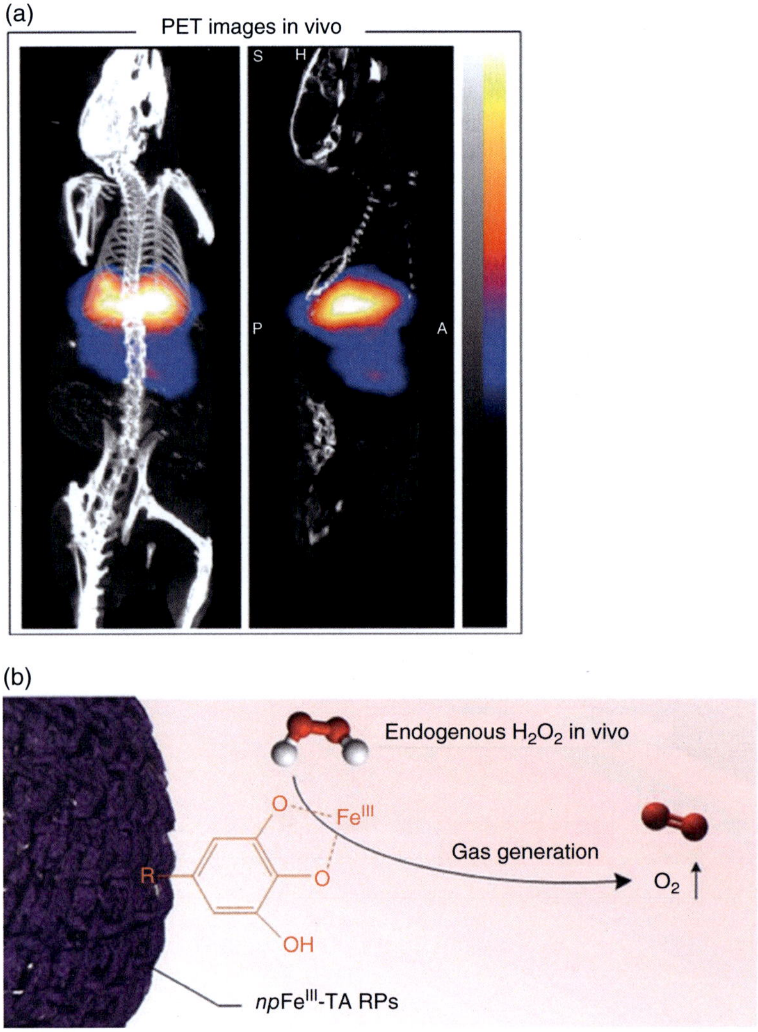

Figure 8.9 (a) PET/computed tomography image of an *in vivo* model (small animal) 30 minutes after injection of 1 MBq $^{64}Cu^{II}$-TA capsules (maximum intensity projection (right), sagittal view (left)). The color scale for the PET image data shows radiotracer uptake with white corresponding to the highest activity and blue to the lowest activity. *Source:* Guo et al. (2014), Reproduced with permission John Wiley & Sons. (b) Interfacial reaction of H_2O_2, splitting to H_2O and O_2 with the catalytic assistance of chelated Fe^{3+} in the nanoporous Fe^{III}-TA replica particles. *Source:* Guo et al. (2015), Reproduced with permission of Wiley.

(a)

(b)

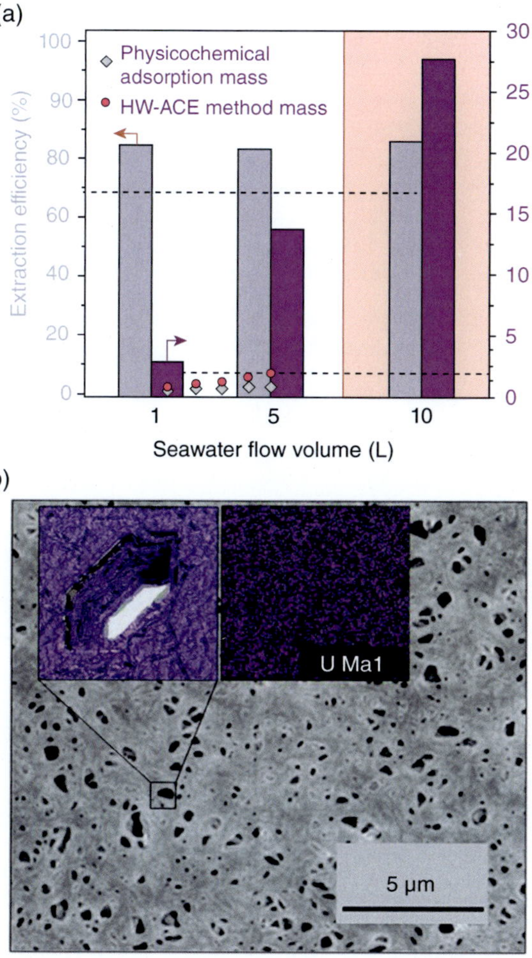

Figure 8.10 (a) Extraction efficiency and mass of uranium from seawater by three layers of MPN-based membranes. *Source:* Luo et al. (2019), figures 3 and 4. Reproduced with permission of the Royal Society of Chemistry. (b) SEM and energy-dispersive X-ray spectroscopy mapping of membranes following uranium deposition. *Source:* Luo et al. (2019), Reproduced with permission Royal Society of Chemistry.

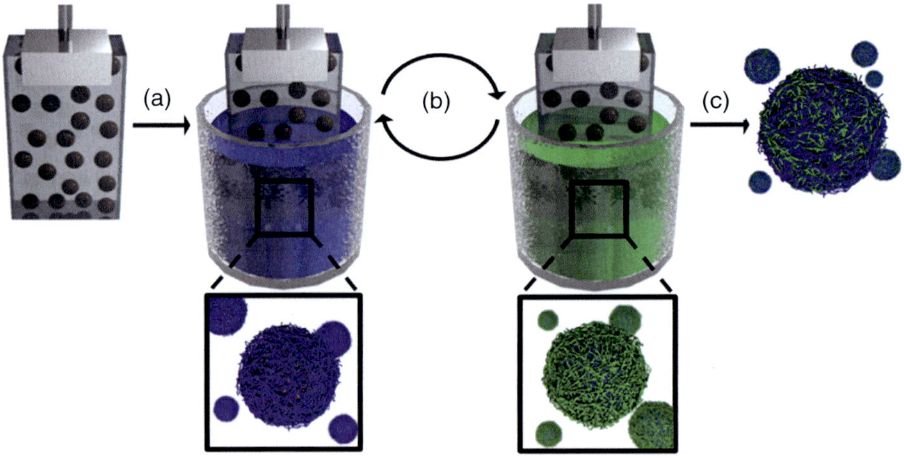

Figure 8.12 Template particle coating via immersive polymer assembly: (a) immobilized particles are immersed in a polymer solution, followed by washing steps (immersion in water); (b) subsequent immersion in a solution of a complementary polymer, followed by water washing steps (the washing steps are not shown for simplicity; this process is repeated until the desired number of layers is attained); and (c) the layered particles are recovered from the polymer agarose, and the template particles can be dissolved to yield capsules. *Source:* Richardson et al. (2013), Reproduced with permission John Wiley & Sons.

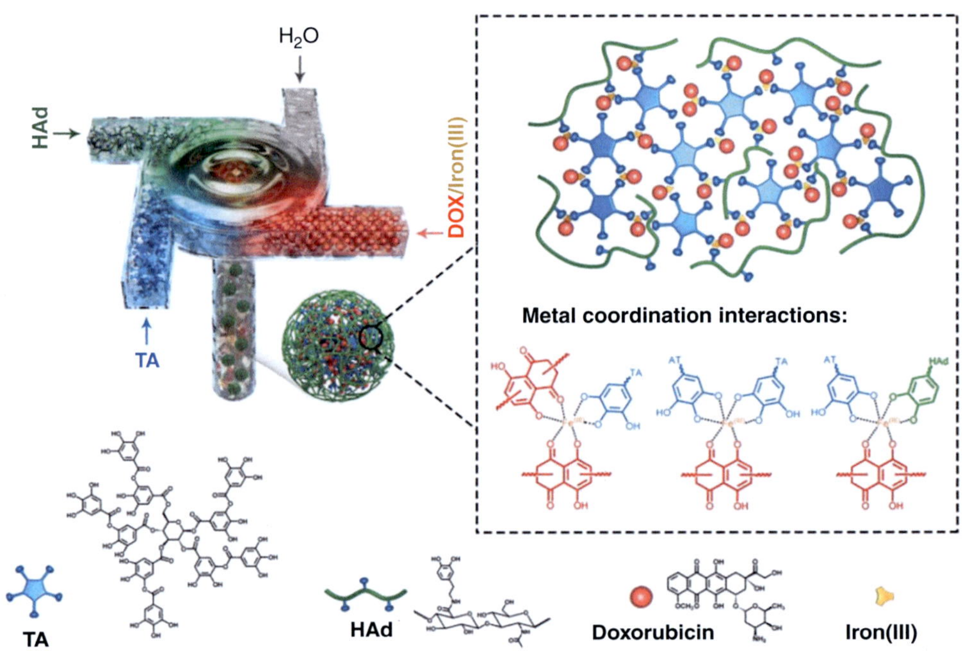

Figure 8.14 MPN formation via flash complexation, with complex formation facilitated via a small, turbulent mixing device into which reagent solutions are fed. *Source:* Liu et al. (2019b), Reproduced with permission Royal Society of Chemistry. Note: The actual composition of TA consists of a complex mixture of different galloylated glucoses (see Sylla et al. 2015).

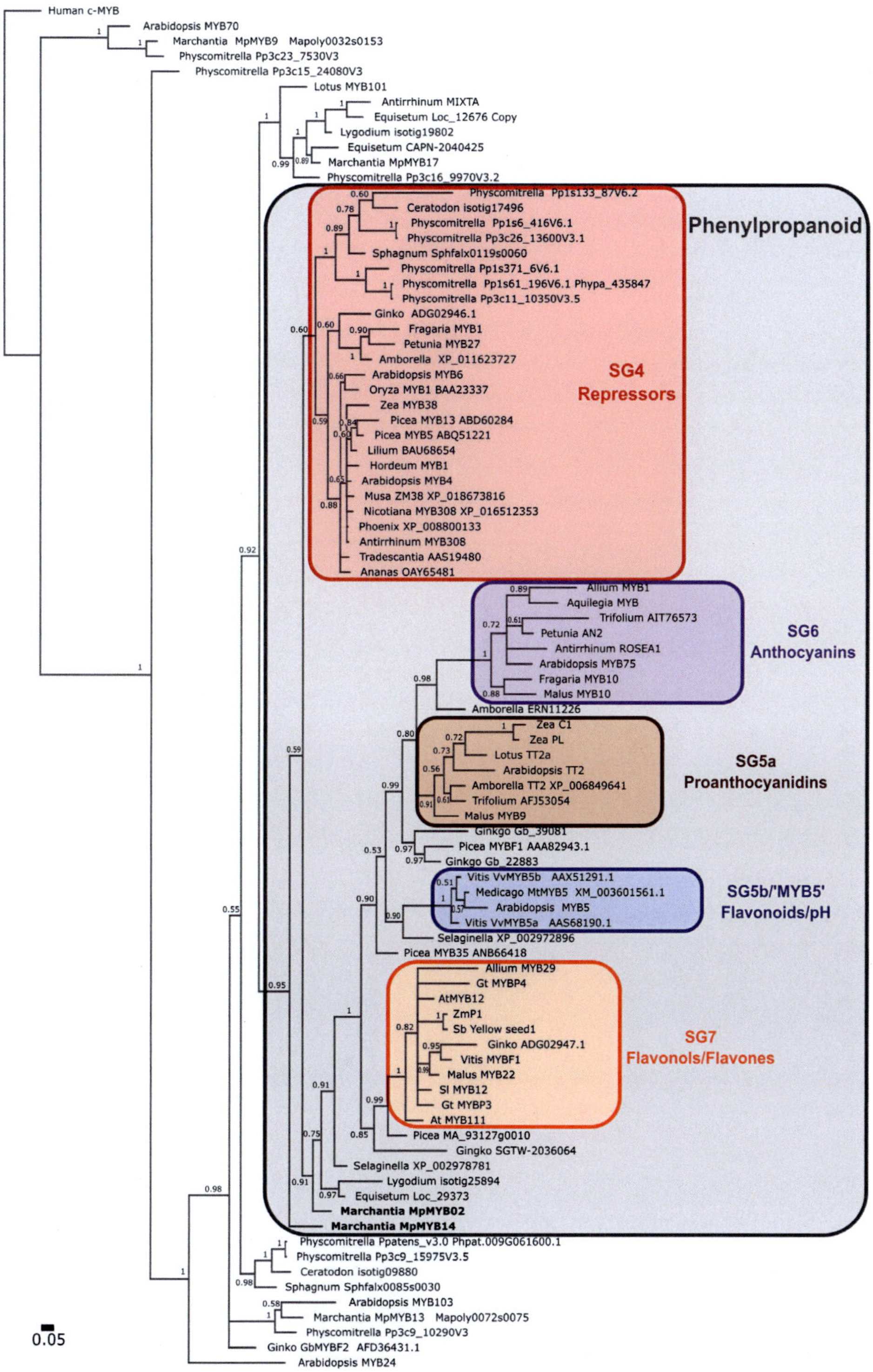

Figure 10.2 Phylogenetic relationships of phenylpropanoid-related R2R3MYB TFs from different plant species. The amino acid sequences of the R2R3MYB domain were aligned in ClustalW and then used to generate a phylogenetic tree with MrBayes in the Geneious suite of software, using default parameters, with the posterior probabilities of the branches shown. The human cMYB sequence (NP_001155129) was used for the outgroup, and a selection of non-flavonoid-related

Figure 10.2 (Continued) R2R3MYBs was included for comparison. The top Blast hits for At*MYB75* (sub-group 6) and Mp*MYB14* from species representing the major plant groups are included. The clade highlighted in grey includes regulators of the general phenylpropanoid pathway, including SG4-containing transcriptional repressors of phenylpropanoid genes (in red), SG5a-containing activators of anthocyanin/proanthocyanidin production (in brown), SG5b-containing activators of proanthocyanidin/hyperacidification (in blue), SG6-containing activators of anthocyanin production (in purple), and SG7-containing activators of flavone/flavonol production (in yellow). Sequence details are: the liverwort *M. polymorpha* Myb02 (Mapoly0006s0226), Myb09 (Mapoly0032s0153), Myb13 (Mapoly0072s0075), Myb14 (Mapoly0073s0038), Myb17 (Mapoly0096s0058); the mosses *Ceratodon purpureus* isotig09880, isotig17496; *Physcomitrella patens* Pp1s133_87V6.2, Pp1s6_416V6.1, Pp1s371_6V6.1,Pp1s61_196V6.1; Pp3c9_15975V3.5, Pp3c9_10290V3, Pp3c11_1035 0V3.5,Pp3c15_24080V3, Pp3c16_9970V3.2, Pp3c23_7530V3, Pp3c26_13600V3.1; *Sphagnum fallax* XP_002978781, SphFalx011950060, SphFalx008550030; the ferns/fern allies *Equisetum diffusum* CAPN-2040425; *Equisetum giganteum* Loc_12676, Loc_29373; *Lygodium japonicum* isotig19802, isotig25894; *Selaginella moellendorffii* XP_002972896, XP_002978781; the gymnosperms *Gingko biloba* SGTW-2036064, Gb_22883, Gb_39081, GbMYBP2 AFD36431.1, ADG02946.1; *Picea mariana* MYBF1 (AAA82943); *Picea glauca* MYB5 (ABQ51221), MYB13 (ABD60284); *Picea abies* MYB35 (ANB66418), MA_93127g0010; and the angiosperms *Allium cepa* MYB1 (KX785130), MYB29 (KX785133); *Amborella trichopoda* ERN11226, TT2 (XP_006849641), ZM38 (XP_011623727); *Ananas comosus* OAY65481; *Antirrhinum majus* Rosea1 (ABB83826), Mixta (CAA55725), Myb308 (P81393); *Aquilegia formosa* MYB1 (ACQ82820); *Arabidopsis thaliana* MYB4 (At4g38620), MYB5 (At3g13540), MYB6 (XP_002872444), MYB12 (ABB03913), MYB24 (AF175987), MYB70 (NP_179910), MYB75/PAP1 (AAG42001), MYB103 (AED96722), MYB111 (At5g49330.1) MYB123/TT2 (CAC40021); *Fragaria × ananassa* MYB1 (AAK84064); *Gentian triflora* GtMYBP4 AB280446, GtMYBP3 AB733016; *Hordeum vulgare* MYB1 (P20026); *Lilium* BAU68654; *Lotus japonicus* MYB101 (BAC75671), TT2a (BAG12893); *Malus × domestica* MYB9 (NP_001280749), MYB10 (ACQ45201), MYB22 (AAZ20438); *Musa acuminata* subsp. *malaccensis* XP_018673816; *Nicotiana tabacum* MYB308 (XP_016512353); *Oryza sativa* MYB1 (BAA23337); *Petunia × hybrida* AN2 (AAF66727), MYB27 (AHX24372); *Phoenix dactylifera* XP_008800133; *Solanum lycopersicum* MYB12 EU419748; Sorghum bicolor yellowseed 1, AY860968; *Tradescantia fluminensis* AAS19480; *Trifolium arvense* AFJ53054, AIT76573; *Vitis vinifera* MybF1 (FJ948477), VvMYB5a AA56868190.1, VvMYB5b AAX51291.1; *Zea mays* C1 (AAA33482), Pl (AAA19821), P1 (GRMZM2G084799), Myb38 (P20025).

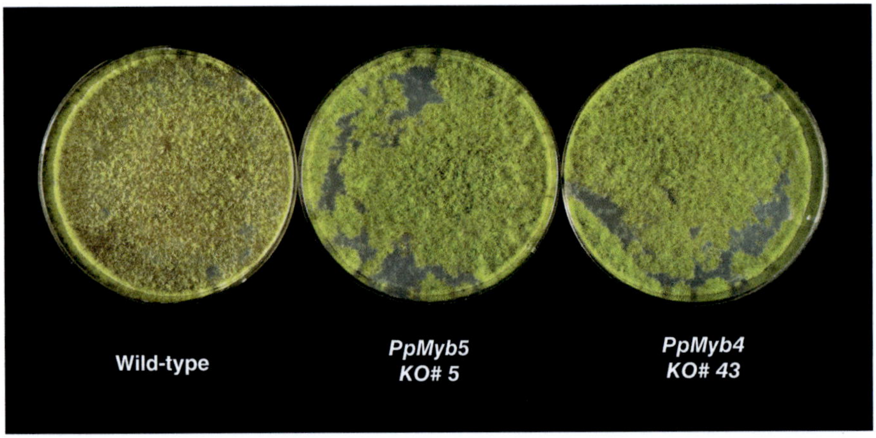

Figure 10.3 Response of *Physcomitrella patens* wild-type and knockout mutants of *PpMyb4* and *PpMyb5* to UV-B irradiation. Both genes encode R2R3Myb subgroup 4 repressors (Figure 10.2). Moss cultures were grown on KNOP medium for seven days at 23 °C, 16 hours light, 8 hours dark and then exposed to UV-B light (260–315 nm) for 20 minutes at a 20 cm distance from plates. Cultures were left to recover for eight days before being photographed. Clearly the *PpMyb4* and *PpMyb5* knockout mutants both survived and grew better than the wild type due to their elevated content of phenolics. *Source:* Jiménez (2010).

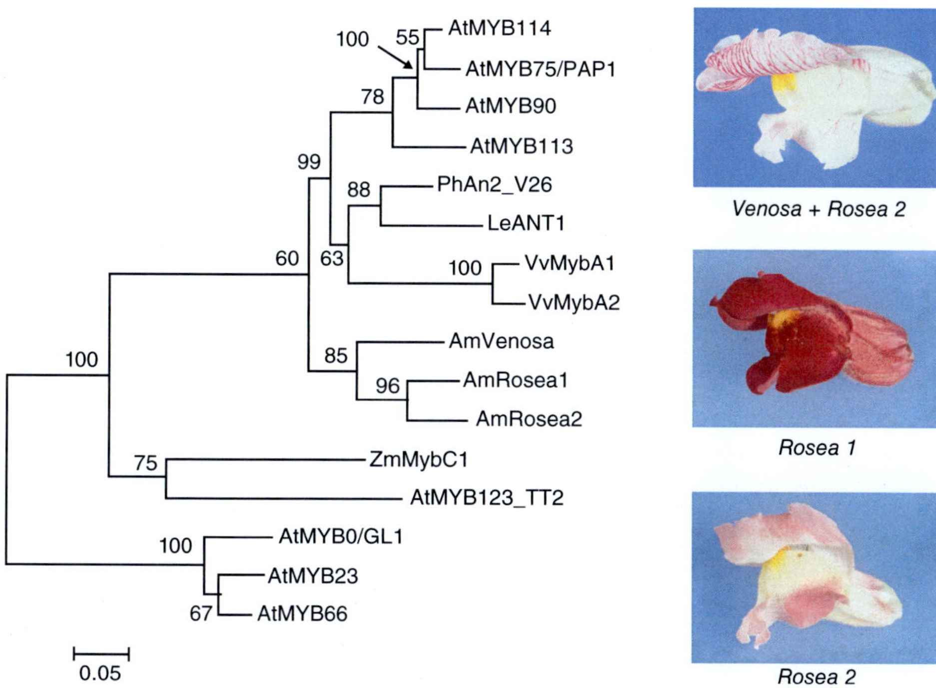

Figure 10.4 Phylogenetic tree comparing the amino acid sequences of the DNA binding domains of recently duplicated families R2R3 Myb subgroup 6 TFs from: *Arabidopsis thaliana* (At) *Antirrhinum majus* (Am) *Vitis vinifera* (Vv) An2 from *Petunia hybrida* (Ph), and LeANT1 from tomato. Together with the anthocyanin regulatory proteins, C1 and Pl from maize. Protein sequences were manually aligned using MacClade 4.08 (D.R. Maddison and W.P. Maddison, Sinauer Associates). Phylogenetic analysis was performed with PAUP* 4.0b10, using only the Myb domain of each protein. One thousand bootstrapped data sets were used to estimate the confidence of each tree clade. The phenotypes of the recently duplicated subgroup 6 R2R3Myb proteins in the flowers of *A.majus* are shown to the right of the figure.

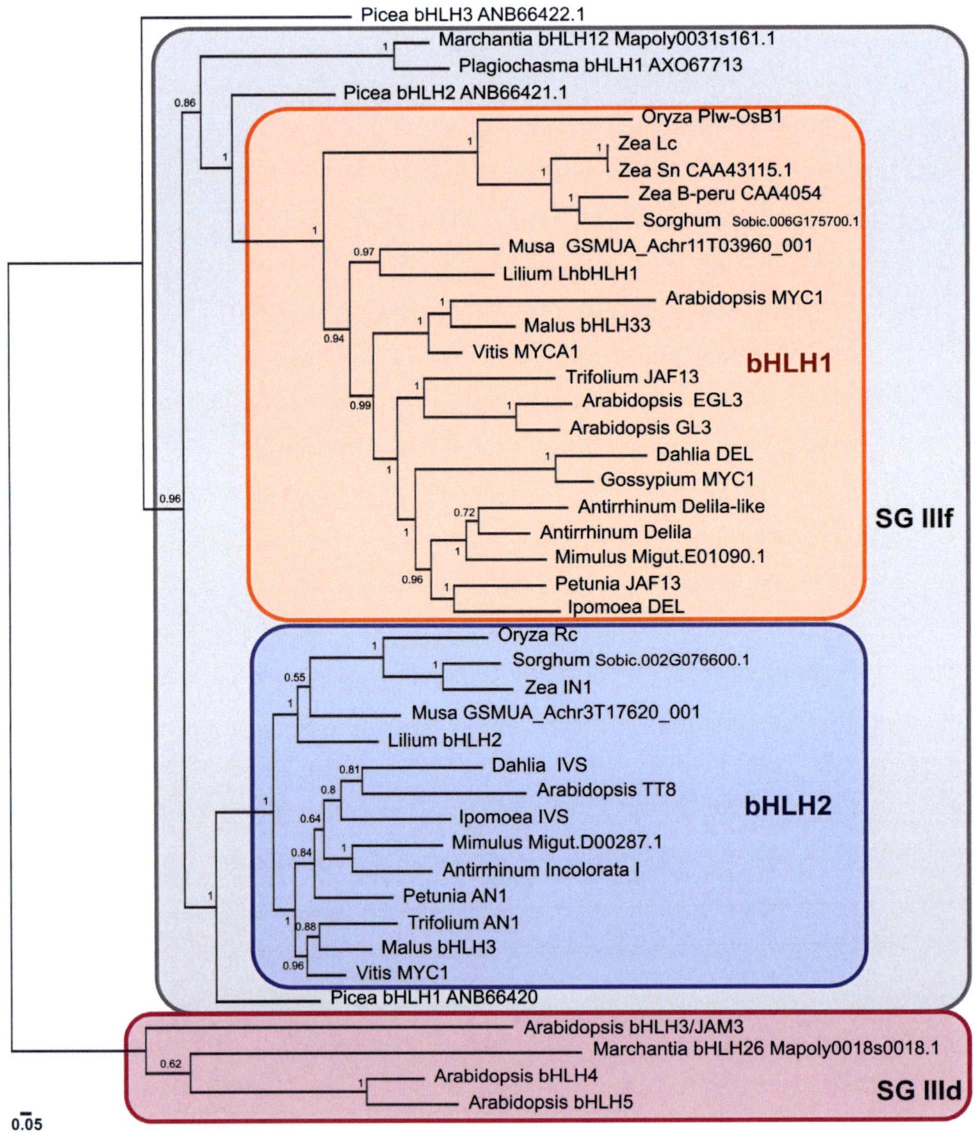

Figure 10.5 Phylogenetic analysis of plant bHLH proteins belonging to subgroup IIIf. *Source:* Adapted from Heim et al. (2003). Deduced amino acid sequences were aligned using MUSCLE to generate a maximum likelihood phylogenetic tree using PhyML with 1000 bootstrap replicates using Geneious (10.0.9) software, using default parameters. Gene IDs/accession numbers of sequences used: *Antirrhinum majus* Antirrhinum Delila (AAA32663.1; Am02g33340); Antirrhinum Delila-like (Am02g28470); Antirrhinum Incolorata I (Am02g53780) (A. majus genome, v.3); *Arabidopsis thaliana* Arabidopsis bHLH3/JAM3 (NP_193376); Arabidopsis EGL3 (OAP12509); Arabidopsis GL3 (NP_680372); Arabidopsis TT8 (Q9FT81); Arabidopsis MYC1 (NP_001154194); Arabidopsis bHLH4 (O49687); Arabidopsis bHLH5 (Q9FIP9); *Dahlia variabilis* Dahlia Del (BAJ33516); Dahlia IVS (BAJ33515); *Gossypium hirsutum* Gossypium MYC1 (CAA07615); *Ipomoea nil* Ipomoea Del (XP_019171149); Ipomoea IVS (XP_019197480); *Lilium hybrida* Lilium bHLH1 (BAE20057); Lilium bHLH2 (BAE20058); *Malus domestica* Malus MdbHLH3 (ADL36597); Malus bHLH33 (ABB84474); *Marchantia polymorpha* Marchantia bHLH12 (Mapoly0031s161.1); Marchantia bHLH26

Figure 10.5 (Continued) (Mapoly0018s0018.1); *Mimulus gattatus* (Erythranthe guttata) Mimulus Migut.E01090.1; Mimulus Migut.D00287.1; *Musa acuminata* GSMUA_Achr11T03960_001; GSMUA_Achr3T17620_001; *Oryza sativa* Oryza Plw-OsB1 (BAB64301); Oryza Rc (ADK36625); *Petunia hybrida* Petunia AN1 (AAG25928); Petunia JAF13 (AAC39455); *Picea abies* Picea bHLH1 ANB66420; Picea bHLH2 ANB66421.1; Picea bHLH3 ANB66422.1; *Sorghum bicolor* Sobic.006G175700.1; Sobic.002G076600.1; *Trifolium repens* Trifolium AN1 (AIT76559) Trifolium JAF13 (AIT76563); *Vitis vinifera* Vitis MYC1 (NM_001281253); Vitis MYCA1 (NP_001267954); *Zea mays* Zea Lc (AAA33504); Zea In1 (AAB03841); Zea Sn (CAA43115.1); Zea B-peru (CAA4054).

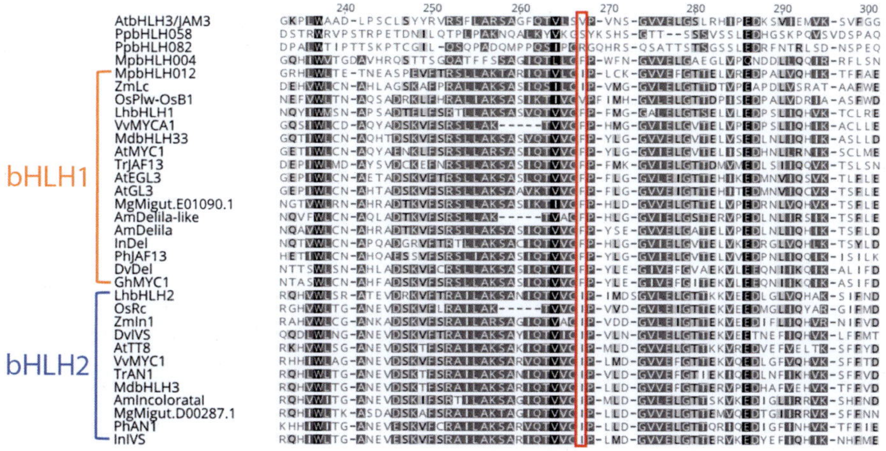

Figure 10.6 Muscle alignment of bHLH proteins. Gene IDs/accession numbers of sequences used: *Antirrhinum majus* AmDelila (AAA32663.1; Am02g33340); AmDelila-like (Am02g28470); AmIncolorata I (Am02g53780) (*A. majus* genome, v.3); *Arabidopsis thaliana* AtbHLH3/JAM3 (NP_193376); AtEGL3 (OAP12509); AtGL3 (NP_680372); AtTT8 (Q9FT81); AtMYC1 (NP_001154194); *Dahlia variabilis* DvDel (BAJ33516); DvIVS (BAJ33515); *Gossypium hirsutum* GhMYC1 (CAA07615); *Ipomoea nil* InDel (XP_019171149); InIVS (XP_019197480); *Lilium hybrida* LhbHLH1 (BAE20057); LhbHLH2 (BAE20058); *Malus domestica* MdbHLH3 (ADL36597); MdbHLH33 (ABB84474); *Marchantia polymorpha* MpbHLH012 (Mapoly0031s0161, M. polymorpha genome MpTak_v6.1); MpbHLH004 (Mapoly0028s0058; M. polymorpha genome MpTak_v6.1); *Mimulus gattatus* (Erythranthe guttata) MgMigut.E01090.1; MgMigut.D00287.1; Oryza sativa OsPlw-OsB1 (BAB64301); OsRc (ADK36625); *Petunia hybrida* PhAN1 (AAG25928); PhJAF13 (AAC39455); *Physcomitrium patens* PpbHLH058 (Pp3c6_3190V3.2); PpbHLH082 (Pp3c16_21680V3.1); *Trifolium repens* TrAN1 (AIT76559) TrJAF13 (AIT76563); *Vitis vinifera* VvMYC1 (NM_001281253); VvMYCA1 (NP_001267954); *Zea mays* ZmLc (AAA33504); ZmIn1 (AAB03841). bHLH-1 proteins are bracketed in orange, bHLH-2 proteins in blue. Residue 177 in At GL3 is boxed in red as this has been shown to influence binding to R2R3Myb proteins depending on whether it is F or I (Zhang et al. 2019).

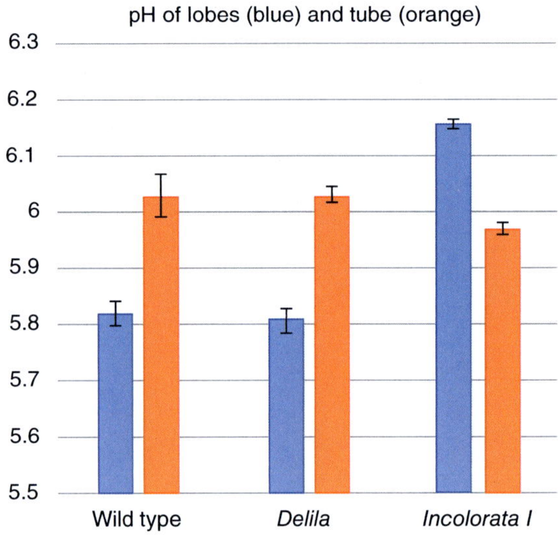

Figure 10.7 The effect of mutations in the genes encoding bHLH-1 (Delila) and bHLH-2 (Incolorata I) on the pH of lobe and tube petal tissue in *Antirrhinum majus*. Petal tissue from five flowers was ground with a pestle and mortar in 4 mL of distilled water, and the pH of the homogenate was measured immediately with a normal pH electrode. Delila is active in both lobes (blue) and tubes (orange), but Incolorata I is active only in lobes (Albert et al. 2020). Error bars show sem $n = 5$.

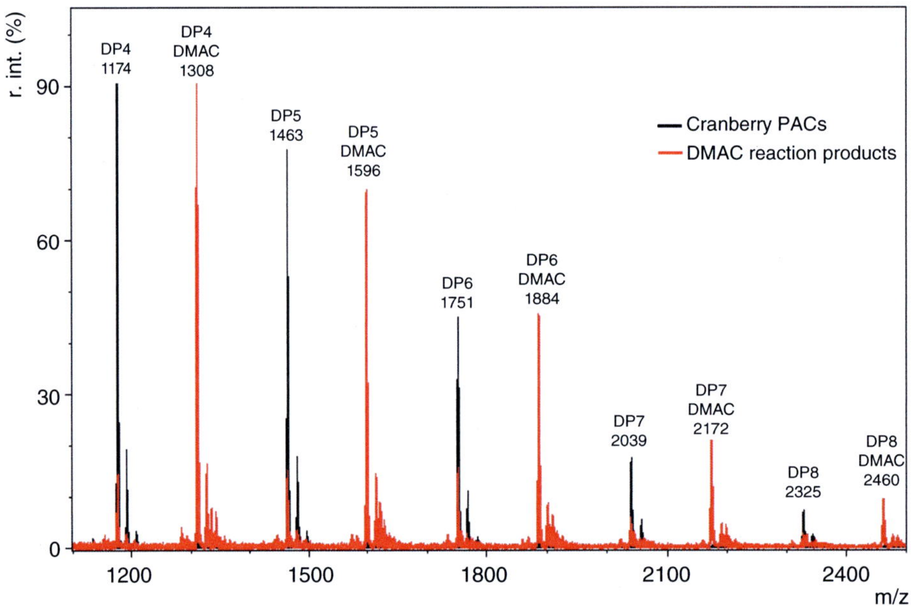

Figure 11.5 MALDI-TOF mass spectrum in positive reflectron mode for unreacted PAC (black) and PAC-DMAC reaction products (red). The degrees of polymerization (DP) for unreacted PAC are labeled from DP4 to DP8, while PAC-DMAC reaction products are labeled from DP4-DMAC to DP8-DMAC. Numbers under the label correspond to the mass of the sodium adduct for PACs and the mass of the respective cations for the DMAC reaction products. As indicated in Figures 11.6 and 11.7, the isotope patterns of the peaks for the DMAC reaction products indicate that each oligomer has two reaction products that correspond to the mass of the open ring cation and the pyrylium cation.

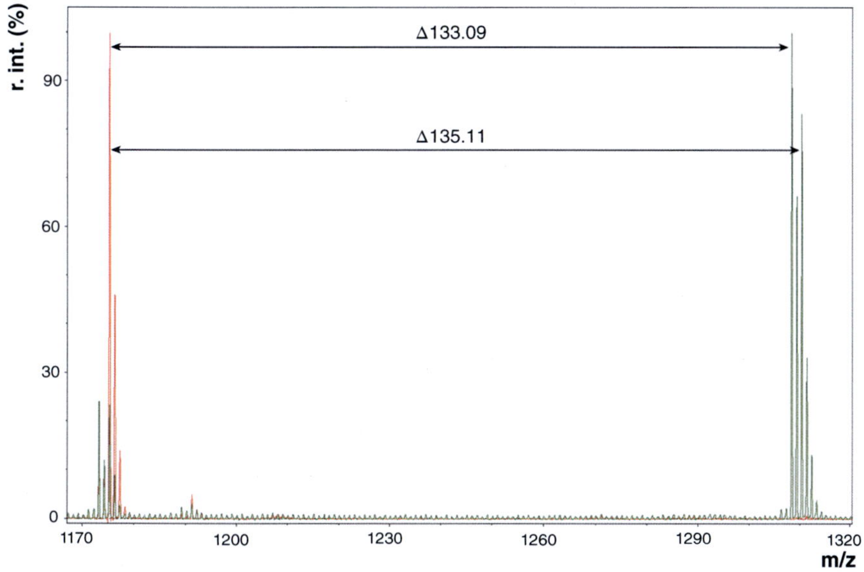

Figure 11.6 MALDI-TOF mass spectrum in positive reflectron mode for unreacted PAC (red) and PAC-DMAC reaction products (green), which show the PAC tetramer and the additive masses for the DMAC condensation products.

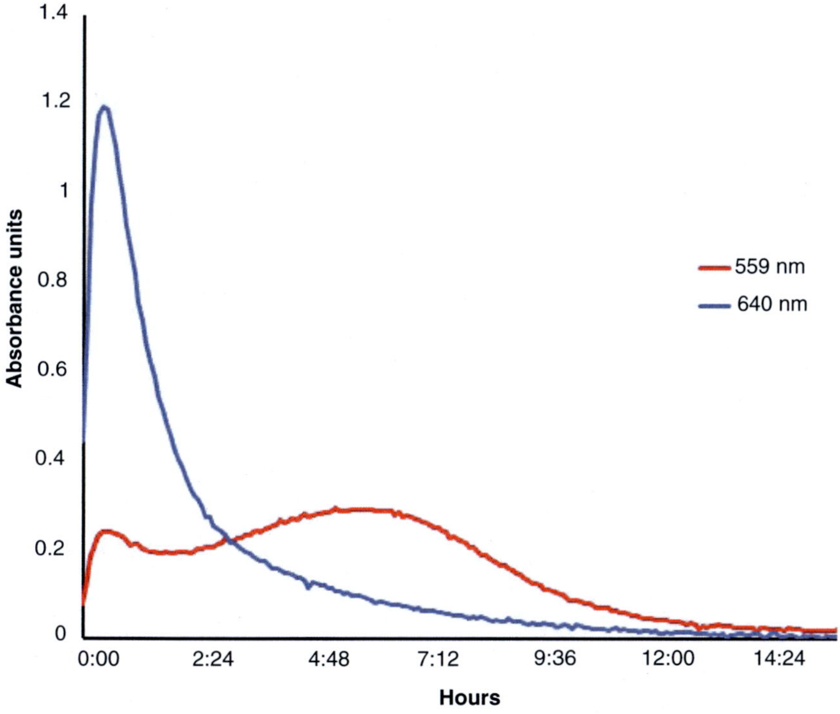

Figure 11.9 Change in absorbance at 559 and 640 nm during the DMAC reaction of procyanidin A2 in the presence of hydrochloric acid (3.24 M), which shows a rapid increase in absorbance at 640 nm (open ring PAC-DMAC cation) and the slow appearance of absorbance at 559 nm (PAC-DMAC pyrylium).

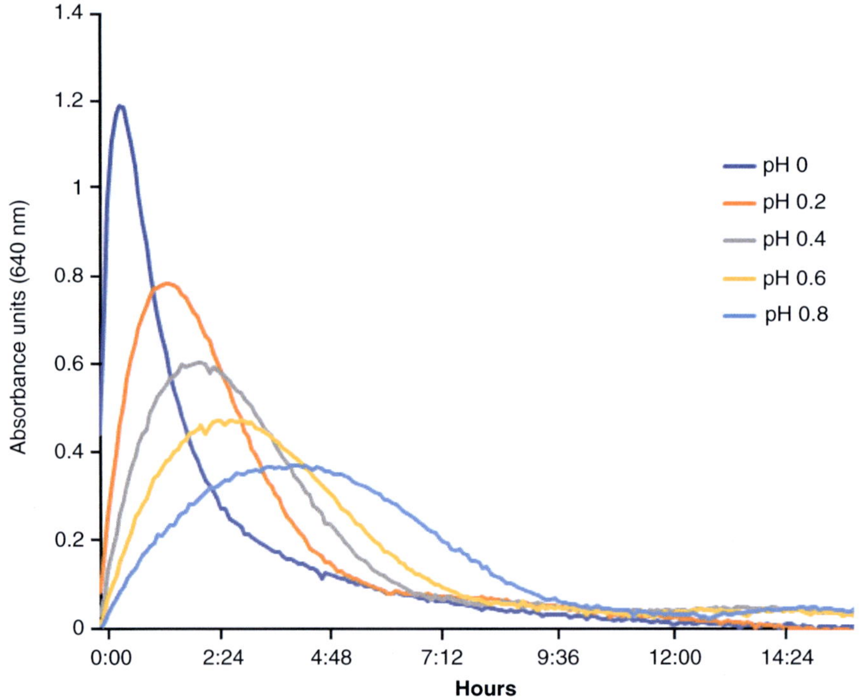

Figure 11.10 Change in absorbance at 640 nm during the DMAC reaction of procyanidin A2 at five different pH values.

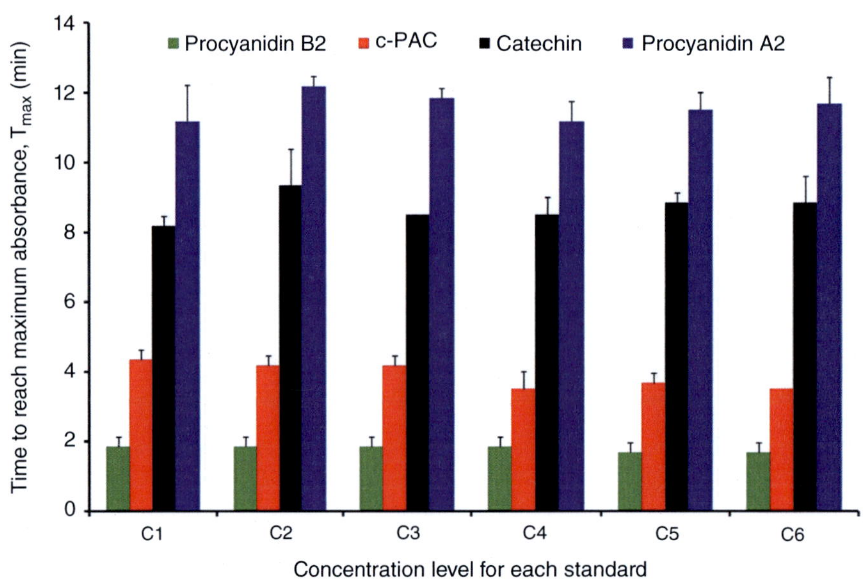

Figure 11.12 Variation of time required to reach maximum absorbance (t_{max}) using six increasing concentrations of catechin, procyanidin A2, procyanidin B2, and c-PAC. Stock solutions were diluted with methanol to produce six levels of concentrations (C1–C6): 2.6, 5.1, 7.7, 10.2, 12.8, and 15.3 mg/L for catechin; 4.8, 9.6, 14.4, 19.2, 24.0, and 28.8 mg/L for procyanidin A2; 5.0, 10.1, 15.1, 20.2, 25.2, and 30.2 mg/L for procyanidin B2; and 16.3, 32.6, 48.9, 65.2, 81.5, and 97.8 for c-PAC.

9

Highly Efficient Production of Dihydroflavonol 4-Reductases in Tobacco Cells and Refinement of the BuOH-HCl Enzymatic Assay

Lingping Zhu[1], *Saku Mattila*[1], *Roosa Matomäki*[1], *Lorenzo Mollo*[1,2], *Sharmin Ahamed*[1], *Sara M. Abdou*[1], *Hany Bashandy*[1,3], and *Teemu H. Teeri*[1]

[1] *Department of Agricultural Sciences, Viikki Plant Science Centre, University of Helsinki, Helsinki, Finland*
[2] *Department of Environmental and Life Sciences, Laboratory of Plant and Algae Physiology, Università politecnica delle Marche, Ancona, Italy*
[3] *Department of Genetics, Cairo University, Giza, Egypt*

9.1 Introduction

The biochemical pathway to anthocyanins is one of the best-known secondary metabolism pathways in plants. The pathway is a branch of the flavonoid pathway that also produces flavones, flavonols and proanthocyanidins (condensed tannins) in all land plants (Stafford 1991). Some plant taxa have specific branches in the flavonoid pathway, such as legumes producing isoflavonoids or snapdragon producing aurones. The flavonoid pathway itself is a branch of the phenylpropanoid pathway shared by land plants and their likely precursors (Fellenberg and Vogt 2015; Wen et al. 2020), where, in vascular plants, massive metabolite flux leads to lignin, the second most abundant polymer in the terrestrial biosphere after cellulose (Boerjan et al. 2003). The connection of the phenylpropanoid pathway to primary metabolism is through phenylalanine (or tyrosine), making the shikimic acid pathway responsible for the heavy flux of carbon into the slowly recycling form of lignin – the nitrogen in the amino acid phenylalanine is however recycled in the plant cells as it is removed in the first step of the phenylpropanoid pathway by the enzyme phenylalanine ammonia-lyase (Barros and Dixon 2020).

Anthocyanins, the name referring to the "blue" of "flowers" in Greek, are responsible for most of the orange, red, and blue colors we see in flowers and fruits. Some plant species color their flowers and fruits (or other tissues) with carotenoids or betalains (Tanaka et al. 2008), but anthocyanins are by far the most common of these pigments. In addition to providing pleasure in the form of beauty in flowers (on which the ornamental plant industry relies heavily), anthocyanins in fruits are considered a healthy component of our diets (Martin 2018). It is believed that the colored anthocyanins evolved as a way to

Recent Advances in Polyphenol Research, Volume 8, First Edition. Edited by Juha-Pekka Salminen, Kristiina Wähälä, Victor de Freitas, and Stéphane Quideau.

cope with the abiotic stresses of terrestrial life (Berland et al. 2019) of plants but in flowering plants, they have further acquired roles in communication with the animal kingdom and in establishing a relationship that benefits both partners (pollination in exchange of food).

The biosynthesis of anthocyanins, within the flavonoid pathway, begins with the action of the enzyme dihydroflavonol 4-reductase (DFR) on the intermediate dihydroflavonol. DFR competes for this substrate with flavonol synthase, an oxidative enzyme that forms flavonols, important pale yellow metabolites in many plant tissues. The product of DFR is leucoanthocyanidin, an unstable intermediate that is converted to anthocyanidin by anthocyanidin synthase (and, as a competing reaction, to catechins by leucoanthocyanidin reductase). Anthocyanidins are unstable in the cellular environment and need glycosylation for their stability. Glycosylation usually starts by adding a glucose moiety to the hydroxyl group in carbon number 3, but can, in different species, extend to other positions and with several different sugar moieties. Additional methylation of the hydroxyl groups in the anthocyanidin backbone and acylation of the hydroxyl groups in the sugar moieties leads to a large number of different anthocyanins in the plant kingdom. This variation, together with control of pH in the vacuole where anthocyanins finally reside (after biosynthesis in the cytosol) and with interactions with colorless flavonoids (copigments) leads to the wonderful color variation we see in flowers and fruits (Koes et al. 2005).

In spite of the large spectrum of colors in flowers and fruits, the main anthocyanin aglycone molecules (anthocyanidins) are relatively few. The most important ones are pelargonidin, cyanidin, and delphinidin, together with the methylated forms of the two latter, peonidin, petunidin and malvidin. The difference between the first three is in the hydroxylation pattern of the B-ring that originates from the aromatic group of phenylalanine. Pelargonidin harbors a single hydroxyl group in the B-ring position 4′, originating from the second enzymatic reaction of the phenylpropanoid pathway where cinnamate (deaminated phenylalanine) is converted to *p*-coumarate by cinnamate 4-hydroxylase (C4H). Cyanidin has a second hydroxyl at position 3′, added by flavonoid 3′-hydroxylase (F3′H), and delphinidin has altogether three B-ring hydroxyl groups, the final two added by flavonoid 3′,5′-hydroxylase (F3′5′H). While all land plants have C4H, without which lignin does not form (Schilmiller et al. 2009), the presence of F3′H and F3′5′H varies depending on the species and the genotype within species (e.g. between cultivars of ornamental plants). As delphinidin (and its methylated derivatives, petunidin and malvidin) are responsible for purple and blue colors in petals, the lack of F3′5′H in, e.g. roses, carnations and chrysanthemums explains the lack of blue colors in these ornamental plants. While this has been spectacularly fixed with genetic engineering (transfer of a gene that encodes F3′5′H), the difficulty in obtaining intense blue in these ornamentals shows that the type of the aglycone is not the only factor determining petal color (Tanaka et al. 1998; Katsumoto et al. 2007; Noda et al. 2017). Similarly, the sky-blue color of the cornflower petals is due to an anthocyanin based on cyanidin – the anthocyanidin that is usually associated with red and pink petal color (Shiono et al. 2005).

Orange and scarlet red anthocyanins are pelargonidin derivatives. While it is easy to explain why some plant species lack delphinidin-based anthocyanins, it is more mysterious why others lack pelargonidin-based pigments. Pelargonidin is the simplest of the anthocyanidins and biosynthesis of cyanidin and delphinidin proceeds through the pelargonidin precursor dihydrokaempferol (DHK) that is hydroxylated to dihydroquercetin (DHQ) and dihydromyricetin (DHM) by F3′H and F3′5′H, respectively. The cause is in substrate specificity of DFR – the enzyme in some species reacts poorly with DHK. Petunia was the first and most famous example. Orange or scarlet red petunia cultivars do not exist, except as genetically modified cultivars that express a gene for DFR from a different species (Meyer et al. 1987; Helariutta et al. 1993). It was recently found that, apparently unintentionally, such modified cultivars had been infiltrated into commercial petunia breeding programs producing novel but illegal colors in several cultivars (Bashandy and Teeri 2017; Haselmair-Gosch et al. 2018; Fraiture et al. 2019).

The interesting differences in substrate specificity of different DFR enzymes have been studied since 1980s but, as yet, which amino acids are responsible for the specificity of the enzyme have not been well defined. DFR is reported as an unstable enzyme (Petit et al. 2007) which is challenging to produce in microbial cells and frequently ends up in insoluble inclusion bodies in bacteria. Fusion with soluble carrier proteins in modern expression vectors helps a lot and DFR has been successfully produced in both bacteria and yeast (Martens et al. 2002; Gosch et al. 2014; Miosic et al. 2014). An alternative is to produce the enzyme in plant cells, but the yields have been low (Martens et al. 2002). In this paper, we show that using the hypertranslatable plant vector pEAQ-HT (Peyret and Lomonossoff 2013), remarkably high yields of DFR can be obtained in tobacco cells after infiltration with agrobacterium and transient expression. The protein is stable and allows easy determination of its substrate preferences. The traditional BuOH-HCl assay for determining DFR activity is refined allowing simple quantitative measurement of activity in terms of the product (leucoanthocyanidin) formed.

9.2 Results

9.2.1 Transient Expression from Hypertranslatable Vectors

The pEAQ-HT series of vectors (Peyret and Lomonossoff 2013) contain (modified) 5′ and 3′ UTR sequences from the Cowpea mosaic virus and yield several-fold more protein product than similar vectors without the viral UTRs. The vectors contain the silencing suppressor P19 as part of the construct, allowing protein expression to proceed for about one week in infiltrated *Nicotiana benthamiana* leaves yielding in sometimes extremely high amounts of foreign protein (e.g. for the petunia DFRA, a band in the gel that matches Rubisco LS in intensity; Figure 9.1). To limit the amount of DFR activity in infiltrated leaf samples, the leaves were harvested three days after infiltration for most assays shown here. For purification of His-tagged proteins, expression was allowed to proceed for six days, at which time the leaves started to show some yellowing.

The crude and cleared extracts contain small molecules from tobacco leaves, which were removed by buffer exchange using gel filtration. However, the level of activity and patterns of substrate use were the same for both the crude and the buffer exchanged extracts, and for the results shown later, simple crude extracts were used. Extracts from petals where the anthocyanin pathway was active were gel filtrated to remove a

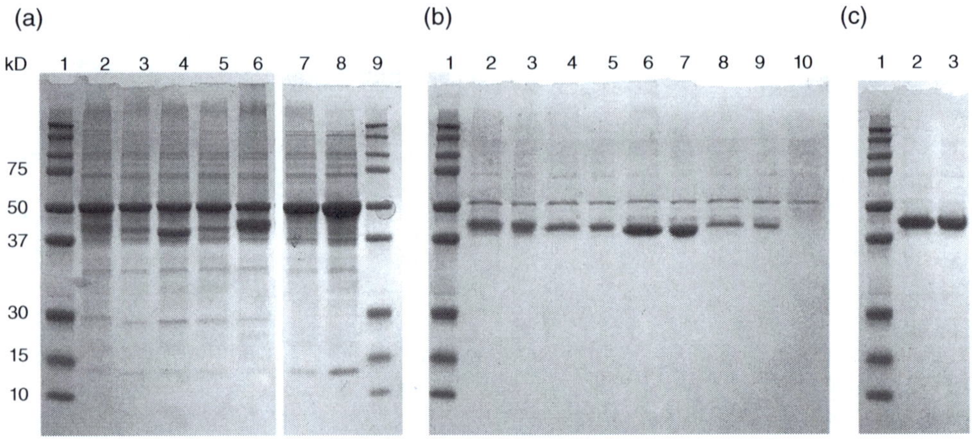

Figure 9.1 Crude extracts (a) of agroinfiltrated *N. benthamiana* leaves expressing His-tagged DFR proteins and nickel column purified proteins (b, c). a1 size standard, a2 His_6-GDFR1-2, a3 His_6-GDFR1-3, a4 GDFR1-2-His_6, a5 GDFR1-3-His_6, a6 His_6-DFRA, a7 empty agrobacterium, a8 not infiltrated, a9 size standard; b1 size standard, b2-3 His_6-GDFR1-2 (H and G), b4-5 His_6-GDFR1-3 (H and G), b6-7 GDFR1-2-His_6 (H and G), b8-9 GDFR1-3-His_6 (H and G), b10 empty agrobacterium (H); c1 size standard, c2-3 His_6-DFRA (H and G). H = Eluted from His SpinTrap, G = Desalted in G25 column. Gel loading corresponds to 1 mg (a) or 2.4 mg (b, c) of leaf tissue fresh weight. Sizes of GDFR1-2, GDFR1-3, and petunia DFRA with the His_6 tag are 41.9, 42.0, and 42.2 kD, respectively.

considerable background due to endogenous metabolites. Stored at −80 °C in 20% glycerol, the samples were relatively stable. After nineteen months of storage, more than 50% of activity remained in the samples. Keeping the extracts overnight on ice, more than 90% of activity was retained.

Most assays shown here were done with crude protein extracts. To test the purification of the proteins, the His-tagged versions were bound to the nickel column, extracted with 500 mM imidazole, and analyzed after buffer exchange. Based on SDS-PAGE, the proteins were practically void of contaminating proteins (except for a thin band migrating at the position of Rubisco LS; Figure 9.1) and they showed similar substrate use profiles as the corresponding crude extracts.

9.2.2 BuOH-HCl Assay Revisited

A simple colorimetric assay for detection of leucoanthocyanidins, the product of DFR activity, is based on assays of condensed tannins (proanthocyanidins), which upon acid hydrolysis in *n*-BuOH-HCl become oxidized and butanoylated to colored anthocyanidin derivatives (Swain and Hillis 1959). Porter et al. (1986) showed that the yield of the process is critically dependent on trace metal-ion impurities and the method used here was based on their suggestion to include ferric iron salt in the mixture. The butanoylated pelargonidin, cyanidin, and delphinidin products have slightly different absorption maxima in BuOH-HCl, and absorbance was measured at the optimum wavelengths of 535 nm, 550 nm, ad 560 nm, respectively, for the three products.

The BuOH-HCl assay for DFR has been much used but also criticized, for example, for differential extractability of the leucoanthocyanidins in ethyl acetate during their isolation from the aqueous reaction mix (Gosch et al. 2014). Furthermore, actual amount of substrate used (or product formed) is not calculated in the traditional BuOH-HCl assay. By addressing these points, we developed the BuOH-HCl assay into a reliable and robust assay, with the aim of studying the substrate specificity of DFR enzymes from different sources. The gerbera enzyme GDFR1-3 does not strongly differentiate between DHK, DHQ, and DHM as its substrate, and a crude extract of GDFR1-3 produced in *N. benthamiana* leaves was used to address the assay parameters. The standard assay contained 1 mM NADPH, 100 μM dihydroflavonol, and ca. 150 μg/mL of tobacco leaf protein in 100 mM MOPS-KOH pH 7.0. The substrate was dissolved in DMSO, which was at a 1% concentration in the final assay. Increasing the DMSO concentration to 2% retained 90% of the activity. The effect of assay pH between 6 and 7 was negligible for the three gerbera enzymes (GDFR1-1, GDFR1-2 and GDFR1-3), but the activity declined when the pH increased above 7.4 (Figure 9.2a). A pH of 7.0 was chosen for the assay based on the fact that the pH of plant cytoplasm, where anthocyanin biosynthesis takes place, is usually neutral to slightly basic (Martinière et al. 2013).

In order to evaluate the linearity of the assay with respect to time and amount of substrate, a simple background corrected absorbance should be sufficient. However, to further assess the fraction of the substrate consumed in the reaction, absorbances were converted to μM concentrations using factors described in the methods section, the derivation of which is described further here.

With the standard amount of protein in the assay, the reaction rate declined after 10 minutes (Figure 9.3a). Apparently, the decline was not due to the stability of the enzyme since a reduced amount of tobacco extract allowed the reaction to continue up to one hour linearly (Figure 9.3b). During the standard assay time of 10 minutes, the reaction remained linear (Figure 9.3c).

The reaction rate was dependent on substrate concentration in the range 33–100 μM used for the assay (Figure 9.2b). The usual K_m of DFR for dihydroflavonols is in the range of 10–20 μM (Fischer et al. 2003; Halbwirth et al. 2003; Gosch et al. 2014), although values as low as 0.4 μM have been reported (Miosic et al. 2014). An NADPH concentration of 1 mM was expected to be saturated (Ruhnau and Forkmann 1988).

Partitioning of leucoanthocyanidins, the DFR products, between acidified aqueous buffer and ethyl acetate was dependent on the degree of hydroxylation of the B-ring, and leucodelphinidin, in particular, extracts poorly (Gosch et al. 2014). In order to take the differential extractability into account, a serial extraction of the stopped reaction mix was done, first with 163 μL ethyl acetate to compensate for the loss due to its solubility in water, and subsequently with 150 μL.

As shown in the graph (Figure 9.2c), leucocyanidin had much lower extractability than leucopelargonidin. Calculated from the linear trend lines, 67%, 46%, and 30% of leucopelargonidin, leucocyanidin, and leucodelphinidin, respectively, were partitioned into ethyl acetate in one extraction step. With two extraction steps in the standard assay, this meant that a correction factor of 1.12, 1.41, and 1.96 for the three reaction products, respectively, was necessary to compensate for the differential extraction.

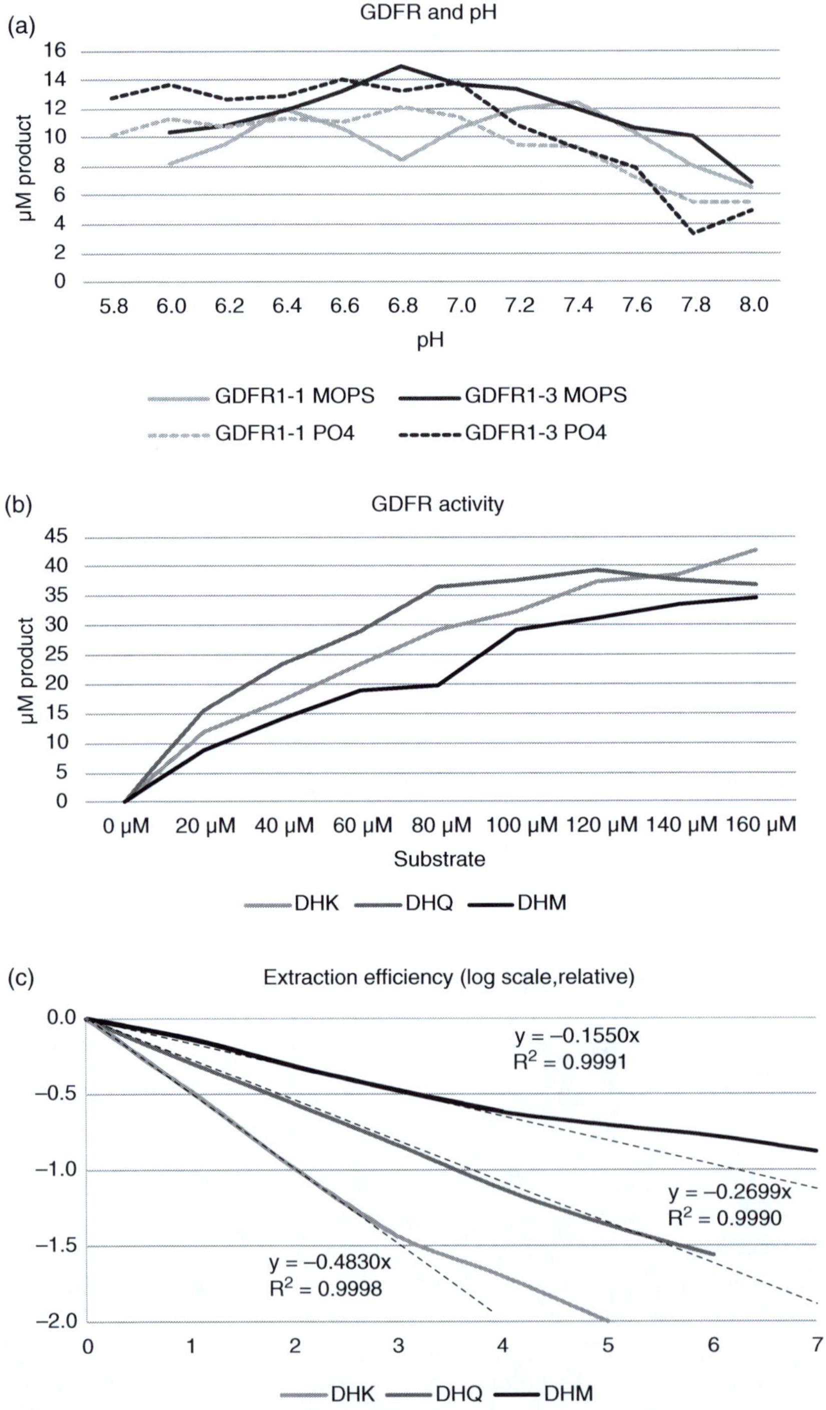

Figure 9.2 pH dependence of two gerbera DFRs (a), dependence on substrate concentration of GDFR1-3 (b), and extractability of DFR products to ethyl acetate after the enzymatic reaction, assayed with BuOH-HCl (c).

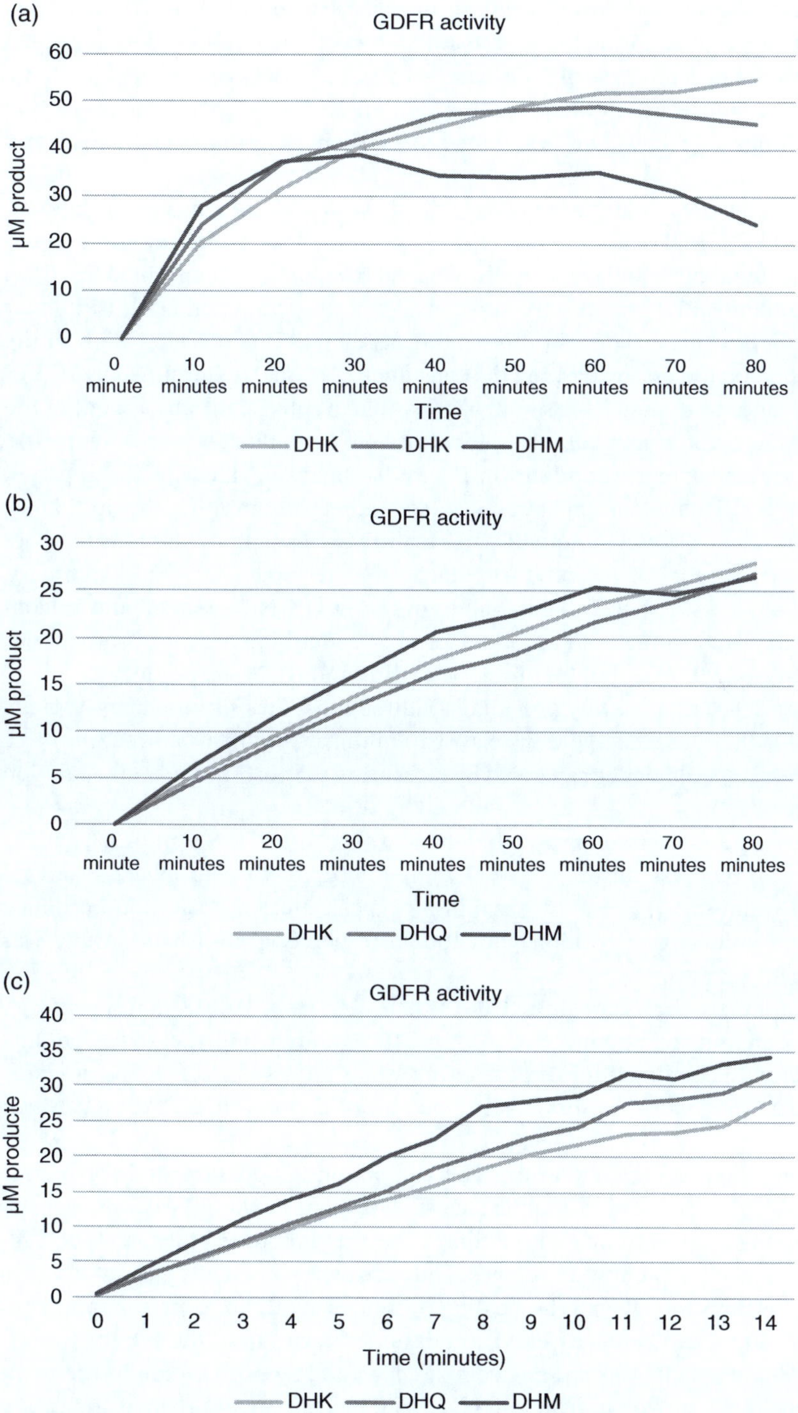

Figure 9.3 Linearity of the GDFR1-3 reaction with 10 µL of tobacco extract (a), 2 µL of tobacco extract (b), and 10 µL of tobacco extract for a shorter time, representing the 10 minutes reaction time of the standard assay (c).

In order to establish the relationship between the absorbance in the BuOH-HCl assay and the amount of product formed, molar extinction coefficients of the butanoylated anthocyanidins were needed. These are however not available and even the molar extinction coefficients of unsubstituted anthocyanidins found in the literature are not useful due to varying conditions (solvents and wavelengths) in their determination (Giusti and Wrolstad 2001).

In order to determine the amount of product formed, we determined instead the amount of substrate consumed with the assumption that this corresponded to the amount of product formed. The residual dihydroflavonol in the stopped reaction was determined by HPLC (without extraction into ethyl acetate). We also considered this as a potential substitute for the standard assay, but the variance in this kind of assay was much higher than in the BuOH-HCl assay. Nevertheless, by plotting the consumed substrate (determined by HPLC) against the absorbance of the BuOH-assay, it was possible to convert the absorbance of the butanoylated product to the amount of substrate consumed (or product formed) (Figure 9.4).

The slopes of the linear regression lines in Figure 9.4 hold the values of the apparent molar extinction coefficients for the butanoylated anthocyanidins, which are 15,894, 13,063, and 10,906 for Bu-Pg, Bu-Cy, and Bu-Dp, respectively. Note that this value includes the differential extractability of the products that were determined earlier. In summary, based on this relationship, the absorbance values of the BuOH-HCl assays could be converted to μM substrate consumed (and product formed) by multiplying the absorbance of the standard assay by 62.9, 76.6, and 91.7 for DHK, DHQ, and DHM, respectively.

We wanted to simplify the assay by providing all three substrates simultaneously. Here, we chose to provide them each at 33 μM instead of 100 μM. One problem was that the three butanoylated products could not be separated by measuring the absorbance and separation by HPLC was necessary.

Butanoylation of Pg and Cy lead to two different peaks on HPLC (at 8.9 and 12.3, and 8.1 and 11.4 minutes in our system, respectively) while for Dp, three peaks were separated (5.0, 7.3 and 10.3 minutes). By simply summing the peak areas for each butanoylated anthocyanidin, an excellent correlation with the corresponding absorbance value was obtained (Figure 9.5).

Using the slope of these regression lines (0.001374, 0.001367, and 0.001812, respectively), the summed peak areas could be converted to absorbances and then further to μM product formed. The assay with combined substrates, however, was not a substitute for assays done with each substrate separately. As shown later, we observed strong interference between the substrates.

Leucoanthocyanidins are known to be unstable compounds that oxidize and polymerize easily (Gosch et al. 2014). In order to establish at which stage of the BuOH-HCl assay instability of the reaction products could cause errors in the assay, we tested the effects of a 24-hour pause in the procedure at different stages. The assay was paused after the 10 minutes reaction (by inactivating the enzyme by heating 5 minutes at 95 °C), after stopping with glacial acetic acid, after extraction with ethyl acetate and after drying of the extract. In each case, more than 90% of activity was present, compared to an assay that was conducted without pauses. In fact, the first two interruptions caused an increase in absorbance, indicating that 5 minutes at 95 °C or lowering the pH by glacial acetic acid does not, in fact, completely or immediately destroy the DFR activity. Nevertheless, the experiment shows that there are no problems of product decay at any stage of the assay. This does not mean that

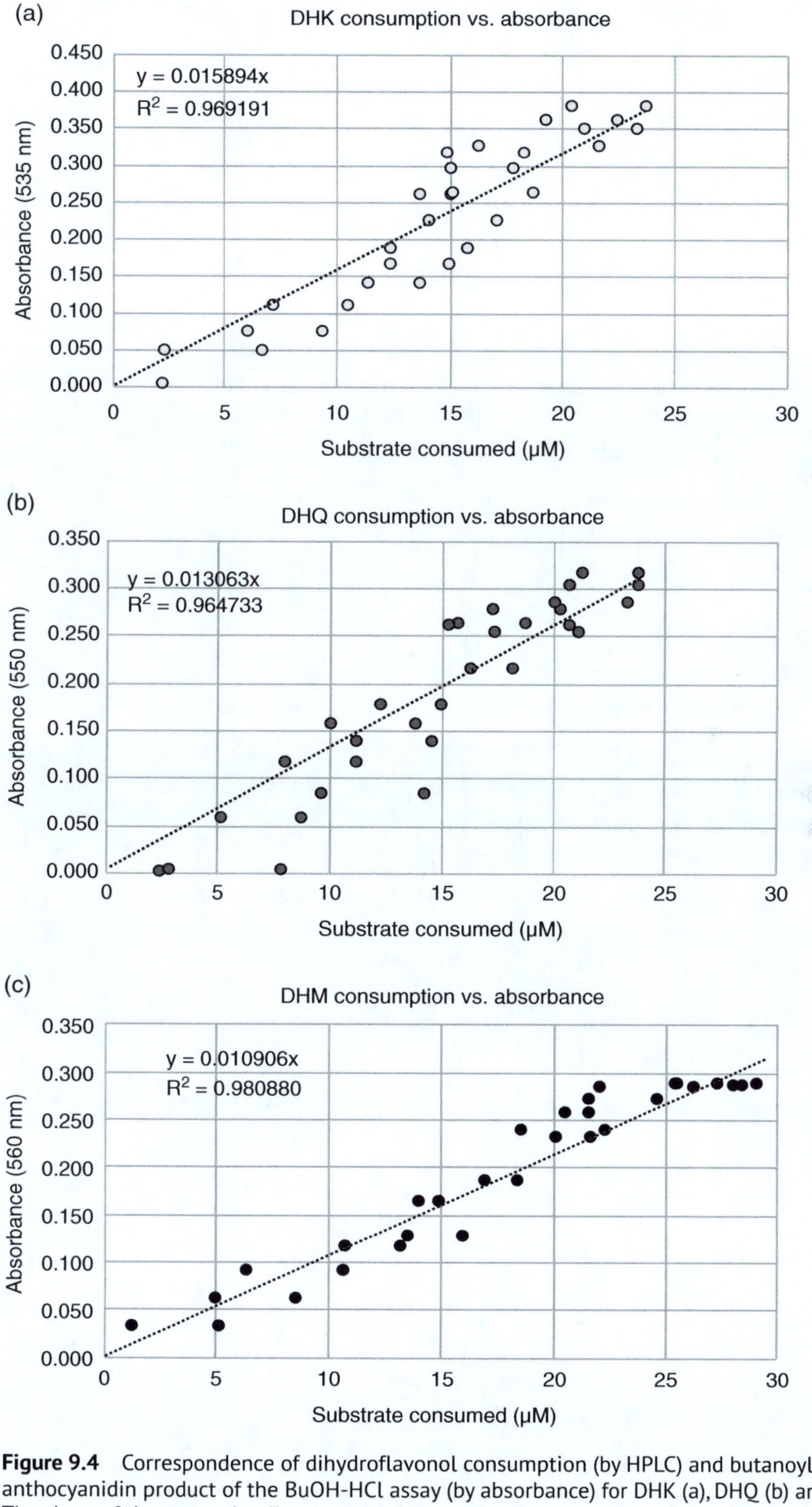

Figure 9.4 Correspondence of dihydroflavonol consumption (by HPLC) and butanoylated anthocyanidin product of the BuOH-HCl assay (by absorbance) for DHK (a), DHQ (b) and DHM (c). The slope of the regression line contains information for conversion of the absorbance value to substrate consumed and product formed in μM.

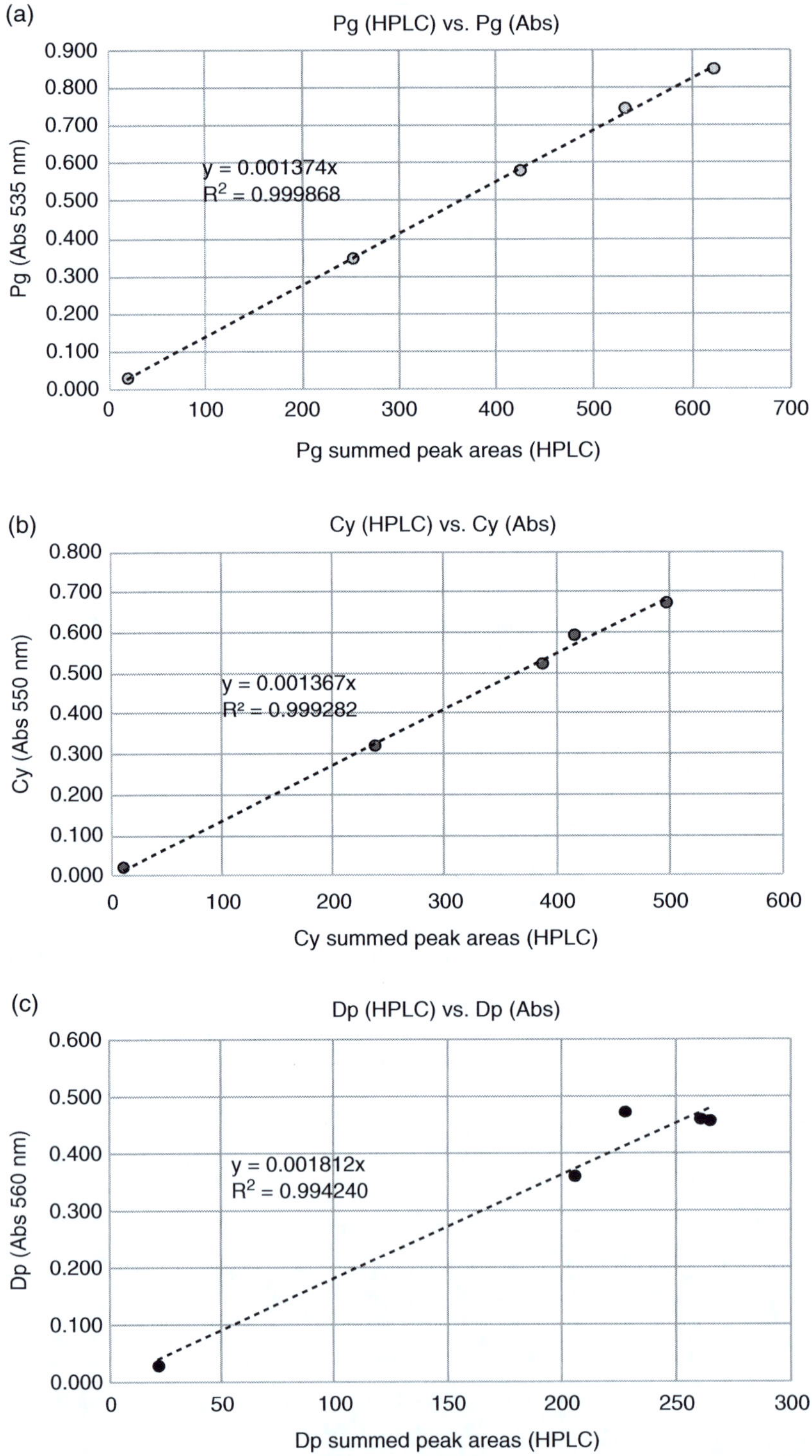

Figure 9.5 Correspondence of the peak area sum of butanoylated Pg (a), Cy (b) and Dp (c) in HPLC and absorbance value of the BuOH-HCl assay allowing commensuration of mixed substrate assay by HPLC to single substrate assay by BuOH-HCl.

oxidation or polymerization of the products does not occur. The BuOH-HCl assay is an oxidative assay that was originally developed to hydrolyze and derivatize condensed tannins (into butanoylated anthocyanidins), which are polymerized leucoanthocyanidin derivatives. The red color of the final assay product is however light-sensitive and decays to half when left for a few hours on the laboratory bench in ambient light. In the dark, the color is stable for several days at room temperature (and for several weeks at −20 °C).

9.2.3 Substrate Profiles of Different DFRs

We have identified three allelic forms of gerbera DFR that are represented in two cultivars with different colors (Bashandy et al. 2015a). The cultivar "Regina" accumulates orange pelargonidin-derived anthocyanins in the petals and is heterozygous for two alleles, GDFR1-1/GDFR1-2. The cultivar "President" with magenta-colored flowers accumulates cyanidin-derived anthocyanins and is homozygous for GDFR1-3. In 1993, we showed that expression of GDFR1-1 in the (nearly) white petunia laboratory strain RL-01 led to intense orange pigmentation (Helariutta et al. 1993). The famous escaped orange GMO petunia was made several years earlier by Meyer et al. (1987) using the maize DFR gene A1, and subsequently crossed with ornamental varieties to yield stable and bright orange cultivars by Oud et al. (1995).

Expressed in *N. benthamiana*, the substrate specificity of these five DFR enzymes is shown in Figure 9.6a. DFR1-1 and GDFR1-2 have similar preferences for the three dihydroflavonols; DHK is strongly preferred and activity with DHQ and DHM is minimal. GDFR1-3, on the other hand, has no particular preference and all three substrates are efficiently reduced. The petunia DFRA shows the well-known pattern of not reducing DHK and showing the highest activity with DHM (Haselmair-Gosch et al. 2018). The maize A1 DFR that is responsible for pelargonidin biosynthesis in the orange petunia cultivars shows high activity with DHK, but also accepts DHQ and DHM.

Substrate preference of enzymes extracted from petals expressing these DFR variants shows similar profiles as the tobacco-produced enzymes (Figure 9.6c). The engineered orange petunia "African Sunset" shows a combination of the maize A1 and petunia DFRA profiles.

When all three substrates are provided simultaneously, the picture changes completely (Figure 9.6b). The four enzymes active with DHK are not able to convert this substrate when all three substrates are given simultaneously. Use of DHQ and DHM is affected less, if at all.

9.3 Materials and Methods

9.3.1 Plant Material and Chemicals

Nicotiana benthamiana plants used for agroinfiltration were sown on peat:vermiculite mixture (1 : 1) and grown for six weeks, until a ten-leaved stage, under fluorescent light before infiltration with agrobacterium (Bashandy et al. 2015b). *Gerbera hybrida* cultivars "Regina" and "President" were grown in standard greenhouse conditions (Elomaa and Teeri 2001). An undefined cultivar of *Petunia hybrida* with dark blue flowers was

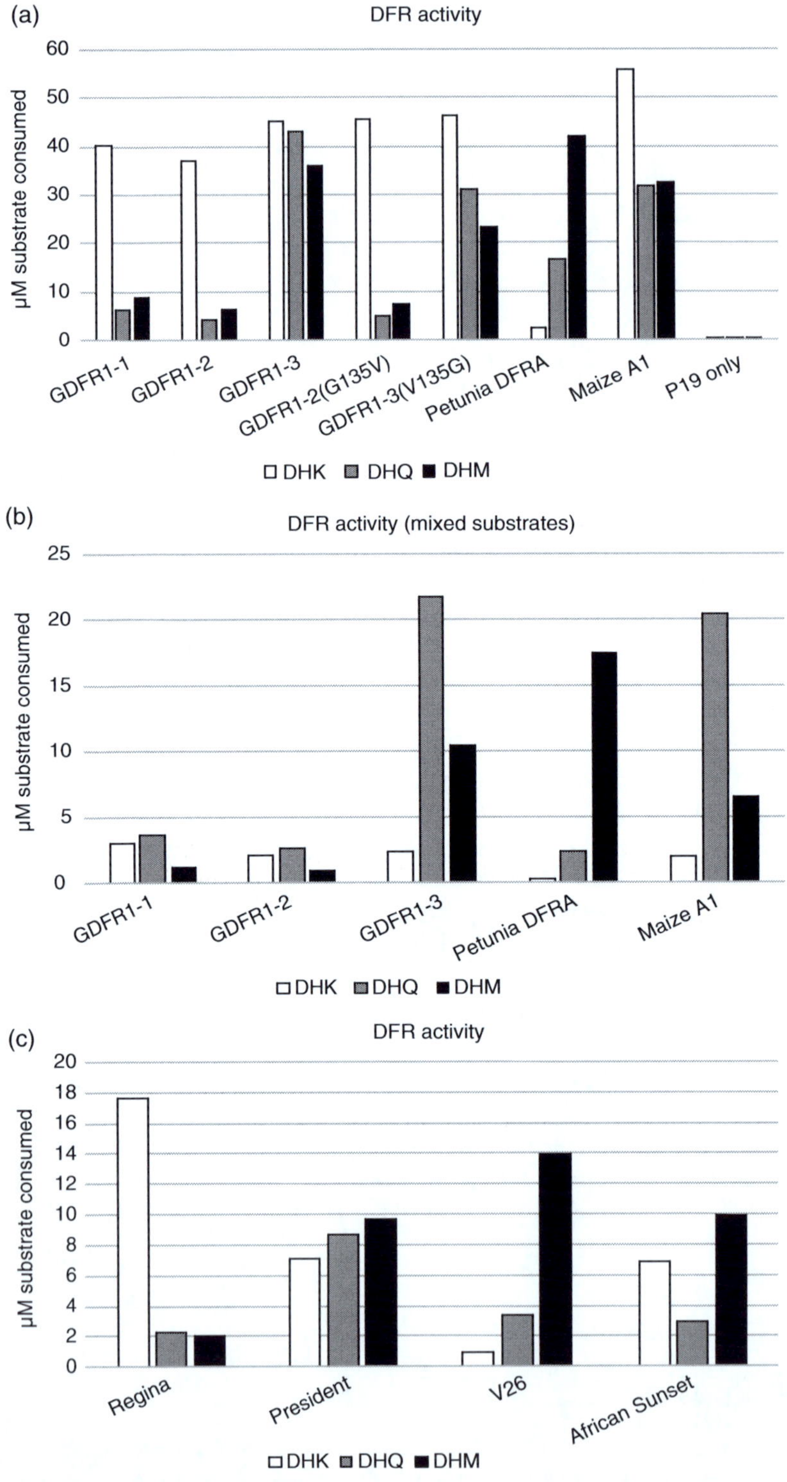

Figure 9.6 Patterns of substrate use by DFR enzymes heterologously produced in *N. benthamiana* with separately (a) or simultanously (b) offered substrates. The substrate preference patterns in petal extracts (c) correspond well with the heterologously produced enzymes. The gerbera cultivar "Regina" is heterozygous for GDFR1-1 and GDFR1-2, and "President" is homozygous for GDFR1-3. V26 is a non-transgenic laboratory line of petunia with intense blue flowers, "African Sunset" is an orange engineered petunia cultivar transgenic for the maize DFR gene A1.

purchased from a local market for isolation of the petunia DFRA cDNA. Petals of the inbred petunia line V26 (Napoli et al. 1990) and the commercial F1 cultivar "African Sunset" (Bashandy and Teeri 2017) were extracted for the enzyme assays. (2R,3R)- DHK, (2R,3R)-dihydroquercetin (DHQ), and (2R,3R)-dihydromyricetin (DHM) were purchased from TransMIT PlantMetaChem (Giessen, Germany) and other chemicals from Sigma-Aldrich.

9.3.2 Isolation of DFR Encoding Sequences and Plasmid Construction

Total RNA was extracted by the method of Chang et al. (1993) from gerbera and petunia petals at a stage where the distal half of the petal had gained visible pigmentation. Poly-A RNA was converted to cDNA using oligo-dT priming and the SuperScript® III Reverse Transcriptase kit (Invitrogen, Carlsbad, CA, USA). DFR coding sequences were amplified from cDNA or preexisting plasmids by two successive PCR reactions, first with gene-specific primers and subsequently with Gateway attB adapter primers. PCR fragments were purified with a High Pure PCR Product Purification kit (Roche) and used in a BP reaction with the vector pDONR221 according to the Gateway manual (Invitrogen). Mutagenesis was done at this stage using primers carrying the intended changes (Zheng et al. 2004). After verifying the inserts by sequencing, they were transferred to the destination vector pEAQ-HT-DEST1, pEAQ-HT-DEST2, or pEAQ-HT-DEST3 in an LR reaction according to the Gateway manual. The latter two vectors code for N-terminal and C-terminal His-tags, respectively, and the first vector has no tag. The resulting expression vectors were verified with restriction enzymes and transformed to *Agrobacterium tumefaciens* C58C1(pGV2260) (Deblaere et al. 1985) using the freeze-thaw method (Wise et al. 2006). The final verification of the agrobacterium strains used in this study was done by amplifying the DFR insert from a bacterial sample and by sequencing the fragment with the amplification primers (separately). Accession numbers for GDFR1-1, GDFR1-2, GDFR1-3, DFRA, and A1 are Z17221, KP765769, KP765770, MW929212, and KY964325, respectively. All primers used are listed in Table 9.1.

9.3.3 Protein Extraction and Purification

Plant tissue was frozen in liquid nitrogen and milled into a fine powder (Oscillating Mill MM400, Retsch). 4.5 volumes of extraction buffer were added to frozen tissue powder (room temperature buffer into powder on dry ice thaws the tissue just right), and the homogenate was kept on ice for 15 minutes and cleared twice in a microcentrifuge (full speed, +4 °C). One-quarter volume of 100% glycerol was added, and the samples were snap frozen in liquid nitrogen in aliquots and stored at −80 °C until used.

Extraction buffer for tobacco tissue was 100 mM Tris-Cl pH 7.5, supplemented with 0.2% 2-mercaptoethanol and a cocktail of protease inhibitors (Complete Mini EDTA free, Roche). For gerbera and petunia petal tissue, rich in polyphenols and prone to intense browning due to polyphenol oxidases, the buffer was degassed by boiling and cooled before adding mercaptoethanol and the protease inhibitors. 50 mg/mL anion exchange matrix Dowex 1 × 2–400 (equilibrated in degassed 100 mM Tris-Cl pH7.5) was included in order to bind polyphenols from the homogenate (Schijlen et al. 2007).

Table 9.1 Primers.

Name	Sequence	Application
GER787	AAAAAGCAGGCTCCATGGAAGAGGATTCTCCG	Forward primer for GDFR1-1 and GDFR1-2
GER788	AAAAAGCAGGCTCCATGGAAGAGGATTATCCA	Forward primer for GDFR1-3
GER789	AAAAAGCAGGCTCCATGCCCCTTCACCTCCG	Forward primer for DFRA
GER790	AGAAAGCTGGGTCTAGACTTCAACATTGC	Reverse primer for DFRA
ATTB2DFRsrv	AGAAAGCTGGGTCCTATTGGCCTTCTTTTGAAC	Reverse primer for GDFR1-1, GDFR1-2 and GDFR1-3 (includes the stop codon)
GER1768	AGAAAGCTGGGTCTTGGCCTTCTTTTGAACAAC	Reverse primer for GDFR1-1, GDFR1-2 and GDFR1-3 (without the stop codon)
GER1592	ACCGTGAACGGACAAGAGAAAC	GDFR1-3 V135G forward primer
GER1593	GTTTCTCTTGTCCGTTCACGGT	GDFR1-3 V135G reverse primer
GER1594	ACCGTGAACGTACAAGAGAAAC	GDFR1-2 G135V forward primer
GER1595	GTTTCTCTTGTACGTTCACGGT	GDFR1-2 G135V reverse primer
GER1567	AAAAAGCAGGCTCCATGGAGGGAGGTGCCGGT	Forward primer for A1
GER1568	AGAAAGCTGGGTTAAGCGCCAATCGTCGC	Reverse primer for A1
GER29	GGGGACAAGTTTGTACAAAAAAGCAGGCT	Gateway attB1 adapter primer
GER30	GGGGACCACTTTGTACAAGAAAGCTGGGT	Gateway attB2 adapter primer
GER1620 GER1621	CTTTGAAACAGAGTTTTCCCGTGG CATAGAAATGCACACCGAATAACAG	Primer pair and sequencing primers for insert amplification from expression vectors

Extraction buffer for the His-tagged DFR proteins was, according to the His SpinTrap manual (GE Healthcare), 20 mM Na-PO_4 pH 7.4, 500 mM NaCl, supplemented with 2-mercaptoethanol and the protease inhibitor cocktail as above, and with 20 mM imidazole. After clearing of the extract, it was applied to the His SpinTrap column, washed once with the extraction buffer (without 2-mercaptoethanol and protease inhibitors) and eluted twice with 250 µL 20 mM Na-PO_4 pH 7.4, 500 mM NaCl, 500 mM imidazole.

Protein amount was measured using the Bradford reagent kit (BioRad) with BSA as standard and buffer change, when necessary, into 100 mM Tris-Cl pH 7.5 was done using G25 size exclusion columns (Amersham), according to the manufacturer's instructions.

9.3.4 BuOH-HCl Assay

The BuOH-HCl assay for DFR enzymatic activity is based on Porter et al. (1986). The assay was performed in a 100 µL volume (79 µL 100 mM MOPS-KOH pH7.0, 10 µL 10 mM NADPH in water, 1 µL 10 mM dihydroflavonol (DHK, DHQ, DHM, or their mixture) in DMSO and 10 µL protein extract in 100 mM Tris-Cl pH7.5, with which the reaction was started). After 10 minutes incubation at 30 °C, the reaction was stopped by adding 20 µL glacial acetic acid, extracted with 163 µL and 150 µL ethyl acetate and the combined extracts were dried in a vacuum centrifuge. The dried extract was dissolved in BuOH-HCl (95 parts *n*-butanol and 5 parts 37% HCl, supplemented with 1/30 volume of 2% $NH_4Fe(SO_4)_2$ $12H_2O$ in 2 N HCl). After 30 minutes incubation at 95 °C, the tubes were allowed to cool down and absorbance was measured at 535 nm, 550 nm and 560 nm for products from DHK, DHQ and DHM, respectively, and at 660 nm that was subtracted as background. Corrected absorbances were converted to µM substrate used (or product formed) by multiplying with 62.9, 76.6, and 91.7 for DHK, DHQ and DHM, respectively.

9.3.5 HPLC

In order to follow the use of the substrates in the DFR assay by HPLC, the reaction was stopped, and proteins were precipitated by adding three volumes of methanol. To separate butanoylated anthocyanidins in the mixed substrate assay, a sample of BuOH-HCl was directly applied to HPLC. HPLC was performed as described in Pietiäinen et al. (2016), detection was at 525 nm for the butanoylated anthocyanidins and 290 nm for the dihydroflavonol substrates.

9.4 Discussion

The gold standard assay for DFR activity uses radioactive substrates and separation of the leucoanthocyanidin products by chromatography for quantification (Kristiansen 1986; Fischer et al. 1988; Ruhnau and Forkmann 1988; Miosic et al. 2014). However, radioactive substrates are not commercially available, and many laboratories have used alternatives. Leucoanthocyanidins have been directly observed in HPLC (Xie et al. 2004; Li et al. 2017) allowing use of nonradioactive substrates. The standard method to assay NADPH dependent reduction reactions is to follow consumption of NADPH spectrophotometrically at 340 nm. However, the DFR substrates (dihydroflavonols) and products (leucoanthocyanidins) also absorb UV-light at this wavelength, making the assay possible but challenging to follow

(Trabelsi et al. 2011). Several papers report they have assayed DFR by following absorbance at 340 nm in a reaction setup with 4 mM NADPH (Hayashi et al. 2005; Chu et al. 2014, 2015). With molar absorptivity of NADPH at this wavelength being 6,220, this is an impossible task.

The classical and still much used assay for leucoanthocyanidins relies on oxidative butanoylation of the compounds into colored anthocyanidins in HCl. The assay has been developed for analysis of condensed tannins (proanthocyanidins) that are polymers of flavan-3-ols (catechin and epicatechin). The criticism of this assay by Gosch et al. (2014) for its variability, in fact, refers to variance in hydrolysis of the polymer (Smathers and Charley 1967). For assay of monomeric leucoanthocyanidins (products of the enzymatic reaction by DFR), we showed here that the assay is robust and dependable. With single substrates, measurement of absorbance at 535–560 nm (depending on the product) is sufficient. With several substrates provided simultaneously, the butanoylated products must be separated by HPLC. Extractability of the reaction products into ethyl acetate differs and products with more hydroxyl groups (leuco-Cy and leuco-Dp) extract less efficiently that leuco-Pg. This can be taken into account, and by taking into account also the apparent molar absorptivity of the butanoylated anthocyanidins, the absorbance values can be converted to moles of substrate consumed and product formed.

The amino acids determining substrate specificity has been studied since 1980s (Beld et al. 1989). In addition to petunia, several other species harbor DFR enzymes incapable of converting the simplest substrate DHK to leuco-Pg and supporting biosynthesis of orange pelargonins in the plant. The Cymbidium orchid DFR does not accept DHK (Johnson et al. 1999) and in strawberry two different isoforms of DFR have different substrate preferences in this respect (Miosic et al. 2014). Also the arabidopsis DFR strongly prefers DHQ and DHM over DHK (Leonard et al. 2008). Johnson et al. (2001) used complementation of a petunia line lacking F3′H and F3′5′H (W80, similar to RL01) to assay *in vivo* the capability of fusions between the petunia DFRA and the gerbera GDFR1-1 to support pelargonin biosynthesis. They pinpointed a 26 amino acid long region (132–157, gerbera coordinates) that is important for substrate choice. In this region, the amino acid at position 134 has been suggested to divide DFRs between DHK users (N) and nonusers (D), but it has been shown that many DFR enzymes do not follow this rule (Nitarska et al. 2018).

The gerbera DFR isoenzymes analyzed here show very distinct patterns of substrate preference. Still, GDFR1-3 that converts all three dihydroflavonols to the corresponding leucoanthocyanidins differs only at eight amino acid positions from GDFR1-1 and GDFR1-2 that strongly prefer DHK. Only one of these is in the region 132–157 (G or V at 135), but swapping that amino acid between the gerbera enzymes did not change the substrate specificity dramatically (Figure 9.6a).

In assaying substrate preferences of enzymes, a choice assay where several substrates are given simultaneously is a standard procedure. When all three substrates are given to any of the three gerbera enzymes, or to the maize A1 DFR, they stop using DHK. Haselmair-Gosch et al. (2018) report that the maize A1 DFR used to engineer the pelargonin producing orange petunia cultivars, as well as petal extracts of the orange cultivars, are very poor in using DHK. This is in stark contrast to the results we found here (Figure 9.6). A possible explanation is that the Michaelis constant K_m of A1 DFR for DHK is higher than for the other substrates. This would lead to a comparatively low activity in an assay with low substrate concentration (1 μM in Haselmair-Gosch et al. 2018; 100 μM in this work).

Why is it so difficult to make orange pelargonidins? Orange and scarlet colors in flowers are attractive to some pollinators (particularly birds, and to humans), but perhaps they do not serve other functions that anthocyanins have in the plant. Even the gerbera cultivar "Regina" with orange pelargonin petals in the showy inflorescence, accumulates cyanins in vegetative tissues (Deng et al. 2014).

Acknowledgements

THT wishes to acknowledge Heidi Halbwirth and Cristian Molitor (Technische Universität Wien, Austria) for fruitful discussion about assays for DFR substrate specificity.

References

Barros, J. and Dixon, R.A. (2020). Plant phenylalanine/tyrosine ammonia-lyases. *Trends in Plant Science*, **25**: 66–79.

Bashandy, H. and Teeri, T.H. (2017). Genetically engineered orange petunias on the market. *Planta*, **246**: 277–280.

Bashandy, H., Jalkanen, S., and Teeri, T.H. (2015a). Within leaf variation is the largest source of variation in agroinfiltration of *Nicotiana benthamiana*. *Plant Methods*, **11**: 1–7.

Bashandy, H., Pietiäinen, M., Carvalho, E., et al. (2015b). Anthocyanin biosynthesis in gerbera cultivar 'Estelle'and its acyanic sport 'Ivory'. *Planta*, **242**: 601–611.

Beld, M., Martin, C., Huits, H., et al. (1989). Flavonoid synthesis in *Petunia hybrida*: partial characterization of dihydroflavonol-4-reductase genes. *Plant Molecular Biology*, **13**: 491–502.

Berland, H., Albert, N.W., Stavland, A., et al. (2019). Auronidins are a previously unreported class of flavonoid pigments that challenges when anthocyanin biosynthesis evolved in plants. *Proceedings of the National Academy of Sciences*, **116**: 20232–20239.

Boerjan, W., Ralph, J., and Baucher, M. (2003). Lignin biosynthesis. *Annual Review of Plant Biology*, **54**: 519–546.

Chang, S., Puryear, J., and Cairney, J. (1993). A simple and efficient method for isolating RNA from pine trees. *Plant Molecular Biology Reporter*, **11**: 113–116.

Chu, Y., Pan, J., Wu, A., et al. (2014). Molecular cloning and functional characterization of dihydroflavonol-4-reductase gene from *Calibrachoa hybrida*. *Scientia Horticulturae*, **165**: 398–403.

Chu, Y.X., Chen, H.R., Wu, A.Z., et al. (2015). Expression analysis of dihydroflavonol 4-reductase genes in *Petunia hybrida*. *Genetics and Molecular Research*, **14**: 5010–5021.

Deblaere, R., Bytebier, B., Greve, D., et al. (1985). Efficient octopine Ti plasmid-derived vectors for Agrobacterium-mediated gene transfer to plants. *Nucleic Acids Research*, **13**: 4777–4788.

Deng, X., Bashandy, H., Ainasoja, M., et al. (2014). Functional diversification of duplicated chalcone synthase genes in anthocyanin biosynthesis of *Gerbera hybrida*. *New Phytologist*, **201**: 1469–1483.

Elomaa, P. and Teeri, T.H. (2001). Transgenic gerbera. In: *Biotechnology in Agriculture and Forestry, Transgenic Crops III. Vol.* **48**, (ed. Bajaj, Y.P.S.), 139–154. Berlin: Springer.

Fellenberg, C. and Vogt, T. (2015). Evolutionarily conserved phenylpropanoid pattern on angiosperm pollen. *Trends in Plant Science*, **20**: 212–218.

Fischer, D., Stich, K., Britsch, L., and Grisebach, H. (1988). Purification and characterization of (+) dihydroflavonol (3-hydroxyflavanone) 4-reductase from flowers of *Dahlia variabilis*. *Archives of Biochemistry and Biophysics*, **264**: 40–47.

Fischer, T.C., Halbwirth, H., Meisel, B., et al. (2003). Molecular cloning, substrate specificity of the functionally expressed dihydroflavonol 4-reductases from *Malus domestica* and *Pyrus communis* cultivars and the consequences for flavonoid metabolism. *Archives of Biochemistry and Biophysics*, **412**: 223–230.

Fraiture, M.A., Ujhelyi, G., Ovesná, J., et al. (2019). MinION sequencing technology to characterize unauthorized GM petunia plants circulating on the European Union market. *Scientific Reports*, **9**: 1–8.

Giusti, M.M. and Wrolstad, R.E. (2001). Characterization and measurement of anthocyanins by UV-visible spectroscopy. In: *Current Protocols in Food Analytical Chemistry*, (ed. Wrolstad, R.E.), 1–13. New York: John Wiley & Sons.

Gosch, C., Nagesh, K.M., Thill, J., et al. (2014). Isolation of dihydroflavonol 4-reductase cDNA clones from *Angelonia x angustifolia* and heterologous expression as GST fusion protein in *Escherichia coli*. *PLoS One*, **9**: e107755.

Halbwirth, H., Martens, S., Wienand, U., et al. (2003). Biochemical formation of anthocyanins in silk tissue of *Zea mays*. *Plant Science*, **164**: 489–495.

Haselmair-Gosch, C., Miosic, S., Nitarska, D., et al. (2018). Great cause—small effect: undeclared genetically engineered orange petunias harbor an inefficient dihydroflavonol 4-Reductase. *Frontiers in Plant Science*, **9**: 149.

Hayashi, M., Takahashi, H., Tamura, K., et al. (2005). Enhanced dihydroflavonol-4-reductase activity and NAD homeostasis leading to cell death tolerance in transgenic rice. *Proceedings of the National Academy of Sciences*, **102**: 7020–7025.

Helariutta, Y., Elomaa, P., Kotilainen, M., et al. (1993). Cloning of cDNA coding for dihydroflavonol-4-reductase (DFR) and characterization of dfr expression in the corollas of *Gerbera hybrida* var. Regina (Compositae). *Plant Molecular Biology*, **22**: 183–193.

Johnson, E.T., Yi, H., Shin, B., et al. (1999). *Cymbidium hybrida* dihydroflavonol 4-reductase does not efficiently reduce dihydrokaempferol to produce orange pelargonidin-type anthocyanins. *The Plant Journal*, **19**: 81–85.

Johnson, E.T., Ryu, S., Yi, H., et al. (2001). Alteration of a single amino acid changes the substrate specificity of dihydroflavonol 4-reductase. *The Plant Journal*, **25**: 325–333.

Katsumoto, Y., Fukuchi-Mizutani, M., Fukui, Y., et al. (2007). Engineering of the rose flavonoid biosynthetic pathway successfully generated blue-hued flowers accumulating delphinidin. *Plant and Cell Physiology*, **48**: 1589–1600.

Koes, R., Verweij, W., and Quattrocchio, F. (2005). Flavonoids: a colorful model for the regulation and evolution of biochemical pathways. *Trends in Plant Science*, **10**: 236–242.

Kristiansen, K.N. (1986). Conversion of (+)-dihydroquercetin to (+)-2, 3-trans-3, 4-cis-leucocyanidin and (+)-catechin with an enzyme extract from maturing grains of barley. *Carlsberg Research Communications*, **51**: 51–60.

Leonard, E., Yan, Y., Chemler, J., et al. (2008). Characterization of dihydroflavonol 4-reductases for recombinant plant pigment biosynthesis applications. *Biocatalysis and Biotransformation*, **26**: 243–251.

Li, Y., Liu, X., Cai, X., et al. (2017). Dihydroflavonol 4-reductase genes from Freesia hybrida play important and partially overlapping roles in the biosynthesis of flavonoids. *Frontiers in Plant Science*, **8**: 428.

Martens, S., Teeri, T., and Forkmann, G. (2002). Heterologous expression of dihydroflavonol 4-reductases from various plants. *FEBS Letters*, **531**: 453–458.

Martin, C. (2018). A role for plant science in underpinning the objective of global nutritional security? *Annals of Botany*, **122**: 541–553.

Martinière, A., Bassil, E., Jublanc, E., et al. (2013). in vivo intracellular pH measurements in tobacco and Arabidopsis reveal an unexpected pH gradient in the endomembrane system. *The Plant Cell*, **25**: 4028–4043.

Meyer, P., Heidmann, I., Forkmann, G., and Saedler, H. (1987). A new petunia flower colour generated by transformation of a mutant with a maize gene. *Nature*, **330**: 677–678.

Miosic, S., Thill, J., Milosevic, M., et al. (2014). Dihydroflavonol 4-reductase genes encode enzymes with contrasting substrate specificity and show divergent gene expression profiles in *Fragaria* species. *PLoS One*, **9**: e112707.

Napoli, C., Lemieux, C., and Jorgensen, R. (1990). Introduction of a chimeric chalcone synthase gene into petunia results in reversible co-suppression of homologous genes in trans. *The Plant Cell*, **2**: 279–289.

Nitarska, D., Haselmair-Gosch, C., Miosic, S. et al. (2018). Reinvestigating substrate specificity of dihydroflavonol 4-reductase of Petunia× hybrid. *XXIXth International Conference on Polyphenols*, Madison, USA (16–20 July 2018).

Noda, N., Yoshioka, S., Kishimoto, S., et al. (2017). Generation of blue chrysanthemums by anthocyanin B-ring hydroxylation and glucosylation and its coloration mechanism. *Science Advances*, **3**: e1602785.

Oud, J.S., Schneiders, H., Kool, A.J., and van Grinsven, M.Q. (1995). Breeding of transgenic orange *Petunia hybrida* varieties. *Euphytica*, **84**: 175–181.

Petit, P., Granier, T., d'Estaintot, B.L., et al. (2007). Crystal structure of grape dihydroflavonol 4-reductase, a key enzyme in flavonoid biosynthesis. *Journal of Molecular Biology*, **368**: 1345–1357.

Peyret, H. and Lomonossoff, G.P. (2013). The pEAQ vector series: the easy and quick way to produce recombinant proteins in plants. *Plant Molecular Biology*, **83**: 51–58.

Pietiäinen, M., Kontturi, J., Paasela, T., et al. (2016). Two polyketide synthases are necessary for 4-hydroxy-5-methylcoumarin biosynthesis in *Gerbera hybrida*. *The Plant Journal*, **87**: 548–558.

Porter, L.J., Hrstich, L.N., and Chan, B.G. (1986). The conversion of procyanidins and prodelphinidins to cyanidin and delphinidin. *Phytochemistry*, **25**: 223–230.

Ruhnau, B. and Forkmann, G. (1988). Flavan-3, 4-diols in anthocyanin biosynthesis, enzymatic formation with flower extracts from *Callistephus chinensis*. *Phytochemistry*, **27**: 1035–1039.

Schijlen, E.G., de Vos, C.R., Martens, S., et al. (2007). RNA interference silencing of chalcone synthase, the first step in the flavonoid biosynthesis pathway, leads to parthenocarpic tomato fruits. *Plant Physiology*, **144**: 1520–1530.

Schilmiller, A.L., Stout, J., Weng, J.K., et al. (2009). Mutations in the cinnamate 4-hydroxylase gene impact metabolism, growth and development in Arabidopsis. *The Plant Journal*, **60**: 771–782.

Shiono, M., Matsugaki, N., and Takeda, K. (2005). Structure of the blue cornflower pigment. *Nature*, **436**: 791–791.

Smathers, R.U. and Charley, H. (1967). Problems in measuring leucoanthocyanin content of pears. *Journal of Food Science*, **32**: 310–314.

Stafford, H.A. (1991). Flavonoid evolution: an enzymic approach. *Plant Physiology*, **96**: 680–685.

Swain, T. and Hillis, W.E. (1959). The phenolic constituents of *Prunus domestica*. I.—The quantitative analysis of phenolic constituents. *Journal of the Science of Food and Agriculture*, **10**: 63–68.

Tanaka, Y., Tsuda, S., and Kusumi, T. (1998). Metabolic engineering to modify flower color. *Plant and Cell Physiology*, **39**: 1119–1126.

Tanaka, Y., Sasaki, N., and Ohmiya, A. (2008). Biosynthesis of plant pigments: anthocyanins, betalains and carotenoids. *The Plant Journal*, **54**: 733–749.

Trabelsi, N., d'Estaintot, B.L., Sigaud, G., et al. (2011). Kinetic and binding equilibrium studies of dihydroflavonol 4-reductase from *Vitis vinifera* and its unusually strong substrate inhibition. *Journal of Biophysical Chemistry*, **2**: 332–344.

Wen, W., Alseekh, S., and Fernie, A.R. (2020). Conservation and diversification of flavonoid metabolism in the plant kingdom. *Current Opinion in Plant Biology*, **55**: 100–108.

Wise, A.A., Liu, Z., and Binns, A.N. (2006). Three Methods for the Introduction of Foreign DNA into *Agrobacterium*. *Methods in Molecular Biology*, **343**: 43–53.

Xie, D.Y., Jackson, L.A., Cooper, J.D., et al. (2004). Molecular and biochemical analysis of two cDNA clones encoding dihydroflavonol-4-reductase from *Medicago truncatula*. *Plant Physiology*, **134**: 979–994.

Zheng, L., Baumann, U., and Reymond, J.L. (2004). An efficient one-step site-directed and site-saturation mutagenesis protocol. *Nucleic Acids Research*, **32**: e115.

10

A Long and Winding Road

The Evolution of Transcriptional Regulation of Polyphenol Biosynthesis

Cathie Martin[1], *Jie Li*[1], *and Nick W. Albert*[2]

[1] *Department of Biochemistry and Metabolism, John Innes Centre, Norwich Research Park, Colney, Norwich, UK*
[2] *Plant & Food Research, Palmerston North, New Zealand*

10.1 Introduction

Polyphenols are not well defined but, in general, they are natural products which are abundant in plants derived from shikimate and/or polyketide pathways and carrying more than one phenolic ring (Quideau et al. 2011). Their multiple phenolic rings carry multiple hydroxyl groups. Polyphenols broadly fall into three major classes: flavonoids, tannins, and lignans, and for the purposes of this review will be discussed together with phenolic acids which contain a single phenolic ring. They are of importance because of the strength and waterproofing properties lignans give to wood and vascular tissues of all vascular plants, because of the antioxidant properties they confer in our diets, because tannins are used as dyes and for tanning garments, because of the organoleptic qualities they confer to wines and because the flavonoids confer many of the beautiful colors to our flowers and fruits. Polyphenols are synthesized through phenylpropanoid metabolism which starts with phenylalanine and comprises a network of specialized metabolic pathways that gives rise to diverse polyphenolic products with diverse distributions and functions in different plant tissues (Figure 10.1). Flux along these specialized pathways is determined principally, but by no means exclusively, by transcriptional regulation.

Understanding of the regulation of polyphenol (or phenylpropanoid) biosynthesis began with the identification of the transcriptional regulators of anthocyanin and phlobaphene biosynthesis in corn, based on the extensive mutant collections of Rollins A. Emerson, Barbara McClintock, Peter Peterson, and Ed Coe (Cone et al. 1986; Paz-Ares et al. 1986, 1987; Lechelt et al. 1989; Ludwig et al. 1989). However, the origin of this research field in angiosperms may have biased our comprehension of the regulation of polyphenol biosynthesis. The increasingly common practice of identifying genes by homology may limit the identification of new/novel regulators, especially where functional identification is unsupported by accompanying genetic analyses.

Recent Advances in Polyphenol Research, Volume 8, First Edition. Edited by Juha-Pekka Salminen, Kristiina Wähälä, Victor de Freitas, and Stéphane Quideau.

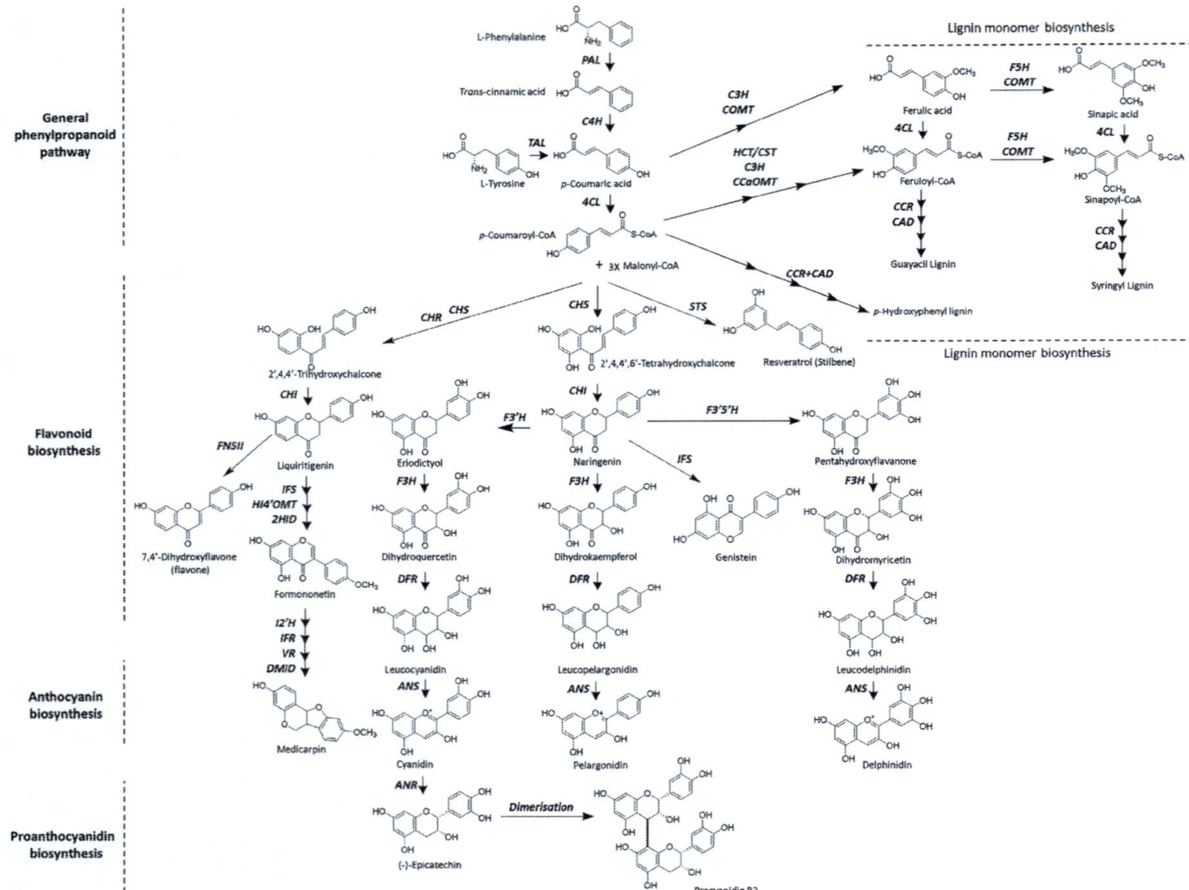

Figure 10.1 Simplified overview of phenylpropanoid metabolic pathway in plants. 2HID, 2-hydroxyisoflavanone dehydratase; 4CL, 4-coumarate CoA ligase; ANR, anthocyanidin reductase; ANS, anthocyanidin synthase; C3H, *p*-coumaroyl shikimate/quinate 3-hydroxylase; C4H, cinnamate 4-hydroxylase; CAD (hydroxy) cinnamyl alcohol dehydrogenase; CCaOMT, caffeoyl CoA *O*-methyltransferase; CCR (hydroxy)cinnamoyl CoA reductase; COMT, caffeic acid/5-hydroxyferulic acid *O*-methyl-transferase; DFR, dihydroflavonol 4-reductase; DMID, 7,2′-dihydroxy-4′-methoxy-isoflavonol dehydratase; F3′5′H, flavonoid 3′, 5′-hydroxylase; F3H, flavanone 3-hydroxylase; F3′H, flavonoid 3′-hydroxylase; F5H ferulate 5-hydroxylase; CHS, chalcone synthase; FNSII, flavone synthase II; HCT/CST, hydroxycinnamoyl CoA:shikimate/quinate hydroxycinnamoyltransferase; I2′H, isoflavone 2′-hydroxylase; I4′OMT, 2-hydroxyisoflavanone 4′-*O*-methyltrasferase; IFR, isoflavone reductase; IFS, isoflavone synthase; PAL, phenylalanine ammonia lyase; STS, stilbene synthase; CHI, chalcone isomerase; TAL, tyrosine ammonia lyase; VR, vestitone reductase.

10.2 The Importance of R2R3Myb Transcription Factors (TFs) in the Regulation of Phenylpropanoid Metabolism in Plants

Central to the regulation of phenylpropanoid metabolism are R2R3Myb TFs. R2R3Myb TFs are specific to the plant lineage, but were founded as variants of a more ancient class of Myb TFs, which have three repeats of about 52 amino acids responsible for DNA-binding, and regulating the cell cycle, albeit by different mechanisms, in all major eucaryotic lineages. R2R3Myb TFs evolved through the loss of one Myb repeat (R1) but are still able to bind DNA, probably as monomers, although they may also bind in complexes with other TFs. They consist of highly conserved DNA-binding domains and highly variable C-terminal domains that modulate the rate of transcriptional initiation. In Solanaceous plants, R2R3Myb TFs regulate parts of the shikimate pathway and aromatic amino acid biosynthesis (EOB I-V, Ph4, Odorant 1, Verdonk et al. 2005; Spitzer-Rimon et al. 2010, 2012; Van Moerkercke et al. 2011; Cna'ani et al. 2015a, 2015b) implying that regulation of phenylpropanoid metabolism (Figure 10.1) probably started with the regulation of the synthesis of specialized phenolic metabolites, which evolved as plants moved onto the land. Indeed, it has been suggested that phenylpropanoid metabolism evolved as a means of protecting terrestrial plant life from the ravages of UV irradiation in a low ozone atmosphere (Weng and Chapple 2010), and in reducing water loss in terrestrial environments, possibly through waterproofing the plant cuticle (Renault et al. 2019). However, an outstanding recent analysis of R2R3Myb genes suggests that the family may have its origins in oceanic green algae, where phenolics might have facilitated the subsequent emergence of photosynthetic organisms onto land (Jiang and Rao 2020). The association of R2R3Myb proteins with the transcriptional regulation of phenylpropanoid metabolism has been confirmed by recent studies of two R2R3Myb genes in the liverwort, *Marchantia polymorpha* (Figure 10.2). One, MpMYB02, encodes a protein that controls the production of marchantin, a cyclic bisbenzyl compound (Takahashi and Asakawa 2017; Kubo et al. 2018), and the other (MpMYB14) the synthesis of ricconidin A/auronidin, a colored aurone derivative that provides protection to the organism from abiotic stress and pathogens (Albert et al. 2018; Berland et al. 2019; Carella et al. 2019). The other R2R3Myb TFs in *Marchantia polymorpha* appear to have functions unrelated to phenylpropanoid biosynthesis.

A number of other R2R3Myb TFs that regulate monolignol and consequently lignin synthesis have been described in angiosperms and gymnosperms. These are either master regulators of secondary cell wall production, controlling hemicellulose and pectin formation (AtMYB46 and 83 in Arabidopsis) as well as lignin, or more direct regulators of monolignol biosynthetic genes (AtMYB58, 63, 85, and 105 in Arabidopsis). These seem to have their evolutionary origin in the last common ancestor of vascular plants and bryophytes (Ohtani and Demura 2019) although broader sampling, claimed to be less biased, suggests an older origin (Jiang and Rao 2020). However, it is agreed that these R2R3Myb genes were/are around in plants which did/do not synthesize lignin (Renault et al. 2019) suggesting that their functionality in angiosperms may not reflect their original functionality, and their regulation of branches of phenylpropanoid metabolism may have been a characteristic acquired in vascular plants.

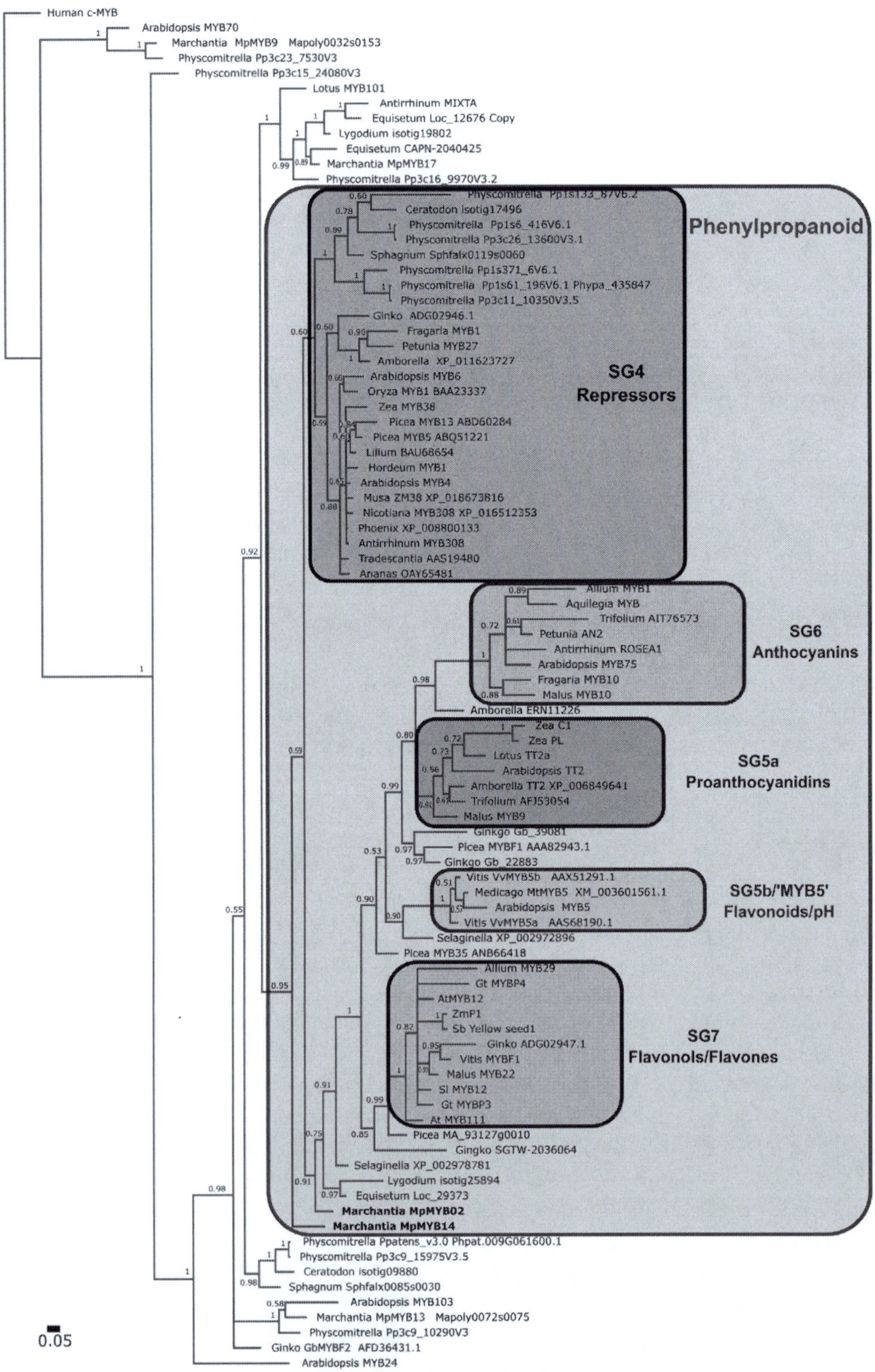

Figure 10.2 Phylogenetic relationships of phenylpropanoid-related R2R3MYB TFs from different plant species. The amino acid sequences of the R2R3MYB domain were aligned in ClustalW and then used to generate a phylogenetic tree with MrBayes in the Geneious suite of software, using

10.2.1 R2R3Myb TFs Regulate Specialized Branches of Polyphenol Metabolism

From the R2R3Myb regulators of the early steps in phenylpropanoid biosynthesis, which have been maintained in many angiosperms, more specialized functions have become associated with specific sub-clades of R2R3Myb TFs. For example, subgroup 4 (terminology based on nomenclature for Arabidopsis R2R3Myb TFs; Kranz et al. 1998; Stracke et al. 2001; Dubos et al. 2010) includes repressors of flavonoid, general phenylpropanoid and monolignol biosynthesis, and subgroups 5, 6 and 7 generally serve as activators of different branches of flavonoid biosynthesis (Figures 10.1 and 10.2).

Subgroup 7 consists of R2R3Myb TFs which broadly regulate flavonol and flavone biosynthesis in angiosperms. They function as activators of transcriptional initiation without other known partners. Some have been shown to upregulate genes encoding enzymes of primary and intermediary metabolism as well as the specialized pathways within flavonoid biosynthesis (Zhang et al. 2015). In the pericarp of maize the P1 R2R3Myb TF, belonging to subgroup 7, regulates the biosynthesis of red phlobaphene pigments through its control of the genes encoding the enzymes required for synthesis of 3-deoxyanthocyanidins, as well as synthesis of *C*-glycosyl flavones (Lechelt et al. 1989; Grotewold et al. 1994; Zhang et al. 2003).

Figure 10.2 (Continued) default parameters, with the posterior probabilities of the branches shown. The human cMYB sequence (NP_001155129) was used for the outgroup, and a selection of non-flavonoid-related R2R3MYBs was included for comparison. The top Blast hits for At*MYB75* (sub-group 6) and Mp*MYB14* from species representing the major plant groups are included. The clade highlighted in grey includes regulators of the general phenylpropanoid pathway, including SG4-containing transcriptional repressors of phenylpropanoid genes (in red), SG5a-containing activators of anthocyanin/proanthocyanidin production (in brown), SG5b-containing activators of proanthocyanidin/hyperacidification (in blue), SG6-containing activators of anthocyanin production (in purple), and SG7-containing activators of flavone/flavonol production (in yellow). Sequence details are: the liverwort *M. polymorpha* Myb02 (Mapoly0006s0226), Myb09 (Mapoly0032s0153), Myb13 (Mapoly0072s0075), Myb14 (Mapoly0073s0038), Myb17 (Mapoly0096s0058); the mosses *Ceratodon purpureus* isotig09880, isotig17496; *Physcomitrella patens* Pp1s133_87V6.2, Pp1s6_416V6.1, Pp1s371_6V6.1,Pp1s61_196V6.1; Pp3c9_15975V3.5, Pp3c9_10290V3, Pp3c11_1035 0V3.5,Pp3c15_24080V3, Pp3c16_9970V3.2, Pp3c23_7530V3, Pp3c26_13600V3.1; *Sphagnum fallax* XP_002978781, SphFalx011950060, SphFalx008550030; the ferns/fern allies *Equisetum diffusum* CAPN-2040425; *Equisetum giganteum* Loc_12676, Loc_29373; *Lygodium japonicum* isotig19802, isotig25894; *Selaginella moellendorffii* XP_002972896, XP_002978781; the gymnosperms *Gingko biloba* SGTW-2036064, Gb_22883, Gb_39081, GbMYBP2 AFD36431.1, ADG02946.1; *Picea mariana* MYBF1 (AAA82943); *Picea glauca* MYB5 (ABQ51221), MYB13 (ABD60284); *Picea abies* MYB35 (ANB66418), MA_93127g0010; and the angiosperms *Allium cepa* MYB1 (KX785130), MYB29 (KX785133); *Amborella trichopoda* ERN11226, TT2 (XP_006849641), ZM38 (XP_011623727); *Ananas comosus* OAY65481; *Antirrhinum majus* Rosea1 (ABB83826), Mixta (CAA55725), Myb308 (P81393); *Aquilegia formosa* MYB1 (ACQ82820); *Arabidopsis thaliana* MYB4 (At4g38620), MYB5 (At3g13540), MYB6 (XP_002872444), MYB12 (ABB03913), MYB24 (AF175987), MYB70 (NP_179910), MYB75/PAP1 (AAG42001), MYB103 (AED96722), MYB111 (At5g49330.1) MYB123/TT2 (CAC40021); *Fragaria × ananassa* MYB1 (AAK84064); *Gentian triflora* GtMYBP4 AB280446, GtMYBP3 AB733016; *Hordeum vulgare* MYB1 (P20026); *Lilium* BAU68654; *Lotus japonicus* MYB101 (BAC75671), TT2a (BAG12893); *Malus × domestica* MYB9 (NP_001280749), MYB10 (ACQ45201), MYB22 (AAZ20438); *Musa acuminata* subsp. *malaccensis* XP_018673816; *Nicotiana tabacum* MYB308 (XP_016512353); *Oryza sativa* MYB1 (BAA23337); *Petunia × hybrida* AN2 (AAF66727), MYB27 (AHX24372); *Phoenix dactylifera* XP_008800133; *Solanum lycopersicum* MYB12 EU419748; Sorghum bicolor yellowseed 1, AY860968; *Tradescantia fluminensis* AAS19480; *Trifolium arvense* AFJ53054, AIT76573; *Vitis vinifera* MybF1 (FJ948477), VvMYB5a AA56868190.1, VvMYB5b AAX51291.1; *Zea mays* C1 (AAA33482), Pl (AAA19821), P1 (GRMZM2G084799), Myb38 (P20025). (*See insert for color representation of the figure.*)

Subgroup 5 consists of R2R3Myb proteins involved in the regulation of proanthocyanidin biosynthesis. These proteins function as part of a complex, comprised of Myb, bHLH, and WD Repeat proteins, the MBW complex and, at least from data in Arabidopsis, all three proteins need to be functional for the complex to induce transcriptional activation of target genes. Subgroup 5 R2R3Myb TFs, as defined by Stracke et al. (2001), can be divided into those most similar to the Arabidopsis TF TT2/AtMYB123, which we will refer to here as subgroup 5a and those most similar to AtMYB5, which we will refer to as subgroup 5b. Subgroup 6 TFs also participate in MBW complexes principally concerned with activation of anthocyanin biosynthesis, although, in the Caryophillaceae, a gene encoding a subgroup 6 R2R3Myb TF appears to have been co-opted from anthocyanin biosynthesis to regulate betalain biosynthesis (Hatlestad et al. 2015). It appears that the R2R3Myb proteins that regulate betalain biosynthesis do not need to interact with bHLH and WDR proteins to promote transcriptional activation.

10.2.2 R2R3Myb Transcriptional Repressors Controlling Phenylpropanoid Metabolism

Within the group of R2R3Myb proteins that regulate polyphenol metabolism, it appears that transcriptional repressors may have evolved early, with structural homologues having been identified in Picea (gymnosperm), the ferns, Salvinia, Azola, and Selaginella. Their presence has also been suggested in the bryophyte, *Physcomitrella patens*, where knockouts of a *Myb* gene *PpMyb4* give rise to elevated levels of soluble polyphenols (Jiménez 2010). Importantly, knockouts of either *PpMyb4*, or a closely related gene *PpMyb5*, gave increased resistance to UV irradiation in *P. patens* (Figure 10.3, Jiménez 2010) in a manner very similar

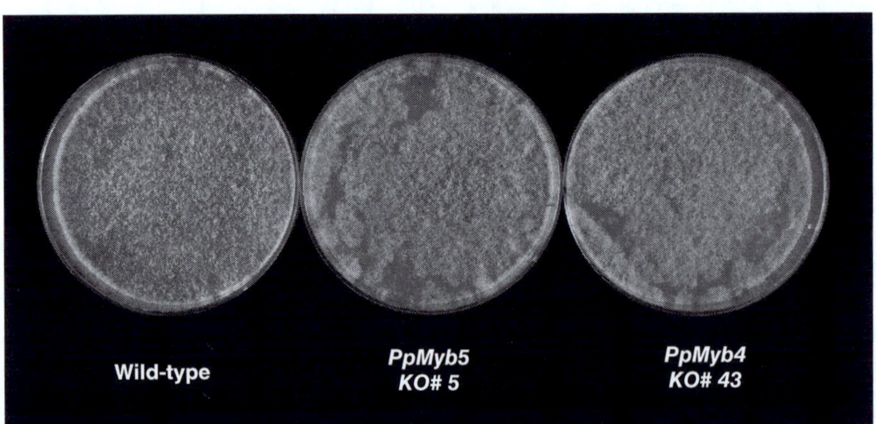

Figure 10.3 Response of *Physcomitrella patens* wild-type and knockout mutants of *PpMyb4* and *PpMyb5* to UV-B irradiation. Both genes encode R2R3Myb subgroup 4 repressors (Figure 10.2). Moss cultures were grown on KNOP medium for seven days at 23 °C, 16 hours light, 8 hours dark and then exposed to UV-B light (260–315 nm) for 20 minutes at a 20 cm distance from plates. Cultures were left to recover for eight days before being photographed. Clearly the *PpMyb4* and *PpMyb5* knockout mutants both survived and grew better than the wild type due to their elevated content of phenolics. *Source:* Jiménez (2010). (*See insert for color representation of the figure.*)

to the knockout mutant of the R2R3Myb repressor, AtMYB4, in Arabidopsis (Jin et al. 2000). These data confirm that such repressors were present early in land plant evolution, and hint that their original functions may have been concerned with regulating polyphenol biosynthesis in response to UV light (Jin et al. 2000; Jiménez 2010; Bowman et al. 2017; Hernández-Hernández et al. 2021).

10.2.3 Stand-Alone R2R3Myb Transcriptional Activators

Transcriptional repressors in eukaryotes work effectively in apposition to transcriptional activators. While some of the subgroup 4 R2R3Myb repressors have domains clearly associated with transcriptional repression (such as the EAR motif believed to inhibit the Mediator Complex of RNA polymerase II), others do not carry such domains and may work as repressors by competing for the same DNA binding motifs as R2R3Myb activators. This implies that a set of R2R3Myb transcriptional activators likely evolved around the same time as the repressors in subgroup 4. Indeed, the R2R3Myb proteins belonging to subgroup 7 appear to have diverged at about the same time as subgroup 4 repressors emerged, supported by the presence of members of the clade in Selaginella, Lycopodium, and Physcomitrella (Bowman et al. 2017). Following this divergence, R2R3Myb proteins participating in MBW complexes evolved, possibly from a common ancestor with subgroup 7 R2R3Myb TFs (Figure 10.2). It has been claimed that all subgroup 4 repressors work within MBW complexes (LaFountain and Yuan 2021) but this conclusion is based on bioinformatic analyses, which depend on the proteins containing the bHLH interaction signature motif within R3 of their DNA binding domains (Zimmermann et al. 2004). While some subgroup 4 proteins do interact with MBW components (PhMYB27, Albert et al. 2014), the subgroup as a whole appears to have evolved before the R2R3Myb activators that operate through the MBW complex. It is likely that those subgroup 4 repressors that interact with MBW complexes have acquired this functionality by convergence, probably because the bHLH interaction signature motif lies in a highly conserved region of R2R3Myb proteins (R3) involved in DNA recognition and consequently, acquisition of the ability to interact with bHLH proteins may be relatively straightforward if the conserved binding domain defining DNA binding specificity carries sequences similar to the bHLH interaction signature motif. Functional tests of subgroup 4 proteins from *A. majus* (AmMyb308, AmMyb330) failed to provide any evidence for interaction with bHLH partners (Tamagnone et al. 1998) and the first functionally characterized subgroup 4 repressor, AtMYB4, failed to interact with bHLH proteins even in yeast 2-hybrid assays in its intact functional form although the interaction was detected when AtMYB4 was truncated (Jin et al. 2000; Zimmermann et al. 2004).

A double negative logic has been proposed to explain why there are so many repressors of anthocyanin biosynthesis (LaFountain and Yuan 2021). However, the extensive range of R2R3Myb repressors influencing phenylpropanoid metabolism generally may also be a feature of their ability to divert flux within phenylpropanoid metabolism. For example, subgroup 4 R2R3Myb TFs are commonly highly expressed in petals of mature flowers and, in species like *Antirrhinum majus*, may be concerned with diverting flux from anthocyanin biosynthesis, which occurs principally in developing petals, to the production of phenolic volatiles like methyl benzoate, which are synthesized and released from petal epidermal cells once the flower has opened, to attract pollinators. Similar arguments could be used to explain the activity of R2R3Myb repressors in redirecting flux in monolignol biosynthesis (Bedon et al. 2010; Legay et al. 2010).

10.2.4 R2R3Myb TFs Working in MBW Complexes to Regulate Phenylpropanoid Metabolism

Two subgroups of R2R3Myb proteins that operate in association with bHLH proteins in MBW complexes are involved in regulating condensed tannin biosynthesis from proanthocyanidins. It is unclear whether the R2R3Myb proteins most similar to AtMYB5 (subgroup 5b) or those most similar to AtMYB123 (TT2) (subgroup 5a) evolved first, but both appear to be involved in regulating proanthocyanidin formation and the production of condensed tannins which may be present in the vacuoles of cells (particularly in seed coats) and also in the cell walls of certain tissues. It is also unclear whether the synthesis of condensed tannins in cell walls is regulated differently from the synthesis of condensed tannins in vacuoles, since Arabidopsis (which carries most of the mutants in this pathway that have been investigated) produces condensed tannins only in the vacuoles of the cells of the seed coat. What is apparent in Arabidopsis is that there is redundancy as well as divergence in the target genes of the MBW complexes composed of the R2R3Myb proteins closest to AtMYB5 (subgroup 5b) and those closest to TT2/AtMYB123 (subgroup 5a) (Xu et al. 2014, 2015). The separation of target genes between these two types of MBW complex has allowed additional functions to be acquired in some angiosperms, such as hyperacidification of petal cells of *Petunia hybrida*, which is associated with the activity of Ph4 (an R2R3Myb TF belonging to subgroup 5b from petunia), and work within an MBW complex with the bHLH protein An1 and the WDR protein An11 (Koes et al. 2005).

A major subgroup of R2R3Myb TFs, subgroup 6, regulates anthocyanin biosynthesis within angiosperms, also working within the context of an MBW complex. The point at which subgroup 6 R2R3Myb TFs diverged from those belonging to subgroups 5a and 5b is not completely clear, but very likely occurred after the divergence of subgroup 5 R2R3Myb TFs from subgroup 7 R2R3Myb transcriptional activators (Figure 10.2). It seems likely that regulators of the anthocyanin pathway might have evolved from regulators of proanthocyanin biosynthesis, because the synthesis of catechin and epicatechin involves the activity of most of the enzymes involved in anthocyanin biosynthesis, requiring only the co-option of an enzyme catalyzing 3-*O*-glycosylation and possibly proteins able to transport anthocyanins specifically to the vacuole to secure the minimum requirements for the regulation of anthocyanin biosynthesis. It is currently unclear whether gymnosperms encode proteins functionally equivalent to subgroup 6 R2R3Myb TFs. They certainly produce anthocyanins and encode proteins belonging to R2R3Myb subgroup 5, associated functionally with MBW complexes, but these appear to be associated predominantly with proanthocyanin biosynthesis (Jiang and Rao 2020).

The potential ease of transition between regulating proanthocyanidin biosynthesis and anthocyanin biosynthesis is emphasized by the fact that the R2R3Myb regulation of anthocyanin biosynthesis appears to be polyphyletic. The R2R3Myb TFs regulating anthocyanin biosynthesis in the Poales, including C1 and Pl in maize and OsC1 and OsMyb3 in rice are related more closely, structurally, to the R2R3Myb TFs of subgroup 5 than those of subgroup 6 (Zheng et al. 2021). However other monocots, such as freesia, lily, onion, and some (but not all) orchids encode R2R3Myb TFs that regulate anthocyanin biosynthesis, and which clearly belong to subgroup 6 R2R3Myb regulators (Yamagishi et al. 2010; Davies et al. 2012; Schwinn et al. 2016; Li et al. 2020) together with those from dicots.

So, starting with the evolution and ascendancy of lignin biosynthesis in vascular plants, reinforced by the evolution of C-type lignin and proanthocyanins/condensed tannins as protectants in seed coats of seed plants, the diversification of anthocyanin production in terms of colors and patterns in flowering plants for pollinator and animal disperser attraction, is associated with repeated species-level amplification of R2R3Myb genes that regulate these specific aspects of phenylpropanoid metabolism. A good example of this is in the regulation of anthocyanin biosynthesis in angiosperm flowers, where species-level gene duplication may underpin many of the patterns of anthocyanin production in flowers. A good example of this is seen in *Antirrhinum majus* (Schwinn et al. 2006; Shang et al. 2011; Whitney et al. 2013, Figure 10.4) where three closely related subgroup 6 R2R3Myb genes regulate the patterning of anthocyanin production in flowers, influencing pollinator

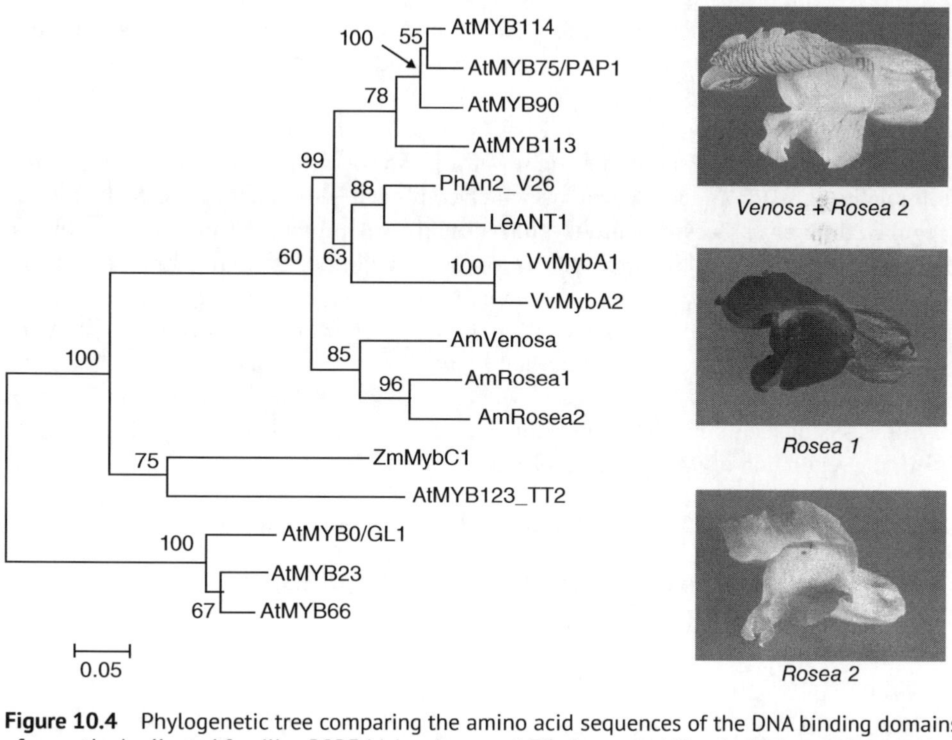

Figure 10.4 Phylogenetic tree comparing the amino acid sequences of the DNA binding domains of recently duplicated families R2R3 Myb subgroup 6 TFs from: *Arabidopsis thaliana* (At) *Antirrhinum majus* (Am) *Vitis vinifera* (Vv) An2 from *Petunia hybrida* (Ph), and LeANT1 from tomato. Together with the anthocyanin regulatory proteins, C1 and Pl from maize. Protein sequences were manually aligned using MacClade 4.08 (D.R. Maddison and W.P. Maddison, Sinauer Associates). Phylogenetic analysis was performed with PAUP* 4.0b10, using only the Myb domain of each protein. One thousand bootstrapped data sets were used to estimate the confidence of each tree clade. The phenotypes of the recently duplicated subgroup 6 R2R3Myb proteins in the flowers of *A.majus* are shown to the right of the figure. (*See insert for color representation of the figure.*)

attraction. Similar gene expansion has been observed in *Petunia hybrida* where duplicate subgroup 6 R2R3Myb TFs control where in vegetative and floral tissue anthocyanins are produced (Albert et al. 2011; Zhang et al. 2019). In petunia, some *R2R3Myb* gene duplication occurred before the formation of the cultivated hybrid and the activity of functionally redundant subgroup 6 R2R3Myb TFs may have been under considerable selection during the breeding of new patterns and intensities of color in this important decorative garden plant (Zhang et al. 2019). Investigation of the Solanaceae more broadly suggests that duplication of subgroup 6 R2R3Myb genes occurred at the level of the genus Solanum conferring differences in pigmentation of flowers and fruits within the different species (Powell et al. 2020). Similar duplications appear to have occurred in the amplification of subgroup 4 R2R3Myb repressors in gymnosperms, perhaps influencing the cell/tissue distribution of flavonoid responses to stress in pine trees (Bedon et al. 2010).

Because of the preponderance of data arising from the studies of anthocyanin and proanthocyanin biosynthesis in angiosperms, the literature is dominated by reports on the roles of the "MBW complex" in regulating phenylpropanoid metabolism. Of course, there are good examples, especially in early land plants where polyphenol metabolism is not regulated by MBW complexes as well as evidence of the importance of R2R3Myb TFs regulating branches of phenylpropanoid metabolism without bHLH TF or WDR partners. The concept of the MBW complex was advanced following studies in Arabidopsis of a number of distinct processes controlled by the pleiotropic *TTG1* locus which encodes a WDR protein (Payne et al. 2000; Baudry et al. 2004; Koes et al. 2005; Ramsay and Glover 2005; Lloyd et al. 2017). Coming from these important opinion/review articles the idea emerged of a commonality of bHLH and WDR proteins which combined with specific R2R3Myb proteins to regulate different processes; anthocyanin biosynthesis, proanthocyanidin biosynthesis, cell hyperacidification and, in Arabidopsis, trichome and non-root hair formation. It was even hinted that there may be many other pathways/processes regulated by the generic units of bHLH and WDR proteins in combination with specific R2R3Myb proteins in different species (Spelt et al. 2002). Time has suggested that such models may have overestimated the reach of MBW complexes but their importance in the regulation of polyphenol biosynthesis can be appreciated best by considering the roles of bHLH TFs and WDR proteins in the regulation of phenylpropanoid metabolism.

10.3 The Role of bHLH Proteins in the Regulation of Phenylpropanoid Metabolism

To date, only one specific clade of bHLH TFs has been shown to have a direct role in regulating the transcription of genes encoding enzymes in polyphenol biosynthesis. This clade has been defined by analysis of bHLH proteins in angiosperms and is the clade IIIf (Heim et al. 2003, Figure 10.5). bHLH proteins of subgroup IIIf share a highly conserved basic domain (which binds DNA), a protein-protein interaction domain (HLH domain) and an N-terminal domain which is believed to interact with R2R3Myb proteins in the MBW complex (Goff et al. 1992; Zimmermann et al. 2004). Although it has been claimed that bHLH proteins of subgroup IIIf exist in bryophytes such as *Marchantia polymorpha* (Catarino et al. 2016), this phylogenetic analysis was based only on comparisons of the bHLH domains of the proteins. MpbHLH12 identified by Catarino et al. (2016) as a member of subgroup IIIf contains an N-terminal domain that partially resembles that of the

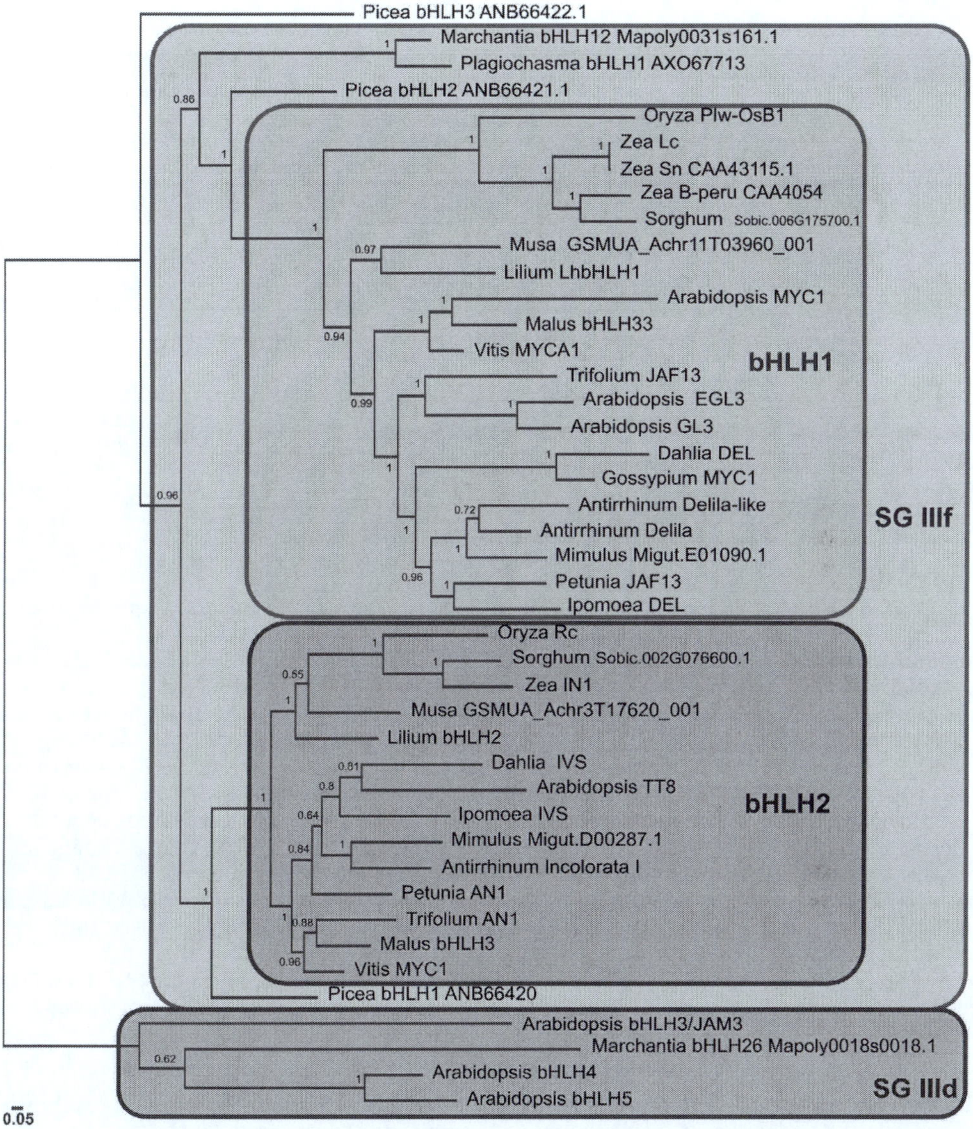

Figure 10.5 Phylogenetic analysis of plant bHLH proteins belonging to subgroup IIIf. *Source:* Adapted from Heim et al. (2003). Deduced amino acid sequences were aligned using MUSCLE to generate a maximum likelihood phylogenetic tree using PhyML with 1000 bootstrap replicates using Geneious (10.0.9) software, using default parameters. Gene IDs/accession numbers of sequences used: *Antirrhinum majus* Antirrhinum Delila (AAA32663.1; Am02g33340); Antirrhinum Delila-like (Am02g28470); Antirrhinum Incolorata I (Am02g53780) (A. majus genome, v.3); *Arabidopsis thaliana* Arabidopsis bHLH3/JAM3 (NP_193376); Arabidopsis EGL3 (OAP12509); Arabidopsis GL3 (NP_680372); Arabidopsis TT8 (Q9FT81); Arabidopsis MYC1 (NP_001154194); Arabidopsis bHLH4 (O49687); Arabidopsis bHLH5 (Q9FIP9); *Dahlia variabilis* Dahlia Del (BAJ33516); Dahlia IVS (BAJ33515); *Gossypium hirsutum* Gossypium MYC1 (CAA07615); *Ipomoea nil* Ipomoea Del (XP_019171149); Ipomoea IVS (XP_019197480); *Lilium hybrida* Lilium bHLH1 (BAE20057); Lilium bHLH2 (BAE20058); *Malus domestica* Malus MdbHLH3 (ADL36597); Malus bHLH33 (ABB84474); *Marchantia polymorpha* Marchantia bHLH12 (Mapoly0031s161.1); Marchantia bHLH26 (Mapoly0018s0018.1); *Mimulus gattatus* (Erythranthe guttata) Mimulus Migut.E01090.1; Mimulus Migut.D00287.1; *Musa acuminata* GSMUA_Achr11T03960_001; GSMUA_Achr3T17620_001; *Oryza sativa* Oryza Plw-OsB1 (BAB64301); Oryza Rc (ADK36625); *Petunia hybrida* Petunia AN1 (AAG25928); Petunia JAF13 (AAC39455); *Picea abies* Picea bHLH1 ANB66420; Picea bHLH2 ANB66421.1; Picea bHLH3 ANB66422.1; *Sorghum bicolor* Sobic.006G175700.1; Sobic.002G076600.1; *Trifolium repens* Trifolium AN1 (AIT76559) Trifolium JAF13 (AIT76563); *Vitis vinifera* Vitis MYC1 (NM_001281253); Vitis MYCA1 (NP_001267954); *Zea mays* Zea Lc (AAA33504); Zea In1 (AAB03841); Zea Sn (CAA43115.1); Zea B-peru (CAA4054). (*See insert for color representation of the figure.*)

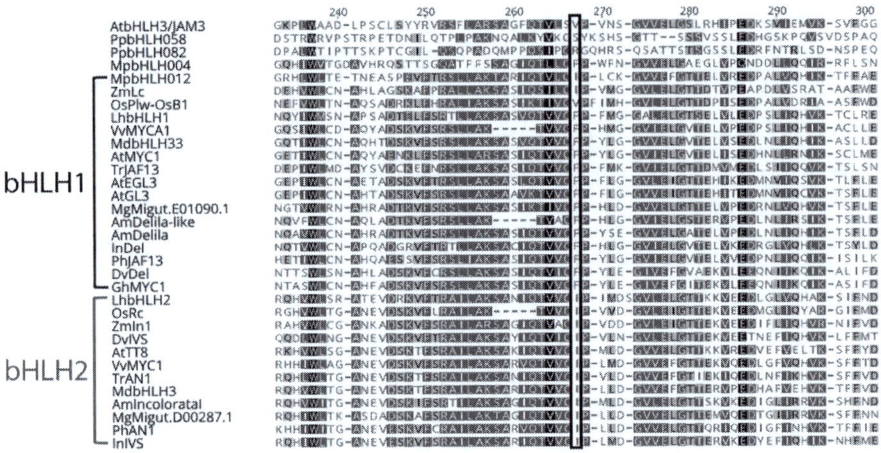

Figure 10.6 Muscle alignment of bHLH proteins. Gene IDs/accession numbers of sequences used: *Antirrhinum majus* AmDelila (AAA32663.1; Am02g33340); AmDelila-like (Am02g28470); AmIncolorata I (Am02g53780) (*A. majus* genome, v.3); *Arabidopsis thaliana* AtbHLH3/JAM3 (NP_193376); AtEGL3 (OAP12509); AtGL3 (NP_680372); AtTT8 (Q9FT81); AtMYC1 (NP_001154194); *Dahlia variabilis* DvDel (BAJ33516); DvIVS (BAJ33515); *Gossypium hirsutum* GhMYC1 (CAA07615); *Ipomoea nil* InDel (XP_019171149); InIVS (XP_019197480); *Lilium hybrida* LhbHLH1 (BAE20057); LhbHLH2 (BAE20058); *Malus domestica* MdbHLH3 (ADL36597); MdbHLH33 (ABB84474); *Marchantia polymorpha* MpbHLH012 (Mapoly0031s0161, M. polymorpha genome MpTak_v6.1); MpbHLH004 (Mapoly0028s0058; M. polymorpha genome MpTak_v6.1); *Mimulus gattatus* (Erythranthe guttata) MgMigut.E01090.1; MgMigut.D00287.1; Oryza sativa OsPlw-OsB1 (BAB64301); OsRc (ADK36625); *Petunia hybrida* PhAN1 (AAG25928); PhJAF13 (AAC39455); *Physcomitrium patens* PpbHLH058 (Pp3c6_3190V3.2); PpbHLH082 (Pp3c16_21680V3.1); *Trifolium repens* TrAN1 (AIT76559) TrJAF13 (AIT76563); *Vitis vinifera* VvMYC1 (NM_001281253); VvMYCA1 (NP_001267954); *Zea mays* ZmLc (AAA33504); ZmIn1 (AAB03841). bHLH-1 proteins are bracketed in orange, bHLH-2 proteins in blue. Residue 177 in At GL3 is boxed in red as this has been shown to influence binding to R2R3Myb proteins depending on whether it is F or I (Zhang et al. 2019). (*See insert for color representation of the figure.*)

conserved N-terminal domain of angiosperm members of this clade (Figure 10.6) but it does not seem to interact with R2R3Myb TFs regulating flavonoid metabolism in Marchantia (Albert et al. 2018). Clearly, knockout mutants of MpbHLH012, generated by CRISPR/Cas 9 gene editing, would address whether bryophytes acquired Myb-bHLH interactions regulating polyphenol metabolism. In contrast, in gymnosperms, which make both proanthocyanins and anthocyanins, there are bHLH proteins which clearly belong to subgroup IIIf (Nemesio-Gorriz et al. 2017) suggesting that the MBW complex evolved for the regulation of proanthocyanin/anthocyanin biosynthesis and that other roles, such as the regulation of trichome formation in Arabidopsis, are derived functionalities (Serna and Martin 2006).

10.3.1 Roles of bHLH-1 and bHLH-2 Clades in Regulating Anthocyanin Biosynthesis

Subgroup IIIf proteins in angiosperms clearly divide into two separate clades which have been termed bHLH-1 and bHLH-2 (Figure 10.5). The proteins in these two clades are distinct structurally and the genes that encode them have distinguishing intron/exon structures supporting the view that they represent two evolutionarily distinct clades. Members of both clades are present in all angiosperm species that have been investigated (Albert et al. 2020). The maintenance of members of two similar but structurally distinct clades in most species suggests that these proteins have distinct functionalities. Indeed, in petunia, only the bHLH-2 protein (encoded by *An1*) appears to regulate anthocyanin biosynthesis directly (Spelt et al. 2000, 2002). Research on An1 in petunia showed that it also regulates hyper-acidification of petal cells and seed coat morphology, and analysis of mutations of another bHLH-2 protein (Noemi) in citrus showed that it controls proanthocyanin production in seed coats, anthocyanin production in leaves and flowers and acidification of the juice sacs of the fruit (Spelt et al. 2002; Butelli et al. 2019). In neither of these examples is there evidence that the bHLH-1 protein participates in an MBW complex that regulates the transcription of these genes directly. Instead, it has been proposed that the bHLH-1 protein in petunia, Jaf13, is required for the activation of An1 and the gene encoding the WDR protein, in a hierarchical transcriptional regulatory cascade (Albert et al. 2014). In other species, bHLH-1 and bHLH-2 proteins seem to have overlapping functions in terms of regulation of anthocyanin biosynthesis, as, for example, in Arabidopsis where bHLH-2 (TT8) can regulate both anthocyanin and proanthocyanin biosynthesis in the tissues in which it is expressed, provided the appropriate R2R3Myb proteins are also present in the same tissues in Arabidopsis. The bHLH-1 proteins, EGL3, and GL3, can also regulate anthocyanin and proanthocyanin biosynthesis within MBW complexes with the appropriate R2R3Myb proteins, although it should be noted that the evidence that bHLH-1 proteins in Arabidopsis do regulate proanthocyanin biosynthesis is limited to studies with target genes linked to reporters in protoplast assays where the levels of the TFs may be very high and not reflect physiological conditions, since mutations in *egl3* and *gl3* do not impact seed coat pigmentation (Xu et al. 2014).

The partial functional redundancy of bHLH-1 and bHLH-2 proteins is demonstrated nicely in *Antirrhinum majus* where the production of anthocyanin in the flower tube is dependent on Delila (Del; a bHLH-1 protein), but the presence of anthocyanin in the lobes of the flower is dependent on the activity of Incolorata I (Inc I; a bHLH-2 protein). This overlapping functional redundancy in the direct control of transcription of anthocyanin biosynthetic genes has been demonstrated by the use of unstable mutations, which show that *Inc I* complements *del* mutations in the lobes of the flower and that *Del* complements *Inc I* mutations in the lobes of the flower. In *A. majus* there is also evidence for the hierarchical regulation of the bHLH-2 gene expression by the bHLH-1 proteins, which has been invoked to explain the bicolored patterning of *del* mutant flowers (Albert et al. 2020).

10.3.2 Roles of bHLH-1 and bHLH-2 Clades in the Regulation of Proanthocyanidin Biosynthesis

It is not possible to discern whether both bHLH-1 (Delila) and bHLH-2 (Incolorata I) function in the MBW complex regulating proanthocyanin biosynthesis in *A. majus* since no proanthocyanins have been reported in this species, which also does not appear to have a gene encoding anthocyanidin reductase (ANR), the key gene in proanthocyanidin biosynthesis (Jun et al. 2021), in its genome (Li et al. 2019). Mutants of *A. majus* can be used to test the roles of Del (bHLH-1) and Inc I (bHLH-2) in the hyperacidification of petal cells.

Mutations of *Inc I* do show an increase in the pH of petal lobe cells compared to wild-type petal lobes, whereas *del* mutations do not affect the pH of the tube tissues (Figure 10.7). These results suggest that while Inc I participates in other MBW complexes, Del is restricted to the MBW complex regulating anthocyanin biosynthesis, either directly or through activating the genes encoding bHLH-2 proteins.

This distinction in functionality between bHLH-1 and bHLH-2 TFs may also explain the activity of the Intensifier TF in maize. *Intensifier* (*In*) has been characterized by the phenotype of its active allele which gives red aleurone in some genetic backgrounds or paler anthocyanin pigmentation in others, and that of a number of knockout and dominant mutations which give dark (black) aleurones. It was therefore reported as a suppressor of anthocyanin accumulation in maize (Fraser 1924). However, the wild-type gene encodes an apparently intact and functional bHLH-2 protein of subgroup IIIf (Burr et al. 1996). Another interesting attribute of the *In* phenotype is that the paler aleurone pigmentation conferred by the wild-type *In* allele is non-diffusible but the darker aleurone pigmentation of the recessive *in* allele is diffusible as well as being more intense (Reddy and Peterson 1978). In rice, the red pericarp is caused by the accumulation of proanthocyanidins and is controlled by a gene, *Rc*, which encodes a bHLH-2 protein, structurally most closely related to Intensifier in maize (Sweeney et al. 2006, Furukawa et al. 2007). In rice, anthocyanins in the pericarp are controlled by a bHLH-1 type protein, Rb (closely related structurally to

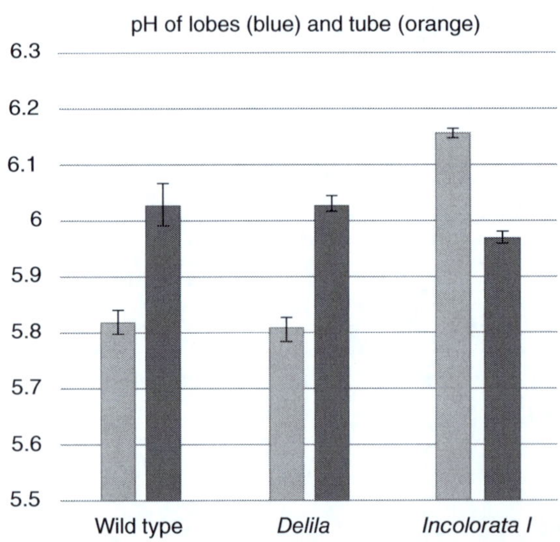

Figure 10.7 The effect of mutations in the genes encoding bHLH-1 (Delila) and bHLH-2 (Incolorata I) on the pH of lobe and tube petal tissue in *Antirrhinum majus*. Petal tissue from five flowers was ground with a pestle and mortar in 4 mL of distilled water, and the pH of the homogenate was measured immediately with a normal pH electrode. Delila is active in both lobes (blue) and tubes (orange), but Incolorata I is active only in lobes (Albert et al. 2020). Error bars show sem $n = 5$. (*See insert for color representation of the figure.*)

the bHLH -1 proteins R/B/Lc/Sn in maize) interacting in an MBW complex with a subgroup 5a R2R3Myb proteins closely related to C1 in maize, OsMyb3 in the pericarp (Zheng et al. 2021) and OsC1 in other tissues (Takahashi 1957; Saitoh et al. 2004). This suggests very strongly that the bHLH-1 proteins in rice regulate anthocyanin biosynthesis and the bHLH-2 proteins regulate proanthocyanin biosynthesis. Probably both proteins work through MBW complexes, and while the R2R3Myb TFs controlling anthocyanin biosynthesis in rice are known to be closely related to C1 (Saitoh et al. 2004; Zheng et al. 2021), the identity of the R2R3Myb protein(s) controlling proanthocyanidin biosynthesis in rice has not yet been confirmed. Mapping and molecular characterization have identified two R2R3Myb genes on Chromosome 7, and a locus *Rp* on chromosome 7 associated with variation in the red color of rice pericarp (Furukawa et al. 2007; Tong et al. 2021), although *Rc* is also located on Chromosome 7. A similar distinction between the MBW complex controlling anthocyanin biosynthesis in maize (in aleurone C1, R and Pac 1) and proanthocyanidin biosynthesis in aleurone (an R2R3Myb, very likely C1, In and Pac 1 (Fraser 1924)) could explain the apparent intensifying effect of loss of function alleles of *Intensifier*. If Intensifier activity switches on proanthocyanidin biosynthesis in maize aleurone, this could produce red or colorless non-diffusible pigmented proanthocyanins (their color being dependent on their degree of polymerization). If Intensifier can induce the production of catechins for proanthocyanidin biosynthesis its activity would reduce the accumulation of anthocyanins as demonstrated by overexpression of ANR in tobacco flowers (Xie et al. 2003). The phenotype that wild-type *In* alleles have reduced accumulation of anthocyanins without synthesizing anything else would be the result of In stimulating the synthesis of colorless (low degree of polymerization) or red-colored proanthocyanins which might be strongly associated with cell walls and therefore not readily extractable (or visually non-diffusible) from maize aleurone (Reddy and Peterson 1978; Chen et al. 2017; Chatham et al. 2019).

Perhaps the most interesting implication for the regulation of phenylpropanoid metabolism of this idea, is that only bHLH-2 proteins may regulate proanthocyanin biosynthesis while bHLH-1 proteins regulate only anthocyanin biosynthesis, at least in the Poales. The functionalities of bHLH-1 and bHLH-2 proteins in MBW complexes may also differ between orders of plants as suggested by the differences between bHLH-1 functions between the Solanaceae (typified by petunia, potato and tomato) and the Plantaginaceae (typified by *A. majus*).

Of considerable interest in this respect are transgenic experiments using bHLH-1 proteins to induce anthocyanin biosynthesis in Solanaceous plants. These seem to work well when *Lc* has been expressed in petunia (Bradley et al. 1998) and *Del* in tomato (Mooney et al. 1995) inducing strong anthocyanin biosynthesis in vegetative tissues of both species, despite the homologous proteins (Jaf13 and SlJaf13, respectively) apparently playing no direct role in anthocyanin biosynthesis. Clearly, bHLH-1 proteins have the potential to regulate anthocyanin production directly in Solanaceous species, especially when combined with a compatible subgroup 6 R2R3Myb protein (Butelli et al. 2008; Albert et al. 2009). Their lack of activity in regulating anthocyanin biosynthesis directly in these species is most easily explained by low levels or an absence of expression of the bHLH encoding genes in the cells that produce anthocyanins.

10.4 The Role of the WDR in the MBW Complex in the Regulation of Polyphenol Metabolism

WD Repeat proteins are defined by the presence of four or more repeating units containing a conserved core of approximately 40 amino acids that usually ends with tryptophan-aspartic acid (WD). WD-repeat proteins are thought to form circularized beta-propeller structures without catalytic activity on which protein complexes assemble. In plants, the MBW complex was defined originally (Payne et al. 2000; Baudry et al. 2004; Ramsay and Glover 2005; Lloyd et al. 2017) following analysis of the phenotype of mutants of *TTG1*, which encodes the WD Repeat protein involved in MBW complexes in Arabidopsis (Walker et al. 1999). Consequently, the WD Repeat protein has been viewed as an essential component of MBW complexes, although that belief may have to be modified as more is discovered about the regulation of polyphenol metabolism in a broad range of plant species. The identification of the role of TTG1/WDR in MBW complexes in Arabidopsis was followed by the discovery that the gene *An11* required for anthocyanin biosynthesis in *Petunia hybrida* encodes a highly conserved WD Repeat (de Vetten et al. 1997). The description of the *pac1* mutation in maize and its identification as affecting a WDR protein involved in anthocyanin accumulation seemed to establish the ubiquity of the WDR proteins in MBW complexes (Selinger and Chandler 1999). However, unlike in Arabidopsis, where *TTG1* is essential for proanthocyanin, anthocyanin and trichome formation, in maize, the knockout mutation, *pac1,* has a pale aleurone phenotype because a little anthocyanin is formed in kernels even in the absence of Pac1 protein (Selinger and Chandler 1999) as does the equivalent mutation of *OsTTG1* in rice (Yang et al. 2021). Although *pac1* mutants show reduced anthocyanin accumulation in root tissue exposed to light, the leaves and other vegetative tissues of *pac1* plants produce normal levels of anthocyanins, suggesting that participation of the WDR protein in the MBW complex controlling anthocyanin biosynthesis is not essential in all tissues in maize (Selinger and Chandler 1999; Carey et al. 2004) or in rice (Yang et al. 2021).

Our understanding of the interactions within the MBW complex has been enhanced significantly by quantitative measurements of the affinities of the different protein components for interaction within the MBW complex, and although these analyses have focussed on the proteins functioning to control trichome development in Arabidopsis, their conclusions appear to be applicable to the proteins of the other MBW complexes controlling phenylpropanoid metabolism in plants (Pesch et al. 2015). These authors showed that the three MBW proteins from Arabidopsis form either GL1 (R2R3Myb)/ GL3 (bHLH-1) or GL3 (bHLH-1)/TTG1 (WDR) dimers. The formation of each dimer with the bHLH protein is counteracted by the respective third protein in a range of in vitro and in vivo assays. In contrast, R2R3Myb proteins like GL1 in Arabidopsis interact positively with WDR (TTG1) and bHLH-2 proteins in regulating their target genes (Pesch et al. 2015; Zhang et al. 2019) and likely operate together in a single MBW complex.

In *A. majus*, the activity of Ros1 (subgroup 6 R2R3Myb) and Del (bHLH-1) in inducing anthocyanin biosynthesis is not enhanced by the presence of a WDR protein, and may, in fact, be inhibited by its presence (Albert et al. 2020). In contrast, the activity of Ros1 and Inc I (bHLH-2) is enhanced by the presence of the WDR protein, meaning that Inc I likely acts in an MBW complex with Ros 1 and WDR, but that the most effective complex of Ros1 and Del in inducing anthocyanin biosynthesis may lack the WDR protein (Albert et al. 2020).

Zhang et al. (2019) also mapped residues important for the interactions between bHLH IIIf proteins with R2R3Myb proteins and showed that within the N-terminal domain conserved within proteins of subgroup IIIf, a phenylalanine residue at position 177 in AtGL3 is conserved in all bHLH-1 proteins and confers significantly greater binding to R2R3Myb proteins than the Isoleucine residue present in bHLH-2 proteins (Figure 10.6). Mutation of F177 to I in AtGL3 removed the competitive interaction between R2R3Myb and WDR proteins. This amino acid lies within the conserved N-terminal domain identified by Goff et al. (1992) in the maize bHLH-1 protein, B, to be necessary for interaction with C1 in the control of anthocyanin biosynthesis. At the very least, these recent results suggest that the WDR proteins are less important to the activity of Myb and bHLH-1 proteins controlling anthocyanin biosynthesis than they are to R2R3Myb and bHLH-2 proteins controlling anthocyanin and proanthocyanidin biosynthesis through the activity of the MBW complex.

10.5 Additional Factors Regulating Transcriptional Control of the MBW Complex

Numerous small Myb proteins may serve as competitive inhibitors of MBW complexes. These are polyphyletic. A group of small Myb proteins carrying just a single Myb repeat (R3) predate the monocot-dicot divide and carry no transcriptional repression domains. They are believed to inhibit the activity of the MBW complex by competing for binding of the bHLH proteins with the R2R3Myb activators, especially those of subgroup 6 controlling anthocyanin biosynthesis (Chen et al. 2019). Their prevalence in angiosperms may reflect the importance of anthocyanin colors and the patterning of flowers to pollinator attraction in flowering plants. The small size of these Myb repressors means that they can diffuse between cells and their importance in determining the patterning of trichome formation has been demonstrated in Arabidopsis where they influence trichome initiation through lateral inhibition of the initiating signal from the MBW complex. Similar patterning functionalities have been suggested for anthocyanin spot development in flowers by Davies et al. (2012) and Albert et al. (2014) and demonstrated to be directed by the counterbalancing activities of the subgroup 6 R2R3Myb transcriptional activator Negan, and the small R3 Myb repressor RTO, in monkey flower (*Mimulus guttatus*) (Ding et al. 2020). Currently, evidence of these passive repressors is associated predominantly with the control of anthocyanin biosynthesis (examples are Atv which represses anthocyanin production in tomato fruit, MYBx in petunia flowers, and TRY-, CPC-, and CPC-like proteins which negatively regulate trichome and non-root hair formation in Arabidopsis). There are additional lineage-specific negative regulators of anthocyanin biosynthesis, such as AtMYBL2, which seem to have evolved from genes encoding R2R3Myb subgroup 4 proteins (Matsui et al. 2008). They have truncated DNA binding domains, which means that they likely do not bind to DNA themselves and, although they may contain repression motifs like the EAR motif, they exert their repression functionality through interaction with bHLH proteins competing with R2R3Myb proteins in MBW complexes. The extent of small Myb protein repression activities of the MBW complexes regulating proanthocyanidin biosynthesis and hyper-acidification of vacuoles is currently not well understood.

10.6 Conclusions

By considering the evolution of transcriptional regulators of polyphenol metabolism, one can see how new metabolic capabilities emerged and how the regulatory networks controlling these capabilities were adjusted. With this holistic view, our considerable knowledge of the workings of the MBW complex in regulating proanthocyanin and anthocyanin biosynthesis can be put in context with the other primary transcriptional regulators of polyphenol biosynthesis. Co-option of components of the MBW complex, as seen in trichome and non-root hair development in the Brassicaceae, and betalain production in the Caryophyllaceae, can also be put in a broader regulatory perspective. An understanding of how the transcriptional regulation of polyphenol metabolism evolved can also facilitate the discovery of new regulators and clarify the functionality of others, already known for some time. Ultimately, understanding these diverged regulatory mechanisms will allow us to breed more effectively for changes in the production of polyphenols, such as improved levels of health-beneficial anthocyanins in food crops and decreased levels of proanthocyanins in forage crops.

Acknowledgments

CM and JL were supported by the Institute Strategic Programme, "Molecules from Nature" (BB/ P012523/1) from the UK Biotechnology and Biological Sciences Research Council and NWA was supported by The Marsden Fund of New Zealand/Te Pūtea Rangahau (contract PAF1501). We thank Andrew Davis at JIC for his photography.

References

Albert, N.W., Lewis, D.H., Zhang, H., et al. (2009). Light-induced vegetative anthocyanin pigmentation in Petunia. *Journal of Experimental Botany*, **60**: 2191–2202.

Albert, N.W., Lewis, D.H., Zhang, H., et al. (2011). Members of an R2R3-MYB transcription factor family in Petunia are developmentally and environmentally regulated to control complex floral and vegetative pigmentation patterning. *The Plant Journal*, **65** (5): 771–784.

Albert, N.W., Davies, K.M., Lewis, D.H., et al. (2014). A conserved network of transcriptional activators and repressors regulates anthocyanin pigmentation in eudicots. *The Plant Cell*, **26** (3): 962–980.

Albert, N.W., Thrimawithana, A.H., McGhie, T.K., et al. (2018). Genetic analysis of the liverwort *Marchantia polymorpha* reveals that R2R3MYB activation of flavonoid production in response to abiotic stress is an ancient character in land plants. *New Phytologist*, **218** (2): 554–566.

Albert, N.W., Butelli, E., Moss, S.M., et al. (2020). Discrete bHLH transcription factors play functionally overlapping roles in pigmentation patterning in flowers of Antirrhinum majus. *New Phytologist*, **231** (2): 849–863 https://doi.org/10.1111/nph.17142.

Baudry, A., Heim, M.A., Dubreucq, B., et al. (2004). TT2, TT8, and TTG1 synergistically specify the expression of BANYULS and proanthocyanidin biosynthesis in Arabidopsis thaliana. *The Plant Journal*, **39** (3): 366–380.

Bedon, F., Bomal, C., Caron, S., et al. (2010). Subgroup 4 R2R3-MYBs in conifer trees: gene family expansion and contribution to the isoprenoid-and flavonoid-oriented responses. *Journal of Experimental Botany*, **61** (14): 3847–3864.

Berland, H., Albert, N.W., Stavland, A., et al. (2019). Auronidins are a previously unreported class of flavonoid pigments that challenges when anthocyanin biosynthesis evolved in plants. *Proceedings of the National Academy of Sciences*, **116** (40): 20232–20239.

Bowman, J.L., Kohchi, T., Yamato, K.T., et al. (2017). Insights into land plant evolution garnered from the *Marchantia polymorpha* genome. *Cell*, **171** (2): 287–304.

Bradley, J.M., Davies, K.M., Deroles, S.C., et al. (1998). The maize Lc regulatory gene up-regulates the flavonoid biosynthetic pathway of Petunia. *The Plant Journal*, **13**: 381–392.

Burr, F.A., Burr, B., Scheffler, B.E., et al. (1996). The maize repressor-like gene intensifier1 shares homology with the r1/b1 multigene family of transcription factors and exhibits missplicing. *The Plant Cell*, **8** (8): 1249–1259.

Butelli, E., Titta, L., Giorgio, M., et al. (2008). Enrichment of tomato fruit with health-promoting anthocyanins by expression of select transcription factors. *Nature Biotechnology*, **26** (11): 1301–1308.

Butelli, E., Licciardello, C., Ramadugu, C., et al. (2019). Noemi controls production of flavonoid pigments and fruit acidity and illustrates the domestication routes of modern citrus varieties. *Current Biology*, **29** (1): 158–164.

Carella, P., Gogleva, A., Hoey, D.J., et al. (2019). Conserved biochemical defenses underpin host responses to Oomycete infection in an early-divergent land plant lineage. *Current Biology*, **29**: 2282–2294 e2285.

Carey, C.C., Strahle, J.T., Selinger, D.A., and Chandler, V.L. (2004). Mutations in the pale aleurone color1 regulatory gene of the Zea mays anthocyanin pathway have distinct phenotypes relative to the functionally similar TRANSPARENT TESTA GLABRA1 gene in Arabidopsis thaliana. *The Plant Cell*, **16** (2): 450–464.

Catarino, B., Hetherington, A.J., Emms, D.M., et al. (2016). The stepwise increase in the number of transcription factor families in the Precambrian predated the diversification of plants on land. *Molecular Biology and Evolution*, **33** (11): 2815–2819.

Chatham, L.A., Paulsmeyer, M., and Juvik, J.A. (2019). Prospects for economical natural colorants: insights from maize. *Theoretical and Applied Genetics*, **132** (11): 2927–2946.

Chen, K., Camberato, J.J., and Vyn, T.J. (2017). Maize grain yield and kernel component relationships to morphophysiological traits in commercial hybrids separated by four decades. *Crop Science*, **57** (3): 1641–1657.

Chen, L., Hu, B., Qin, Y., et al. (2019). Advance of the negative regulation of anthocyanin biosynthesis by MYB transcription factors. *Plant Physiology and Biochemistry*, **136**: 178–187.

Cna'ani, A., Muehlemann, J.K., Ravid, J., et al. (2015a). Petunia × hybrida floral scent production is negatively affected by high-temperature growth conditions. *Plant, Cell & Environment*, **38** (7): 1333–1346.

Cna'ani, A., Spitzer-Rimon, B., Ravid, J., et al. (2015b). Two showy traits, scent emission and pigmentation, are finely coregulated by the MYB transcription factor PH 4 in petunia flowers. *New Phytologist*, **208** (3): 708–714.

Cone, K.C., Burr, F.A., and Burr, B. (1986). Molecular analysis of the maize anthocyanin regulatory locus C1. *Proceedings of the National Academy of Sciences*, **83** (24): 9631–9635.

Davies, K.M., Albert, N.W., and Schwinn, K.E. (2012). From landing lights to mimicry: the molecular regulation of flower colouration and mechanisms for pigmentation patterning. *Functional Plant Biology*, **39** (8): 619–638.

Ding, B., Patterson, E.L., Holalu, S.V., et al. (2020). Two MYB proteins in a self-organizing activator-inhibitor system produce spotted pigmentation patterns. *Current Biology*, **30** (5): 802–814.

Dubos, C., Stracke, R., Grotewold, E., et al. (2010). MYB transcription factors in Arabidopsis. *Trends in Plant Science*, **15** (10): 573–581.

Fraser, A.C. (1924). Heritable characters of maize: XVII—Intensified red and purple aleurone color. *Journal of Heredity*, **15** (3): 119–124.

Furukawa, T., Maekawa, M., Oki, T., et al. (2007). The Rc and Rd genes are involved in proanthocyanidin synthesis in rice pericarp. *The Plant Journal*, **49** (1): 91–102.

Goff, S.A., Cone, K.C., and Chandler, V.L. (1992). Functional analysis of the transcriptional activator encoded by the maize B gene: evidence for a direct functional interaction between two classes of regulatory proteins. *Genes & Development*, **6** (5): 864–875.

Grotewold, E., Drummond, B.J., Bowen, B., and Peterson, T. (1994). The myb-homologous P gene controls phlobaphene pigmentation in maize floral organs by directly activating a flavonoid biosynthetic gene subset. *Cell*, **76** (3): 543–553.

Hatlestad, G.J., Akhavan, N.A., Sunnadeniya, R.M., et al. (2015). The beet Y locus encodes an anthocyanin MYB-like protein that activates the betalain red pigment pathway. *Nature Genetics*, **47** (1): 92–96.

Heim, M.A., Jakoby, M., Werber, M., et al. (2003). The basic helix–loop–helix transcription factor family in plants: a genome-wide study of protein structure and functional diversity. *Molecular Biology and Evolution*, **20** (5): 735–747.

Hernández-Hernández, B., Tapia-López, R., Ambrose, B.A., and Vasco, A. (2021). R2R3-MYB gene evolution in plants, incorporating ferns into the story. *International Journal of Plant Sciences*, **182** (1): 1–8.

Jiang, C.K. and Rao, G.Y. (2020). Insights into the diversification and evolution of R2R3-MYB transcription factors in plants. *Plant Physiology*, **183** (2): 637–655.

Jiménez, S. (2010). Comparative investigations of the regulatory roles of MYB transcription factors across the plant kingdom. PhD thesis. University of East Anglia.

Jin, H., Cominelli, E., Bailey, P., et al. (2000). Transcriptional repression by AtMYB4 controls production of UV-protecting sunscreens in Arabidopsis. *The EMBO Journal*, **19** (22): 6150–6161.

Jun, J.H., Lu, N., Docampo-Palacios, M., et al. (2021). Dual activity of anthocyanidin reductase supports the dominant plant proanthocyanidin extension unit pathway. *Science Advances*, **7** (20): eabg4682.

Koes, R., Verweij, W., and Quattrocchio, F. (2005). Flavonoids: a colorful model for the regulation and evolution of biochemical pathways. *Trends in Plant Science*, **10** (5): 236–242.

Kranz, H.D., Denekamp, M., Greco, R., et al. (1998). Towards functional characterisation of the members of the R2R3-MYB gene family from Arabidopsis thaliana. *The Plant Journal*, **16** (2): 263–276.

Kubo, H., Nozawa, S., Hiwatashi, T., et al. (2018). Biosynthesis of riccionidins and marchantins is regulated by R2R3-MYB transcription factors in Marchantia polymorpha. *Journal of Plant Research*, **131** (5): 849–864.

LaFountain, A.M. and Yuan, Y.W. (2021). Repressors of anthocyanin biosynthesis. *New Phytologist*, **231** (3): 933–949.

Lechelt, C., Peterson, T., Laird, A., et al. (1989). Isolation and molecular analysis of the maize P locus. *Molecular and General Genetics MGG*, **219** (1): 225–234.

Legay, S., Sivadon, P., Blervacq, A.S., et al. (2010). EgMYB1, an R2R3 MYB transcription factor from eucalyptus negatively regulates secondary cell wall formation in Arabidopsis and poplar. *New Phytologist*, **188** (3): 774–786.

Li, M., Zhang, D., Gao, Q., et al. (2019). Genome structure and evolution of Antirrhinum majus L. *Nature Plants*, **5** (2): 174–183.

Li, Y., Shan, X., Tong, L., et al. (2020). The conserved and particular roles of the R2R3-MYB regulator FhPAP1 from freesia hybrida in flower Anthocyanin Biosynthesis. *Plant and Cell Physiology*, **61** (7): 1365–1380.

Lloyd, A., Brockman, A., Aguirre, L., et al. (2017). Advances in the MYB–bHLH–WD repeat (MBW) pigment regulatory model: addition of a WRKY factor and co-option of an anthocyanin MYB for betalain regulation. *Plant and Cell Physiology*, **58** (9): 1431–1441.

Ludwig, S.R., Habera, L.F., Dellaporta, S.L., and Wessler, S.R. (1989). Lc, a member of the maize R gene family responsible for tissue-specific anthocyanin production, encodes a protein similar to transcriptional activators and contains the myc-homology region. *Proceedings of the National Academy of Sciences*, **86** (18): 7092–7096.

Matsui, K., Umemura, Y., and Ohme-Takagi, M. (2008). AtMYBL2, a protein with a single MYB domain, acts as a negative regulator of anthocyanin biosynthesis in Arabidopsis. *The Plant Journal*, **55** (6): 954–967.

Mooney, M., Desnos, T., Harrison, K., et al. (1995). Altered regulation of tomato and tobacco pigmentation genes caused by the delila gene of Antirrhinum. *The Plant Journal*, **7**: 333–339.

Nemesio-Gorriz, M., Blair, P.B., Dalman, K., et al. (2017). Identification of Norway spruce MYB-bHLH-WDR transcription factor complex members linked to regulation of the flavonoid pathway. *Frontiers in Plant Science*, **8**: 305.

Ohtani, M. and Demura, T. (2019). The quest for transcriptional hubs of lignin biosynthesis: beyond the NAC-MYB-gene regulatory network model. *Current Opinion in Biotechnology*, **56**: 82–87.

Payne, C.T., Zhang, F., and Lloyd, A.M. (2000). GL3 encodes a bHLH protein that regulates trichome development in Arabidopsis through interaction with GL1 and TTG1. *Genetics*, **156** (3): 1349–1362.

Paz-Ares, J., Wienand, U., Peterson, P.A., and Saedler, H. (1986). Molecular cloning of the c locus of Zea mays: a locus regulating the anthocyanin pathway. *The EMBO Journal*, **5** (5): 829–833.

Paz-Ares, J., Ghosal, D., Wienand, U., et al. (1987). The regulatory c1 locus of Zea mays encodes a protein with homology to myb proto-oncogene products and with structural similarities to transcriptional activators. *The EMBO Journal*, **6** (12): 3553–3558.

Pesch, M., Schultheiß, I., Klopffleisch, K., et al. (2015). TRANSPARENT TESTA GLABRA1 and GLABRA1 compete for binding to GLABRA3 in Arabidopsis. *Plant Physiology*, **168** (2): 584–597.

Powell, A.F., Courtney, L.E., Schmidt, M.H.W., et al. (2020). A Solanum lycopersicoides reference genome facilitates biological discovery in tomato. *BioRxiv.* https://doi.org/10.1101/2020.04.16.039636.

Quideau, S., Deffieux, D., Douat-Casassus, C., and Pouységu, L. (2011). Plant polyphenols: chemical properties, biological activities, and synthesis. *Angewandte Chemie International Edition*, **50** (3): 586–621.

Ramsay, N.A. and Glover, B.J. (2005). MYB–bHLH–WD40 protein complex and the evolution of cellular diversity. *Trends in Plant Science*, **10** (2): 63–70.

Reddy, A.R. and Peterson, P.A. (1978). The action of the intensifier (In) gene in flavonoid production in aleurone tissue of maize. *Canadian Journal of Genetics and Cytology*, **20** (3): 337–347.

Renault, H., Werck-Reichhart, D., and Weng, J.K. (2019). Harnessing lignin evolution for biotechnological applications. *Current Opinion in Biotechnology*, **56**: 105–111.

Saitoh, K., Onishi, K., Mikami, I., et al. (2004). Allelic diversification at the C (OsC1) locus of wild and cultivated rice: nucleotide changes associated with phenotypes. *Genetics*, **168** (2): 997–1007.

Schwinn, K., Venail, J., Shang, Y., et al. (2006). A small family of MYB-regulatory genes controls floral pigmentation intensity and patterning in the genus Antirrhinum. *The Plant Cell*, **18** (4): 831–851.

Schwinn, K.E., Ngo, H., Kenel, F., et al. (2016). The onion (Allium cepa L.) R2R3-MYB gene MYB1 regulates anthocyanin biosynthesis. *Frontiers in Plant Science*, **7**: 1865.

Selinger, D.A. and Chandler, V.L. (1999). A mutation in the pale aleurone color1 gene identifies a novel regulator of the maize anthocyanin pathway. *The Plant Cell*, **11** (1): 5–14.

Serna, L. and Martin, C. (2006). Trichomes: different regulatory networks lead to convergent structures. *Trends in Plant Science*, **11** (6): 274–280.

Shang, Y., Venail, J., Mackay, S., et al. (2011). The molecular basis for venation patterning of pigmentation and its effect on pollinator attraction in flowers of Antirrhinum. *New Phytologist*, **189** (2): 602–615.

Spelt, C., Quattrocchio, F., Mol, J.N., and Koes, R. (2000). Anthocyanin1 of petunia encodes a basic helix-loop-helix protein that directly activates transcription of structural anthocyanin genes. *The Plant Cell*, **12** (9): 1619–1631.

Spelt, C., Quattrocchio, F., Mol, J., and Koes, R. (2002). ANTHOCYANIN1 of petunia controls pigment synthesis, vacuolar pH, and seed coat development by genetically distinct mechanisms. *The Plant Cell*, **14** (9): 2121–2135.

Spitzer-Rimon, B., Marhevka, E., Barkai, O., et al. (2010). EOBII, a gene encoding a flower-specific regulator of phenylpropanoid volatiles' biosynthesis in petunia. *The Plant Cell*, **22** (6): 1961–1976.

Spitzer-Rimon, B., Farhi, M., Albo, B., et al. (2012). The R2R3-MYB–like regulatory factor EOBI, acting downstream of EOBII, regulates scent production by activating ODO1 and structural scent-related genes in petunia. *The Plant Cell*, **24** (12): 5089–5105.

Stracke, R., Werber, M., and Weisshaar, B. (2001). The R2R3-MYB gene family in Arabidopsis thaliana. *Current Opinion in Plant Biology*, **4** (5): 447–456.

Sweeney, M.T., Thomson, M.J., Pfeil, B.E., and McCouch, S. (2006). Caught red-handed: Rc encodes a basic helix-loop-helix protein conditioning red pericarp in rice. *The Plant Cell*, **18** (2): 283–294.

Takahashi, M.E. (1957). Analysis on apiculus color genes essential to anthocyanin coloration rice. *Journal of the Faculty of Agriculture, Hokkaido University*, **50** (3): 266–362.

Takahashi, H. and Asakawa, Y. (2017). Transcriptome analysis of Marchantin biosynthesis from the liverwort Marchantia polymorpha. *Natural Product Communications*, **12** (8): 1934578X1701200831.

Tamagnone, L., Merida, A., Parr, A., et al. (1998). The AmMYB308 and AmMYB330 transcription factors from Antirrhinum regulate phenylpropanoid and lignin biosynthesis in transgenic tobacco. *The Plant Cell*, **10** (2): 135–154.

Tong, J., Han, Z., and Han, A. (2021). Genetic analysis and molecular mapping of Rp, a mutant gene encoding red pericarp in rice (Oryza sativa L.). *Czech Journal of Genetics and Plant Breeding*, **57** (2): 51–57.

Van Moerkercke, A., Haring, M.A., and Schuurink, R.C. (2011). The transcription factor EMISSION OF BENZENOIDS II activates the MYB ODORANT1 promoter at a MYB binding site specific for fragrant petunias. *The Plant Journal*, **67** (5): 917–928.

Verdonk, J.C., Haring, M.A., Van Tunen, A.J., and Schuurink, R.C. (2005). ODORANT1 regulates fragrance biosynthesis in petunia flowers. *The Plant Cell*, **17** (5): 1612–1624.

de Vetten, N., Quattrocchio, F., Mol, J., and Koes, R. (1997). The an11 locus controlling flower pigmentation in petunia encodes a novel WD-repeat protein conserved in yeast, plants, and animals. *Genes & Development*, **11** (11): 1422–1434.

Walker, A.R., Davison, P.A., Bolognesi-Winfield, A.C., et al. (1999). The TRANSPARENT TESTA GLABRA1 locus, which regulates trichome differentiation and anthocyanin biosynthesis in Arabidopsis, encodes a WD40 repeat protein. *The Plant Cell*, **11** (7): 1337–1349.

Weng, J.K. and Chapple, C. (2010). The origin and evolution of lignin biosynthesis. *New Phytologist*, **187** (2): 273–285.

Whitney, H.M., Milne, G., Rands, S.A., et al. (2013). The influence of pigmentation patterning on bumblebee foraging from flowers of Antirrhinum majus. *Naturwissenschaften*, **100** (3): 249–256.

Xie, D.Y., Sharma, S.B., Paiva, N.L., et al. (2003). Role of anthocyanidin reductase, encoded by BANYULS in plant flavonoid biosynthesis. *Science*, **299** (5605): 396–399.

Xu, W., Grain, D., Bobet, S., et al. (2014). Complexity and robustness of the flavonoid transcriptional regulatory network revealed by comprehensive analyses of MYB–b HLH–WDR complexes and their targets in A rabidopsis seed. *New Phytologist*, **202** (1): 132–144.

Xu, W., Dubos, C., and Lepiniec, L. (2015). Transcriptional control of flavonoid biosynthesis by MYB–bHLH–WDR complexes. *Trends in Plant Science*, **20** (3): 176–185.

Yamagishi, M., Shimoyamada, Y., Nakatsuka, T., and Masuda, K. (2010). Two R2R3-MYB genes, homologs of petunia AN2, regulate anthocyanin biosyntheses in flower tepals, tepal spots and leaves of Asiatic hybrid lily. *Plant and Cell Physiology*, **51** (3): 463–474.

Yang, X., Wang, J., Xia, X., et al. (2021). OsTTG1, a WD40 repeat gene, regulates anthocyanin biosynthesis in rice. *Plant Journal*, **107** (1): 198–214 https://doi.org/10.1111/tpj.15285.

Zhang, P., Wang, Y., Zhang, J., et al. (2003). A maize QTL for silk maysin levels contains duplicated Myb-homologous genes which jointly regulate flavone biosynthesis. *Plant Molecular Biology*, **52** (1): 1–15.

Zhang, Y., Butelli, E., Alseekh, S., et al. (2015). Multi-level engineering facilitates the production of phenylpropanoid compounds in tomato. *Nature Communications*, **6** (1): 1–11.

Zhang, B., Chopra, D., Schrader, A., and Hülskamp, M. (2019). Evolutionary comparison of competitive protein-complex formation of MYB, bHLH, and WDR proteins in plants. *Journal of Experimental Botany*, **70** (12): 3197–3209.

Zhang, H., Koes, R., Shang, H., et al. (2019). Identification and functional analysis of three new anthocyanin R2R3-MYB genes in Petunia. *Plant Direct*, **3** (1): e00114.

Zheng, J., Wu, H., Zhao, M., et al. (2021). OsMYB3 is a R2R3-MYB gene responsible for anthocyanin biosynthesis in black rice. *Molecular Breeding*, **41** (8): 1–15.

Zimmermann, I.M., Heim, M.A., Weisshaar, B., and Uhrig, J.F. (2004). Comprehensive identification of Arabidopsis thaliana MYB transcription factors interacting with R/B-like BHLH proteins. *The Plant Journal*, **40** (1): 22–34.

11

Analysis of Proanthocyanidins in Food Ingredients by the 4-(Dimethylamino)cinnamaldehyde Reaction

Daniel Esquivel-Alvarado[1], Emilia Alfaro-Viquez[1], Andrew Birmingham[2], Abigail Kramschuster[2], Christian G. Krueger[1,2], and Jess D. Reed[1,2]

[1] *Reed Research Group, Department of Animal and Dairy Sciences, University of Wisconsin-Madison, Madison, WI, USA*
[2] *Complete Phytochemical Solutions LLC, Cambridge, WI, USA*

11.1 Introduction

Food ingredients that contain proanthocyanidins (PACs) are increasingly used to manufacture dietary supplements because PACs have many putative health benefits, which are related to antimicrobial, antioxidant, and anti-inflammatory properties. Intake of PACs in foods and beverages is associated with decreased risk of a number of diseases such as cardiovascular disease (Schoonees et al. 2012; Sahebkar 2014; Zhang et al. 2016), urinary tract infections (Feliciano et al. 2015a), protection against periodontal disease (Nawrot-Hadzik et al. 2021), stomach ulcers (Saito et al. 1998), and treatment of ulcerative colitis in remission (Głąbska et al. 2020). The association with decreased risk of disease is stimulating increased demand in the dietary supplement market for ingredients that contain PACs. Sources of PACs that are used to formulate dietary supplements include extracts of the food industry such as grape seeds, cranberry extracts, extracts from the fruit of *Aronia*, and many other fruits and by-products of fruit processing such as skins, press cakes, seeds, and hulls. Botanicals that contain PACs, such as pine bark and hawthorn, are also used as ingredients in dietary supplements. Unfortunately, the content and nature of PACs in these ingredients are poorly characterized because the market is not stringently regulated, which creates problems, such as PACs, not being present in detectable amounts and false label claims regarding the presence of ingredients.

One of the better-known health benefits of PACs is the relationship between A-type PACs in cranberries and reduced risk of urinary tract infections. This association has led to the creation of a large market for dietary supplements that contain cranberry by-products and extracts and many of these products are standardized on PAC content as estimated by the DMAC method. Clinical studies have found a positive association between the consumption

Recent Advances in Polyphenol Research, Volume 8, First Edition. Edited by Juha-Pekka Salminen, Kristiina Wähälä, Victor de Freitas, and Stéphane Quideau.

of cranberry juice and cranberry dietary supplements with a decrease in the number of symptomatic cases (Burleigh et al. 2013). However, a meta-analysis of clinical studies found that while the association seemed strong, results were inconclusive because the studies used differing doses and products, lacked compositional analysis of bioactives at study, and lacked mechanistic guidance for selection of subjects and study design (Jepson and Craig 2007; Jepson and Craig 2008; Jepson et al. 2012). Therefore, the ingredient market for the manufacturing of dietary supplements has serious problems related to standardization, authenticity, and efficacy, yet the market is quite large with US consumers spending over $9.6 billion on supplements annually (Smith et al. 2020). The herbal ingredient market is over $800M/year and the botanical market is $37B/year (Smith et al. 2015).

11.2 Background on the 4-(Dimethylamino)cinnalmaldehyde (DMAC) Reaction with PACs

Application of the acid-catalyzed reaction of DMAC for analysis of PACs has the potential to help solve the problem of detection and quantification of PACs in these ingredients. The method was first published by Thies and Fischer (1971), as a colorimetric method for the quantification of catechin in solution, and not for the analysis of PACs. Under the strong acid conditions used by Thies and Fischer (1971), the chromophore develops rapidly. This property made the DMAC reaction attractive as a post-column derivatization method for the analysis of catechin and PACs by high-performance liquid chromatographic as developed by Treutter (1989). Subsequent research showed that the DMAC reaction is a useful and relatively simple colorimetric method for the detection and quantification of PACs. For instance, the DMAC assay has been used to determine the content of PACs in selected chocolate and confectionery products (Payne et al. 2010), cranberry products (Prior et al. 2010; Birmingham et al. 2020), and forage legumes (Li et al. 1996). In addition, by using the DMAC assay, our group has determined the content of PACs in three *Vaccinium* species from Costa Rica, peanut skins from the Spanish, Valencia, and Virginia varieties, and the content of PACs from the Costa Rican guava (*Psidium friedrichsthaliaum*) (Esquivel-Alvarado et al. 2019; Muñoz-Arrieta et al. 2021a; Muñoz-Arrieta et al. 2021b). The chromophore formed in the initial reaction has a maximum absorbance of 640 nm. This property avoids interference from anthocyanins, especially in fruits and fruit products that also contain high levels of anthocyanins, which is advantageous over the vanillin-HCl or butanol-HCl methods that produce chromophores with similar absorbance spectra to anthocyanins.

As mentioned earlier, PACs react with vanillin under acid conditions and develop a red color with maximum wavelength absorptions from 500 to 540 nm (Price et al. 1978). The maximum wavelength in the vanillin-HCl method is strongly dependent of the solvent and source of PACs (Makkar and Becker 1993; Mitsunaga et al. 1998). The vanillin moiety reacts at the C_6 or C_8 position of the upper A-ring of PACs (Mitsunaga et al. 1998), which are also the reactive carbons in the DMAC reaction. Although the vanillin-HCl method has a similar mechanism to the DMAC reaction, the vanillin-HCl method is less specific, more labor-intensive and time-consuming, has lower throughput, and requires highly trained personnel for repeatability (Hümmer and Schreier 2008; Wallace 2010; Herald et al. 2014).

The DMAC reaction appears to have greater sensitivity than either of the vanillin-HCl or the butanol-HCl methods (Hümmer and Schreier 2008; Wallace 2010). Published research determined the optimal condition for the DMAC reaction and showed that the method may be used for quantifying PACs in foods, feeds, and botanicals (Sintara et al. 2018; Birmingham et al. 2020). In practice, the method does have good repeatability and precision (Prior et al. 2010; Wallace and Giusti 2010; Wang et al. 2016; Birmingham et al. 2020). However, the method has serious problems because PAC degree of polymerization and structural variation, control of reaction kinetics, and choice of standards influence estimates of sample concentration (Wallace and Giusti 2010; Feliciano et al. 2012; Wang et al. 2016; Birmingham et al. 2020). In this chapter, we discuss the mechanism of reaction, share our recent research on reaction kinetics and products, and discuss the problem of standards.

11.3 Mechanism of the Acid-Catalyzed DMAC Reaction with PACs

The cinnamaldehydes are a group of naturally occurring phenylpropanoids, of which DMAC (a) is a synthetic derivative. The parent molecule is cinnamaldehyde ((2E)-3-phenylprop-2-enal); (b) (Figure 11.1), which lacks substitution on the phenyl group. The cinnamaldehydes are characterized by an aldehyde at C_1 and a *trans* double bond between C_2 and C_3, and a phenyl group at C_3. Cinnamaldehyde occurs in the bark of cinnamon trees and other species of the genus *Cinnamomum* and is the main component of cinnamon's flavor and odor. Cinnamon also contains proanthocyanidins with A-type interflavan bonds (Anderson et al. 2004). Condensation reactions between cinnamaldehyde and proanthocyanidins occur in cinnamon and these reactions create the dark red-brown colors of the spice (Tanaka et al. 2008). Other naturally occurring cinnamaldehydes, such as sinapaldehyde (c) and coniferyl aldehyde (d), that occur in oak wood and bark (Figure 11.1), also undergo condensation reactions with PACs during the aging of wines in oak barrels to form flavylium pigments (De Freitas et al. 2004; Sousa et al. 2005).

Proanthocyanidins (PACs) are oligomeric flavans and flavan-3-ols that have much structural heterogeneity as related to the type of interflavan bonds, degree of polymerization, and substitutions on the flavan monomers (Cheynier 2005; Reed et al. 2005). For instance, Figure 11.1 shows the structure of a PAC trimer based on epicatechin monomers **(e)** that were isolated from cranberries (Foo et al. 2000). The absolute structure was determined by a combination of ^{13}C NMR, mass spectrometry, and chemical derivatization methods after painstaking chromatographic separations from the large number of PAC oligomers that occur in cranberries. This trimer has a B-type interflavan bond between the upper epicatechin and middle epicatechin ($4\beta \rightarrow 8$) and an A-type interflavan bond between the middle epicatechin and the terminal epicatechin ($4\beta \rightarrow 8$ and $2\beta \rightarrow O \rightarrow 7$), this is only one of the three isolated trimers (Foo et al. 2000). Our group, using matrix-assisted laser desorption/ionization time-of-flight mass spectrometry (MALDI-TOF MS), has detected cranberry

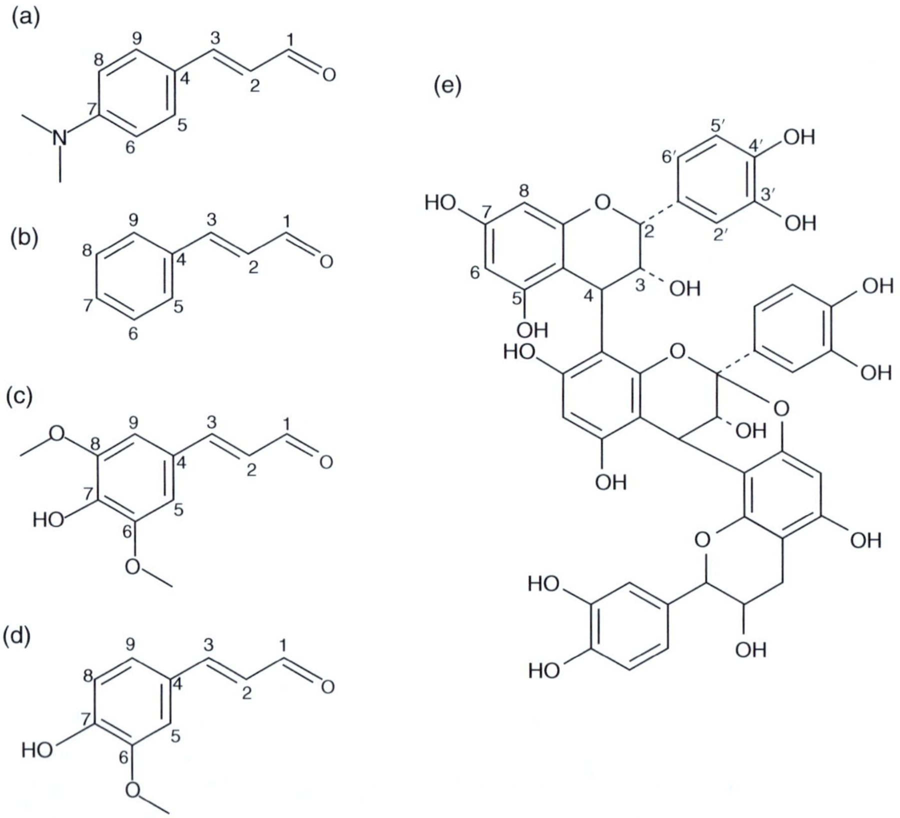

Figure 11.1 Chemical structures of 4-(dimethylamino)cinnamaldehyde (a), cinnamaldehyde (b), sinapaldehyde (c), coniferyl aldehyde (d), and PAC trimer based on epicatechin moieties (e).

PACs with a degree of polymerization of up to 36 with multiple A-type interflavan bonds (Alfaro-Viquez et al. 2019). In addition, our MS spectra indicate that cranberry PACs also contain gallocatechin moieties as originally suggested by Foo et al. (2000), (Reed et al. 2005; Alfaro-Viquez et al. 2018; Esquivel-Alvarado et al. 2021a, 2021b). Therefore, the number of isomers in cranberry PAC is unknown, but by extrapolation from the identified dimers and trimers to higher degrees of polymerization would indicate that a conservative estimate would number in thousands. The formula “$X^{N}Y^{(N-1)}$” estimates the number of individual molecules in a series of PAC oligomers, where X is the number of monomers, i.e. (epi)catechin, (epi)gallocatechin, amongst others, Y is the type of interflavan bonds, and *N* is the degree of polymerization (Cheynier 2005). For B-type proanthocyanidins composed of catechin and epicatechin, this formula yields 8 dimers, 32 trimers, and 128 tetramers. However, PACs in ingredients are more complex and vary in the types of monomers, contain both A- and B-type interflavan bonds, and may have different types of substitutions on their flavan monomers (Reed et al. 2005). In addition, PACs undergo reactions during food processing to form covalent linkages to anthocyanins and other phenolic compounds such as cinnamaldehydes to form derived compounds. This large structural heterogeneity presents a serious challenge to developing rapid colorimetric methods for the accurate determination of PACs in food ingredients.

A proposed reaction mechanism of DMAC with PACs is presented in Figure 11.2, which illustrates the reaction with the trimer shown in Figure 11.1. Much of this mechanism was proposed by De Freitas et al. (2004) for the reaction of sinapaldehyde with catechin to form a new catechin-pyrylium pigment. The first step is the generation of a carbocation in the aldehyde at C_1 carbon (f) in the presence of acid. As discussed later, pH greatly influences the rate of formation of the initial chromophore. The upper C_8 flavan-3-ol unit is the most reactive site for the electrophilic attack by the DMAC carbocation because C_8 is more electronegative. This electrophilic attack generates the initial condensation reaction. However, C_6 of the A ring is also electronegative and may participate in the reaction. Indeed, in the

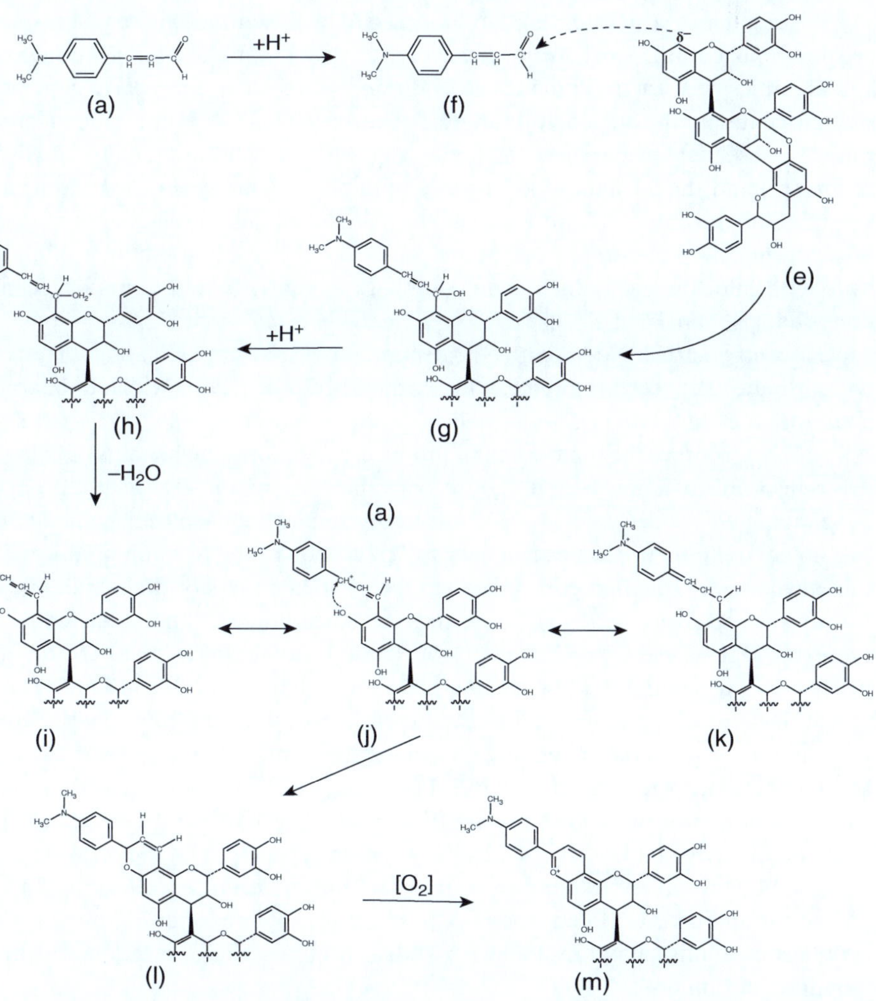

Figure 11.2 Proposed reaction mechanism for the formation of the open ring PAC-DMAC cation (i–k) and a stable PAC-DMAC pyrylium cation (m) obtained from DMAC (a) and a PAC trimer (e).

reaction between cinnamaldehyde and epicatechin, more complex condensation reactions occur in which cinnamaldehyde condenses with both C_8 and C_6 and these condensation reactions continue to crosslink catechin to form oligomeric structures (Tanaka et al. 2008). The product of the initial condensation reaction (g) is unstable and generates an oxonium cation in the presence of strong acid (h) after the initial product dehydrates. This carbocation (i) is probably the first DMAC reaction product that is stable, because, unlike the other cinnamaldehydes that have been studied, this positive charge can delocalize across several different resonance structures (i–k). In the initial condensation reaction, there is a net loss of 17 amu (1 oxygen, 2 electrons, and 1 proton) from the reactants (PACs and DMAC) during the generation of the carbocation. The second proton in the generation of water during the dehydration step comes from the acid catalyst.

The possible resonance structures for the DMAC reaction differ from the reaction of natural cinnamaldehydes because the charge can delocalize to the amino nitrogen (i–k) and this resonance structure has been proposed to be the chromophore that is measured at 640 nm (Wallace and Giusti 2010). However, we are not aware of a publication that isolates and identifies this reaction product. In addition, the resonance structure generates a new reactive center for the nucleophilic attack by the oxygen on C_7 of the A ring to form a pyran ring (l). Subsequent oxidation of the pyran ring leads to the formation of a stable pyrylium cation (m) and the formation of a new chromophore. This type of pyrylium cation was identified as the chromophore produced in the acid-catalyzed reaction of sinapaldehyde and catechin (De Freitas et al. 2004), and similar to the reaction products of cinnamaldehyde with catechin and PACs in cinnamon bark along with more complicated condensation products (Tanaka et al. 2008).

We recently reacted DMAC with epicatechin, procyanidin A2, and oligomeric cranberry PACs, isolated the reaction products by chromatographic techniques and determined the main products by MALDI-TOF MS in positive reflectron mode (Esquivel-Alvarado et al. 2021c). Because the main reaction products are cations, the benefits of using the positive reflectron mode in MALDI-TOF MS are that cations readily desorb, do not require ionization, and are detected at mass unit resolution as their molecular mass without the addition of a proton or monovalent cation. The spectra of the reaction product of catechin show two masses, the first corresponds to the predicted mass of the pyrylium cation (n), and the second mass, which is 2 amu heavier, corresponds to the predicted mass of the open ring DMAC cation (o) (Figure 11.3). These results suggest that both the open ring structure, in which the positive charge can delocalize to the amino group (NH_2), and the pyrylium cation are products in the reaction of DMAC with catechin. These two products are stable enough to be isolated by chromatography and characterized by MALDI-TOF MS. Furthermore, the MALDI-TOF MS spectra of DMAC reaction with procyanidin A2, a dimeric procyanidin, also show two reaction products (Figure 11.4). In fact, the predicted masses and observed masses for both the pyrylium cation (p) and the open ring cation (q) of the DMAC reaction products are within one-hundredth of a mass unit. Although the addition of the DMAC moiety at the C_8 position is shown in both Figures 11.3 and 11.4, there is a possibility that the addition of DMAC occurs at the C_6 position, resulting in the formation of isomers.

MALDI-TOF MS spectra of cranberry PACs, which were isolated through a series of chromatographic steps (Esquivel-Alvarado et al. 2021a), show a series of oligomeric PACs

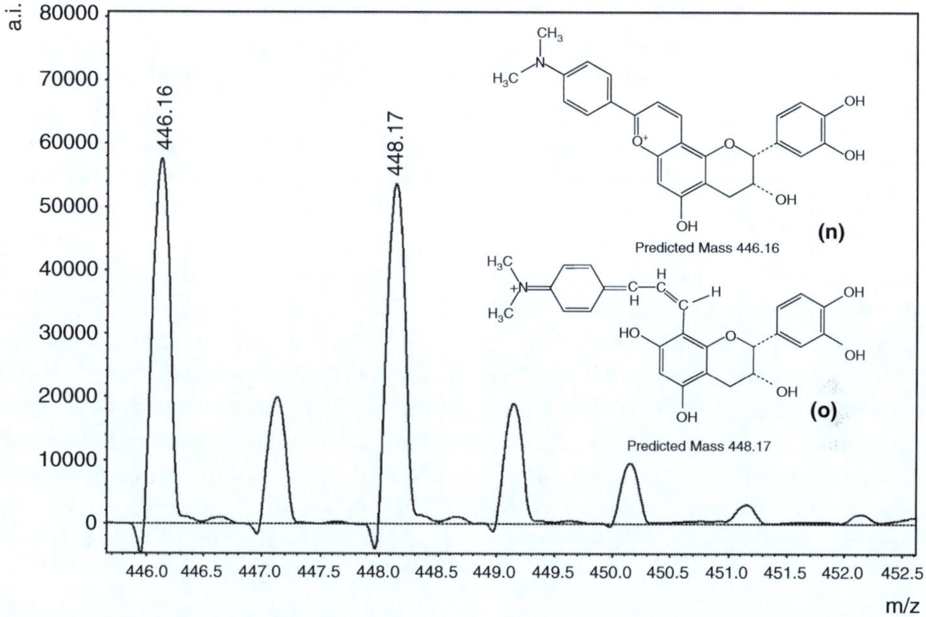

Figure 11.3 MALDI-TOF mass spectrum in positive reflectron mode for the reaction of 4-(dimethylamino)cinnamaldehyde (a) with epicatechin to form the epicatechin-DMAC pyrylium (n) and the open ring epicatechin-DMAC cation (o).

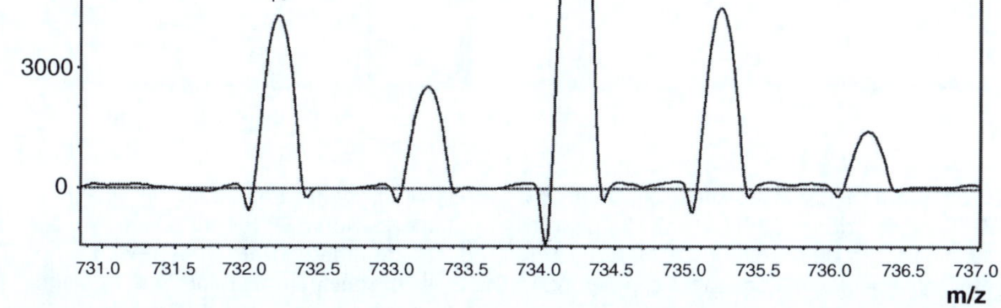

Figure 11.4 MALDI-TOF mass spectrum in positive reflectron mode for the reaction of 4-(dimethylamino)cinnamaldehyde (a) with procyanidin A2 to form the procyanidin A2-DMAC pyrylium (p) and the open ring procyanidin A2-DMAC cation (q).

with one or more A-type interflavan bonds (Figure 11.5), with masses as sodium adducts that correspond to oligomers with a degree of polymerization from tetramers to octamers. Figure 11.5 also shows spectra of the DMAC reaction products of the cranberry PAC oligomers overlayed with the spectra of the unreacted PACs. DMAC reaction products and PACs up to 8 degrees of polymerization were detected in positive reflectron MALDI-TOF MS. The mass difference between PACs and the DMAC reaction products corresponds to the additive mass of DMAC. The enlarged spectra for the PAC tetramer show the mass of the sodium adduct of a tetramer with 1 A-type interflavan bond and the additive masses for the DMAC condensation cations (Figure 11.6), 135.11 amu is the additive mass for the DMAC open ring condensation product where the charge is delocalized within the open ring structure and putatively on the amino group and 133.09 amu is the additive mass for the DMAC pyrylium condensation product. Thus, MALDI-TOF MS is suitable for the detection of stable products of the reaction of DMAC with oligomeric PACs (PAC-DMAC pyrylium cation and the open ring PAC-DMAC cation). The enlarged spectra of the DMAC reaction products with the tetramer show the two overlapping isotope patterns that are typical for masses that differ by 2 amu in MALDI TOF MS (Figure 11.7). The observed and predicted masses of open ring DMAC reaction product and pyrylium DMAC reaction product with the tetramer are within one-hundredth of a mass unit (Figure 11.7).

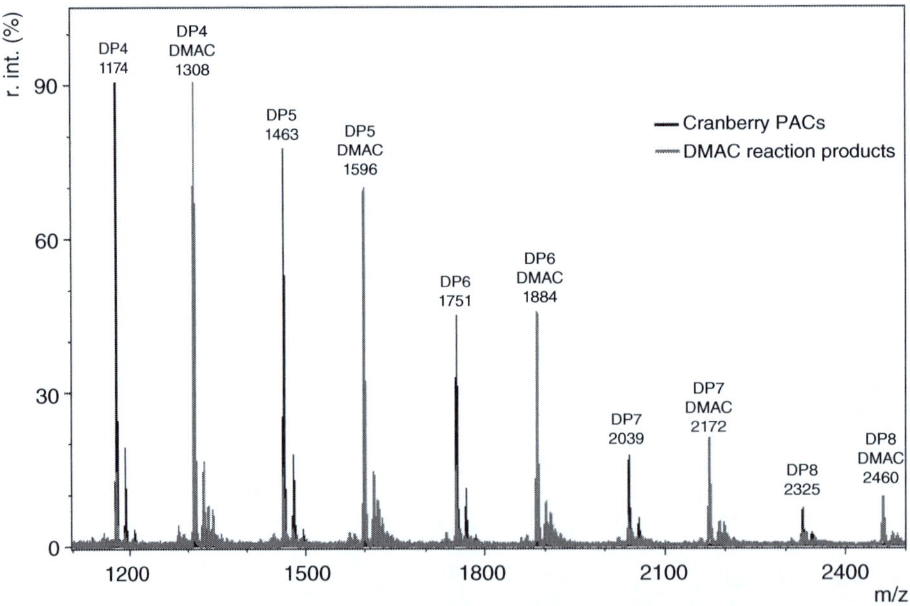

Figure 11.5 MALDI-TOF mass spectrum in positive reflectron mode for unreacted PAC (black) and PAC-DMAC reaction products (red). The degrees of polymerization (DP) for unreacted PAC are labeled from DP4 to DP8, while PAC-DMAC reaction products are labeled from DP4-DMAC to DP8-DMAC. Numbers under the label correspond to the mass of the sodium adduct for PACs and the mass of the respective cations for the DMAC reaction products. As indicated in Figures 11.6 and 11.7, the isotope patterns of the peaks for the DMAC reaction products indicate that each oligomer has two reaction products that correspond to the mass of the open ring cation and the pyrylium cation. (*See insert for color representation of the figure.*)

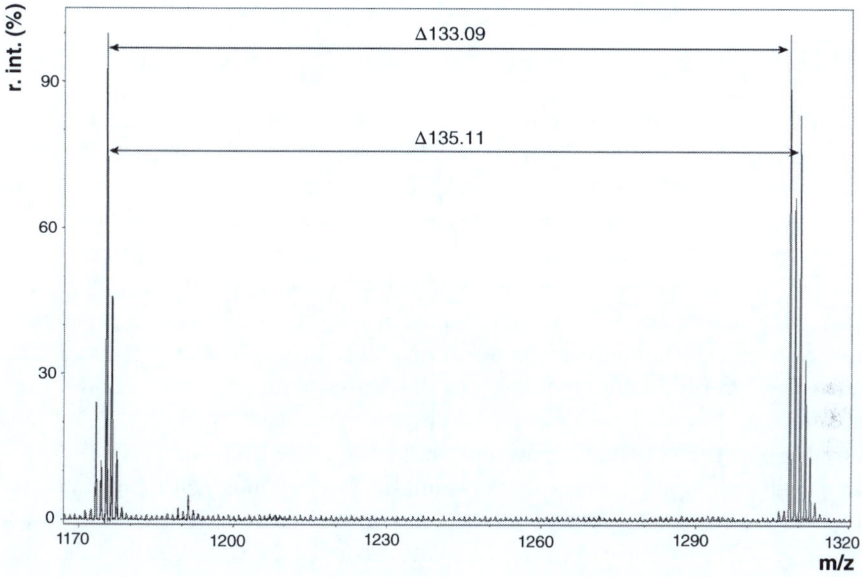

Figure 11.6 MALDI-TOF mass spectrum in positive reflectron mode for unreacted PAC (red) and PAC-DMAC reaction products (green), which show the PAC tetramer and the additive masses for the DMAC condensation products. (*See insert for color representation of the figure.*)

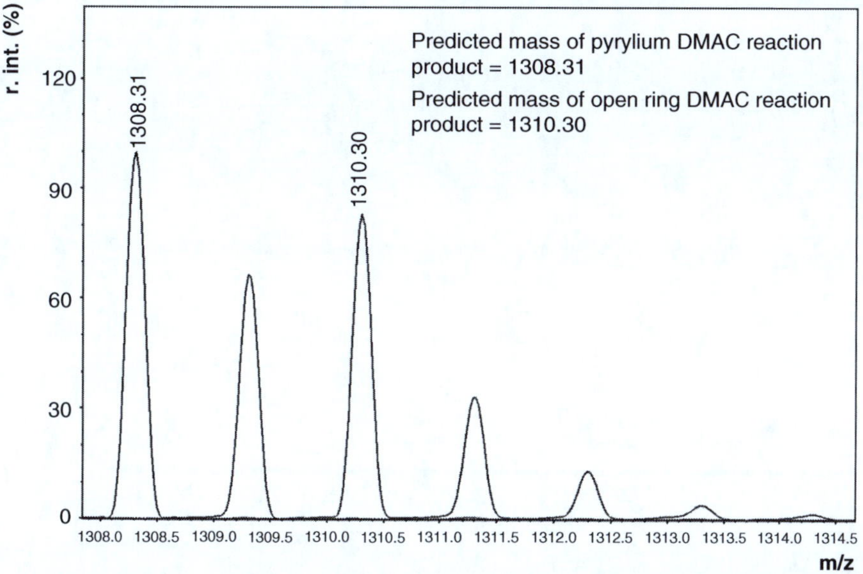

Figure 11.7 MALDI-TOF mass spectrum in positive reflectron mode for the tetramer of the DMAC reaction products, which shows the two overlapping isotope patterns for the PAC-DMAC pyrylium and the open ring PAC-DMAC cation.

11.4 Absorption and Emission Spectra of the DMAC Reaction Products

As described earlier, the products of the DMAC reaction with naturally occurring PACs are complex and primarily consist of a single DMAC moiety covalently linked to the PAC oligomers. Our research indicates that this DMAC condensation product consists of both the open ring cation (i, j, and k) and the pyrilium cation (m). The UV/visible spectrum of the PAC-DMAC reaction products indicates that their absorbance maximum (559 nm) is at a different wavelength than used in the DMAC method (640 nm) (Figure 11.8). The products were isolated after neutralizing the reaction media and chromatographic separation. These results led us to determine the effect of pH on absorbance at 559 and 640 nm during the DMAC reaction. In the DMAC method, the DMAC reagent (0.1% w/v dissolved in a solution mixture containing ethanol, water, and hydrochloric acid 75 : 12.5 : 12.5; v/v/v, respectively) has a pH of zero. In our experiment, six additional DMAC solutions at 0.1% w/v were prepared by adjusting the volume of hydrochloric acid to obtain solutions at pH 0.0, 0.2, 0.4, 0.6, and 0.8, and adding ethanol with no hydrochloric acid. Reactions were performed in a 96-well plate placed in a microplate reader at 25 °C and absorbance was measured at wavelengths of 520, 559, and 640 for 16 hours with a reading every five minutes. Figure 11.9 shows the DMAC reaction of procyanidin A2 under the conditions of our

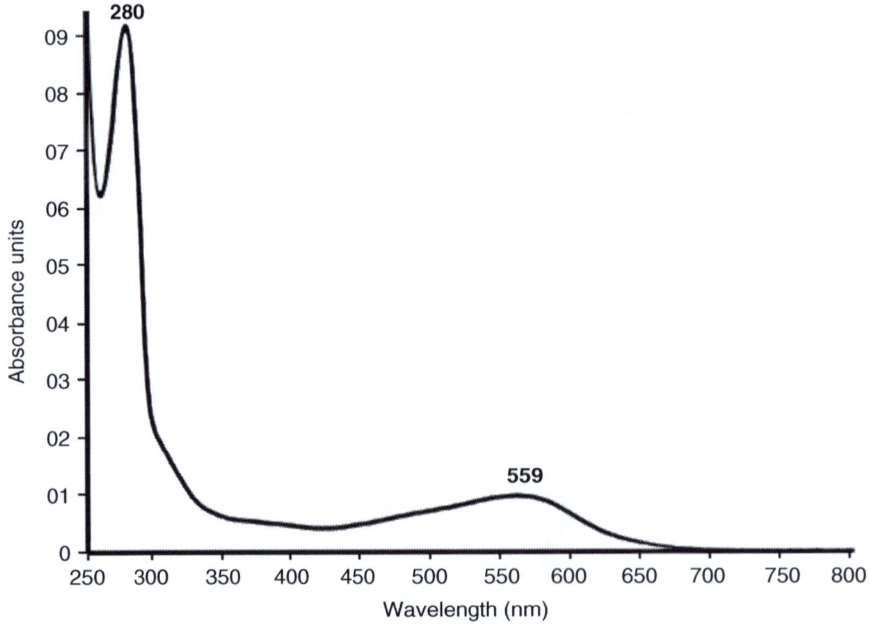

Figure 11.8 UV-visible spectrum of isolated DMAC reaction products, showing the maximum absorbance at 559 nm recorded at pH 7.

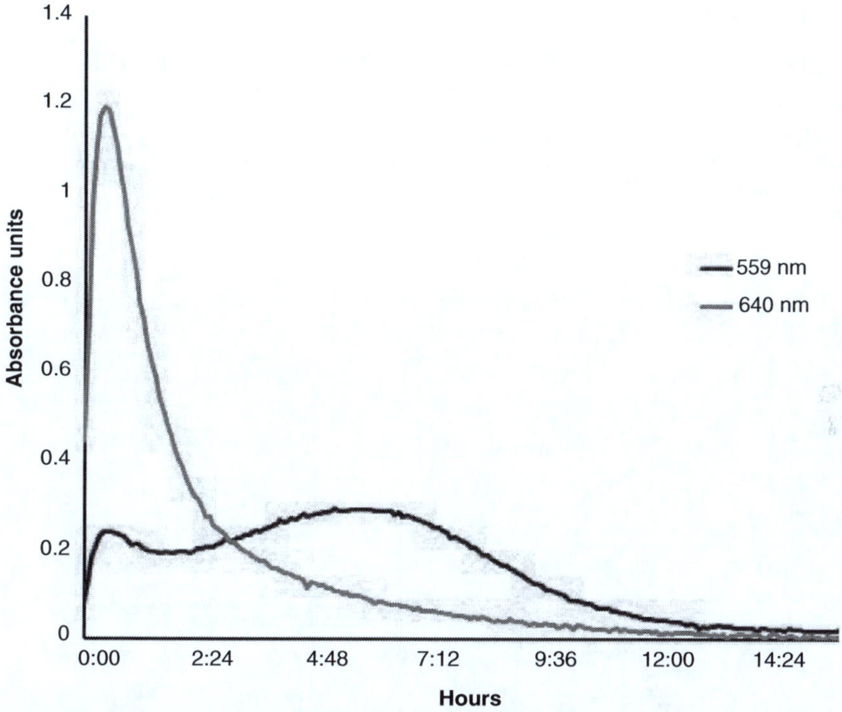

Figure 11.9 Change in absorbance at 559 and 640 nm during the DMAC reaction of procyanidin A2 in the presence of hydrochloric acid (3.24 M), which shows a rapid increase in absorbance at 640 nm (open ring PAC-DMAC cation) and the slow appearance of absorbance at 559 nm (PAC-DMAC pyrylium). (*See insert for color representation of the figure.*)

published DMAC method (Birmingham et al. 2020), which is in the presence of strong acid (HCl; 3.24 M). The strong acid environment leads to a rapid increase in absorbance at 640 nm and a slow appearance of absorbance at 559 nm. These results suggest that the initial chromophore formed in the DMAC reaction may be the open ring structure (i–k), which is the published putative structure (Wallace and Giusti 2010), that has a UV visible maximum at 640 nm. The open ring structure may be slowly converted to the pyrylium cation (m), which may have a lower absorbance maximum. On the other hand, the chromophore with a maximum of 640 nm may be an intermediate in the earlier reaction because absorbance at 640 nm decreases rapidly after reaching its maximum, indicating that this chromophore is unstable. For procyanidin A2, the initial chromophore starts to disappear after about 15 to 20 minutes. For catechin, procyanidin B2, and PACs, the reaction reaches a maximum at 640 nm much earlier than 15 to 20 minutes. However, absorbance at 559 nm increases during this time, which is the maximum of the stable chromophores that we isolated and detected by mass spectrometry. More research on the rate of reaction and the structure of the chromophores is required to improve the usefulness of the DMAC reaction for quantifying PACs.

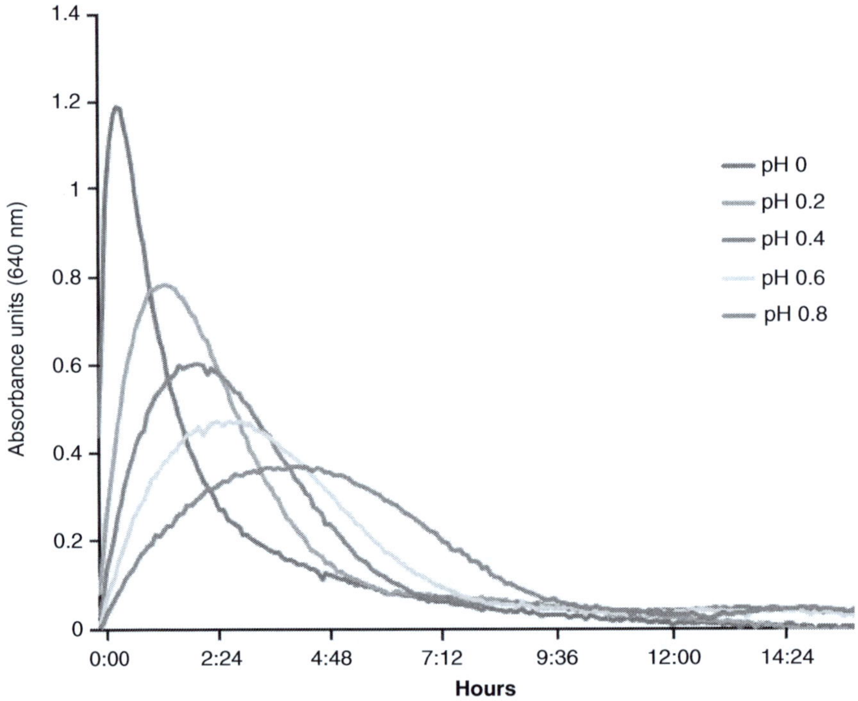

Figure 11.10 Change in absorbance at 640 nm during the DMAC reaction of procyanidin A2 at five different pH values. (*See insert for color representation of the figure.*)

The pH of the reaction has a large effect on absorbance at 640 nm (Figure 11.10). For procyanidin A2 at a low pH, the reaction is rapid, but the maximum absorbance is delayed to over 4.5 hours at a pH of 0.8. Acid is a catalyst to generate the DMAC carbocation that reacts with the terminal C_8 position of PACs to form the initial condensation product. More research is needed to understand the initial reaction kinetics and formation of the chromophore with a maximum absorbance of 640 nm. In contrast, we had observed that the effect of pH on rate and maximum absorbance at 559 nm is much lower than for 640 nm, which suggests separate reactions occur over a longer time frame to form the more stable chromophores. The dehydration of the initial condensation product to form the stable open ring cation (i–k) and the subsequent oxidation to form the pyrylium cation (m) may occur over a longer time frame. Indeed, the condensation reaction of cinnamaldehyde with catechin occurs by heating the reaction mixture at 100 °C for 50 minutes without acid and the reaction with procyanidin B1 occurs in the dark at room temperature after eight days, again without acid (Tanaka et al. 2008).

The PAC-DMAC reaction products have a stable purple color. When a fluorescence emission spectrum was acquired using a fluorimeter at the excitation wavelength of 559 nm, we discovered that the DMAC reaction products exhibit fluorescence at 613 nm. We also reacted four other cinnamaldehydes (cinnamaldehyde, 4-methylcinnamaldehyde, 4-hydroxy-3-methoxycinnamaldehyde, and 4-hydroxy-3,5-dimethoxycinnamaldehyde) with PACs in order to study the spectral properties of the reaction products in comparison to the DMAC products. Reaction products between these cinnamaldehydes and PACs produced unique compounds with different and useful absorption and emission wavelengths. For instance, the product of the reaction between (i) cinnamaldehyde with PACs exhibited a maximum absorption at 404 nm and a maximum emission at 473 nm, (ii) 4-methylcinnamaldehyde with PACs exhibited a maximum absorption at 416 nm and a maximum emission at 480 nm, (iii) 4-hydroxy-3-methoxycinnamaldehyde with PACs exhibited a maximum absorption at 443 nm and a maximum emission at 524 nm, and (iv) 4-hydroxy-3,5-dimethoxycinnamaldehyde with PAC exhibited a maximum absorption at 460 nm and a maximum emission at 536 nm.

11.5 Standards for the DMAC Reaction and Accuracy of the Method

The use of commercially available standards for the DMAC reaction to quantify PACs creates several problems related to the kinetics and stoichiometry of the reaction in relationship to PAC structure. Food ingredients have large differences in PAC structure and processing also affects PAC structure (Reed et al. 2005). For instance, the kinetics of reaction for potential standards such as catechin, procyanidin A2, and procyanidin B2, which can be purchased from chemical supply vendors, may be used to develop a standard curve between absorbance in the DMAC reaction and concentration of PACs in a dietary ingredient. Unfortunately, the reaction kinetics and the time required for the reaction to reach maximum absorbance (t_{max}) for each of these potential standards are substantially different from the PACs in food ingredients (Figure 11.11). Concentration has little effect on t_{max},

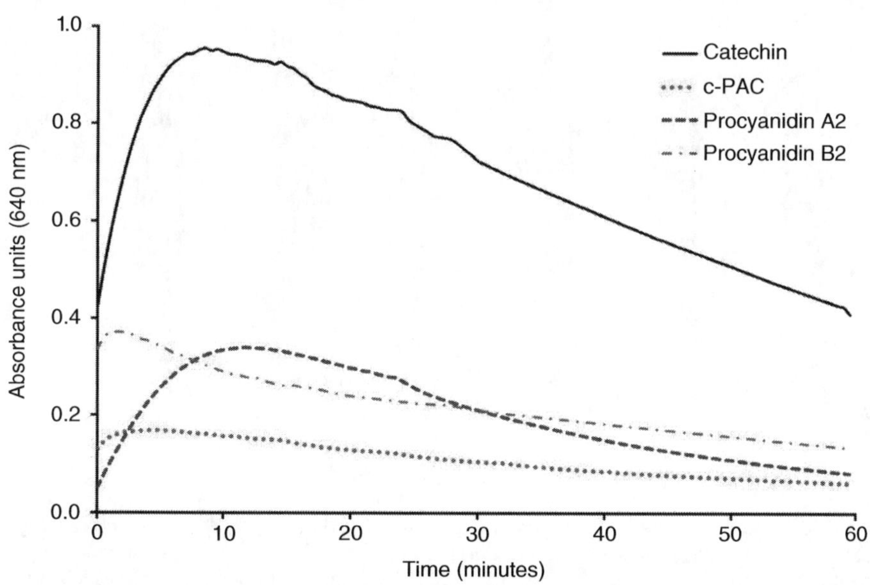

Figure 11.11 Absorbance at 640 nm over 60 minutes after reaction of 4-(dimethylamino) cinnamaldehyde with four different standards: catechin (1.07 μg), procyanidin A2 (1.09 μg), procyanidin B2 (1.06 μg), and c-PAC (1.14 μg). *Source:* The DMAC method is explained in Feliciano *et al.* (2012)/with permission of American Chemical Society.

but there is a greater than five-fold difference in t_{max} among the different standards. Procyanidin A2 had the highest and procyanidin B2 had the lowest t_{max} (Figure 11.12). The t_{max} for cranberry PACs was closer to procyanidin B2, although procyanidin A2 has been accepted as the most appropriate standard for cranberry PACs. Procyanidin B2, procyanidin A2, and cranberry PACs will generate a different standard curve depending on the time chosen for measuring absorbance and therefore give a different estimate of concentration in the sample under study. This problem may be solved by using the t_{max} absorbance value as shown in Figure 11.11. However, using the t_{max} absorbance values for catechin, procyanidin B2, procyanidin A2, and other commercially available standards does not solve the problem of accuracy that is created by the different reactivity between naturally occurring PACs and the standards (Feliciano et al. 2012; Birmingham et al. 2020).

The weight-absorbance relationship for standard curves generated at t_{max} for catechin, procyanidin A2, and procyanidin B2, and the cranberry PACs are substantially different, with a seven-fold difference in slopes (Figure 11.13) (Feliciano et al. 2012). The slopes for procyanidin dimers A2 and B2 were similar, although a subsequent publication by Wang et al. (2016) indicates that their slopes are different. The contradiction between these two publications may be related to differences in the methods and also illustrates the problem associated with the development of standards for the DMAC reaction. For instance, Feliciano et al. (2012) took absorbance reading at 640 nm every 30 seconds for one hour. Conversely, Wang et al. (2016) took only one reading at 640 nm at nine minutes. Vidal-Casanella et al. (2021) also demonstrated that the sensitivity for the procyanidin B2 and

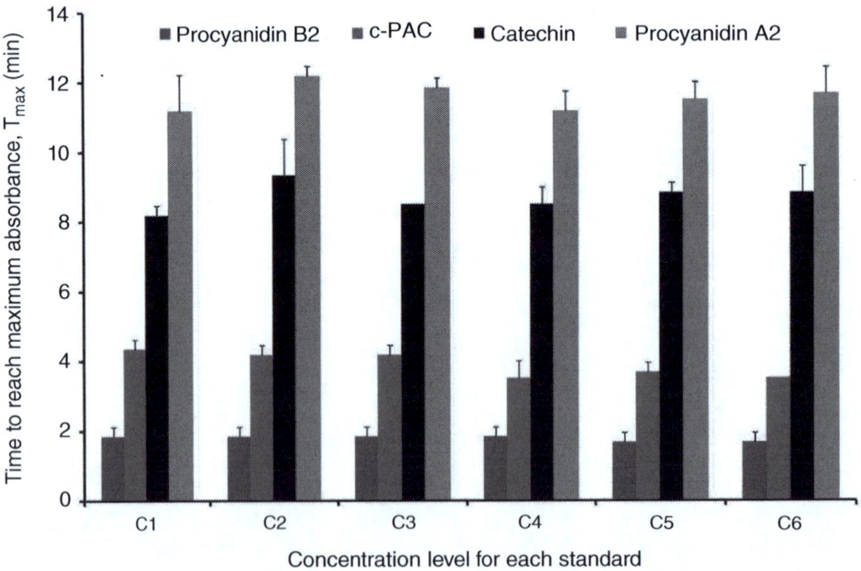

Figure 11.12 Variation of time required to reach maximum absorbance (t_{max}) using six increasing concentrations of catechin, procyanidin A2, procyanidin B2, and c-PAC. Stock solutions were diluted with methanol to produce six levels of concentrations (C1–C6): 2.6, 5.1, 7.7, 10.2, 12.8, and 15.3 mg/L for catechin; 4.8, 9.6, 14.4, 19.2, 24.0, and 28.8 mg/L for procyanidin A2; 5.0, 10.1, 15.1, 20.2, 25.2, and 30.2 mg/L for procyanidin B2; and 16.3, 32.6, 48.9, 65.2, 81.5, and 97.8 for c-PAC. (*See insert for color representation of the figure.*)

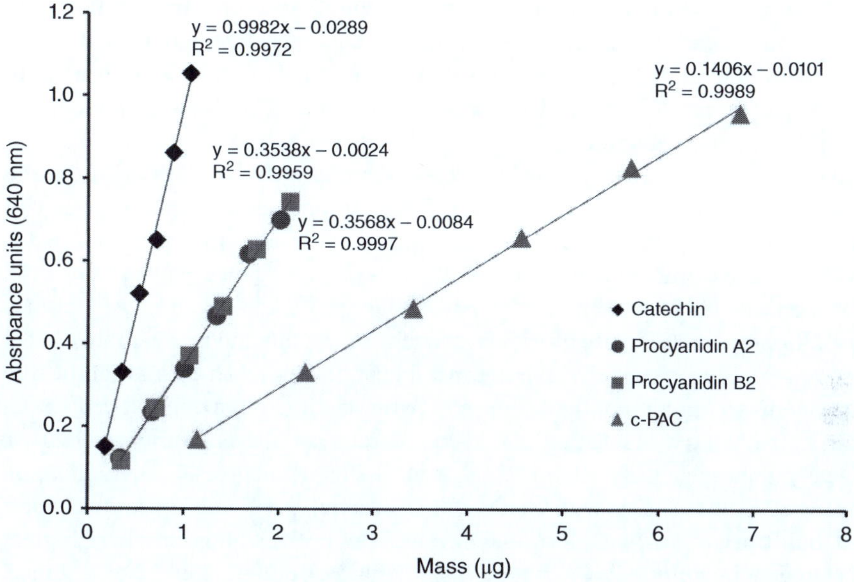

Figure 11.13 Calibration curves for catechin, procyanidin A2, procyanidin B2, and c-PAC after reaction with 4-(dimethylamino)cinnamaldehyde.

trimer C1, which contains only B-type interflavan bonds, is higher than procyanidin A2, which contains A-type interflavan bonds when the DMAC assay is used. These results suggest that the reactivity of PACs does not depend exclusively on the concentration but also on the type of interflavan bonds in the oligomers (Feliciano et al. 2012). If the standard curve for procyanidin A2 in Figure 11.13 is used to estimate the content of cranberry PACs in various cranberry products, then the amount of PAC is two to three-fold lower in comparison to using isolated cranberry PACs as the standard (Table 11.1). Unfortunately,

Table 11.1 Content of PACs determined using the 4-(dimethylamino) cinnamaldehyde method in cranberry powders and cranberry juices using procyanidin A2 and c-PAC as standards[a].

	Procyanidin A2	c-PAC
Powder 1	198.8±3.5	421.8±51.2
Powder 2	123.0±13.6	271.7±37.0
Powder 3	126.2±9.1	316.9±22.1
Juice 1	174.3±16.5	381.1±11.3
Juice 2	777.2±79.9	1716±107.6
Juice 3	95.5±9.1	207.4±3.1

[a] Results were expressed as the mean of three independent replicates and the standard deviation in mg/g for powders and mg/L for juices.

procyanidin A2 is the recommended standard for use in the analysis of cranberry PACs and the use of this standard leads to an underestimate of PAC content (Birmingham et al. 2020). This is a problem for the use of food ingredients that are enriched in PACs because an accurate measurement is essential to quantify composition for standardization, labeling, and formulation of dietary supplements.

In summary, our research on reaction kinetics shows that standards differ in time to reach maximum absorbance and using a fixed time point for measuring absorbance will lead to very poor precision and accuracy. We recommend using a microplate reader that is capable of kinetic measurements for obtaining the t_{max} absorbance values. However, although this approach will improve precision, it will not improve accuracy because accuracy is a function of the standard. The use of standards such as catechin and procyanidin dimers will lead to serious inaccuracy, which can be a three-fold or greater underestimate of the true value. This problem is a function of the degree of polymerization of the oligomeric PACs as related to the stoichiometry of reaction of the commercial standards (Wang et al. 2016). More specifically, the molar reactivity of PACs decreases as the degree of polymerization increases (Wang et al. 2016) because the main DMAC reaction product is presumed to be a single DMAC substitution on the C_8 terminal flavan of the PAC. In addition, there is great structural heterogeneity among PACs in food ingredients (Reed et al. 2005). For instance, in derived PACs, which occur in grapes and cranberries (Reed et al. 2005), anthocyanins form new covalent bonds to PACs that often involve the C_8 upper flavan-3-ol unit of the PAC (Salas et al. 2003). The formation of this bond would make the PAC much less reactive to DMAC because the C_8 upper flavan-3-ol is the putative reactive site. Cinnamon is a fascinating case because the PACs in cinnamon bark are already highly complexed with cinnamaldehyde, and so the DMAC reaction will not occur in PACs that have already undergone the condensation reaction with cinnamaldehyde, as would be the case for PACs from wines aged in oak barrels that have undergone reactions with sinapaldehyde and syringaldehyde. Therefore, our research indicates that better accuracy will be achieved by using PAC standards that are similar to those of the extract being studied (Birmingham et al. 2020). However, isolation of such extracts is not an easy task and may also introduce other problems such as contamination of the standard with non-PAC components, modification of the degree of polymerization due to loss of lower or higher molecular weight species in the isolation procedure, and structural modifications that occur during the extraction and purification of the standard. Our research on the development of a cranberry PAC standard used both qualitative and quantitative analyses of the standard to ensure that it was not contaminated with other components (Feliciano et al. 2012; Birmingham et al. 2020).

11.6 Interaction of PAC-DMAC Reaction Products with Extra-Intestinal Pathogenic *Escherichia coli*

The intake of cranberry products is associated with the prevention of urinary tract infections because cranberry PACs prevent the adhesion of extra-intestinal pathogenic *Escherichia coli* (ExPEC) to uroepithelial cells (Howell et al. 1998). Atomic force microscopy measures nanoscale adhesion forces between fimbriae-like structures of *E. coli* and uroepithelial cells and this technique was used to demonstrate that cranberry juice disrupts the binding of bacterial ligands to uroepithelial cell receptors (Liu et al. 2010). PACs from cranberries and apples, containing A- and B-type interflavan bonds, respectively, agglutinate and prevent ExPEC invasion of gut epithelial cells (Feliciano et al. 2014). The interaction between cranberry PACs and ExPEC was studied using PACs that were tagged with 5-([4,6-dichlorotriazin-2-yl]amino (DTAF) fluorescein, allowing the visualization of PAC-DTAF by fluorescent microscopy (Feliciano et al. 2015b). By using PAC-DTAF, it was possible to demonstrate that dendritic cells recognize and actively internalize PAC complexed with helminth antigens, suggesting that a combination therapy with helminth products and PAC may be a novel future avenue for autoimmune treatments (Williams et al. 2017). However, autofluorescence at the excitation wavelength of fluorescein (494 nm) produces difficulties in distinguishing signal from noise in fluorescent microscopy when PAC-DTAF is used. The PAC-DMAC reaction products produce fluorescence that is similar to Texas Red; a useful wavelength for fluorescent microscopy because the excitation and emission filters are available on most fluorescent microscopes. Therefore, we used the fluorescent properties of the PAC-DMAC reaction products to study the effects of PACs on ExPEC. Fluorescent and differential interference contrast (DIC) microscopy were used to visualize the complexation of ExPEC strain CFT073, which was modified to express a green fluorescent protein (GFP) by the PAC-DMAC reaction products (Esquivel-Alvarado et al. 2021c). Three treatments were visualized by fluorescent microscopy of the bacteria. The first treatment was ExPEC alone, the second was the PAC-DMAC reaction products alone, and the third treatment was the PAC-DMAC reaction product with the ExPEC. Fluorescent micrographs of the first treatment showed that the ExPEC were dispersed across the field of view and visualized using the GFP filter and DIC microscope but were not visualized using the Texas Red filter. Fluorescent micrographs of the second treatment showed that the DMAC reaction products with PAC were dispersed across the field of view and visualized using the Texas Red filter and DIC microscope but were not detected using the GFP filter. Finally, fluorescent micrographs of the third treatment showed that the DMAC reaction products with PAC and the ExPEC were overlapping and visualized using the GFP and Texas Red filters and DIC microscope. These results indicate that PAC-DMAC reaction products may have a greater ability to agglutinate ExPEC than untagged PACs (Feliciano et al. 2015b; Esquivel-Alvarado et al. 2021c).

11.7 Conclusion

Although the DMAC method overcomes some of the challenges when compared to previous methods such as the vanillin-HCl or the butanol-HCl, the DMAC method still has limitations. For instance, results from the DMAC method are affected by both DMAC reaction conditions and the source of PACs. Additionally, the use of commercially available standards for the DMAC reaction to quantify PACs creates several problems related to the kinetics of the reaction in relationship to PAC structure. For instance, we demonstrated that catechin, procyanidin A2, procyanidin B2 and PACs differ in the kinetics of chromophore formation so that the time at which absorbance is measured will influence the slope of the calibration curve. In addition, the use of catechin or procyanidin dimers as standards will grossly underestimate PAC concentration, which suggests that accurate results may only be obtained by using PACs that have been isolated from the plant material of interest as standards for the DMAC reaction. Our group also found that the isolated PAC-DMAC reaction products are stable with an absorbance max at 559 nm and these chromophores may not be the reaction products that are typically measured at 640 nm in the DMAC method. Our group also discovered that the DMAC reaction products with PAC exhibited fluorescence at useful wavelengths for fluorescent microscopy. More research is required to determine the structure of stable DMAC-PAC reaction products, especially in regard to their use for quantifying PACs by means of fluorescence spectrophotometry. In addition, more research on the use of the fluorescent properties of PAC-DMAC reaction products in studies on PAC bioactivity, such as the interaction of PACs with bacteria, mammalian cells, and other tissues may improve our understanding of PAC in dietary supplements consumed for health-related benefits.

References

Alfaro-Viquez, E., Esquivel-Alvarado, D., Madrigal-Carballo, S., et al. (2018). Cranberry proanthocyanidin-chitosan hybrid nanoparticles as a potential inhibitor of extra-intestinal pathogenic *Escherichia coli* invasion of gut epithelial cells. *International Journal of Biological Macromolecules*, **111**: 415–420.

Alfaro-Viquez, E., Esquivel-Alvarado, D., Madrigal-Carballo, S., et al. (2019). Proanthocyanidin-chitosan composite nanoparticles prevent bacterial invasion and colonization of gut epithelial cells by extra-intestinal pathogenic *Escherichia coli*. *International Journal of Biological Macromolecules*, **135**: 630–636.

Anderson, R., Broadhurst, C., Polansky, M., et al. (2004). Isolation and characterization of polyphenol type-A polymers from cinnamon with insulin-like biological activity. *Journal of Agricultural and Food Chemistry*, **52**: 65–70.

Birmingham, A., Esquivel-Alvarado, D., Maranan, M., et al. (2020). Inter-laboratory validation of 4-(dimethylamino)cinnamaldehyde (DMAC) assay using cranberry proanthocyanidin standard for quantification of soluble proanthocyanidins in cranberry foods and dietary supplements, first action official method SM: 2019.06. *Journal of AOAC International*, **104**: 216–222.

Burleigh, A., Benck, S., Mcachran, S., et al. (2013). Consumption of sweetened, dried cranberries may reduce urinary tract infection incidence in susceptible women–a modified observational study. *Nutrition Journal*, **12**: 1–7.

Cheynier, V. (2005). Polyphenols in foods are more complex than often thought. *The American Journal of Clinical Nutrition*, **81**: 223S–229S.

De Freitas, V., Sousa, C., Silva, A., et al. (2004). Synthesis of a new catechin-pyrylium derived pigment. *Tetrahedron Letters*, **45**: 9349–9352.

Esquivel-Alvarado, D., Muñoz-Arrieta, R., Alfaro-Viquez, E., et al. (2019). Composition of anthocyanins and proanthocyanidins in three tropical *Vaccinium* species from Costa Rica. *Journal of Agricultural and Food Chemistry*, **68**: 2872–2879.

Esquivel-Alvarado, D., Alfaro-Viquez, E., Krueger, C., et al. (2021a). Classification of proanthocyanidin profiles using matrix-assisted laser desorption/ionization time-of-flight mass spectrometry (MALDI-TOF MS) spectra data combined with multivariate analysis. *Food Chemistry*, **336**: 127667.

Esquivel-Alvarado, D., Alfaro-Viquez, E., Krueger, C., et al. (2021b). Identification of A-type proanthocyanidins in cranberry-based foods and dietary supplements by matrix-assisted laser desorption/ionization time-of-flight mass spectrometry, First Action Method: 2019.05. *Journal of AOAC International*, **104**: 223–231.

Esquivel-Alvarado, D., Alfaro-Viquez, E., Polewski, M., et al. (2021c). Synthesis of fluorescent proanthocyanidin-cinnamaldehydes pyrylium products for microscopic detection of interactions with extra-intestinal pathogenic *Escherichia coli*. *Journal of Agricultural and Food Chemistry*, **69**: 10700–10708.

Feliciano, R., Shea, M., Shanmuganayagam, D., et al. (2012). Comparison of isolated cranberry (*Vaccinium macrocarpon* Ait.) proanthocyanidins to catechin and procyanidins A2 and B2 for use as standards in the 4-(dimethylamino) cinnamaldehyde assay. *Journal of Agricultural and Food Chemistry*, **60**: 4578–4585.

Feliciano, R., Meudt, J., Shanmuganayagam, D., et al. (2014). Ratio of "A-type" to "B-type" proanthocyanidin interflavan bonds affects extra-intestinal pathogenic *Escherichia coli* invasion of gut epithelial cells. *Journal of Agricultural and Food Chemistry*, **62**: 3919–3925.

Feliciano, R., Krueger, C., and Reed, J. (2015a). Methods to determine effects of cranberry proanthocyanidins on extraintestinal infections: relevance for urinary tract health. *Molecular Nutrition & Food Research*, **59**: 1292–1306.

Feliciano, R.P., Heintz, J., Krueger, C., et al. (2015b). Fluorescent labeling of cranberry proanthocyanidins with 5-([4, 6-dichlorotriazin-2-yl] amino) fluorescein (DTAF). *Food Chemistry*, **166**: 337–345.

Foo, Y., Lu, Y., Howell, A., and Vorsa, N. (2000). A-Type proanthocyanidin trimers from cranberry that inhibit adherence of uropathogenic p-fimbriated *Escherichia coli*. *Journal of Natural Products*, **63**: 1225–1228.

Głąbska, D., Guzek, D., Gałązka, K., and Lech, G. (2020). Therapeutic potential of proanthocyanidins in ulcerative colitis in remission. *Journal of Clinical Medicine*, **9**: 771.

Herald, T., Gadgil, P., Perumal, R., et al. (2014). High-throughput micro-plate HCl–vanillin assay for screening tannin content in sorghum grain. *Journal of the Science of Food and Agriculture*, **94**: 2133–2136.

Howell, A., Vorsa, N., Marderosian, A., and Foo, L. (1998). Inhibition of the adherence of P-fimbriated *Escherichia coli* to uroepithelial-cell surfaces by proanthocyanidin extracts from cranberries. *New England Journal of Medicine*, **339**: 1085–1086.

Hümmer, W. and Schreier, P. (2008). Analysis of proanthocyanidins. *Molecular Nutrition & Food Research*, **52**: 1381–1398.

Jepson, R. and Craig, J. (2007). A systematic review of the evidence for cranberries and blueberries in UTI prevention. *Molecular Nutrition & Food Research*, **51**: 738–745.

Jepson, R. and Craig, J. (2008). Cranberries for preventing urinary tract infections. *Cochrane Database of Systematic Reviews*, **10** (10): Cd001321.

Jepson, R.G., Williams, G., and Craig, J.C. (2012). Cranberries for preventing urinary tract infections. *Cochrane Database of Systematic Reviews*, **10** (10): CD001321.

Li, Y.G., Tanner, G., and Larkin, P. (1996). The DMACA–HCl protocol and the threshold proanthocyanidin content for bloat safety in forage legumes. *Journal of the Science of Food and Agriculture*, **70**: 89–101.

Liu, Y., Pinzón-Arango, P., Gallardo-Moreno, A., and Camesano, T. (2010). Direct adhesion force measurements between E. coli and human uroepithelial cells in cranberry juice cocktail. *Molecular Nutrition & Food Research*, **54**: 1744–1752.

Makkar, H. and Becker, K. (1993). Vanillin-HCl method for condensed tannins: effect of organic solvents used for extraction of tannins. *Journal of Chemical Ecology*, **19**: 613–621.

Mitsunaga, T., Doi, T., Kondo, Y., and Abe, I. (1998). Color development of proanthocyanidins in vanillin-hydrochloric acid reaction. *Journal of Wood Science*, **44**: 125–130.

Muñoz-Arrieta, R., Esquivel-Alvarado, D., Alfaro-Viquez, E., et al. (2021a). Nutritional and bioactive composition of Spanish, Valencia, and Virginia type peanut skins. *Journal of Food Composition and Analysis*, **98**: 103816.

Muñoz-Arrieta, R., Esquivel-Alvarado, D., Alfaro-Viquez, E., et al. (2021b). Nutritional and chemical composition of the Costa Rican guava (*Psidium friedrichsthalianum* [O. Berg] Nied): an underexploited edible fruit with nutritional and industrial potential. *ACS Food Science & Technology*, **1**: 1970–1978.

Nawrot-Hadzik, I., Matkowski, A., Kubasiewicz-Ross, P., and Hadzik, J. (2021). Proanthocyanidins and flavan-3-ols in the prevention and treatment of periodontitis—immunomodulatory effects, animal and clinical Studies. *Nutrients*, **13**: 239.

Payne, M., Hurst, W., Stuart, D., et al. (2010). Determination of total procyanidins in selected chocolate and confectionery products using DMAC. *Journal of AOAC International*, **93**: 89–96.

Price, M., Van Scoyoc, S., and Butler, L. (1978). A critical evaluation of the vanillin reaction as an assay for tannin in sorghum grain. *Journal of Agricultural and Food Chemistry*, **26**: 1214–1218.

Prior, R., Fan, E., Ji, H., et al. (2010). Multi-laboratory validation of a standard method for quantifying proanthocyanidins in cranberry powders. *Journal of the Science of Food and Agriculture*, **90**: 1473–1478.

Reed, J., Krueger, C., and Vestling, M. (2005). MALDI-TOF mass spectrometry of oligomeric food polyphenols. *Phytochemistry*, **66**: 2248–2263.

Sahebkar, A. (2014). A systematic review and meta-analysis of the effects of pycnogenol on plasma lipids. *Journal of Cardiovascular Pharmacology and Therapeutics*, **19**: 244–255.

Saito, M., Hosoyama, H., Ariga, T., et al. (1998). Antiulcer activity of grape seed extract and procyanidins. *Journal of Agricultural and Food Chemistry*, **46**: 1460–1464.

Salas, E., Fulcrand, H., Meudec, E., and Cheynier, V. (2003). Reactions of anthocyanins and tannins in model solutions. *Journal of Agricultural and Food Chemistry*, **51**: 7951–7961.

Schoonees, A., Visser, J., Musekiwa, A., and Volmink, J. (2012). Pycnogenol® (extract of French maritime pine bark) for the treatment of chronic disorders. *Cochrane Database of Systematic Reviews*.

Sintara, M., Li, L., Cunningham, D., et al. (2018). Single-laboratory validation for determination of total soluble proanthocyanidins in cranberry using 4-dimethylaminocinnamaldehyde. *Journal of AOAC International*, **101**: 805–809.

Smith, T., Lynch, M.E., Johnson, J., et al. (2015). Herbal dietary supplement sales in US increase 6.8% in 2014. *HerbalGram*, **107**: 52–59.

Smith, T., May, G., Eckl, V., and Reynolds, C. (2020). US sales of herbal supplements increase by 8.6% in 2019. *HerbalGram*, **127**: 54–69.

Sousa, C., Mateus, N., Perez-Alonso, J., et al. (2005). Preliminary study of oaklins, a new class of brick-red catechinpyrylium pigments resulting from the reaction between catechin and wood aldehydes. *Journal of Agricultural and Food Chemistry*, **53**: 9249–9256.

Tanaka, T., Matsuo, Y., Yamada, Y., and Kouno, I. (2008). Structure of polymeric polyphenols of cinnamon bark deduced from condensation products of cinnamaldehyde with catechin and procyanidins. *Journal of Agricultural and Food Chemistry*, **56**: 5864–5870.

Thies, M. and Fischer, R. (1971). Über eine neue farbreaktion zum mikrochemischen nachweis und zur quantitativen bestimmung von catechinen. *Microchimica Acta*, **59**: 9–13.

Treutter, D. (1989). Chemical reaction detection of catechins and proanthocyanidins with 4-dimethylaminocinnamaldehyde. *Journal of Chromatography A*, **467**: 185–193.

Vidal-Casanella, O., Nuñez, O., Hernández-Cassou, S., and Saurina, J. (2021). Assessment of experimental factors affecting the sensitivity and selectivity of the spectrophotometric estimation of proanthocyanidins in foods and nutraceuticals. *Food Analytical Methods*, **14**: 485–495.

Wallace, T. (2010). Analysis of procyanidins and anthocyanins in food products using chromatographic and spectroscopic techniques. The Ohio State University.

Wallace, T. and Giusti, M. (2010). Evaluation of parameters that affect the 4-dimethylaminocinnamaldehyde assay for flavanols and proanthocyanidins. *Journal of Food Science*, **75**: C619–C625.

Wang, Y., Singh, A., Hurst, W., et al. (2016). Influence of degree-of-polymerization and linkage on the quantification of proanthocyanidins using 4-dimethylaminocinnamaldehyde (DMAC) assay. *Journal of Agricultural and Food Chemistry*, **64**: 2190–2199.

Williams, A., Klaver, E., Laan, L., et al. (2017). Co-operative suppression of inflammatory responses in human dendritic cells by plant proanthocyanidins and products from the parasitic nematode Trichuris suis. *Immunology*, **150**: 312–328.

Zhang, H., Liu, S., Li, L., et al. (2016). The impact of grape seed extract treatment on blood pressure changes: a meta-analysis of 16 randomized controlled trials. *Medicine*, **95**: e4247.

12

Reactions of Ellagitannins Related to Their Metabolism in Higher Plants

Takashi Tanaka

Graduate School of Biomedical Sciences, Nagasaki University, Nagasaki, Japan

12.1 Introduction

Ellagitannins are hydrolyzable tannins that produce ellagic acid following hydrolysis of their hexahydroxydiphenoyl (HHDP) groups (Figure 12.1). Therefore, the HHDP moiety is the most representative acyl group characterizing ellagitannins (Haslam 1982; Okuda et al. 1995; Okuda 2005; Niemetz and Gross 2005; Quideau et al. 2011). In addition, hydrolyzable tannins containing acyl groups that are biogenetically related to HHDP have been isolated from higher plants, and these tannins are also sometimes referred to as ellagitannins even if they do not yield ellagic acid upon hydrolysis. Ellagitannins are derived from the oxidative metabolism of galloyl glucose. Different regio- and stereoselective couplings between two galloyl (3,4,5-trihydroxybenzoyl) groups bound to a glucopyranose core can generate numerous isomers, so the structural diversity has attracted considerable attention from organic chemists. Pioneering work involving ellagitannin chemistry began in the early twentieth century (Freudenberg 1919), and structures of ellagitannins found in commercially important plant extracts were determined by Schmidt and Mayer in the 1960s (Schmidt et al. 1965; Mayer et al. 1967; Schmidt et al. 1967; Mayer et al. 1969). Since then, many reports describing structural features (Okuda et al. 1995), biological activities (Yoshida et al. 2010; Okuda and Ito 2011), gut microbiota metabolism (Larrosa et al. 2010; Tomas-Barberan et al. 2017), ecological significance (Hagerman 2012; Salminen 2014), and total syntheses of ellagitannins (Feldman 2005; Quideau et al. 2011; Hirokane et al. 2014; Ashibe et al. 2017; Richieu et al. 2017; Yamada et al. 2017; Shibayama et al. 2021) have been published. However, the detailed chemical mechanisms involved in the biosynthesis and degradation of ellagitannins in higher plants, which give rise to the vast structural diversity, still require clarification. A chemotaxonomical overview of ellagitannin structures showed that taxonomically related plants tend to contain structurally related ellagitannins,

---

This chapter is dedicated to the deceased Professor Hidetoshi Yamada.

Recent Advances in Polyphenol Research, Volume 8, First Edition. Edited by Juha-Pekka Salminen, Kristiina Wähälä, Victor de Freitas, and Stéphane Quideau.

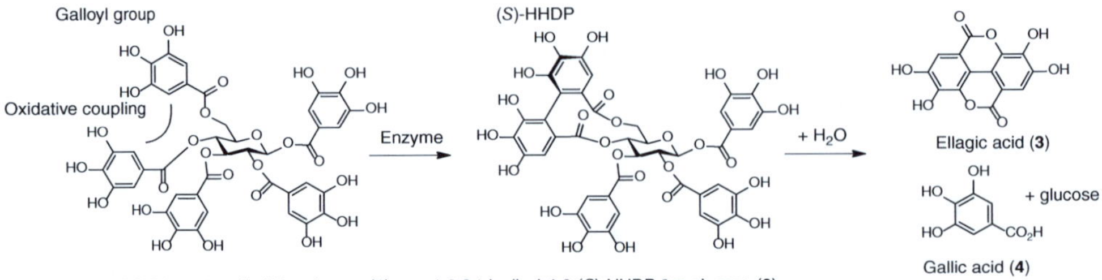

Figure 12.1 Enzymatic production of an ellagitannin (**2**) from pentagalloyl glucose (**1**) (Niemetz et al. 2001), and generation of ellagic acid (**3**) by hydrolysis. *Source:* Adapted from Niemetz et al. (2001).

indicating that regioselective couplings of galloyl groups in ellagitannin biosynthesis are precisely controlled in each plant species (Haddock et al. 1982a, 1982b; Okuda et al. 1992, 1993). Therefore, intramolecular coupling of galloyl groups within galloyl glucose molecules under ordinary physiological conditions must be carried out enzymatically. Niemetz et al. characterized the enzyme capable of converting penta-*O*-galloyl-β-D-glucose to a 4,6-(*S*)-HHDP analog (Figure 12.1), and they also described the production of a naturally occurring dimeric ellagitannin (Niemetz and Gross 2003a, 2003b; Niemetz et al. 2001, 2003). On the other hand, studies on ellagitannin total synthesis using unique reagents and protecting groups in organic solvents have provided important information regarding the regio- and stereochemical nature of biphenyl bond formation (Khanbabaee and Lötzerich 1998; Feldman 2005; Ashibe et al. 2017; Yamada et al. 2017; Tanaka et al. 2011). However, there is currently limited understanding of the mechanisms governing the species-specific, regio-, and stereoselective couplings of galloyl groups or the metabolism of HHDP and dehydrohexahydroxydiphenoyl (DHHDP), which is an oxidized form of HHDP, in higher plants. This chapter discusses the structures and reactions of ellagitannin acyl groups, focusing primarily on the non-enzymatic redox disproportionation of DHHDP. Various structural alterations of ellagitannins, which occur during the seasonal changes of ellagitannin composition in plant leaves, are also introduced.

12.2 Structural Variety of Ellagitannin Acyl Groups

The structural diversity of ellagitannins primarily originates from the acyl groups and their location on the glucose core, although intermolecular couplings, which afford oligomeric molecules, are also important (Tanaka et al. 1996b; Okuda 2005; Yoshida et al. 2005). The most typical acyl groups are HHDP and DHHDP, and the HHDP moiety on the glucopyranose in reported ellagitannins is located at the 1,6- (Haddock et al. 1982a, 1982b; Lee et al. 1990; Nonaka et al. 1992), 2,3- (Schmidt et al. 1965), 2,4- (Zhang et al. 2001), 3,4- (Tanaka et al. 2001), 3,6- (Jochims et al. 1968), or 4,6-positions (Schmidt et al. 1965). In contrast, the DHHDP moiety is typically located at the 1,3- (Schmidt et al. 1967a, 1967b;

Lee et al. 1991), 1,4- (Orejola et al. 2020), 2,4- (Okuda et al. 1982b), 3,6- (Okuda et al. 1982a; Foo 1993), or 4,6-positions (Schmidt et al. 1967b; Nonaka et al. 1981) of D-glucopyranose. There are biogenetically related acyl groups that are found in nature, in addition to the HHDP and DHHDP groups; the carboxylic acid forms of these acyl groups are shown in Figure 12.2. A comparison between the molecular formula of each analog with that of the HHDP acid is shown in parentheses under each structure. These details suggest differences in the oxidation state of the acyl groups relative to HHDP acid by denoting their H-atom deficiencies.

Chebulic acid (the 2,4-acyl group of chebulagic acid) (Yoshida et al. 1980; 1982) and the acyl groups of penthorumnin A (Era et al. 2019) and mallotusinin (Saijo et al. 1989a) are in the same oxidation state as the HHDP acid. The oxidation states of the acyl groups of repandusinic acid A (Saijo et al. 1989b), acalyphidin M2 (Amakura et al. 1999), mallotinin (Saijo et al. 1989b), geraniinic acids B and C (Ito et al. 1999), phyllanthusiins A and B (Yoshida et al. 1992), phyllanemblinin C (Zhang et al. 2001) and amariinic acid (Foo 1995) are the same as that of DHHDP acid. Phyllanthusiin E (Yoshida et al. 1992), brevifolin carboxylic acid (Schmidt and Eckert 1958), and the acyl group of phyllanthusiin C (Yoshida et al. 1992) represent decarboxylated products, and their oxidation states are also considered to be the same as that of DHHDP acid. In contrast, hydrated biscyclohexenetrione dicarboxylic (HBCHT) acid (the acyl group of carpinins A) (Nonaka et al. 1990), and the acyl groups of penthorumnin C (Era et al. 2019), euphormisin M2 (Yoshida et al. 1994), and pelargoniin E (Latte et al. 2008) are in higher oxidation states compared with the DHHDP acid.

12.3 Reactions of the DHHDP Group

Some of the acyl groups in Figure 12.2 can be chemically derived from DHHDP (Figure 12.3). For example, subjecting geraniin (**5**) to weakly alkaline conditions yields repandusinic acid A (Figure 12.3a). In contrast, treating **5** with triethyl amine in CH_3CN generates a product with a brevifolin carboxylic moiety (Figure 12.3b) (Tanaka et al. 1990), which was also detected in the hydrolysis products of **5** (Okuda et al. 1977). Interestingly, heating **5** in the presence of pyridine in CH_3CN affords reduction products, mallotusinin and 1-*O*-galloyl-2,4:3,6-bis-HHDP glucose (Figure 12.3c) (Tanaka et al. 1990; Era et al. 2020). Furthermore, the reaction between **5** and cysteine methyl ester yields an adduct, **5a**, which can be subsequently reduced with Raney Ni to afford chebulagic acid (Figure 12.3d) (Tanaka et al. 1996a). Although Schmidt et al. proposed that the chebuloyl group of chebulagic acid is derived directly from the HHDP group via hydrolytic ring fission of the aromatic ring (Schmidt et al. 1967), it can also be produced in the reductive metabolism of DHHDP (Yoshida et al. 1982). These chemical conversions of DHHDP to produce naturally occurring ellagitannin acyl groups suggest that the ellagitannin acyl groups depicted in Figure 12.2 can be generated from DHHDP and/or HHDP.

As described earlier, the DHHDP group of **5** was reduced after heating with pyridine in CH_3CN, and this reaction produced a mixture of tetrahydroxydibenzofuran dicarboxyl and (*R*)-HHDP esters (Tanaka et al. 1990; Era et al. 2020). A similar reduction of DHHDP

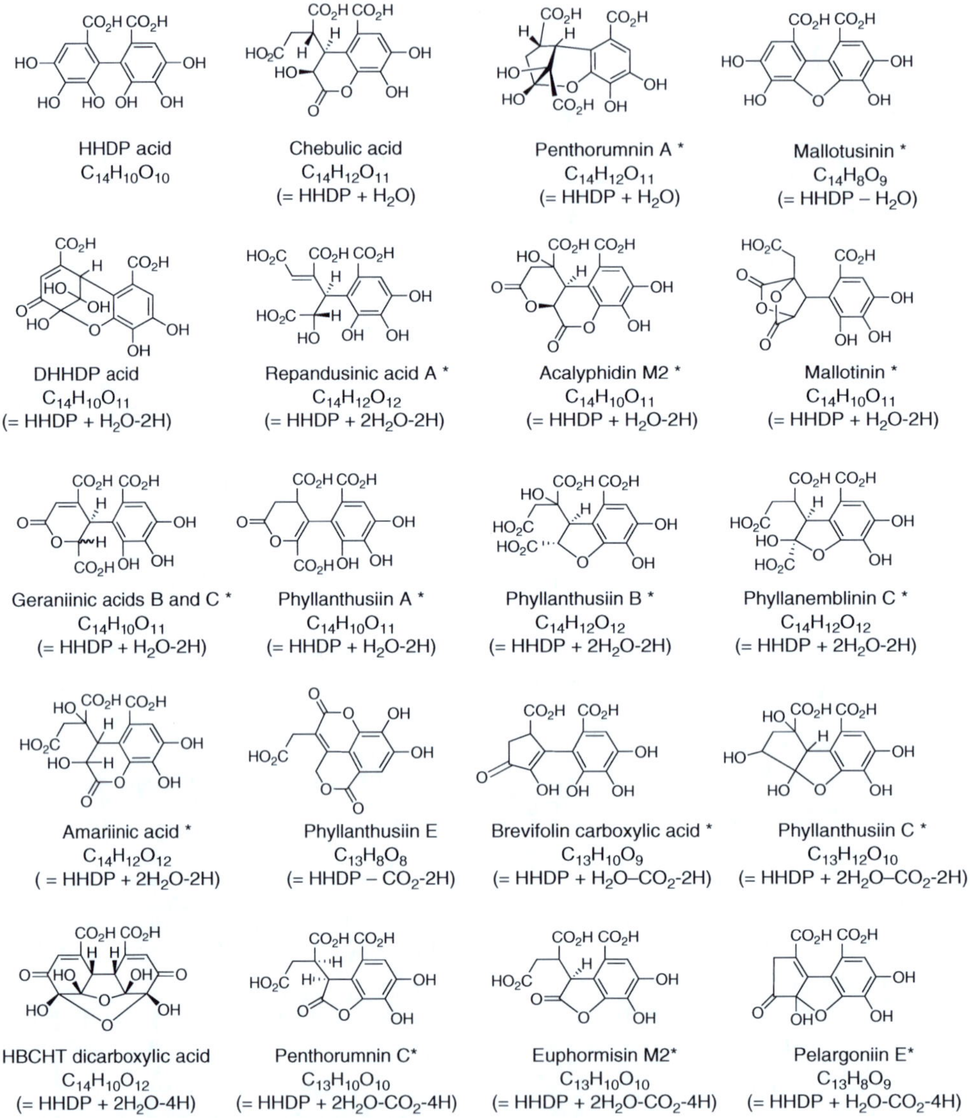

Figure 12.2 Carboxylic acid forms of ellagitannin acyl groups. Comparisons between their molecular formulae and that of HHDP acid are given in parentheses. A name with a "*" indicates that the carboxylic acid moiety represents the characteristic acyl group of the ellagitannin.

to HHDP was first reported by Schmidt et al. in 1967. In that study, hydrolysis of 1,3:4,6-bis-DHHDP glucose (brevilagin I) in hot water afforded ellagic acid (**3**) and brevifolin carboxylic acid (**6**). The authors speculated that the reaction involved a redox-disproportionation, where an enediol compound derived by decarboxylation of DHHDP acts as the reductant for converting DHHDP to HHDP (Figure 12.4) (Schmidt et al. 1967b; Schmidt et al. 1967c). Meanwhile, this enediol compound is a precursor of **6**.

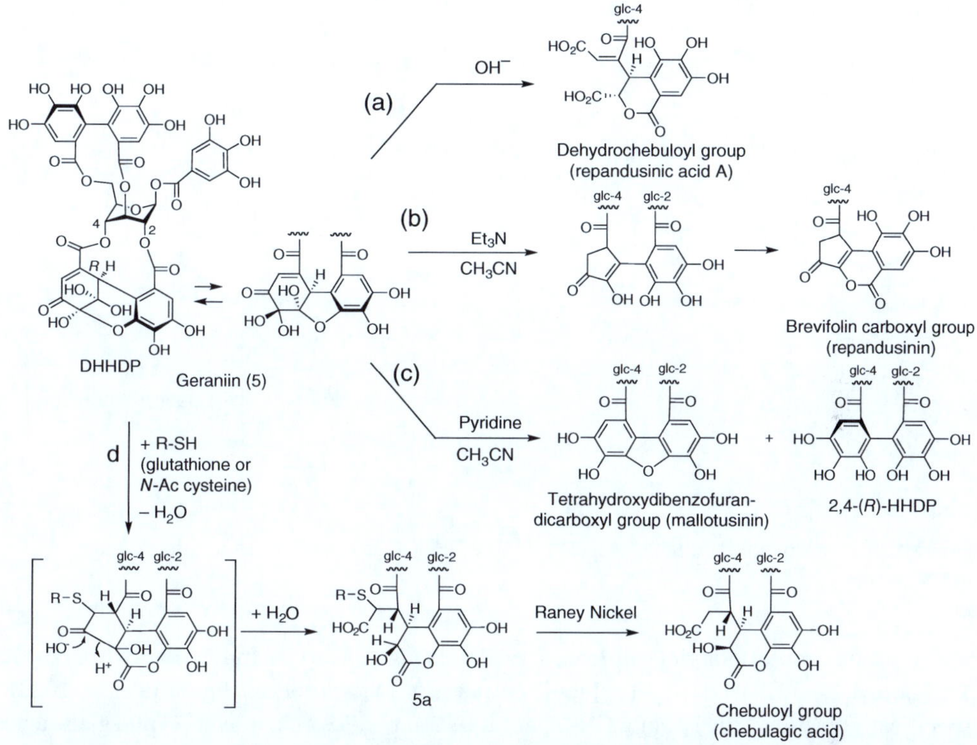

Figure 12.3 Chemical conversions of geraniin (**5**) to produce naturally occurring ellagitannins.

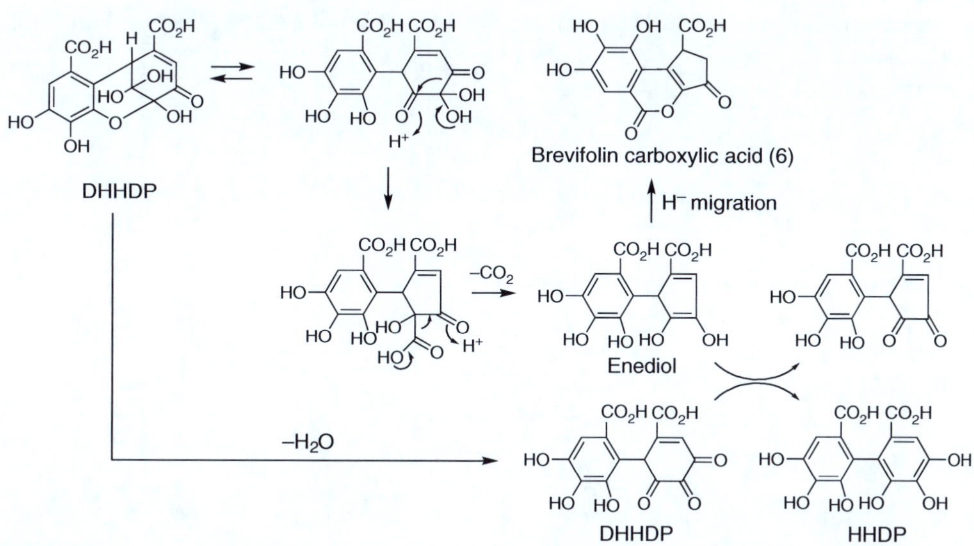

Figure 12.4 Mechanism of DHHDP redox disproportionation. *Source:* Adapted from Schmidt et al. (1967).

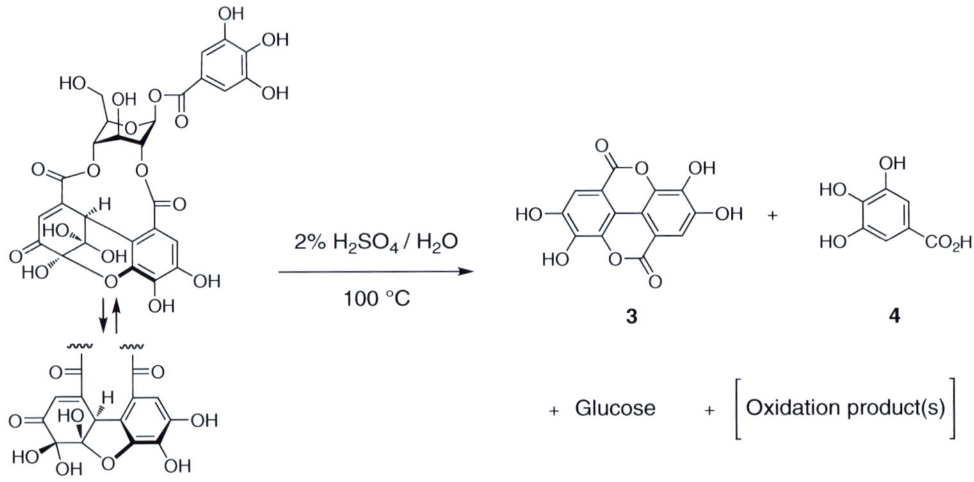

Figure 12.5 Production of ellagic acid (**3**) from the DHHDP group of furosin (**7**) upon acid hydrolysis.

The production of ellagic acid from DHHDP is also observed following hydrolysis of 1-*O*-galloyl-2,4-(*R*)-DHHDP-β-D-glucose (furosin, **7**) (Yazaki et al. 1989; Era et al. 2020); specifically, heating **7** in H_2O or 2% H_2SO_4 at 100 °C generates ellagic acid (**3**) and gallic acid (**4**) (Figure 12.5). Brevifolin carboxylic acid (**6**) was not detected in the reaction mixture after these experiments, and a limited amount of **6** was observed when **7** was heated in pH 4 or pH 5 phosphate-citrate buffer. This result is consistent with the mechanism proposed by Schmidt et al. (Figure 12.4). In addition, another potential mechanism describing the production of **3** from DHHDP also can be constructed (Figure 12.6). This mechanism is

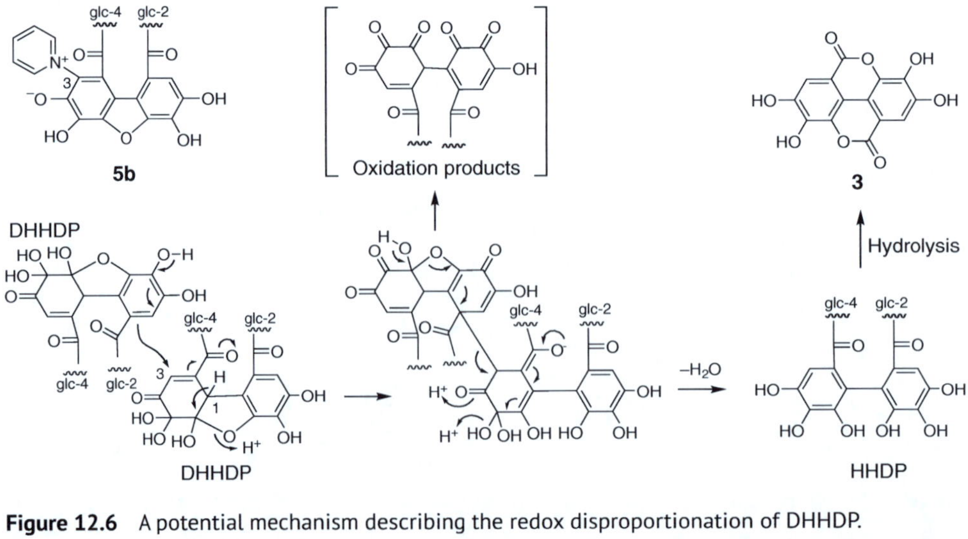

Figure 12.6 A potential mechanism describing the redox disproportionation of DHHDP.

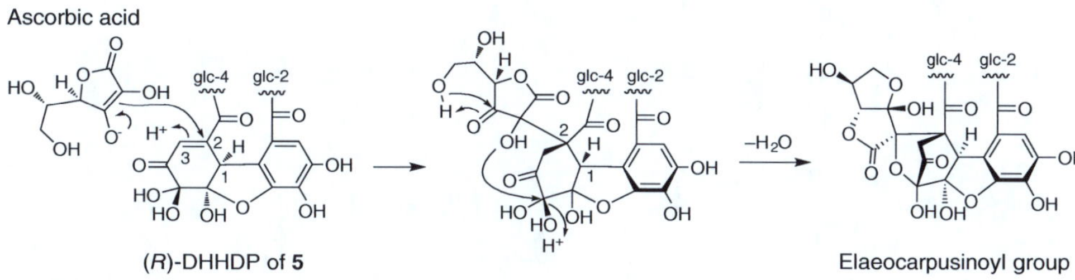

Figure 12.7 A potential mechanism for the reaction of ascorbic acid with an (*R*)-DHHDP group.

supported by the production of a minor product, **5b** (Figure 12.6), along with the reduction products generated when **5** is heated with pyridine (Figure 12.3c) (Era et al. 2020). As illustrated in Figure 12.6, intermolecular electron transfer from a DHHDP pyrogallol ring to an electron-deficient cyclohexenetrione ring of another molecule's DHHDP group leads to covalent bond formation at C-3 of the DHHDP group. Subsequent elimination converts the DHHDP cyclohexenetrione ring to a pyrogallol ring. Efforts are currently underway to identify the missing oxidation product(s) that are generated simultaneously in this reaction. The presented reaction mechanism may be related to the addition of ascorbic acid to the (*R*)-DHHDP group of geraniin (**5**). This addition occurs at C-2 of the DHHDP moiety to give an ascorbic acid adduct known as an elaeocarpusinoyl group (Figure 12.7) (Okuda et al. 1986; Tanaka et al. 1986).

12.4 Decomposition of 1,4-DHHDP-α-D-glucose

Recently, the first ellagitannin with a DHHDP group at the 1,4-positions of α-D-glucose was isolated from the rhizome of *Nuphar japonicum* (Nymphaeaceae) and given the trivial name nupharanin (**8**) (Orejola et al. 2020). This tannin gradually decomposes in a neutral aqueous solution to yield a product with a unique acyl group (**8a**). A potential reaction mechanism involving oxidative ring fission is proposed in Figure 12.8.

12.5 Amariin as a Precursor of Geraniin

Amariin (**9**) was first isolated from *Phyllanthus amarus* (Euphorbiaceae) by Foo in 1993 and characterized as 1-*O*-galloyl-2,4:3,6-bis-(*R*)-DHHDP-β-D-glucose (Figure 12.9) (Foo 1993). They demonstrated that heating **8** in aqueous EtOH generated the reduction products, geraniin (**5**), ellagic acid (**3**), gallic acid (**4**), and brevifolin carboxylic acid (**6**). The authors also highlighted the fact that the characteristic reaction of **9** was similar to that of the partially characterized ellagitannin, isogeraniin, which was previously reported by Haddock and co-workers (Haddock et al. 1982a, 1982b). Geraniin (**5**) is the most common DHHDP-bearing ellagitannin (dehydroellagitannin), and it is found in plants

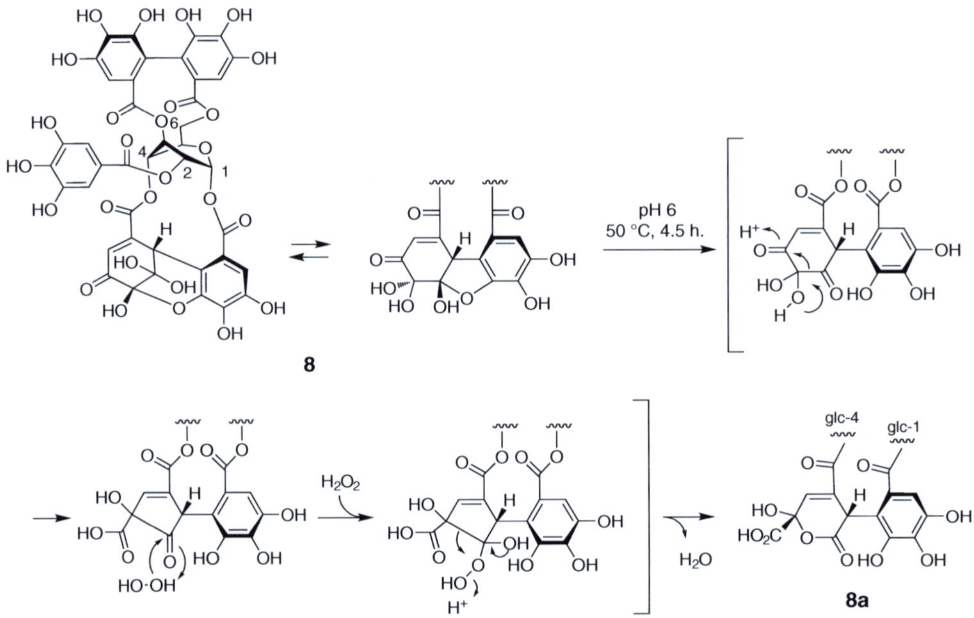

Figure 12.8 Degradation of the DHHDP group of nupharanin (**8**) in pH 6 citrate-phosphate buffer.

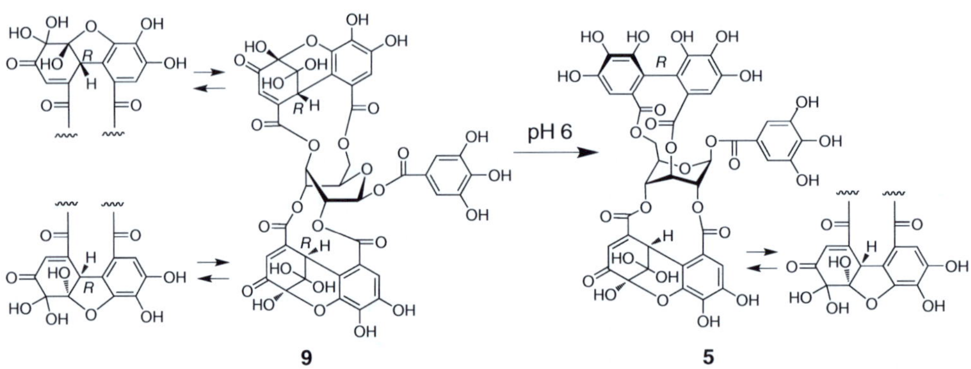

Figure 12.9 Redox disproportionation of the 3,6-DHHDP group of amariin (**9**) in pH 6 citrate-phosphate buffer solution to produce geraniin (**5**).

belonging to the Euphorbiaceae, Geraniaceae, and Aceraceae, among others. Importantly, Haddock et al. pointed out that isogeraniin coexists with geraniin in most geraniin-rich plants. Therefore, if isogeraniin is identical to amariin, the coexistence of geraniin and isogeraniin in those plants suggests that the characteristic reaction involving the 3,6-DHHDP moiety of **9** may be related to the biosynthesis of the 3,6-HHDP group of **5**.

The leaves of *Carpinus laxiflora* (Betulaceae) (Nonaka et al. 1992) and *Triadica sebifera* (Euphorbiaceae) accumulate **5** in their mature summer leaves (Era et al. 2020); however, **9** is the major tannin in the young spring leaves of these plants. As mentioned earlier, Foo described the conversion of **9** to **5** upon heating in aqueous EtOH. Similarly, **9** gradually decomposes in pH 6 citrate-phosphate buffer to give **5** as the major reduction product, indicating that the reaction is a redox disproportionation (Era et al. 2020). However, despite a detailed analysis of the reaction mixture, oxidation products of **9** have not yet been identified. The reactivity of **9** and the seasonal changes observed in the aforementioned plants suggest that the 3,6-HHDP group of **5** is produced via redox disproportionation of the DHHDP group. The 2,4-DHHDP group is more stable in these experiments relative to the unstable 3,6-DHHDP group. The difference in their reactivities is likely caused by a difference in strain between the glucose core and the DHHDP groups; the glucopyranose of **9** adopts a boat conformation, and that of **5** adopts a $^{1}C_{4}$ chair conformation.

12.6 Triterpenoid HHDP Esters in *Castanopsis sieboldii*

The DHHDP group was also determined to be the biogenetic precursor of HHDP in the young spring leaves of *Castanopsis sieboldii* (Fagaceae) (Wakamatsu et al. 2019). The mature leaves are reported to contain HHDP esters of 28-*O*-glucosyl 2α,3β,23,24-tetrahydroxyolean- and urs-12-en-28-oic acids (Ageta et al. 1988). However, high-performance liquid chromatography (HPLC) analysis of the young spring leaves at the tops of the twigs revealed the presence of unstable phenolic metabolites. Following extraction of the leaves with a solvent containing 1,2-phenylenediamine, these metabolites were determined to be DHHDP esters. The isolated phenazine derivative (**10a**) suggested that the original compound was the 3,24-DHHDP ester of 28-*O*-glucosyl 2α,3β,23,24-tetrahydroxyoleanen-28-oic acid (**10**) (Figure 12.10). This DHHDP ester, **10**, was detected as broad peaks in the HPLC chromatogram obtained by aqueous CH_3CN extraction of the

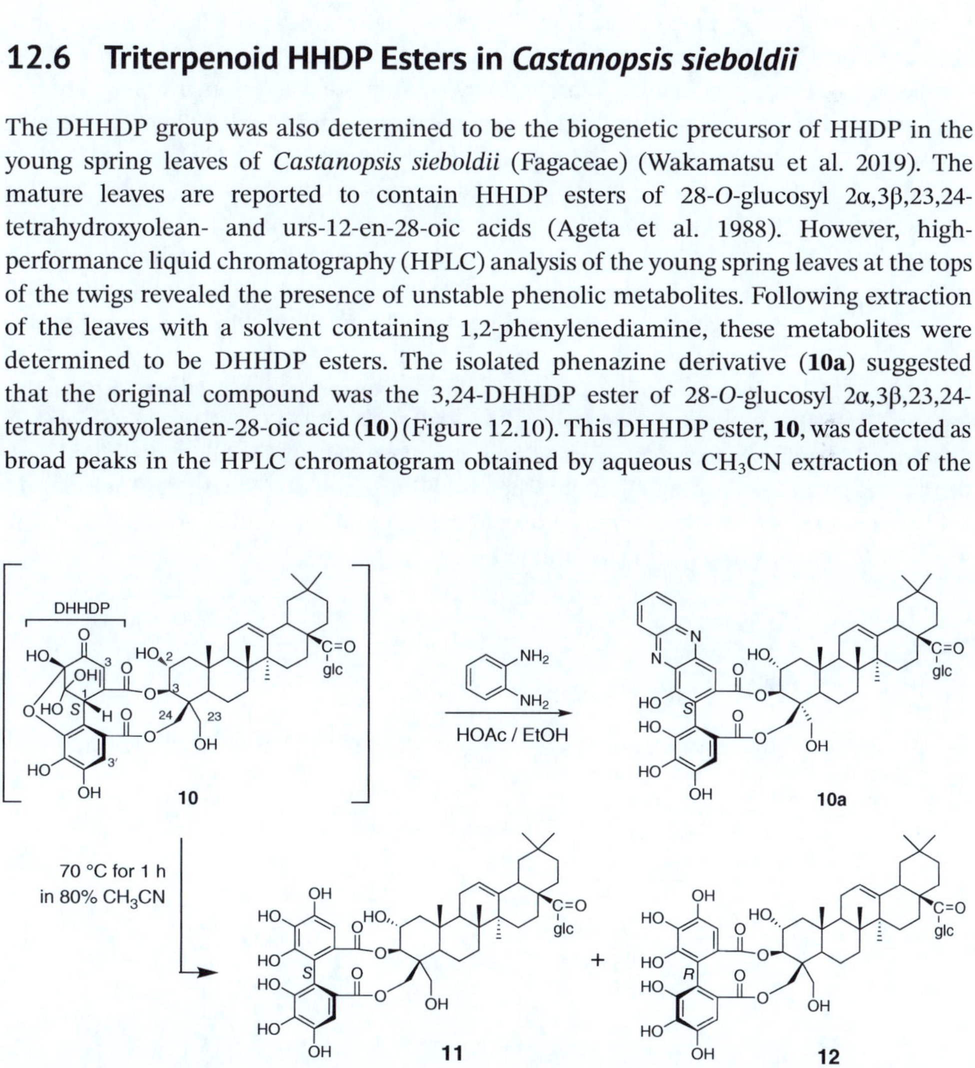

Figure 12.10 Reactions of DHHDP esters of 28-*O*-glucosyl 2α,3β,23,24-tetrahydroxyoleanen-28-oic acid (**10**). Compounds **10–12** coexist with ursane-type isomers, but only the oleanane-type triterpenoids are shown in this figure.

young leaves; however, after heating the extract, **10** disappeared and the (*S*)-HHDP (**11**) and (*R*)-HHDP (**12**) esters were detected. The HHDP esters **11** and **12** are constituents of the mature summer leaves, but they were not detected in the young spring leaves. Interestingly, **11** (containing (*S*)-HHDP) was the major product when the young leaf extract was heated at 70 °C. In contrast, treatment of the same extract with a reducing agent, such as dithiothreitol, afforded **12** (containing (*R*)-DHHDP).

12.7 Highly Oxidized Ellagitannins in *Carpinus japonica*

Leaves of *Carpinus japonica* (Betulaceae) contain highly oxidized ellagitannins with a characteristic HBCHT group at the 4,6-positions of glucose (Figures 12.2 and 12.11) (Nonaka et al. 1990). The mature summer leaves contain the diastereomeric ellagitannins, carpinins A (**14**) and isocarpinins A (**15**), which bear (*S*)- and (*R*)-HHDP groups, respectively, at the 2,3-positions of their glucose cores. However, carpinins F (**13**), which contains a 2,3-(*S*)-DHHDP moiety, is the major ellagitannin found in the young spring leaves. This observation suggests that the production of the (*R*)- and (*S*)-HHDP esters from (*S*)-DHHDP occurs during the seasonal change in ellagitannin composition in the leaves (Kojima et al. 2020). Reduction of **13** using ascorbic acid exclusively afforded **14**, and heating of **13** in dilute H_2SO_4 also yielded the reduction product, **14**, together with the partial hydrolysate, 4,6-HBCHT-D-glucose (**14a**), ellagic acid (**3**), and brevifolin carboxylic acid (**6**) (Figure 12.11). In contrast, heating **13** in H_2O afforded the reduction product, **15** [containing an (*R*)-DHHDP group], along with 2,3-(*R*)-HHDP-4,6-(*S*)-DHHDP-D-glucose (**16**). Similarly, **13** gradually decomposed in pH 6 citrate-phosphate buffer solution at room temperature to give **15** as the major product. The instability of the 2,3-(*S*)-DHHDP moiety of **13** is likely responsible for the observed seasonal change in ellagitannin composition in the leaves of *C. japonica* from spring to summer.

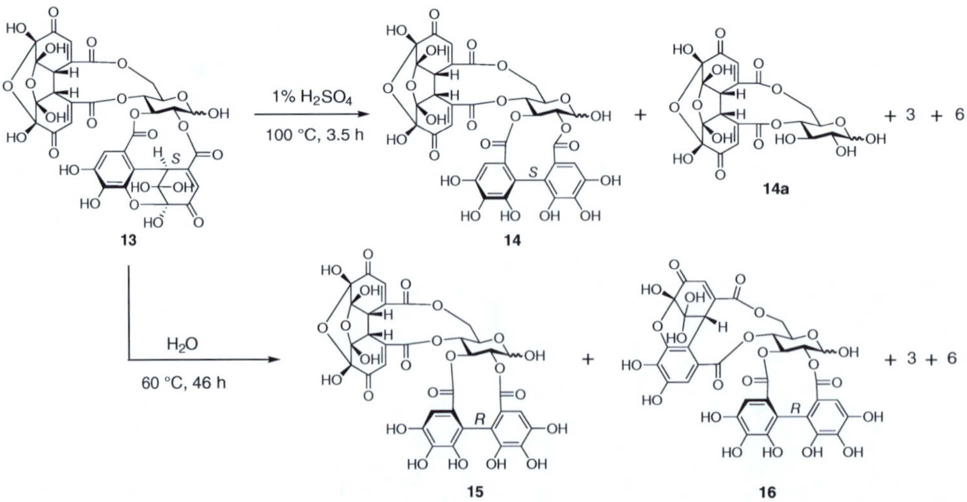

Figure 12.11 Reactions of carpinins F (**13**) in 1% H_2SO_4 or H_2O at elevated temperature.

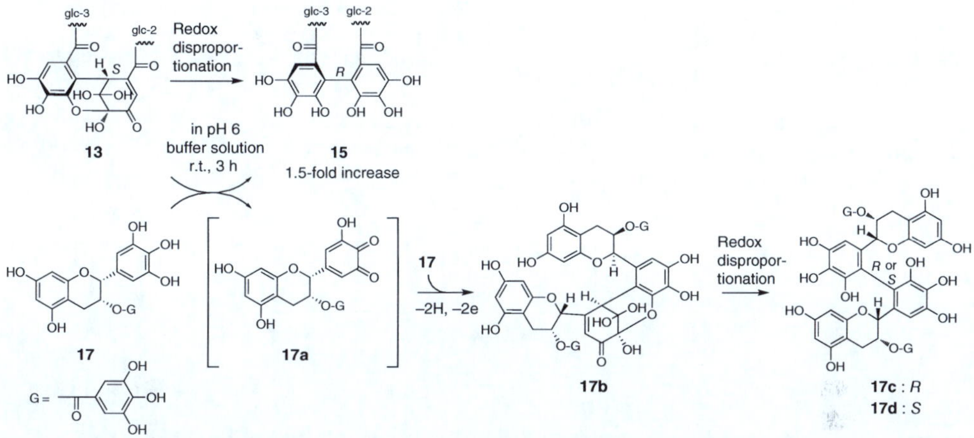

Figure 12.12 Increased production of isocarpinins A (**15**) from carpinins F (**13**) via redox disproportionation in the presence of epigallocatechin-3-*O*-gallate (**17**), and simultaneous oxidation of **17** to dehydrotheasinensin A (**17b**) and theasinensins A (**17c**) and D (**17d**) (in pH 6 phosphate-citrate buffer solution at room temperature for three hours). Compound **17** alone was stable under these experimental conditions.

The reduction of **13** to **15** in H_2O or pH 6 buffer solution also involves redox disproportionation of the DHHDP group. This conclusion was supported by a simple experiment that demonstrated that the production of **15** from **13** in pH 6 citrate-phosphate buffer at ambient temperature increased upon adding an equimolar amount of epigallocatechin-3-*O*-gallate (**17**). Specifically, the yield of **15** increased by about 50% compared with that obtained with **13** alone. Interestingly, the oxidation products of **17** (i.e. dehydrotheasinensin A (**17b**) and theasinensins A (**17c**) and D (**17d**)) were also detected in the reaction product mixture (Figure 12.12) (Tanaka et al. 2003; Matsuo and Tanaka 2017). Because **17** alone is stable under the applied conditions, the observed composition of the product mixture indicates that oxidation of **17** was coupled to the reduction of **13**, and further, that reduction of **13** is promoted by electron transfer from **17**. These results support the proposal that redox disproportionation of **13** proceeds via electron transfer from a DHHDP pyrogallol-ring to a hydrated cyclohexenetrione ring of the DHHDP group in another molecule, following a bimolecular mechanism analogous to that illustrated in Figure 12.6.

12.8 Similarity of Catechin Oxidation to Oxidation of Methyl Gallate

Theasinensin A (**17c**) in Figure 12.12 and the desgalloyl analog are representative black tea polyphenols, and their structures closely resemble that of HHDP. Black tea is produced by curling and crushing withered fresh leaves, followed by a tea-fermentation process, in which catechins are oxidized by polyphenol oxidase enzymes. It is clear that **17c** is produced from oxidative coupling of the most abundant tea catechin, **17.** However, **17c** was not detected in crushed fresh tea leaves and was only identified when the tea leaves were first dried using heat (Tanaka et al. 2002b). Analysis of the crushed leaves revealed

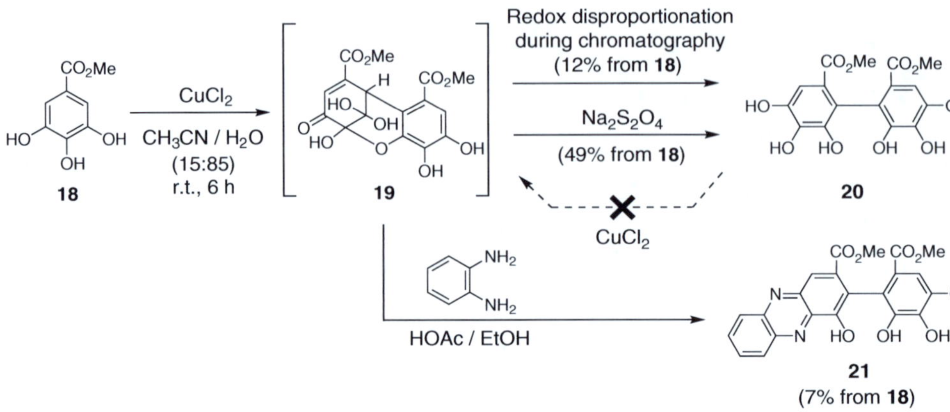

Figure 12.13 Oxidation of methyl gallate (**18**) with $CuCl_2$, and subsequent reactions of the unstable DHHDP dimethyl ester (**19**).

accumulation of the unstable quinone dimer, **17b**. A detailed study of the degradation of **17b** indicated that the black tea polyphenol, **17c**, is produced via redox disproportionation of **17a** (Tanaka et al. 2003), and a complex mixture of oxidation products was observed in addition to reduction product **17c**. These results are related to the previously mentioned reactions involving DHHDP; the similar structures and reactivities of **17b** and DHHDP suggest that the HHDP group can be produced from galloyl esters via a DHHDP group.

The enzymes responsible for the relevant oxidation reactions (i.e. tyrosinase and catechol oxidase) contain a binuclear copper active site and are ubiquitous in plants (Decker and Tuczek 2000). These enzymes perform *o*-diphenol oxidation to the corresponding *o*-quinone, coupled with the reduction of molecular oxygen to water. To mimic the enzymatic oxidation, **17** was treated with a mild oxidizing agent, $CuCl_2$, and the reaction afforded **17b** as the major product (Shii et al. 2011). Similarly, oxidation of methyl gallate (**18**) with $CuCl_2$ in aqueous CH_3CN afforded dimethyl DHHDP (**19**) (Figure 12.13) (Yamashita et al. 2021). The structure of the product was confirmed based on spectroscopic data analysis, reduction with $Na_2S_2O_4$ to dimethyl HHDP (**20**), and derivatization to a phenazine derivative (**21**). The ester, **19**, was unstable, and gradually disappeared during chromatographic separation, and instead, its reduction product, dimethyl HHDP (**20**), was detected. Furthermore, treatment of **20** with $CuCl_2$ did not afford **19**. These results support the hypothesis that an HHDP group can be produced via redox disproportionation of a DHHDP group.

12.9 Production Mechanism of DHHDP and HHDP

Results of several studies described earlier prompted us to propose a new production mechanism of HHDP: (i) oxidative coupling of two galloyl groups affords a DHHDP group, and (ii) subsequent redox disproportionation of the DHHDP moieties generates HHDP (Figure 12.14, solid arrow). The new mechanism is an alternative to the conventional route (Figure 12.14, dashed arrow), in which HHDP is produced directly from galloyl groups via oxidative coupling and further oxidation of one of the HHDP aromatic

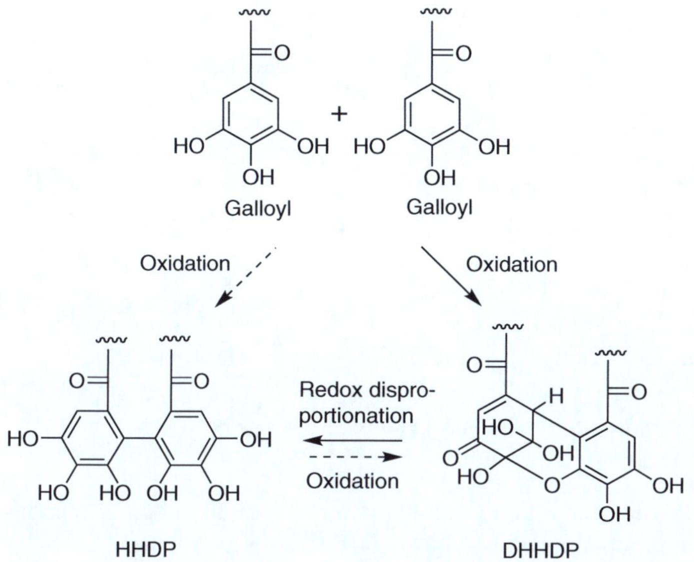

Figure 12.14 Two possible routes for producing HHDP from galloyl groups. Dashed arrow: conventional route. Solid arrow: a new route including redox disproportionation.

rings generates DHHDP. In ellagitannin biosynthesis, the HHDP is produced under species-specific enzymatic regulation, and the detailed mechanisms at the active center of the enzymes remain completely unknown. It is possible that both the new and conventional production mechanisms in Figure 12.14 are involved in ellagitannin biosynthesis. Many synthetic studies showed that simple oxidative couplings of protected galloyl groups directly afford modified HHDP moieties. Therefore, plant enzymes may catalyze the direct conversion of galloyl groups to the HHDP moiety. However, it is also possible that both oxidation of galloyl groups to DHHDP and reduction of DHHDP to HHDP are successively catalyzed by plant enzymes, and the leaves of *C. japonica* may lack the ability to reduce DHHDP to HHDP. As exemplified by geraniin (**5**), the DHHDP moiety at the 2,4-positions of $^4C_1$ glucose is more stable compared to the 3,6- and 2,3-DHHDP analogues, and many ellagitannins with a 2,4-DHHDP glucose unit have been isolated (Okuda et al. 1995; Lee et al. 1991); however, ellagitannins with 2,4-HHDP glucose is rare in nature (Zhang et al. 2001). This fact may be implicated in the difficulty of reduction of the 2,4-DHHDP unit in ellagitannin metabolism.

12.10 Oxidative Degradation of Ellagitannins

In addition to the variation of acyl groups, intramolecular C-glycosidation, condensation with coexisting compounds, and oligomerization also give rise to the large structural diversity of ellagitannins (Okuda et al. 1995; Yoshida et al. 2010; Quideau et al. 2011; Koga et al. 2017). On the other hand, degradation of ellagitannins in plant tissues and in ecological systems are also important issues in ellagitannin chemistry.

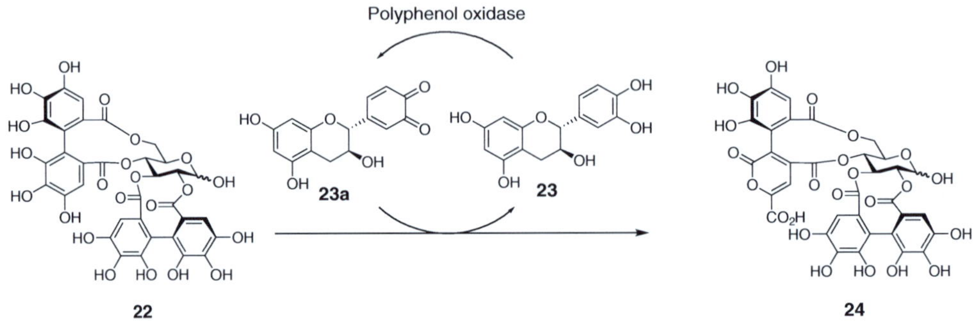

Figure 12.15 Enzymatic oxidation of pedunculagins (**22**) in the presence of catechin (**23**).

12.10.1 Degradation of Pedunculagins in the Leaves of Common Camellia Species

Newly grown young, soft leaves of common Camellia plants collected in May (*C. japonica*, Theaceae) contain α- and β-pedunculagins (**22**) and catechin (**23**). These two polyphenols are not detected in leaves obtained in July, but rather, epicatechin and proanthocyanidins become predominant in the mature, hard leaves. Although polyphenol oxidase alone does not directly oxidize **22**, the enzyme immediately oxidizes **22** in the presence of **23** to produce **24** as one of the initial oxidation products (Figure 12.15) (Tsujita et al. 2017). In this reaction, the enzyme oxidizes the catechol-type B-ring of **23**, and the resulting *o*-quinone (**23a**) oxidizes the pyrogallol ring of **22**. The mechanism is similar to that proposed for the oxidation of tea catechins to black tea polyphenols (Tanaka et al. 2002a). However, the oxidation of **22** is not regioselective, so a complex mixture of oxidation products is generated.

12.10.2 Degradation of Vescalagin in the Leaves of Japanese Blue Oak

A similar oxidative degradation of an ellagitannin was observed in the leaves of *Quercus glauca* (Fagaceae) (Wakamatsu et al. 2020). Vescalagin (**25**) is a representative *C*-glycosidic ellagitannin, which is a known component of oak barrels used for aging wines and spirits (Viriot et al. 1993; Jourdes et al. 2011; Petit et al. 2019). Vescalagin (**25**) is also the major ellagitannin present in the young spring leaves of *Q. glauca*; however, the concentration of **25** in the leaves is very low during the summer. The major polyphenols of the summer leaves are catechin (**23**) and proanthocyanidins, and this seasonal change was similar to that observed for European oak leaf (Feeny 1970). As observed in the reactions involving **22**, **25** cannot be oxidized with polyphenol oxidase alone. However, in the presence of **23**, the enzyme oxidizes **25** to afford **26**, which has a cyclopentendione structure (Figure 12.16). HPLC analysis of the young leaves of *Q. glauca* confirmed the presence of **26**, and this result was further supported by the detection of a quinoxaline derivative, **26a**, which can be prepared by treating the extract with 1,2-phenylenediamine. Furthermore, it was demonstrated that the dioxo moiety of **26** spontaneously reacted with the phloroglucinol-A-ring of **23** to yield the adduct, **27**. This suggested that the oxidative degradation of **25** in the *Q. glauca* leaf may involve a reaction with coexisting catechin and/or proanthocyanidins.

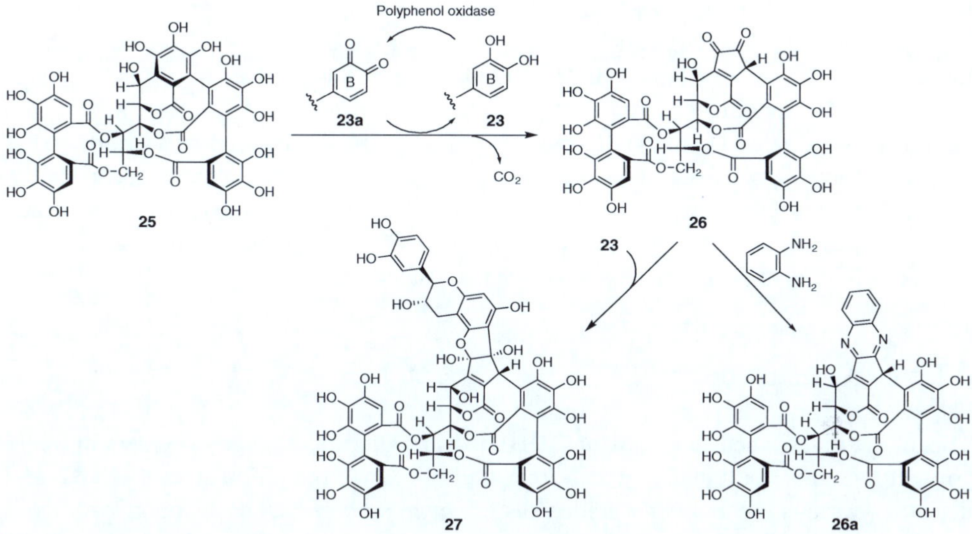

Figure 12.16 Enzymatic oxidation of vescalagin (**25**) in the presence of catechin (**23**), and subsequent conjugation of product **26** with **23**. Preparation of the quinoxaline derivative (**26a**) is also possible.

In contrast to the reactivity observed for **22**, oxidation of **25** seems to be regioselective, and the pyrogallol-ring attached at C-1 of the glucose unit in **25** is most susceptible to oxidation. Similar regioselective oxidation of **25** has also been observed following its autoxidation in aqueous ethanol (Fujieda et al. 2008; Wakamatsu et al. 2020).

12.10.3 Degradation of Vescalagin with Wood-Decaying Fungi

Traditional cultivation of shiitake mushrooms (*Lentinula edodes*) in Japan preferentially involves using the wood of Quercus species containing **25**. Interestingly, when **25** was added to the experimental culture medium of *L. edodes*, the oxidation product, **26**, was detected as the initial metabolite, and later, **28** and **29** were identified in the medium (Figure 12.17) (Wakamatsu et al. 2020). Production of **28** indicated an oxidative ring

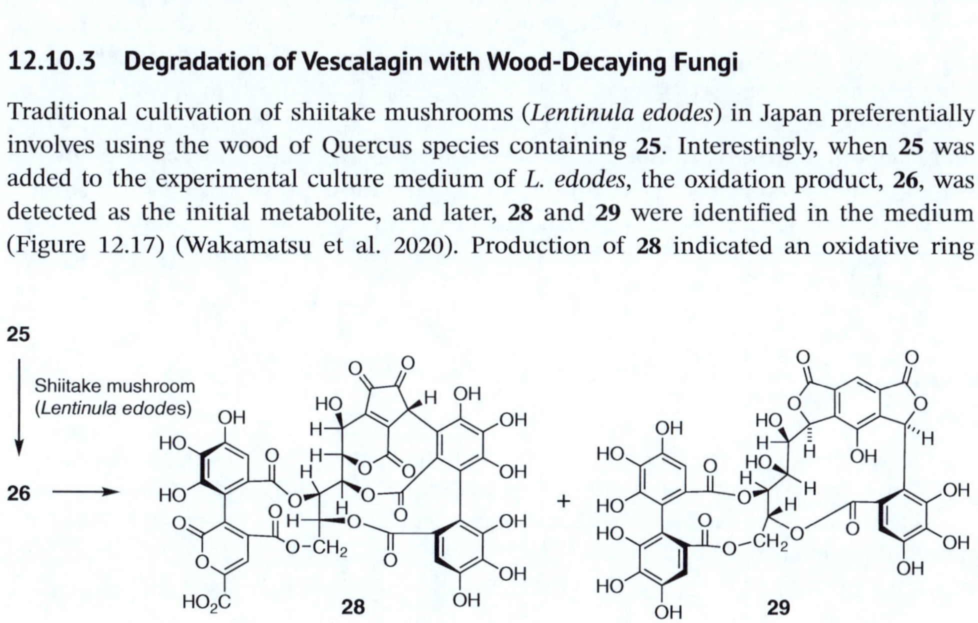

Figure 12.17 Oxidative degradation of vescalagin (**25**) by *Lentinula edodes*.

fission, similar to that observed for **22**. The unusual metabolite, **29**, was first isolated from the sapwood of *Quercus crispula* (Omar et al. 2014), and its production from **25** suggests an unusual complex rearrangement of **26**. It is likely that **26** is produced by metabolic processes involving endophytes in the living wood. However, it is also possible that the degradation of **25** was caused by fungi during the lumbering and drying processes applied to the wood. Although it is commonly accepted that wood ellagitannins are produced as defensive agents against microorganisms (Taguri et al. 2006), enzymes in wood-decaying fungi, such as laccases, lignin peroxidase, and manganese peroxidase, degrade the ellagitannins, thus contributing to the overall forest ecological system.

12.11 Conclusions

This chapter described reactions of DHHDP and structural change of ellagitannins in young leaves of several plants, which importantly include the production of HHDP, and discusses various mechanisms. Redox disproportionation of DHHDP upon hydrolysis was first proposed by Schmidt (Schmidt et al. 1967c), and later, Foo demonstrated the non-enzymatic production of HHDP from the DHHDP group in amariin (**9**) upon simple heating (Foo 1993). In addition, a related redox disproportionation reaction is involved in the production of black tea polyphenols, theasinensins (**17c** and **17d**), from epigallocatechin-3-*O*-gallate (**17**) (Figure 12.12). A review of black tea chemistry published in 1962 mentioned the possibility of redox disproportionation of quinone dimers during tea fermentation (Roberts 1962). Such insights, in combination with recent experimental results, suggest that the reaction mechanisms involved in ellagitannin metabolism and catechin oxidation for the production of black tea polyphenols are similar. Furthermore, as demonstrated by the reaction of DHHDP in **10** and **13**, the generation of (*R*)- and (*S*)-HHDP from (*S*)-DHHDP may be related to the coexistence of diastereomeric ellagitannins containing (*R*)- and (*S*)-HHDP moieties in *Nuphar japonicum* (Ishimatsu et al. 1989a, 1989b), *Castanopsis sieboldii* (Nonaka et al. 1989; Yamaguchi et al. 2013), and *Penthorum chinense* (Era et al. 2015). The results described in this chapter are not sufficient to conclude that redox disproportionation is the dominant mechanism governing HHDP production. Therefore, further studies from biochemical and organic synthesis perspectives are required to gain a comprehensive understanding of ellagitannin metabolism.

References

Ageta, M., Nonaka, G., and Nishioka, I. (1988). Tannins and related compounds. LXVII. Isolation and characterization of castanopsinins AH, novel ellagitannins containing a triterpenoid glycoside core, from *Castanopsis cuspidata* var. *sieboldii* Nakai. (3). *Chemical & Pharmaceutical Bulletin*, **36**: 1646–1663.

Amakura, Y., Miyake, M., Ito, H., et al. (1999). Tannins and related polyphenols of euphorbiaceous plants. 16. Acalyphidins M1, M2 and D1, ellagitannins from *Acalypha hispida*. *Phytochemistry*, **50**: 667–675.

Ashibe, S., Ikeuchi, K., Kume, Y., et al. (2017). Non-enzymatic oxidation of a pentagalloylglucose analogue into members of the ellagitannin family. *Angewandte Chemie International Edition*, **56**: 15402–15406.

Decker, H. and Tuczek, F. (2000). Tyrosinase/catecholoxidase activity of hemocyanins: structural basis and molecular mechanism. *Trends in Biochemical Sciences*, **25**: 392–397.

Era, M., Matsuo, Y., Shii, T., et al. (2015). Diastereomeric ellagitannin isomers from *Penthorum chinense*. *Journal of Natural Products*, **78**: 2104–2109.

Era, M., Matsuo, Y., Saito, Y., et al. (2019). Ellagitannins and related compounds from *Penthorum chinense*. *Journal of Natural Products*, **82**: 129–135.

Era, M., Matsuo, Y., Saito, Y., and Tanaka, T. (2020). Production of ellagitannin hexahydroxydiphenoyl ester by spontaneous reduction of dehydrohexahydroxydiphenoyl ester. *Molecules*, **25**: 1051. doi: 10.3390/molecules25051051.

Feeny, P.P. (1970). Seasonal changes in oak leaf tannins and nutrients as a cause of spring feeding by winter moth caterpillars. *Ecology*, **51**: 565–581.

Feldman, K.S. (2005). Recent progress in ellagitannin chemistry. *Phytochemistry*, **66**: 1984–2000.

Foo, L.Y. (1993). Amariin, a di-dehydrohexahydroxydiphenoyl hydrolyzable tannin from *Phyllanthus amarus*. *Phytochemistry*, **33**: 487–491.

Foo, L.Y. (1995). Amariinic acid and related ellagitannins from *Phyllanthus amarus*. *Phytochemistry*, **39**: 217–224.

Freudenberg, K. (1919). Tannins. II. Chebulinic acid. *Berichte der Deutschen Chemischen Gesellschaft*, **52B**: 1238–1246.

Fujieda, M., Tanaka, T., Suwa, Y., et al. (2008). Isolation and structure of whiskey polyphenols produced by oxidation of oak wood ellagitannins. *Journal of Agricultural and Food Chemistry*, **56**: 7305–7310.

Haddock, E.A., Gupta, R.K., Al-Shafi, S.M.K., et al. (1982a). Metabolism of gallic acid and hexahydroxydiphenic acid in plants. Part 4. The metabolism of gallic acid and hexahydroxydiphenic acid in plants: biogenetic and molecular taxonomic considerations. *Phytochemistry*, **21**: 1049–1062.

Haddock, E.A., Gupta, R.K., and Haslam, E. (1982b). The metabolism of gallic acid and hexahydroxydiphenic acid in plants. Part 3. Esters of (*R*) and (*S*)-hexahydroxydiphenic acid and dehydrohexahydroxydiphenic acid with D-glucopyranose ($^1C_4$ and related conformations). *Journal of the Chemical Society, Perkin Transactions*, **1**: 2535–2545.

Hagerman, A.E. (2012). Fifty years of polyphenol-protein complexes. In: Recent Advances in Polyphenol Research. *Vol.* **3**, (eds. Cheynier, V., Sarni-Manchado, P., and Quideau, S.), 71–97. Hoboken, New Jersey: Wiley-Blackwell.

Haslam, E. (1982). The metabolism of gallic acid and hexahydroxydiphenic acid in higher plants. In: Progress in the Chemistry of Organic Natural Products. *Vol.* **41**, (eds. Herz, W., Grisebach, H., and Kirby, G.W.), 1–46. Vienna: Springer.

Hirokane, T., Hirata, Y., Ishimoto, T., et al. (2014). A unified strategy for the synthesis of highly oxygenated diaryl ethers featured in ellagitannins. *Nature Communications*, **5**: 4478/1–4478/10.

Ishimatsu, M., Tanaka, T., Nonaka, G., et al. (1989a). Tannins and related compounds. LXXV. Isolation and characterization of novel diastereoisomeric ellagitannins, nupharins A and B, and their homologs from *Nuphar japonicum* DC. *Chemical & Pharmaceutical Bulletin*, **37**: 129–134.

Ishimatsu, M., Tanaka, T., Nonaka, G., et al. (1989b). Tannins and related compounds. LXXIX. Isolation and characterization of novel dimeric and trimeric hydrolyzable tannins, nuphrins C, D, E and F, from *Nuphar japonicum* DC. *Chemical & Pharmaceutical Bulletin*, **37**: 1735–1743.

Ito, H., Hatano, T., Namba, O., et al. (1999). Constituents of *Geranium thunbergii* Sieb. et Zucc. XY. Modified dehydroellagitannins, geraniinic acids B and C, and phyllanthusiin F. *Chemical & Pharmaceutical Bulletin*, **47**: 1148–1151.

Jochims, J.C., Taigel, G., and Schmidt, O.T. (1968). Natural tannins. XLI. Proton resonance spectra and conformation determination of some natural tannins. *Justus Liebigs Annalen der Chemie*, **717**: 169–185.

Jourdes, M., Michel, J., Saucier, C., et al. (2011). Identification, amounts, and kinetics of extraction of C-glucosidic ellagitannins during wine aging in oak barrels or in stainless steel tanks with oak chips. *Analytical and Bioanalytical Chemistry*, **401**: 1535–1543.

Khanbabaee, K. and Lötzerich, K. (1998). Synthesis of enantiomerically pure unusual ellagitannins 1,4,6-tri-O-galloyl-2,3-(*R*)-hexahydroxydiphenoyl-β-D-glucopyranoside and 4,6-di-*O*-galloyl-2,3-(*R*)-hexahydroxydiphenoyl-D-glucoside. The proposed chemical structures for cercidinin A and B must be revised. *Journal of Organic Chemistry*, **63**: 8723–8728.

Koga, Y., Matsuo, Y., and Tanaka, T. (2017). Conjugation of vescalagin with glucose and phenylpropanoid: reactions related to the insolubilization of oak wood ellagitannins. *Natural Product Communications*, **12**: 1291–1294.

Kojima, D., Shimizu, K., Aritake, K., et al. (2020). Highly oxidized ellagitannins of *Carpinus japonica* and their oxidation-reduction disproportionation. *Journal of Natural Products*, **83**: 3424–3434.

Larrosa, M., García-Conesa, M.T., Espín, J.C., and Tomás-Barberán, F.A. (2010). Ellagitannins, ellagic acid and vascular health. *Molecular Aspects of Medicine*, **31**: 513–539.

Latte, K.P., Kaloga, M., Schaefer, A., and Kolodziej, H. (2008). An ellagitannin, n-butyl gallate, two aryltetralin lignans, and an unprecedented diterpene ester from *Pelargonium reniforme*. *Phytochemistry*, **69**: 820–826.

Lee, S.H., Tanaka, T., Nonaka, G., and Nishioka, I. (1990). Tannins and related compounds. XCV. Isolation and characterization of helioscopinins and helioscopins, four new hydrolyzable tannins from *Euphorbia helioscopia* L. (1). *Chemical & Pharmaceutical Bulletin*, **38**: 1518–1523.

Lee, S.H., Tanaka, T., Nonaka, G., and Nishioka, I. (1991). Tannins and related compounds. CV. Monomeric and dimeric hydrolyzable tannins having a dehydrohexahydroxydiphenoyl group, supinanin, euphorscopin, euphorhelin and jolkianin, from Euphorbia species. *Chemical & Pharmaceutical Bulletin*, **39**: 630–638.

Matsuo, Y. and Tanaka, T. (2017). Oxidation mechanism of polyphenols and chemistry of black tea. In: Recent Advances in Polyphenol Research. *Vol.* **5**, (eds. Yoshida, K., Cheynier, V., and Quideau, S.), 67–88. Hoboken, New Jersey: Wiley-Blackwell.

Mayer, W., Einwiller, A., and Jochims, J.C. (1967). Tanning agents for the wood of *Castanea sativa* and *Quercus sessiliflora*. III. *Structure of castalin, Justus Liebigs Annalen der Chemie*, **707**: 182–189.

Mayer, W., Seitz, H., and Jochims, J.C. (1969). On tanning compounds from the wood of chestnut and oak. IV. *The structure of castalagin, Justus Liebigs Annalen der Chemie*, **721**: 186–193.

Niemetz, R. and Gross, G.G. (2003a). Oxidation of pentagalloylglucose to the ellagitannin, tellimagrandin II, by a phenol oxidase from *Tellima grandiflora* leaves. *Phytochemistry*, **62**: 301–306.

Niemetz, R. and Gross, G.G. (2003b). Ellagitannin biosynthesis: laccase catalyzed dimerization of tellimagrandin II to cornusiin E in *Tellima grandiflora*. *Phytochemistry*, **64**: 1197–1201.

Niemetz, R. and Gross, G.G. (2005). Enzymology of gallotannin and ellagitannin biosynthesis. *Phytochemistry*, **66**: 2001–2011.

Niemetz, R., Schilling, G., and Gross, G.G. (2001). Ellagitannin biosynthesis: oxidation of pentagalloylglucose to tellimagrandin II by an enzyme from *Tellima grandiflora* leaves. *Chemical Communication*: 35–36.

Niemetz, R., Schilling, G., and Gross, G.G. (2003). Biosynthesis of the dimeric ellagitannin, cornusiin E, in *Tellima grandiflora*. *Phytochemistry*, **64**: 109–114.

Nonaka, G., Matsumoto, Y., and Nishioka, I. (1981). Trapain, a new hydrolyzable tannin from Trapa Japonica Flerov. *Chemical & Pharmaceutical Bulletin*, **29**: 1184–1187.

Nonaka, G., Ishimatsu, M., Ageta, M., and Nishioka, I. (1989). Tannins and related compounds. LXXVI. Isolation and characterization of cercidinins A and B and cuspinin, unusual 2,3-(*R*)-hexahydroxydiphenoyl glucoses from *Cercidiphyllum japonicum* and *Castanopsis cuspidata* var. *sieboldii*. *Chemical & Pharmaceutical Bulletin*, **37**: 50–53.

Nonaka, G., Mihashi, K., and Nishioka, I. (1990). Novel metabolites of hexahydroxydiphenic acid esters (ellagitannins) from *Carpinus japonica*. *Journal of the Chemical Society, Chemical Communications*: 790–791.

Nonaka, G., Akazawa, M., and Nishioka, I. (1992). Two new ellagitannin metabolites, carpinusin and carpinusnin from *Carpinus laxiflora*. *Heterocycles*, **33**: 597–606.

Okuda, T. (2005). Systematics and health effects of chemically distinct tannins in medicinal plants. *Phytochemistry*, **66**: 2012–2031.

Okuda, T. and Ito, H. (2011). Tannins of constant structure in medicinal and food plants—Hydrolyzable tannins and polyphenols related to tannins. *Molecules*, **16**: 2191–2217. doi: 10.3390/molecules16032191.

Okuda, T., Yoshida, T., and Nayeshiro, H. (1977). Constituents of *Geranium thunbergii* Sieb. et Zucc. IV. Ellagitannins. (2). Structure of geraniin. *Chemical & Pharmaceutical Bulletin*, **25**: 1862–1869.

Okuda, T., Hatano, T., and Yazaki, K. (1982a). Dehydrogeraniin, furosinin and furosin, dehydroellagitannins from *Geranium thunbergii*. *Chemical & Pharmaceutical Bulletin*, **30**: 1113–1116.

Okuda, T., Yoshida, T., and Hatano, T. (1982b). Constituents of *Geranium thunbergii* Sieb. et Zucc. Part 12. Hydrated stereostructure and the equilibration of geraniin. *Journal of the Chemical Society, Perkin Transactions*, **1**: 9–14.

Okuda, T., Yoshida, T., Hatano, T., and Ikeda, Y. (1986). Biomimetic synthesis of elaeocarpusin. *Heterocycles*, **24**: 1841–1843.

Okuda, T., Yoshida, T., Hatano, T., et al. (1992). Tannins of Rosaceous plants. Part 11. Hydrolysable tannins as chemotaxonomic markers in the Rosaceae. *Phytochemistry*, **31**: 3091–3096.

Okuda, T., Yoshida, T., and Hatano, T. (1993). Classification of oligomeric hydrolyzable tannins and specificity of their occurrence in plants. *Phytochemistry*, **32**: 507–521.

Okuda, T., Yoshida, T., and Hatano, T. (1995). Hydrolyzable tannins and related polyphenols. In: Progress in The Chemistry of Organic Natural Products, (eds. Herz, W., Kirby, G.W., Moore, R.E., et al.), 1–117. Springer-Verlag: New York.

Omar, M., Matsuo, Y., Maeda, H., et al. (2014). New metabolites of C-glycosidic ellagitannin from Japanese oak sapwood. *Organic Letters*, **16**: 1378–1381.

Orejola, J.J., Era, M., Matsuo, Y., et al. (2020). Nupharanin, the first ellagitannin with 1,4-dehydrohexahydroxydiphenoyl-α-D-glucopyranose from *Nuphar japonicum*. *Tetrahedron*, **76**: 131204. doi: 10.1016/j.tet.2020.131204.

Petit, E., Jacquet, R., Pouysegu, L., et al. (2019). Reactivity of wine polyphenols under oxidation conditions: hemisynthesis of adducts between grape catechins or oak ellagitannins and odoriferous thiols. *Tetrahedron*, **75**: 551–560.

Quideau, S., Deffieux, D., Douat-Casassus, C., and Pouységu, L. (2011). Plant polyphenols: chemical properties, biological activities, and synthesis. *Angewandete Chemie International Eddition*, **50**: 586–621.

Richieu, A., Peixoto, P.A., Pouysegu, L., et al. (2017). Bio-inspired total synthesis of (−)-vescalin: a nona-hydroxytriphenoylated C-glucosidic ellagitannin. *Angewandte Chemie, International Edition*, **56**: 13833–13837. doi: 10.1002/anie.201707613.

Roberts, E.A.H. (1962). Economic importance of flavonoid substances; tea fermentation. In: The Chemistry of Flavonoid Compounds, (ed. Geissman, T.A.), 468–512. New York, N.Y: MacMillan Co.

Saijo, R., Nonaka, G., and Nishioka, I. (1989a). Tannins and related compounds. LXXXIV. Isolation and characterization of five new hydrolyzable tannins from the bark of *Mallotus japonicus*. *Chemical & Pharmaceutical Bulletin*, **37**: 2063–2070.

Saijo, R., Nonaka, G., and Nishioka, I. (1989b). Tannins and related compounds. LXXXVII. Isolation and characterization of four new hydrolyzable tannins from the leaves of *Mallotus repandus*. *Chemical & Pharmaceutical Bulletin*, **37**: 2624–2630.

Salminen, J.-P. (2014). The chemistry and chemical ecology of ellagitannins in plant-insect interactions: from underestimated molecules to bioactive plant constituents. In: Recent Advances in Polyphenol Research. *Vol.* **4**, (eds. Romani, A., Lattanzio, V., and Quideau, S.), 83–113. Hoboken, New Jersey: Wiley-Blackwell.

Schmidt, O.T. and Eckert, R. (1958). Natural tannins. XXIX. Constitution of brevifolincarboxylic acid. *Justus Liebigs Annalen der Chemie*, **618**: 71–82.

Schmidt, O.T., Wuertele, L., and Harreus, A. (1965). Natural tannins. XXXIII. Pedunculagin, a 2,3:4,6-bis[(−)-hexahydroxydiphenoyl]-D-glucose from the gallnut. *Justus Liebigs Annalen der Chemie*, **690**: 150–162.

Schmidt, O.T., Schulz, J., and Wurmb, R. (1967). Natural tannins. XXXVI. Terchebin. *Justus Liebigs Annalen der Chemie*, **706**: 169–179.

Schmidt, O.T., Schanz, R., Wurmb, R., and Groebke, W. (1967a). Natural tannnins. XXXV. Brevilagin 2. *Justus Liebigs Annalen der Chemie*, **706**: 154–168.

Schmidt, O.T., Schanz, R., Eckert, R., and Wurmb, R. (1967b). Natural tannins. XXXIV. Brevilagin 1. *Justus Liebigs Annalen der Chemie*, **706**: 131–153.

Schmidt, O.T., Wurmb, R., and Schulz, J. (1967c). Natural tannins. XXXVII. Hexahydroxydiphenic acid (ellagic acid), brevifolincarboxylic acid and chebulic acid as rearrangement products of brevilagins and terchebins. *Justus Liebigs Annalen der Chemie*, **706**: 180–186.

Shibayama, H., Ueda, Y., Tanaka, T., and Kawabata, T. (2021). Seven-step stereodivergent total syntheses of punicafolin and macaranganin. *Journal of the American Chemical Society*, **143**: 1428–1434.

Shii, T., Miyamoto, M., Matsuo, Y., et al. (2011). Biomimetic one-pot preparation of a black tea polyphenol theasinensin A from epigallocatechin gallate by treatment with copper(II) chloride and ascorbic acid. *Chemical & Pharmaceutical. Bulletin*, **59**: 1183–1185.

Taguri, T., Tanaka, T., and Kouno, I. (2006). Antibacterial spectrum of plant polyphenols and extracts depending upon hydroxyphenyl structure. *Biological & Pharmaceutical Bulletin*, **29**: 2226–2235.

Tanaka, T., Nonaka, G., Nishioka, I., et al. (1986). Tannins and related compounds. Part 37. Iisolation and structure elucidation of elaeocarpusin, a novel ellagitannin from *Elaeocarpus sylvestris* var. *ellipticus*. *Journal of the Chemical Society, Perkin Transactions*, **1**: 369–376.

Tanaka, T., Nonaka, G., and Nishioka, I. (1990). Tannins and related compounds. C. Reaction of dehydrohexahydroxydiphenic acid esters with bases, and its application to the structure determination of pomegranate tannins, granatins A and B. *Chemical & Pharmaceutical Bulletin*, **38**: 2424–2428.

Tanaka, T., Kouno, I., and Nonaka, G. (1996a). Glutathione-mediated conversion of the ellagitannin geraniin into chebulagic acid. *Chemical & Pharmaceutical Bulletin*, **44**: 34–40.

Tanaka, T., Ueda, N., Shinohara, H., et al. (1996b). C-Glycosidic ellagitannin metabolites in the heartwood of Japanese chestnut tree (*Castanea crenata* Sieb. et Zucc.). *Chemical & Pharmaceutical Bulletin*, **44**: 2236–2242.

Tanaka, T., Nonaka, G., Ishimatsu, M., et al. (2001). Revised Structure of cercidinin A, a novel ellagitannin having (*R*)-hexahydroxydiphenoyl esters at the 3, 4-positions of glucopyranose. *Chemical & Pharmaceutical Bulletin*, **49**: 486–487.

Tanaka, T., Mine, C., Inoue, K., et al. (2002a). Synthesis of theaflavin from epicatechin and epigallocatechin by plant homogenates and role of epicatechin quinone in the synthesis and degradation of theaflavin. *Journal of Agricultural and Food Chemistry*, **50**: 2142–2148.

Tanaka, T., Mine, C., Watarumi, S., et al. (2002b). Accumulation of epigallocatechin quinone dimers during tea fermentation and formation of theasinensins. *Journal of Natural Products*, **65**: 1582–1587.

Tanaka, T., Watarumi, S., Matsuo, Y., et al. (2003). Production of theasinensins A and D, epigallocatechin gallate dimers of black tea, by oxidation-reduction dismutation of dehydrotheasinensin A. *Tetrahedron*, **59**: 7939–7947.

Tanaka, T., Kouno, I., and Nonaka, G. (2011). Biomimetic synthesis and related reactions of ellagitannins. In: Biomimetic Organic Synthesis. *Vol.* **2**, (eds. Poupon, E. and Nay, B.), 639–675. Weinheim: Wiley-VCH Verlag.

Tomas-Barberan, F.A., Gonzalez-Sarrias, A., Garcia-Villalba, R., et al. (2017). Urolithins, the rescue of “old” metabolites to understand a “new” concept: metabotypes as a nexus among phenolic metabolism, microbiota dysbiosis, and host health status. *Molecular Nutrition & Food Research*, **61**: 1500901. doi: 10.1002/mnfr.201500901.

Tsujita, T., Matsuo, Y., Saito, Y., and Tanaka, T. (2017). Enzymatic oxidation of ellagitannin and a new ellagitannin metabolite from *Camellia japonica* leaves. *Tetrahadron*, **73**: 500–507.

Viriot, C., Scalbert, A., Lapierre, C., and Moutounet, M. (1993). Ellagitannins and lignins in aging of spirits in oak barrels. *Journal of Agricultural and Food Chemistry*, **41**: 1872–1879.

Wakamatsu, H., Tanaka, S., Matsuo, Y., et al. (2019). Reductive metabolism of ellagitannins in the young leaves of *Castanopsis sieboldii*. *Molecules*, **24**: 4279. doi: 10.3390/molecules24234279.

Wakamatsu, H., Matsuo, Y., Omar, M., et al. (2020). Oxidation of the oak ellagitannin, vescalagin. *Journal of Natural Products*, **83**: 413–421.

Yamada, H., Hirokane, T., Ikeuchi, K., and Wakamori, S. (2017). Fundamental methods in ellagitannin synthesis. *Natural Product Communications*, **12**: 1934578X1701200846.

Yamaguchi, S., Hirokane, T., Yoshida, T., et al. (2013). Roxbin B is cuspinin: structural revision and total synthesis. *Journal of Organic Chemistry*, **78**: 5410–5417.

Yamashita, T., Matsuo, Y., Saito, Y., and Tanaka, T. (2021). Formation of dehydrohexahydroxydiphenoyl esters by the oxidative coupling of galloyl esters in an aqueous medium involved in ellagitannin biosynthesis. *Chemistry - An Asian Journal*, **16**: 1735–1740.

Yazaki, K., Hatano, T., and Okuda, T. (1989). Constituents of *Geranium thunbergii* Sieb. et Zucc. Part 14. Structures of didehydrogeraniin, furosinin and furosin. *Journal of the Chemical Society, Perkin Transactions 1*, **1989**: 2289–2296.

Yoshida, T., Fujii, R., and Okuda, T. (1980). Revised structures of chebulinic acid and chebulagic acid. *Chemical & Pharmaceutical Bulletin*, **28**: 3713–3715.

Yoshida, T., Okuda, T., Koga, T., and Toh, N. (1982). Absolute configurations of chebulic, chebulinic and chebulagic acid. *Chemical & Pharmaceutical Bulletin*, **30**: 2655–2658.

Yoshida, T., Itoh, H., Matsunaga, S., et al. (1992). Tannins and related polyphenols of euphorbiaceous plants. IX. Hydrolyzable tannins with $^1C_4$ glucose core from *Phyllanthus flexuosus* Muell. Arg. *Chemical & Pharmaceutical Bulletin*, **40**: 53–60.

Yoshida, T., Amakura, Y., Liu, Y.-Z., and Okuda, T. (1994). Tannins and related polyphenols of euphorbiaceous plants. XI. Three new hydrolyzable tannins and a polyphenol glucoside from *Euphorbia humifusa*. *Chemical & Pharmaceutical Bulletin*, **42**: 1803–1807.

Yoshida, T., Ito, H., and Hipolito, I.J. (2005). Pentameric ellagitannin oligomers in Melastomataceous plants-chemotaxonomic significance. *Phytochemistry*, **66**: 1972–1983.

Yoshida, T., Amakura, Y., and Yoshimura, M. (2010). Structural features and biological properties of ellagitannins in some plant families of the order Myrtales. *International Journal of Molecular Sciences*, **11**: 79–106. doi: 10.3390/ijms11010079.

Zhang, Y.-J., Abe, T., Tanaka, T., et al. (2001). Phyllanemblinins A–F, new ellagitannins from *Phyllanthus emblica*. *Journal of Natural Products*, **64**: 1527–1532.

Index

Note: Page numbers followed by "*f*" and "*t*" indicate figures and tables, respectively.

Recent Advances in Polyphenol Research, Volume 8, First Edition. Edited by Juha-Pekka Salminen, Kristiina Wähälä, Victor de Freitas, and Stéphane Quideau.

d

q

r

u

v